T0291763

CAMBRIDGE LIBRARY COLLECTION

Books of enduring scholarly value

Life Sciences

Until the nineteenth century, the various subjects now known as the life sciences were regarded either as arcane studies which had little impact on ordinary daily life, or as a genteel hobby for the leisured classes. The increasing academic rigour and systematisation brought to the study of botany, zoology and other disciplines, and their adoption in university curricula, are reflected in the books reissued in this series.

Handbook of the New Zealand Flora

Sir Joseph Dalton Hooker (1817–1911), botanist, explorer, and director of the Royal Botanical Gardens at Kew, is chiefly remembered as a close friend and colleague of Darwin, his publications on geographical distribution of plants supporting Darwin's theory of evolution by natural selection. In 1839 Hooker became an assistant surgeon on HMS *Erebus* during Ross' Antarctic expedition. The boat wintered along the New Zealand coast, Tasmania and the Falkland Islands, enabling Hooker to collect over 700 plant species. Drawing heavily on Hooker's illustrated *Flora Novae Zelandiae* (1854–1855), this two-volume work (1864–1867) contains a comprehensive list of New Zealand plant species as well as those of the Chatham, Kermadec, Auckland, Campbell and Macquarrie Islands. As the first major study of New Zealand flora, Hooker's handbook remained the authority on the subject for half a century. Volume 2 continues Hooker's meticulous description and categorization of New Zealand flora.

Cambridge University Press has long been a pioneer in the reissuing of out-of-print titles from its own backlist, producing digital reprints of books that are still sought after by scholars and students but could not be reprinted economically using traditional technology. The Cambridge Library Collection extends this activity to a wider range of books which are still of importance to researchers and professionals, either for the source material they contain, or as landmarks in the history of their academic discipline.

Drawing from the world-renowned collections in the Cambridge University Library, and guided by the advice of experts in each subject area, Cambridge University Press is using state-of-the-art scanning machines in its own Printing House to capture the content of each book selected for inclusion. The files are processed to give a consistently clear, crisp image, and the books finished to the high quality standard for which the Press is recognised around the world. The latest print-on-demand technology ensures that the books will remain available indefinitely, and that orders for single or multiple copies can quickly be supplied.

The Cambridge Library Collection will bring back to life books of enduring scholarly value (including out-of-copyright works originally issued by other publishers) across a wide range of disciplines in the humanities and social sciences and in science and technology.

Handbook of the New Zealand Flora

A Systematic Description of the Native Plants of New Zealand

VOLUME 2

JOSEPH DALTON HOOKER

CAMBRIDGE UNIVERSITY PRESS

Cambridge, New York, Melbourne, Madrid, Cape Town,
Singapore, São Paolo, Delhi, Tokyo, Mexico City

Published in the United States of America by Cambridge University Press, New York

www.cambridge.org
Information on this title: www.cambridge.org/9781108030403

This edition first published 1867
This digitally printed version 2011

ISBN 978-1-108-03040-3 Paperback

HANDBOOK

OF THE

NEW ZEALAND FLORA.

Part I., pp. 1–392, *published* 1864.

Part II., pp. 393 *to end, published* 1867.

HANDBOOK

OF THE

NEW ZEALAND FLORA:

A SYSTEMATIC DESCRIPTION

OF THE

Native Plants

OF

NEW ZEALAND

AND THE

CHATHAM, KERMADEC'S, LORD AUCKLAND'S, CAMPBELL'S, AND MACQUARRIE'S ISLANDS.

BY

J. D. HOOKER, M.D., F.R.S. L.S. & G.S.,

AND HONORARY MEMBER OF THE PHILOSOPHICAL INSTITUTE OF CANTERBURY, NEW ZEALAND.

PUBLISHED

UNDER THE AUTHORITY OF THE GOVERNMENT OF NEW ZEALAND.

PART II.

LONDON:

REEVE & CO., 5, HENRIETTA STREET, COVENT GARDEN.

1867.

Order IV. MUSCI.

Cellular, usually tufted plants, rarely more than a few inches high and usually much less, with distinct erect or prostrate stems, generally branched but sometimes quite simple; cells and tubes composing the stems and leaves without transverse bars or spiral threads, except in *Sphagnum*. Leaves imbricate distichous or 3–8-farious, sometimes of two forms, one larger, the other resembling stipules; never lobed or divided; margin entire or toothed, sometimes thickened. The leaves at the base of the fruitstalk usually differ from the others, and are called *perichætial* leaves. Reproductive organs of two kinds: *capsules*, containing minute spores, produced from archegonia; and *antheridia*, which are minute membranous sacs, by means of whose contents the archegonia are fertilized. I. Capsule (*theca*, *sporangium*) lateral or terminal on the branches, rarely radical, sessile or on a fruitstalk (*seta*), globose oblong-ovoid turbinate or pyriform, terete or angular, equal at the base or with a swelling (*apophysis* or *struma*) at its union with the fruitstalk, indehiscent or bursting by 4 valves at the sides (*Andreæa*), or most commonly by a transversely deciduous cap (*operculum*), covered by a deciduous *calyptra*, which is entire (*mitriform*), or split on one side (*dimidiate* or *cucullate*). The operculum, on falling away, exposes a circular orifice or mouth of the capsule, which is usually open, rarely closed (in *Polytrichum* and its allies) by a transverse membrane; rim of the mouth naked or crowned with 1–3 concentric rows of appendages (*peristome*), consisting, 1. of the *annulus*, a row of loose cells that often curls upon removal; 2. of the *outer peristome*, a row of erect or incurved teeth, often reflexed when dry, separate or variously combined, free or united by transverse bars (*trabeculate*), simple or forked, straight or tortuous; 3. of the *inner peristome*, consisting of a fine membrane split into cilia or teeth; this membrane lines the capsule to its bottom, and is there drawn up as an axis or central column (*columella*). Spores numerous, escaping from the mouth of the capsule, green yellow or brown, simple or combined, without external covering or markings. II. Antheridia: Oblong or linear, stalked, membranous sacs, mixed with jointed filaments (*paraphyses*), surrounded by whorls of leaves (*perigonia*), either axillary or terminal on the stems or branches. The sacs have open mouths, and contain a multitude of cells, each with an enclosed spiral filament (*antherozoid*) endowed with motion. When the antheridia are on the same plant with the capsule, the Moss is called monœcious; when on separate, diœcious; in fewer cases the antheridia and archegonia (which become capsules) are found intermixed in the same inflorescence.

One of the most beautiful tribes of plants, found in all parts of the globe except the driest, abounding in temperate perennially humid latitudes, as the west coast of New Zealand, which is very rich in species.

The capsules, from whose form and structure the main divisional and generic characters are drawn, are thus developed. The female *inflorescence*, as it is called, consists of a terminal or axillary bundle of two or more archegonia, mixed with jointed filaments, and surrounded by perichætial leaves. Archegonia are very slender, erect, cellular, flagon-shaped bodies, tumid and hollow at the base, the hollow being continuous upwards to the tip, which is open; a minute loose cell is contained in the base of their cavity. The process of fertilization has never been observed, but it is believed that one or more of the antherozoids, having escaped from the antheridia, find their way into the cavity of the archegonium, and fertilize its contained cell; it is, however, certain that from this cell the fruitstalk and capsule are developed. The upper part of the archegonium, after fertilization, swells very much into a campanulate body, which breaks away at its base, and is carried up, as the calyptra, by the lengthening fruitstalk; the base of the archegonium, on the other hand, usually becomes a fleshy cylinder supporting the fruitstalk, and is called the *vaginula*. After the fruitstalk has attained some length, its apex begins to swell and to develope into the capsule. When the capsule is ripe it thus dehisces; at the line of dehiscence one row of the cells of its cellular tissue enlarges, and pushes off the upper part or operculum; this row of cells is the future *annulus* of the peristome. The spores are now ready to escape, and their exit is regulated in many species by the hygrometric action of the teeth of the peristome.

In germination, the spore emits cellular filaments, quite like the branches of a *Conferva*, which sometimes form matted, green, velvet-like patches; on these threads the stems and rootlets are formed, the former growing upwards, the latter downwards. Dr. J. B. Hicks describes these filaments as giving rise to *gonidia* (reproductive cells, like those of Lichens). This subject is one of great interest and novelty, but most difficult of investigation, about which I must refer the reader to his valuable paper published in the Linnean Society's Transactions (vol. xxiii. p. 567).

By many muscologists, including Wilson, this Order has been considered as a family, and divided into three—*Andreæaceæ*, *Bryaceæ*, and *Sphagnaceæ*. To me, however, these appear to form one Order, equivalent to one of the average Orders of Flowering Plants. Of the genera, *Andreæa* is, no doubt, one of the most remarkable in the dehiscence of its capsule; but though differing from all other Mosses in this respect, the passage through *Phascum* to these is obvious. It further differs from most Musci in the capsule being seated directly on the vaginula, and in the colour of the foliage. The *Andreæas* have been supposed to form a connecting link with the *Hepaticæ*, and in this respect of the dehiscence of the capsule they are, of all Mosses, those which show the nearest approach to that Order; but the differences are far too great to justify their being regarded as intermediate in the proper sense of that term. *Sphagnum* is an even more peculiar genus; it, too, wants the fruitstalk, and has the capsule supported on the vaginula directly, and it further has truly fascicled branches, no rootlets, peculiar antheridia, and the cellular structure of the leaves is of two kinds, usually furnished with pores and spiral fibres; but besides *Sphagnum* showing in these respects no tendency to other Orders, we find an approach to the same cellular structure in *Leucobryum*, and a few other Mosses of very different affinity. A few muscologists have carried the subdivision of Mosses into Orders much further even than this; but as the terms they use are not the equivalents of those used in other groups of plants, it is impossible to attach any precise value to such divisions. Thus, Dr. Schimper, of Strasburg, the most learned and able living muscologist, divides the Order into, first, *Musci* and *Sphagna*, and the *Musci* successively into Sections, Orders, Tribes, Families, and Genera—a sequence of terms not adopted in other branches of botany.

The Musci were worked up by Mr. Wilson, the most able British muscologist, for the 'Flora of New Zealand;' and I have for the most part adhered to his generic characters and limitations; having myself studied the species which I had collected, both in the Bay of Islands and in Lord Auckland's group, and prepared the analyses of the greater number figured in the 'Flora Antarctica,' we published under our joint names. Since that period many new species have been described by Mr. Mitten, to whom I am indebted for most cordial assistance in the following pages, and who has examined and named for the Hookerian Herbarium, all the specimens that have been received from recent New Zealand collectors. These I have incorporated in the present work, together with a few species which were published by C. Mueller in the 'Flora' (a German periodical) in 1855, a few months before the appearance of the 'New Zealand Flora,' but which were not known to Mr. Wilson or myself at the time of our printing the latter work. I extremely regret that my friend Mr. Wilson's health has prevented him from undertaking a revision of the New Zealand Mosses for this Handbook. A very considerable proportion of the Mosses here described are imperfectly known. In such genera as *Hypnum*, *Bryum*, *Dicranum*, etc., limits scarcely exist between the forms that compose some of the extensive groups; and a special study of a large series of specimens from many parts of the islands is necessary to determine their limits, whether naturally or artificially.

The books that will prove most useful to the student of New Zealand Mosses are:—

Wilson's 'Bryologia Britannica,' 1 vol. 8vo, with 61 plates; Schimper's 'Synopsis Muscorum Europæorum,' 1 vol. 8vo; Hooker's 'Musci Exotici,' 2 vols. 8vo, with 200 plates; C. Mueller's 'Synopsis Muscorum Frondosorum,' 2 vols. 8vo; Hedwig's 'Species Muscorum,' and Schwægrichen's Supplements to the same, in several volumes, 4to, with between 300 and 400 plates. The New Zealand, Tasmanian, and Antarctic Floras, all expensive works in quarto, also contain very numerous plates of New Zealand Mosses.

KEY TO THE GENERA OF NEW ZEALAND MOSSES.

A. **Acrocarpi.**—*Fruitstalk springing from the ends of the branches (lateral in one* Fissidens, *apparently lateral but really terminal in* Sphagnum, Dicnemon, Hedwigia, Mielichoferia, *etc.*).

I. *Leaves inserted all round the stem or branches.* (*Distichous in* Distichium.)

§ *Operculum adherent, never falling away.*

Capsule bursting by 4 lateral slits . . . . 1. ANDREÆA.
Capsule bursting irregularly . . . . 3. PHASCUM.

§§ *Operculum falling away.*

Division 1. ASTOMI.—*Peristome* 0.

* *Capsule sessile on a short white receptacle* . . . . 2. SPHAGNUM.

** *Capsule on a distinct (often very short) red, yellow, or brown fruitstalk.*

† *Calyptra mitriform. Capsule erect.*

Capsule obovate or clavate; operculum conical. Leaf-cells large, lax 43. PHYSCOMITRIUM.
Capsule cylindric; beak of operculum straight. Calyptra very large 21. ENCALYPTA.
Capsule ovoid, grooved. Calyptra plicate, lacerate at the base . . 27. MACROMITRIUM.
Capsule globose, sessile amongst the perichætial leaves . . . . 22. HEDWIGIA.

†† *Calyptra cucullate.*

α. *Capsule globose.*

Stems ½–2 in. long. Fruitstalk short. Operculum beaked . . . 22. HEDWIGIA.
Stems 2–6 in. long. Fruitstalk long, slender. Operculum beaked 23. BRAUNIA.
Stems ⅛–¼ in. long. Fruitstalk 0. Operculum scarcely beaked . 24. LEPTANGIUM.

β. *Capsule not globose.*

Mouth with a membranous ring. Leaves piliferous; cells large, lax 31. LEPTOSTOMUM.
Mouth without a membranous ring. Leaf-cells minute.
Capsule erect, ovoid, not grooved. Operculum with an acute inclined beak . . . . 5. GYMNOSTOMUM.
Capsule erect, ovoid, grooved or striate. Operculum with an acute inclined beak . . . . 30. ZYGODON.
Capsule suberect, ovoid-oblong, smooth. Operculum hardly beaked. Nerve of leaf very thick . . . . 18. DESMATODON.

Division 2. APLOPERISTOMI.—*Peristome single.*

* *Calyptra mitriform, plaited. Capsule erect.*

Capsule ovoid. Calyptra laciniate at the base . . . . 28. MACROMITRIUM.
Capsule pyriform. Calyptra crenate at the base, usually very hairy 29. ORTHOTRICHUM.

** *Calyptra mitriform, not plaited. Capsule erect.*

Calyptra split at the base . . . . 25. GRIMMIA.
Calyptra fringed at the base . . . . 26. RACOMITRIUM.

*** *Calyptra cucullate, mitriform in one* Tortula. (*See* Grimmia, *in* **.)

Peristome a membranous ring . . . . 31. LEPTOSTOMUM.
Peristome of distinct teeth.

α. *Capsule erect or nearly so, not globose. Teeth* 8, *2-fid, or* 16 *equidistant, or* 32 *in pairs, rarely in fours, then short, broad.*

† *Cells of leaf large, lax.*

Teeth erect, when dry . . . . 43. ENTOSTHODON
Teeth recurved when dry . . . . 45. EREMODON.

†† *Cells of leaf minute.*

Teeth recurved when dry. Capsule grooved. . . . . 30. ZYGODON.
Teeth erect or recurved when dry. Capsule not grooved. Perichætial leaves short . . . . 5. WEISSIA.
Teeth incurved when dry. Capsule not grooved. Perichætial leaves very long, sheathing the fruitstalk . . . . 6. SYMBLEPHARIS.

β. *Capsule globose, inclined; mouth small; teeth* 16, *united by their tips.*

Leaves subulate, serrate . . . . 39. CONOSTOMUM.

γ. *Fruitstalk long and slender. Capsule erect or nearly so, narrow ovoid-oblong or cylindrical, rarely slightly curved, quite equal at the base. Teeth* 16, *2–3-fid, or* 32 *in pairs, slender, erect or incurved when dry.*

Leaves broad, soft, papillose at the back; nerve very thick. Teeth 16, 2–3-fid . . . . 18. DESMATODON.
Leaves narrow. Teeth 16 or 32, lanceolate, entire 2-fid or perforate, fragile, irregular . . . . 17. DIDYMODON.
Leaves narrow, distichous. Teeth 16, equidistant . . . . . 19. DISTICHIUM.
Leaves narrow. Teeth 16, perforated, or 32 in pairs, slender, erect, of one series of cells . . . . 15. TRICHOSTOMUM.
Leaves broad or narrow. Teeth 32, slender, twisted, of two series of cells . . . . 16. TORTULA.

δ. *Fruitstalk long or short. Capsule inclined cernuous or pendulous, more or less curved and unequal at the base. Teeth* 16, *equidistant, 2-fid, or* 32 *in pairs.*

† *Fruitstalk erect. Calyptra not fringed at the base.*

Stems creeping. Fruitstalk clothed with long perichætial leaves. Teeth 16, 2-fid . . . . 9. DICNEMON.
Stems erect.
Teeth 32, united by trabeculæ . . . . 20. CERATODON.
Teeth 16, 2-fid. Leaves thick, white, spongy . . . . . . 10. LEUCOBRYUM.
Teeth 16, 2-fid. Leaves of ordinary texture . . . . 11. DICRANUM.

†† *Fruitstalk curved or flexuose.*

Capsule ovoid, furrowed. Calyptra fringed at the base. Teeth 16, 2-fid . . . . 13. CAMPYLOPUS.
Capsule oblong, not furrowed. Calyptra entire at the base. Teeth 16, 2-partite . . . . 12. DICRANODONTIUM.
Capsule pyriform, with a long narrow struma. Calyptra inflated. Teeth 32, granulate . . . . 14. TREMATODON.
Capsule pyriform, with a tapering struma. Calyptra small, narrow. Teeth 16, connate at the base . . . . 35. MIELICHOFERIA.

ε. *Teeth* 32 *or* 64, *very short, equidistant, horny.*

Stout, erect, rigid Mosses. Leaves usually lamellate, brown or lurid green . . . . 46. POLYTRICHUM.

Division 3. DIPLOPERISTOMI.—*Peristome double.*

Capsule horizontal, flat above, convex below. Teeth innumerable, forming a filiform brush . . . . 47. DAWSONIA.
Capsule terete. Teeth short, definite in number.

* *Calyptra mitriform, plicate. Capsule erect; beak of operculum long, erect.*

Calyptra with 4 or more inflexed lobes at the base . . . . . . 27. SCHLOTHEIMIA.
Calyptra not inflexed, but laciniate at the base . . . . 28. MACROMITRIUM.
Calyptra (usually very hairy) not inflexed, crenate or split at the base . . . . 29. ORTHOTRICHUM.

** *Calyptra cucullate.*

a. Leaves long, subulate, serrate; cells minute. Teeth incurved.

Capsule globose, furrowed; operculum nearly flat . . . . 41. BARTRAMIA.
Capsule oblong, erect, hidden by the leaves; operculum convex, apiculate . . . . 40. CRYPTOPODIUM.

β. Leaves short, entire; cells minute. Capsule erect, ovoid, grooved; outer teeth recurved when dry . . . . 30. ZYGODON.

γ. Leaves flaccid; cells large, lax. Capsule pyriform, inclined or pendulous; mouth small; teeth horizontal . . . . . . 42. FUNARIA.

δ. Leaves various. Capsule cylindric, pyriform or clavate, suberect inclined or pendulous; outer teeth erect or spreading.

† *Capsule suberect.*

Capsule narrow, cylindric, furrowed . . . . 32. AULACOMNION.
Capsule oblong, furrowed . . . . 33. ORTHODONTIUM.
Capsule clavate, pyriform, not furrowed . . . . 34. BRACHYMENIUM.
Capsule clavate, pyriform, curved, gibbous, with a small oblique mouth . . . . 38. MEESIA.

†† *Capsule horizontal or drooping, smooth.*

Shoots from the sides or tops of the stems. Operculum shortly beaked or not beaked . . . . 36. BRYUM.
Shoots from the bases of the stems, often creeping. Operculum with a long beak. Leaves large, flaccid . . . . 37. MNIUM.

II. *Leaves distichous or 2–3-farious.* (*See* Distichium *in I.*)

Peristome 0. Leaves 3-farious, of 2 forms, the lateral distichous . 69. CALOMNION.
Teeth 16, 2-fid, spreading when dry. Leaves distichous, equitant . 7. FISSIDENS.

B. **Pleurocarpi.**—*Fruitstalk springing from the sides of the branches.* (*See* Fissidens adiantoides *and* Sphagnum, Dicnemon, Mielichoferia *and* Hedwigia, *which have apparently lateral fruitstalks under* A.)

Division 1. ASTOMI.—*Peristome* 0.

Leaves imbricate all round the stem . . . . 47. ANŒCTANGIUM.
Leaves distichous . . . . 49. AULACOPILUM.

Division 2. APLOPERISTOMI.—*Peristome single.*

* *Leaves distichous or tristichous.*

Calyptra mitriform, hairy, torn at the base. Teeth 16, entire. Leaves complicate, keeled, shining . . . . 58. PHYLLOGONIUM.
Calyptra cucullate. Teeth 16, membranous, united at the base. Leaves piliferous, papillose . . . . 66. HYMENODON.
Calyptra cucullate. Teeth 16, 2-fid, red. Fruitstalk moderate . 7. FISSIDENS.
Calyptra mitriform. Teeth 16, 2-fid, red. Fruitstalk very short . 8. CONOMITRIUM.
Calyptra mitriform, lacerate at the base. Teeth 16, membranous, united at the base. Leaves ciliate . . . . 67. HYPOPTERYGIUM.

** *Leaves not distichous. Calyptra cucullate.*

Teeth 16, short, membranous entire. Calyptra clothed below with long hairs . . . . 52. LEPTODON.
Teeth 16, 2–3-partite, united at the base . . . . 51. LEUCODON.
Teeth 16, in pairs, inflexed, entire . . . . 50. FABRONIA.
Teeth 16, short, free, lacerate at the apex . . . . 57. MESOTUS.

Division 3. Diploperistomi.—*Peristome double.*

* *Calyptra cucullate.*

α. *Leaves strongly distichous, tristichous or bifarious.*

† *Fruitstalk very short.*

Leaves distichous, oblique, broad, shining. Capsule erect . . . 59. NECKERA.

†† *Fruitstalk moderate or long.*

Leaves distichous, not oblique, serrate. Capsule erect . . . . 60. TRACHYLOMA.
Stems prostrate or creeping. Leaves distichous. Capsule cernuous or horizontal . . . . 64. OMALIA.
Stems erect, hardly branched. Leaves distichous, strongly serrate. Fruitstalk from the base of the stem. Capsule inclined . . . . 65. RHIZOGONIUM.
Stems dendroid, much branched. Leaves tristichous. Capsule suberect, cernuous or pendulous . . . . 67. HYPOPTERYGIUM,
Stems creeping or prostrate. Leaves distichous, piliferous, nerveless. Capsule cernuous . . . . 63. HYPNUM.

β. *Leaves imbricate all round the stem, secund or pointing every way, rarely obscurely distichous or bifarious.*

† *Fruitstalk very short.*

Stem erect. Capsule not furrowed. Leaves with a stout nerve . 56. CYRTOPUS.
Rhizome creeping. Capsule erect, furrowed. Leaves nerveless or 2-nerved . . . . 53. CLADOMNION.
Rhizome creeping. Stems pendulous, very long (erect in *M. nitens*). Leaves shining, membranous, concave. Capsule erect . . . 54. METEORIUM.

†† *Fruitstalk moderate or long.*

Habit dendroid. Stems naked below, above pinnately or verticillately branched. Capsule erect, or inclined by the curvature of the fruitstalk . . . . 61. ISOTHECIUM.
Stems creeping or prostrate, pinnately or vaguely branched. Capsule curved, horizontal, or pendulous, not furrowed. Calyptra glabrous . . . . 63. HYPNUM.
Stem compressed, pinnately branched. Capsule erect, straight, cylindric. Leaves entire, nerves 0, or 2, and short . . . . 62. ENTODON.
Stems erect, simple. Leaves strongly serrate. Fruitstalk from the bases of the stems. Capsule inclined . . . . 65. RHIZOGONIUM.
Stem creeping, branched. Leaves piliferous. Capsule curved, furrowed, cernuous. Calyptra hairy . . . . 70. RACOPILUM.

** *Calyptra mitriform.*

α. *Fruitstalk very short.*

Stem erect or pendulous, branched. Leaves imbricate all round the stem. Capsule erect, sunk in the perichætial leaves . . . 55. CRYPHÆA.
Stem erect, unbranched. Leaves tristichous, dorsal small. Capsule erect, almost sessile . . . . 68. CYATHOPHORUM.

β. *Fruitstalk moderate or long.*

Stems creeping, branched. Leaves obscurely tristichous, piliferous. Capsule curved, grooved. Calyptra hairy . . . . 70. RACOPILUM.
Stems dendroid. Leaves tristichous, dorsal smaller. Capsule suberect cernuous or pendulous . . . . 67. HYPOPTERYGIUM.
Stems suberect or creeping. Leaves more or less distichous; cells large, lax. Capsule cernuous. Teeth with transverse bars . . 71. HOOKERIA.
Stems erect, tufted. Leaves imbricate all round. Capsule erect . 72. DALTONIA.

The arrangement adopted in this work is that which Mr. Wilson followed in the Floras of New Zealand and of Tasmania, viz. Dr. Schimper's, with a few unimportant modifications. This arrangement is natural, and the best hitherto devised, but, unfortunately, the characters of the groups do not contrast, and depend to so great a degree upon habit, texture, etc., that they cannot be expressed with sufficient definition for the purposes of a beginner. The following is the order of the Suborders, Tribes, and Genera:—

SUBORDER I. **Andreæeæ.**
1. Andreæa.
SUBORDER II. **Sphagneæ.**
2. Sphagnum.
SUBORDER III. **Bryeæ.**
Sect. 1. ACROCARPI.—*Fruit terminal.*
Tribe I. PHASCEÆ.
3. Phascum.
Tribe II. WEISSIEÆ.
4. Gymnostomum.
5. Weissia.
6. Symblepharis.
Tribe III. FISSIDENTEÆ.
7. Fissidens.
8. Conomitrium.
Tribe IV. DICRANEÆ.
9. Dicnemon.
10. Leucobryum.
11. Dicranum.
12. Dicranodontium.
13. Campylopus.
14. Trematodon.
Tribe V. TRICHOSTOMEÆ.
15. Trichostomum.
16. Tortula.
17. Didymodon.
18. Desmatodon.
19. Distichium.
20. Ceratodon.
Tribe VI. ENCALYPTEÆ.
21. Encalypta.
Tribe VII. HEDWIGIACEÆ.
22. Hedwigia.
23. Braunia.
24. Leptangium.
Tribe VIII. GRIMMIEÆ.
25. Grimmia.
26. Racomitrium.
Tribe IX. ORTHOTRICHEÆ.
27. Schlotheimia.
28. Macromitrium.
29. Orthotrichum.
Tribe X. ZYGODONTEÆ.
30. Zygodon.
Tribe XI. BRYEÆ.
31. Leptostomum.
32. Aulacomnium.
33. Orthodontium.
34. Brachymenium.
35. Mielichoferia.
36. Bryum.
37. Mnium.
38. Meesia.
Tribe XII. BARTRAMIEÆ.
39. Conostomum.
40. Cryptopodium.
41. Bartramia.
Tribe XIII. FUNARIEÆ.
42. Funaria.
43. Entosthodon.
44. Physcomitrium.
Tribe XIV. SPLACHNEÆ.
45. Eremodon.
Tribe XV. POLYTRICHEÆ.
46. Polytrichum.
47. Dawsonia.
Sect. 2. PLEUROCARPI.—*Fruit lateral.*
Tribe XVI. ANŒCTANGIEÆ.
48. Anœctangium.
Tribe XVII. FABRONIEÆ.
49. Aulacopilum.
50. Fabronia.
Tribe XVIII. LEUCODONTEÆ.
51. Leucodon.
52. Leptodon.
53. Cladomnion.
Tribe XIX. PILOTRICHEÆ.
54. Meteorium.
55. Cryphæa.
56. Cyrtopus.
57. Mesotus.
Tribe XX. PHYLLOGONIEÆ.
58. Phyllogonium.
Tribe XXI. NECKEREÆ.
59. Neckera.
60. Trachyloma.
Tribe XXII. ISOTHECIEÆ.
61. Isothecium.
62. Entodon.
Tribe XXIII. HYPNEÆ.
63. Hypnum.
Tribe XXIV. OMALIEÆ.
64. Omalia.
Tribe XXV. RHIZOGONIEÆ.
65. Rhizogonium.
66. Hymenodon.
Tribe XXVI. HYPOPTERYGIEÆ.
67. Hypopterygium.
68. Cyathophorum.
69. Calomnion.
Tribe XXVII. RACOPILEÆ.
70. Racopilum.
Tribe XXVIII. HOOKERIEÆ.
71. Hookeria.
72. Daltonia.

1. ANDREÆA, Ehrhart.

Small, perennial, dark brown, purplish or black, monœcious or diœcious mosses, growing usually on rocks. Leaves often secund; cells dot-like. Fruitstalk terminal, short, seated on an elongated pale receptacle, as in *Sphagnum*. Capsule erect, black, splitting laterally into 4 valves, which are held together by the persistent operculum. Calyptra mitriform, small.

Arctic, alpine, and subalpine mosses, found in all mountainous countries; intermediate in respect of the form of the capsule between *Hepaticæ* and *Musci*.

Leaves nerveless.
  Leaves subulate-lanceolate, gibbous at the base . . . . 1. *A. acutifolia.*
  Leaves ovate-subulate, papillose, spreading, sheathing at the base . . 2. *A. rupestris.*
  Leaves ovate-lanceolate, not papillose, erecto-patent . . . . . . 3. *A. mutabilis.*
  Leaves oblong-ovate, obtuse, shining, erect . . . . 4. *A. nitida.*
Leaves with a broad nerve, ovate-subulate, falcate . . . . 5. *A. subulata.*

1. **A. acutifolia,** *Hook. f. and Wils. Fl. Antarct.* i. 118. *t.* 151. *f.* 2; —*Fl. N. Z.* ii. 57. Stems $\frac{3}{4}$–1 in. high. Leaves reddish or almost black, erecto-patent, incurved, erect when dry, subulate-lanceolate, acute, rigid, concave, nerveless, gibbous at the base; perichætial similar.

Northern Island: Ruahine range, *Colenso*. **Campbell's** Island: on rocks (barren), *J. D. H.* (Kerguelen's Land, Fuegia and Andes of South America).

2. **A. petrophila,** *Ehrhart;—Fl. Tasm.* ii. 161.—*A. rupestris*, Fl. N. Z. ii. 57, not of Linn. Stems very short. Leaves reddish-brown, not glossy, spreading, often subsecund, ovate-subulate or acuminate, base sheathing, back papillose, appressed when dry, nerveless.—Wils. Bryol. Brit.

Northern Island: on the mountains, *Colenso*. (Fuegia, Tasmania, Andes of South America, Europe, and most subalpine regions.)

3. **A. mutabilis,** *Hook. f. and Wils. Fl. Antarct.* i. 119. *t.* 57. *f.* 2. Stems $\frac{1}{2}$–1 in. high, naked below, branched above, slender. Leaves red-brown, yellow at the base, erecto-patent, appressed when dry, rarely falcate and secund, lanceolate or ovate-lanceolate, slightly concave, nerveless, hardly papillose at the back; perichætial longer. elliptic-lanceolate, convolute.

Lord Auckland's group and **Campbell's** Island: on rocks on the hills, *J. D. H.* (Falkland Islands). Near *A. rupestris*, but the leaves are more erect and narrower.

4. **A. nitida,** *Hook. f. and Wils. Fl. Antarct.* i. 118. *t.* 57. *f.* 3. Stems suberect, $\frac{1}{2}$–$\frac{3}{4}$ in. high, sparingly branched. Leaves red-brown or black-purple, crowded, suberect or erecto-patent, scarcely crisped when dry, ovate-oblong, obtuse, apiculate, nerveless, shining, concave; margins reflexed; perichætial longer, erect. Antheridia 6, with a few paraphyses mixed; archegonia 4, without paraphyses.

Lord Auckland's group: on rocks on the hills, rare, *J. D. H.* A very distinct and curious species. (Tasmania).

5. **A. subulata,** *Harvey in Hook. Ic. Pl. t.* 201;—*Fl. Antarct.* i. 119. *t.* 57. *f.* 1. Stems rather branched. Leaves red-brown, falcate, secund, subulate from a dilated base, concave, rigid; nerve very broad and thick; perichætial small, very inconspicuous.

Lord Auckland's group and **Campbell's** Island: rocks on the hills, *J. D. H.* (Tasmania, Australia, Cape of Good Hope, and Fuegia The specimens from South Africa have larger perichætial leaves.

2. SPHAGNUM, Linn.

Flaccid, white or pale-pink, branched, monœcious or diœcious mosses, growing in water bogs or wet woods; branches fascicled or whorled. Leaves of very narrow long cells, forming a network, between whose meshes are larger thin-walled cells, often marked with a spiral line and pierced with large pores. Fruitstalk axillary, at first lateral, but by elongation of the receptacle apparently terminal (*cladocarpous*). Capsule erect, dark-brown, without annulus or teeth. Operculum small, flattish. Calyptra rupturing transversely from the receptacle, the lower part adhering to the fruit; antheridia subspherical, as in *Jungermanniæ*.

A considerable genus of very variable mosses, found in all parts of the world, especially alpine and temperate, contributing extensively to the formation of peat. The species are most difficult of determination, being variable in size and habit, and the specific characters are to a great extent founded on structural peculiarities, the relative constancy of which may be much overrated. Much stress is laid upon the presence or absence of spiral marks within the cells of various parts of the plant; and on the number of series of cells forming the periphery (cortex) of the stem and branches.

The genus itself forms an Order, distinct from that of mosses, in the opinion of many modern Cryptogamists, but though most peculiar in habit and mode of growth, and differing in so many details from all other mosses, I do not think that it should be so regarded, the essential structure of the organs of reproduction being the same, and the range of variation amongst *Musci* in both habit and characters being very wide.

Branches with one or two cortical layers of cells.
Leaves ovate-acuminate . . . . 1. *S. cuspidatum.*
Leaves subovate, obtuse or præmorse . . . . 2. *S. subsecundum.*
Leaves obovate-oblong, obtuse, denticulate . . . . 3. *S. novo-Zelandicum.*
Branches with three or more cortical layers of cells.
Cells of branches with spiral fibres . . . . 4. *S. cymbifolium.*
Cells of branches without spiral fibres.
Stem-leaves fimbriate at the tip . . . . 5. *S. fimbriatum.*
Leaves of branches ovate, obtuse . . . . 6. *S. australe.*
Leaves of branches orbicular-ovate, obtuse . . . . 7. *S. antarcticum.*
Leaves of branches ovate-lanceolate, acuminate . . . . . 8. *S. acutifolium.*

1. **S. cuspidatum,** *Ehrhart;—Fl. N. Z.* ii. 58. Stems long, weak, flaccid; branches remote, fascicled, deflexed, narrowed to the apex. Stem-leaves ovate, acute, spreading; branch-leaves lanceolate-acuminate; margin waved when dry; perichætial acute. Inflorescence diœcious.—Wils. Bryol. Brit. 21. t. 61.

Var. *β. recurvum.* Leaves shorter, recurved when dry.—*S. recurvum,* Pallisot.
Var. *γ. plumosum.* Leaves longer, more attenuated.—*S. plumosum.*

Northern Island, *Colenso.*—Var. *γ*, Oldfield's collection. The above is taken from Wilson. Schimper describes this species as having two layers of cortical cells, Mitten as having but one. (Europe, Fuegia, etc.).

2. **S. subsecundum,** *Nees and Hornsch.?—S. compactum,* var. *ambiguum,* Fl. N. Z. ii. 57. Stems erect, cæspitose, fastigiate; branches crowded, short, erecto-patent; cortical cells in one layer, without spiral fibres. Leaves subovate; apices scarcely incurved, obtuse or premorse.

Northern Island: marshes, Bay of Islands, etc., *Colenso,* etc. A very doubtful plant, differing from *S. compactum* in the cortical cells being in only one layer, referred by Mr. Mitten to *S. subsecundum,* which is a native of Europe, etc.

3. **S. novo-Zelandicum,** *Mitten in Journ. Linn. Soc.* iv. 99. Habit

compact of *S. cymbifolium*. Cortical cells without spiral fibres. Leaves, cauline obovate-lingulate, obtuse, denticulate; branch-leaves ovate acuminate.

New Zealand, Northern Island: *Kerr, Knight*. Description abridged from Mitten's.

4. **S. cymbifolium,** *Dillenius;—Fl. N. Z.* ii. 57. Stems long, stout; branches crowded, short, tumid. Cortical cells of the branches in several series, with spiral fibres. Leaves broadly ovate or orbicular-ovate, obtuse, concave, minutely rough below the tip. Inflorescence diœcious.—Wils. Bryol. Brit. 17. t. 4.

Northern and **Middle** Islands: common, *Colenso, Sinclair, Haast,* etc. (Various temperate climates).

5. **S. fimbriatum,** *Wils. in Fl. Antarct.* 398;—*Fl. N. Z.* ii. 57. Stems long, slender; branches slender, deflexed, with narrow tips. Stem-leaves obovate, obtuse, fimbriate at the tip, others ovate-lanceolate, acuminate; perichætial obovate, obtuse, cucullate. Inflorescence monœcious.—Wils. Bryol. Brit. 21. t. 60; *S. acutifolia,* Mont. Voy. au Pôle Sud, Bot. Crypt. 282.

Middle Island, *Lyall* (a scrap), identified by Wilson. The cauline leaves are loosely reticulated, the proper cells are full of chlorophyll, the interstitial without a spiral fibre (Fuegia, Falkland Islands, Britain, etc.).

6. **S. australe,** *Mitten in Fl. Tasm.* ii. 162.—*S. compactum, γ. ovatum,* Fl. Antarct. 122. Habit, etc., of *S. cymbifolium,* but the cortical layers of cells have no spiral fibres. Leaves of branches ovate, obtuse, of branchlets ovate-lanceolate, obtuse.

Campbell's Island: in bogs, *J. D. H.* (Tasmania).

7. **S. antarcticum,** *Mitten in Journ. Linn. Soc.* iv. 100. Also similar to *S. cymbifolium,* but differing in the cortical cells having no spiral fibre. From *S. australe* it differs in the suborbicular leaves of the branchlets.

Campbell's Island: in bogs, *J. D. H.*

8. **S. acutifolium,** *Ehrhart.* Stem elongate; branches crowded, slender, attenuate, cortical cells in 3 or 4 layers, without spiral fibres. Leaves of stem ovate, erect, of branches erecto-patent, ovate-lanceolate, tapering, subacute, imbricate.—Wils. Bryol. Brit. 20. t. 4.

Chatham Island, *W. Travers;* determined by Mitten. (Europe, etc)

3. PHASCUM, Linn.

Extremely minute, monœcious, green, ephemeral mosses, growing on earth, mudbanks, etc. Fruitstalks terminal, very short. Capsule erect, not dehiscing. Operculum persistent and continuous with the capsule. Calyptra campanulate or cucullate.

A considerable genus in temperate regions, rare in tropical; the species, owing to their most minute size and fugacious character, are often overlooked, and require high powers of the microscope for their study. The genus has been divided by modern Cryptogamists into eight or more.

Stems annual, erect, simple . . . . 1. *P. apiculatum.*
Stems perennial, creeping . . . . 2. *P. nervosum.*

1. **P. (Acaulon) apiculatum,** *Hook. f. and Wils. Fl. N. Z.* ii. 58. *t.* 83. *f.* 1. A minute gregarious annual moss, each plant simple, the size of a grain

of mustard, consisting of a little ball of leaves surrounding a capsule. Leaves few, closely imbricate, two inner the largest, broadly obovate-rotundate, acute or acuminate, very concave, quite entire; nerve excurrent; apex erect; margins not recurved. Capsule immersed, globose, erect; fruitstalk very short. Calyptra very minute.

Northern Island: Hawke's Bay, on the ground, *Colenso*. (Tasmania.) Allied to the British *P. muticum*. Mitten (Journ. Linn. Soc. Bot. iv. 71) has pointed out that the figure of the calyptra in Fl. N. Z. is quite erroneous. It belongs to the subgenus *Acaulon*, which are annual, simple, and have globose immersed capsules with a minute calyptra.

2. **P. (Pleuridium) nervosum,** *Hook. Musc. Exot. t.* 105;—*Fl. N. Z.* ii. 58. A minute, creeping, perennial moss; stems short, nearly simple. Lower leaves ovate; upper elliptic-lanceolate, acuminate, entire or denticulate, appressed; nerve stout, excurrent. Fruitstalk very short, straight or curved. Capsule immersed or exserted, apiculate.

Northern Island: Bay of Islands, in clay hills, *Colenso, J. D. H.* (Australia, South Africa, North America).

4. GYMNOSTOMUM, Hedwig.

Tufted, usually short, green, monœcious or diœcious mosses, of various habits, usually growing on rock or earth. Fruitstalk terminal, slender, rarely short. Capsule erect, rarely inclined; annulus obscure, persistent; teeth 0. Operculum obliquely beaked. Calyptra cucullate.

For other acrocarpous mosses, without peristome, that might be referred here, see in *Weissia, Didymodon, Macromitrium, Zygodon, Leptostomum, Physcomitrion*, and *Hedwigia*.

A very large genus, common in temperate countries, rarer in arctic and tropical; as formerly constituted, it contains mosses of very various affinities, which agreed only in wanting the peristome; recently, however, it has been restricted to mosses which, but for the above character, would be referred to *Weissia*.

Leaves linear-lanceolate; margins flat . . . . 1. *G. calcareum.*
Leaves oblong-lanceolate; margins incurved . . . . 2. *G. tortile.*

1. **G. calcareum,** *Nees and Hornsch.;—Fl. N. Z.* ii. 59. Very minute; stems densely tufted, $\frac{1}{8}$–$\frac{1}{4}$ in. high, branched, very slender. Leaves spreading, linear-lanceolate, rather obtuse, sometimes crisped when dry; margins flat. Fruitstalk very slender, $\frac{1}{8}$ in. long. Capsule oval-oblong; neck short; mouth red; operculum conico-subulate, nearly as long as the capsule.

Northern and **Middle** Islands: Bay of Islands, on clay soil, the form with crisped leaves, on lime in walls, *J. D. H.;* Nelson, *Mantell;* Otago, *Hector and Buchanan.* (Europe, India, and Australia). I follow Mr. Wilson in placing this moss in *Gymnostomum*, on account of the absence of peristome; the European state has 16 equidistant teeth, and is *Seligeria calcarea*, Br. and Sch. (Wils. Bryol. Brit. 54. t. 15).

2. **G. tortile,** *Schwægrichen;—Fl. N. Z.* ii. 59. Stems branched, $\frac{1}{6}$–$\frac{1}{3}$ in. high, pulvinate-cæspitose; branches fastigiate. Leaves crowded, spreading, curved upwards, twisted when dry, oblong-lanceolate, rather obtuse acuminate or apiculate; margins incurved, quite entire; nerve thick. Capsule ovoid, rather thick; operculum with a long beak.—Wils. Bryol. Brit. 45. t. xxxviii.; *Hymenostomum*, Bridel, Bryol. Europ.

Northern Island: Bay of Islands, on clay soil, *Colenso, J. D. H.;* North Cape, *Jol-*

liffe; Auckland, *Knight.* Very similar to *Weissia controversa*, but more robust, foliage firmer, less incurved at the margin, and peristome 0. (Europe, S. Africa.)

5. WEISSIA, Hedwig.

Tufted, green, usually short, monœcious or diœcious, perennial mosses, of various habit. Leaves spreading, curved and crisped when dry, nerved; cells minute. Fruitstalk terminal, usually slender. Capsule erect, oval-oblong; annulus persistent or 0; teeth 16 (rarely 0), in one row, equidistant, free to the base, transversely barred, entire or perforate, apices sometimes 2 fid. Operculum obliquely beaked. Calyptra cucullate.

An extensive genus, especially abundant in temperate regions.

Fruitstalk slender. Capsule exserted.
- Leaves lanceolate, spreading; margin incurved . . . . 1. *W. controversa.*
- Leaves narrow-lanceolate, spreading; margin nearly flat . . . . 2. *W. flavipes.*
- Leaves subulate-lanceolate, secund . . . . 3. *W. crispula.*
- Leaves oblong below, subulate above . . . . 4. *W. irrorata.*

Fruitstalk very short. Capsule immersed . . . . 5. *W. contecta.*

1. **W. controversa,** *Hedwig;—Fl. N. Z.* ii. 59. Tufted, bright green; stems short. Leaves spreading, linear-lanceolate or lanceolate, mucronate; margins quite entire, incurved or involute; nerve excurrent. Fruitstalk $\frac{1}{6}$–$\frac{1}{4}$ in. long. Capsule ovoid, substriate when dry; teeth linear-lanceolate, rather obtuse, nearly entire; operculum conical, beaked.—Wils. Bryol. Brit. 46. t. xv.

Northern Island: Bay of Islands, on clay banks, *J. D. H.;* Auckland, *Knight.* (Tasmania, Europe, and many other countries.)

2. **W. flavipes,** *Hook. f. and Wils. Fl. N. Z.* ii. 59. *t.* 83. *f.* 2. Stems tufted, $\frac{1}{4}$ in. high. Leaves erecto-patent, twisted inwards when dry, linear-lanceolate or linear from a broader base, apiculate; margin nearly flat or subincurved; nerve excurrent. Fruitstalk very slender, $\frac{1}{2}$ in. long, yellow. Capsule cylindric, erect; mouth red; teeth 0 or acute, perforate, denticulate, tapering from the base, red; operculum with a slender beak.

Northern Island: Bay of Islands, on the ground, *J. D. H.; Auckland, Knight, Sinclair* (a form with no teeth). Mr. Wilson, who has examined the Auckland specimens very closely, finds no trace of teeth, and suggests that this may be a different species, for which he proposes the name *Gynostomum patulum.* The specimens are all immature. (Tasmania.)

3. **W. crispula,** *Ludwig;—Fl. Antarct.* i. 127. *t.* 58. *f.* 2. Stems tufted, bright green, $\frac{1}{2}$–$\frac{3}{4}$ in. high. Leaves imbricate, erecto-patent, secund, elongate, lanceolate-subulate, acuminate, channelled, quite entire, more or less crisp when dry; perichætial acuminate, with an excurrent nerve. Fruitstalk $\frac{1}{4}$ in. long. Capsule erect, oblong; teeth entire; operculum with a long slender beak, almost as long as the capsule.—Wils. Bryol. Brit. 48. t. xv.; *Blindia antarctica*, C. Muell. Synops. Musc. i. 344.

Campbell's Island: rocks on the mountains, abundant, *J. D. H.* This is a variety of the European plant (*β. ambigua*, Wils.), which has the leaves shorter, less secund, less crisped when dry, a shorter fruitstalk, and the teeth sometimes approximate in pairs. (Fuegia.)

4. **W. (Eucladium) irroratum,** *Mitten, mss.* Verdigris-green, in-

crusted with calcareous matter; stems about 1 in. high, with crowded shoots. Leaves erecto-patent, below oblong, above narrow linear-subulate, obtuse at the apex; nerve broad, papillose at the back; margin subdenticulate above and crenulate or papillose at the apex; lower cells oblong, green, pale, pellucid, upper shorter, subrotund; perichætial twice as broad, obtuse. Fruitstalk ¾ in. long. Capsule ovoid; teeth short, red (*Mitten*).

Northern Island, *Stephenson*. Mitten describes this as resembling *W. verticillata*, Brid., but having narrower and more obtuse leaves.

5. **W. contecta,** *Hook. f. and Wils. Fl. Antarct.* 127. *t.* 58. *f.* 3. Stems densely tufted, ½–1½ in. high, rigid, fragile, fastigiately branched. Leaves dark-green, crowded, erect, striate, rigid, lanceolate below, narrowed into a long subulate point, channelled; nerve solid, excurrent; margin quite entire, below inflexed; perichætial similar, larger. Fruitstalk very short. Capsule immersed, ovoid; mouth open, exannulate; teeth yellow-red, reflexed when dry, pyramidal, entire or 2-fid; operculum with a long slender oblique beak. Calyptra small, subulate.—*Blindia*, C. Muell.

Campbell's Island: on alpine rocks, *J. D. H.* A remarkable plant, with the habit of some small *Dicrana*, but the teeth entire or nearly so. (Kerguelen's Land.)

6. SYMBLEPHARIS, Montagne.

Tufted perennial mosses, growing on trees or stones. Leaves long, slender; perichætial sheathing the fruitstalk; cells dot-like. Fruitstalk terminal, slender. Capsule erect, subcylindric; mouth small; annulus persistent; teeth 8, in one row, bigeminate, or 32 in fours, short, conniving into a cone when dry. Operculum with a very slender beak. Calyptra long, subulate, dimidiate.

A small genus, confined to the southern hemisphere. The name *Holomitrium* should, in M. Mitten's opinion, be retained for this genus.

Leaves erecto-patent, crisped when dry . . . . 1. *S. perichætialis.*
Leaves falcate-secund, unaltered when dry . . . . 2. *S. pumila.*

1. **S. perichætialis,** *Wils. in Fl. N. Z.* ii. 60. Stem erect, branched, ½–1 in. high. Leaves crowded, erecto-patent, crisped when dry, lanceolate below, narrowed to a long subulate point, quite entire, channelled; nerve continuous; margin flat; perichætial very long, sheathing. Fruitstalk ¼–1 in. long, pale; annulus small, persistent; teeth 32, erect or recurved when moist, dull red, 2-fid at the apex, perforate below; operculum as long as the capsule, almost setaceous. Calyptra twice as long. Male inflor. unknown. —*Trichostomum*, Hook. Musc. Exot. t. 73; *Olomitrium* and *Holomitrium*, Bridel; *Acalyphum cylindricum*, Palisot; *Sprucea*, Hook. f. and Wils. Fl. Antarct. i. 128.

Common throughout the **Northern** and **Middle** Islands, *Menzies*, etc., and **Campbell's** Island, *J. D. H.* (Australia, Tasmania, the Mauritius.)

2. **S. pumila,** *Mitten, mss.* Stems slender, tufted, brownish-green. Leaves all falcate and secund, unaltered when dry, circinately curved, linear-lanceolate, gradually narrowed from the base, quite entire; nerve slender, percurrent; cells at the base large, pale, quadrate. Fruitstalk rather stout. Capsule immature.

Middle Island: dry ground, Otago, *Hector and Buchanan*. Allied to *H. antarcticum*,

C. Mueller, which has distinct masses of alary cells at the contracted bases of the somewhat crisped leaves (*Mitten*).

7. FISSIDENS, Hedwig.

Small, usually bright green, monœcious or diœcious mosses, growing on earth and stones, rarely on roots of trees. Leaves distichous, equitant, alternate, semiamplexicaul, the blade unequal-sided and vertical; cells minute or lax and distinct. Fruitstalk terminal in all but one of the New Zealand species, sometimes on short lateral shoots. Capsule erect inclined or cernuous; annulus 0 or present; teeth 16, in one row, equidistant, with a dark line down the middle, entire 2-fid or 3-fid at the apex, when dry spreading and curling outwards at the base and then inwards. Operculum conical or beaked. Calyptra cucullate, rarely mitriform.

A very large genus of mosses, abounding in all parts of the world except the arctic and alpine. The New Zealand species, I suspect, want a very careful re-examination, with much more numerous specimens than have been hitherto collected. The characters of several are very difficult to detect—if, indeed, they be valid,—especially the denticulation and thickened margin of the leaf, length of stem, and direction of the capsule.

Fruitstalk lateral . . . . 1. *F. adiantoides.*
Fruitstalk terminal.

a. Margin of leaf entire or nearly so, not thickened and hyaline.
Stem ½–1 in. Leaves crisped when dry. Fruitstalk stout . . . 2. *F. asplenioides.*
Stem ½–1 in. Leaves not crisp when dry. Fruitstalk slender . 3. *F. oblongifolius.*
Stem ¼ in. Leaves pale-yellow . . . . 4. *F. pallidus.*

β. Margin of leaf crenulate . . . . 5. *F. tenellus.*

γ. Margin of leaf thickened and hyaline.
Leaves nerveless . . . . 6. *F. dealbatus.*
Leaves furnished with a nerve.
† *Stems* 1–2 *in long* . . . . 7. *F. rigidulus.*

†† *Stems less than* ½ *in. long.*

Leaves broad-lanceolate, apiculate . . . . 8. *F. bryoides.*
Leaves lanceolate, acuminate . . . . 9. *F. viridulus.*
Leaves ovate-acute, cymbiform . . . . 10. *F. brevifolius.*
Leaves verdigris-green, opaque, lanceolate . . . . 11. *F. æruginosus.*

1. **F. adiantoides,** *Hedwig.* Stems 1–2 in. high, elongate, branched. Leaves crowded, crisped and incurved when dry, ovate-lanceolate, denticulate at the apex, minutely serrate below, nerved to the apex. Fruitstalk lateral. Capsule cernuous, ovoid-oblong; operculum beaked.—Wils. Bryol. Brit. 307. t. 16.

Middle Island: Otago, in damp places, *Hector and Buchanan.* (Europe, Tasmania, N. America.)

2. **F. asplenioides,** *Swartz.*—*F. ligulatus,* Hook. f. and Wils. Fl. N. Z. ii. 63. t. 84. f. 1. Stems ½–1 in. high, erect. Leaves very numerous, suberect, rather remote, crisped and involute when dry, ligulate, obtuse; margin subcrenulate or denticulate, not thickened; nerve stout, pellucid, not continuous to the apex; cells minute. Fruitstalk stout. Capsule cernuous, with a large mouth; operculum conic, beaked, half as long as the capsule; male inflorescence on separate stems from the female.

Northern Island: Bay of Islands, near waterfalls, *J. D. H.*, etc.; Auckland, *Knight;*

Wangaroa, *Jolliffe.* (Tasmania, Tristan d'Acunha.) M. Mitten has referred *F. ligulatus* to *F. asplenioides*, Sw.

3. **F. oblongifolius,** *Hook. f. and Wils. Fl. N. Z.* ii. 62. *t.* 83. *f.* 8. Stems $\frac{1}{3}$–$\frac{2}{3}$ in. long, erect. Leaves very numerous, linear or linear-lanceolate, obtuse acute or acuminate, somewhat inflexed when dry; margin not thickened, quite entire except at the crenulate apex; nerve pale, pellucid, not continuous to the apex; cells roundish, opaque. Fruitstalk slender. Capsule very small, ovoid, inclined or subcernuous; operculum not seen. Male inflorescence axillary on the same stems with the female.

Var. α. Leaves linear, ligulate, obtuse.

Var. β. Leaves linear-lanceolate, acuminate.

Northern Island: Bay of Islands, on rocks near waterfalls, both varieties, *J. D. H.* (Tasmania.)

4. **F. pallidus,** *Hook. f. and Wils. Fl. N. Z.* ii. 62. *t.* 83. *f.* 7. Stems tufted, decumbent, very small. Leaves flabellately spreading, pale yellow-green, glossy, inflexed but scarcely altered in drying, linear, hardly lanceolate, acute; margin not thickened, quite entire; nerve continuous to the apex; cells hexagonal, pellucid. Capsule cernuous; operculum longer than the capsule. Calyptra submitriform, inflexed at the base. Male inflorescence not seen.

Northern Island, *Colenso.* (Tasmania.)

5. **F. tenellus,** *Hook. f. and Wils. Fl. N. Z.* ii. 62. *t.* 83. *f.* 6. Minute. Stems very short, decumbent. Leaves few, linear-lanceolate, acuminate, straight; margin not thickened, crenulate; sheathing base denticulate; nerve stout, concolorous, continuous to the apex, excurrent; cells minute. Capsule erect; annulus 0; operculum with a beak nearly as long as the capsule. Calyptra papillose at the apex. Male inflorescence basilar in Tasmanian specimens.

Northern and **Middle** Islands: Bay of Islands, *J. D. H.*; Auckland, *Sinclair*; Thomson's Sound, *Lyall.* (Victoria, Tasmania.)

6. **F. dealbatus,** *Hook. f. and Wils. Fl. N. Z.* i. 63. *t.* 84. *f.* 2. Minute. Stems very short, not $\frac{1}{6}$ in. long, slender, pale greenish-white. Leaves few, linear-oblong, acuminate; margin thickened, quite entire, nerveless; cells large, lax, rhomboid. Fruitstalk short. Capsule suberect, narrow-ovoid, contracted below the mouth; annulus 0; operculum conical, beaked, as long as the capsule. Calyptra red-brown. Male inflorescence on different stems from female.

Northern Island: Bay of Islands, on rocks near waterfalls, *J. D. H.*, growing with other mosses. (Tasmania.)

7. **F. rigidulus,** *Hook. f. and Wils. Fl. N. Z.* ii. 61. *t.* 83. *f.* 3. Stem 1-2 in. long, slender, branched, leafy. Leaves rigid, crisped when dry, lurid green, ovate-lanceolate, subacute; margin broad, much thickened, entire; nerve continuous to the apex, pellucid, stout. Fruitstalk $\frac{1}{8}$–$\frac{1}{4}$ in. long, pale, slender, sometimes several together. Capsule small, ovoid, oblique; operculum shortly beaked. Male inflorescence terminal.

Northern and **Middle** Islands: Bay of Plenty, *Jolliffe*; East Cape, *Sinclair*; Wellington, *Lyall;* Auckland, *Knight;* Southern Alps, *Hector, Travers,* etc.; Chain Hills, Otago, *Lindsay.* (Tasmania.)

8. **F. bryoides,** *Hedwig;—Fl. N Z.* ii. 61. Stems $\frac{1}{6}$–$\frac{1}{2}$ in. high. Leaves broadly lanceolate, apiculate; margin thickened, quite entire; nerve slightly excurrent. Fruitstalk red. Capsule elliptic-oblong, erect; operculum conical, acuminate. Male inflorescence axillary on the stem with the female.—Wils. Bryol. Brit. 304. t. xvi.

Northern Island: Bay of Islands, *J. D. H.* Specimens imperfect. (Europe, N. America.)

9. **F. viridulus,** *Wahlenberg;*—var. *acuminatus,* Fl. N. Z. ii. 61. Stems short, simple, decumbent. Leaves rigid, lanceolate, acuminate; margin much thickened, quite entire, blade on the lower half, not continued to the base; nerve nearly continuous to the apex. Capsule oblong-ovoid, erect; operculum conical, acuminate. Inflorescence monœcious, male terminal or basilar. —Wils. Bryol. Brit. 303. t. liii.

Var. *incurvus.* Capsule cernuous, curved.—Wils. Bryol. Brit. 303. t. 53; *F. incurvus,* Schwæg.—Fl. N. Z. ii. 61.

Northern Island: Bay of Islands, *Colenso* and *J. D. H.* (Europe, N. America, both vars.) Distinguished from *F. bryoides* by the male inflorescence being never axillary.

10. **F. brevifolius,** *Hook. f. and Wils.;—Fl. N. Z.* ii. 61. *t.* 83. *f.* 4. Stems very short, $\frac{1}{2}$ in. long, giving off barren creeping shoots. Leaves on the shoots crowded, very shortly ovate, acute, cymbiform, lower margin not continued to the base; margins not thickened beyond the sheathing part, entire; nerve continuous to the apex; stem-leaves narrower, with a subulate or lanceolate acuminate blade. Capsule suberect or inclined; operculum conic with a short beak. Male inflorescence at the base of the barren shoots (axillary in a Tasmanian state).

Northern Island: *Colenso.* Allied to *F. viridulus,* but differing in bearing barren shoots, in the very short leaves, and in the thickening of the margin of the leaf being confined to the sheathing portion. (Tasmania, extratropical S. Africa, and S. America.)

11. **F. æruginosus,** *Hook f. and Wils.;—Fl. N. Z.* ii. 62. *t.* 83. *f.* 5. Stems very short, decumbent. Leaves crowded, verdigris-green, narrow lanceolate, straight, acuminate, not crisped when dry, opaque; margin not thickened, minutely crenulate towards the apex; nerve pellucid, continuous to the apex; cells dot-like. Capsule not seen. Male inflorescence basal.

Northern Island: *Colenso.* Described from a few fragments only.

8. CONOMITRIUM, Montagne.

Aquatic, slender, branched mosses. Leaves distichous, equitant, alternate, semi-amplexicaul, the blade unequal-sided, vertical. Fruitstalk very short, lateral or terminal. Capsule small, erect, equal, gradually narrowed below; teeth 16, in one row, equidistant, regular or irregular, with no dark line down the centre, truncate, 2-fid at the apex. Operculum conical. Calyptra conical, nearly entire at the base.

A small genus of mosses, of which one species is European.

1. **C. Dillenii,** *Montagne;—Fl. N. Z.* ii. 63. Stems floating, simple

or branched. Leaves oblong-lanceolate, erect, straight, nerve not continuous to the apex. Fruitstalk short, lateral, solitary or 2 together. Capsule ovoid; operculum cuspidate, incurved; teeth regular or irregular.—Montagne in Ann. Sc. Nat. 1837, 250. *Skitophyllum*, La Pylaie. *Octodiceras*, Bridel.

Northern Island: East Coast, by watercourses, *Colenso.* (Australia, S. America.)

9. DICNEMON, Schwægr.

Creeping tufted mosses, with short erect curved branches, growing on trees. Leaves imbricate all round the stems. Fruitstalk terminal on very short lateral branchlets, clothed by the long perichætial leaves which partially hide the capsule also. Capsule elongate, oblique, inclined; teeth 16, in one row, 2-fid to below the middle, the divisions tubercled, incurved. Operculum with a slender oblique beak. Calyptra large, cucullate, rough at the apex, inflated when young.

A genus generally associated with *Leucodon*, but according to Wilson most clearly allied to *Dicranum*, and differing chiefly in the creeping habit.

1. **D. calycinum,** *Wils. and Hook.;—Fl. N. Z.* ii. 64. Stems creeping; branches erect or ascending, divided, 1–1½ in. long, stout, terete, acute. Leaves closely imbricate, pale yellow-green, ovate-lanceolate, concave, quite entire; nerve suddenly ceasing; perichætial large and sheathing. Fruitstalk ¼ in. long. Capsule narrow oblong, subcylindrical, somewhat curved and strumose, tapering into the fruitstalk. Male inflorescence hidden amongst the stem-leaves.—*Leucodon calycinus*, Hook. Musc. Exot. t. 17.

Northern and **Middle** Islands: on trunks of trees, common, *Menzies*, etc.

10. LEUCOBRYUM, Hampe.

Rather large, tufted, white, stout, perennial, monœcious mosses, growing in wet ground, in woods, and on trunks of trees. Leaves spongy, nerveless, glaucous, cells perforated. Fruitstalk terminal, but apparently lateral. Capsule cernuous, strumous, grooved when dry; annulus 0; teeth 16, in one row, subulate, transversely barred, 2-fid at the apex, rough externally. Operculum long-beaked. Calyptra inflated, cucullate, long-beaked.

A considerable tropical genus, rare in temperate regions, remarkable for its white colour, and the spongy texture of the leaves, which are formed of two layers of cells.

1. **L. candidum,** *Hampe;—Fl. N. Z.* ii. 64. Stems erect, tufted, ½–2 in. high, stout, brittle, dichotomously branched. Leaves densely imbricated, erect or falcate and recurved, ovate-lanceolate, concave, wrinkled at the back near the apex. Fruitstalks on short lateral branches. Capsule cernuous, strumose, grooved. Male inflorescence axillary, clustered, on separate stems.—*Dicranum*, Bridel. *L. brachyphyllum*, Hornsch., C. Muell. in Bot. Zeit. 1851, 546.

Common throughout the islands, on decayed trunks, etc., seldom fruiting. (Tasmania, Australia.)

11. DICRANUM, Hedwig.

Tufted, erect, green, perennial, monœcious or diœcious mosses, often large, of various habit. Fruitstalks terminal, slender, sometimes 2–3 together. Capsule inclined or cernuous, often strumose, rather long; teeth 16, subulate, joined at the base, transversely barred, 2-fid (often unequally) to the middle, rarely 3-fid, when dry spreading at the base and then curling upwards. Operculum with a long oblique beak. Calyptra cucullate.

One of the largest genera of Mosses, very abundant in all regions of the globe, but especially the temperate and cold.

I. LEUCOLOMA. Fruitstalk on a very short lateral shoot. Margin of leaf white, composed of long slender flexuous cells.
Upper cauline leaves piliferous . . . . 1. *D. Sieberianum.*
Cauline leaves all acuminate . . . . 2. *D. incanum.*
II. TRIDONTIUM. Teeth usually 3-fid. Leaves oblong-lanceolate, obtuse . . . . 3. *D. Tasmanicum.*
III. DICRANUM. Margin of leaf not pellucid. Teeth 2-fid (rarely 3-fid).
* Leaves with broad subquadrate sheathing bases.
Stems 1–2 in. Leaves squarrose. Capsule suberect . . . 4. *D. vaginatum.*
Stems $\frac{1}{3}$–$\frac{1}{2}$ in. Leaves flexuous Capsule cernuous, strumose 5. *D. Schreberi.*
Stems $\frac{1}{4}$–$\frac{1}{2}$ in. Leaves strict. Capsule suberect, not strumose 6. *D. campylophyllum.*
** Leaves without broad quadrate sheathing bases, usually secund.
Capsule equal, erect . . . . 7. *D. trichopodum.*
Capsule unequal, inclined or cernuous.
† Stems stout. Perichætial leaves long, striate.
+ Nerve stout broad.
Leaves ovate-lanceolate below, setaceous above . . . . . 8. *D. dicarpon.*
Leaves convolute . . . . 9. *D. robustum.*
‡ Nerve more slender.
Stems 1 in. high. Leaves spinulose-serrate . . . . . 10. *D. fasciatum.*
Stems 2–4 in. high. Leaves serrated towards the tip . . 11. *D. Billardieri.*
†† Stem stout or slender. Perichætial leaves short.
Fruitstalk 1 in. long, stout . . . . 12. *D. setosum.*
Fruitstalk hardly longer than the leaves . . . . . 13. *D. Menziesii.*

1. **D. Sieberianum,** *Hornsch.;—Fl. N. Z.* ii. 67. Stems $\frac{1}{2}$–1 in. high, fastigiately branched, $\frac{1}{10}$ in. diam. Leaves yellowish, falcate, convolute, ovate-lanceolate, acuminate, the upper piliferous, papillose at the back, perichætial sheathing, piliferous; margin entire with a narrow pellucid border. Fruitstalks $\frac{1}{2}$–1 in. long, slender, red. Capsule small, ovoid, cernuous; operculum with a long beak.—*Leucodon pallidus*, Hook. Musc. Exot. t. 172.

Var. *β*. Leaves appressed when dry, nearly smooth at the back.

Northern Island: Bay of Islands. *α*. *Cunningham*. *β*. On wet rocks, *J. D. H.* (Australia, Tasmania.)

2. **D. incanum,** *Mitt. mss.* (*Leucoloma*).—Stems $\frac{1}{2}$–1 in. high, creeping below, fastigiately-branched. Leaves pale-yellow-green, erecto-patent, subsecund, rather crisped when dry, ovate-lanceolate, acuminate, convolute, none piliferous, except the perichætial. Fruitstalk $\frac{1}{4}$–$\frac{1}{2}$ in. long, red. Capsule subcylindrical; operculum with a beak as long as the capsule.

Northern Island: *Sinclair*.

3. **D. Tasmanicum,** *Hook. f. in Hook. Ic. Plant. t.* 248 (*Tridontium*); —*Fl. N. Z.* ii. 65. Stems erect, tufted, 1–2 in. high, branched. Leaves pale or dark green or reddish, lax, spreading, lanceolate-oblong, obtuse, channelled,

usually concave at the apex, quite entire; crisped and incurved when dry; nerve vanishing. Fruitstalk stout, $\frac{1}{2}$ in. long. Capsule erect, turbinate; mouth large, teeth large, often 3-fid; annulus 0; operculum with a slender beak, longer than the capsule. Inflorescence diœcious.

Northern and **Middle** Islands: not uncommon on wet stones, rocks, etc. *Colenso*, etc. (Tasmania.)

4. **D. clathratum,** *Mitten, mss.;—D. vaginatum*, var. *clathratum*, Fl. N. Z. ii. 65. Stems 1–2, long, slender, branching. Leaves distant, imbricating, with a broad square sheathing base and long subulate rigid upper portion, the latter is squarrose when dry, almost entire or toothed at the apex, and wholly occupied by the stout nerve. Fruitstalk slender. Capsule ovoid or turbinate, erect or suberect; annulus 0; teeth long, papillose, variously perforate, sometimes 3-fid; operculum with a long beak. Inflorescence diœcious.

Northern Island: wet rocks, Bay of Islands, falls of the Keri Keri river, *J. D. H.*

5. **D. Schreberi,** *Hedwig;—Fl. N. Z.* ii. 65. Stems tufted, short, $\frac{1}{4}$–$\frac{1}{3}$ in. high, nearly simple. Leaves with a broad sheathing base, and lanceolate-subulate, spreading, channelled, nearly entire upper part, which is flexuous when dry; nerve slender. Fruitstalks $\frac{1}{4}$–$\frac{1}{3}$ in. high. Capsule ovoid-oblong, cernuous, strumose; operculum with a short beak as long as the capsule. Inflorescence diœcious.—Wils. Bryol. Brit. 69. t. xxxix.

Northern Island: on moist banks, Bay of Islands, *J. D. H.* (Britain.) Mr. Wilson remarks that this differs from the European form in the entire leaves.

6. **D. campylophyllum,** *Tayl.?—Fl. N. Z.* ii. 65. Closely resembling *D. Schreberi*, but leaves longer and more rigid, less suddenly dilated below. Capsule less oblique, longer, without a strumose operculum.

Northern Island: *Colenso.* Hawke's Bay, *Jolliffe.* (Andes.)

7. **D. trichopodum,** *Mitten, mss.* Stems 1 in. high, tomentose. Leaves subsecund, pale yellowish, erecto-patent, base elliptic-oblong, then setaceous acuminate; nerve rigid, occupying the whole subulate portion of the leaf; margins serrulate at the apex; perichætial convolute, with long slender apices. Fruitstalk slender, $\frac{1}{2}$ in. long, flexuous, yellow. Capsule cylindric-ovoid, equal, erect; mouth small; teeth short, red.—Mitten.

Middle Island: Otago. *Hector and Buchanan.* Foliage of *D. setosum* and capsule of *Holomitrium perichætiale.*

8. **D. dicarpon,** *Hornsch.;—Fl. N. Z.* ii. 66. Stems tufted, robust, 1–2 in. high, densely covered with matted fibrils. Leaves spreading, squarrose or subsecund, linear-cuspidate, from an ovate-lanceolate base, flexuous, striate; margins and back spinulose-serrate; nerve solid. Fruitstalks 2 or more together, sheathed by long perichætial leaves. Capsules curved, cernuous; operculum with a long beak.—Schwægr. Suppl. t. 251. *D. leucolomoides*, C. Muell. in Bot. Zeit. 1851, 550.

Northern and **Middle** Islands, common. (Tasmania and Australia.) *D. spinosum*, Wils. mss., is considered by him a large variety with numerous (3–8) longer setæ.

9. **D. robustum,** *Hook. f. and Wils. Fl. Antarct.* 406. *t.* 152. *f.* 8;—*Fl. N. Z.* ii. 66. Stems forming large tufts, tall and robust. Leaves sub-

erect or falcate and subsecund, flexuous, long subulate-lanceolate, narrowed into very long almost capillary points, convolute below, spinulose-serrate above; nerve variable in width, well defined, excurrent; perichætial longer, sheathing. Fruitstalk 1 inch long. Capsule cylindrical, inclined or nearly erect, substrumose; operculum with a long beak.

Var. β. *pungens.* Perichætial leaves longer, capsule less curved.—*D. pungens*, Hook. f. and Wils. Fl. Antarct. i. 129. t. 59. f. 1.

Middle Island: var. β, Jackson's Bay, *Lyall.* **Lord Auckland's** group and **Campbell's** Island, *J. D. H.* (Norway, Australia, Tasmania, Chile.)

10. **D. fasciatum,** *Hedwig;—Fl. N. Z.* ii. 66. Stems stout, 1 in. high, covered with matted fibrils. Leaves glossy yellow-green, substriated, crowded, secund, subulate-lanceolate, channelled, acuminate, spinulose-serrate; nerve slender; perichætial much longer, often overtopping the capsules, long-acuminate. Fruitstalks usually twin, very short. Capsule almost hidden, oblong, curved, substrumose; annulus 0; operculum with a long beak. Calyptra rough at the apex. Inflorescence monœcious or diœcious.

Northern and **Middle** Islands. Bay of Islands, *Sinclair, J. D. H.;* Auckland, *Jolliffe;* Otago, *Hector and Buchanan.* The figure of Hedwig is indifferent, and Wilson thinks may indicate a different species.

11. **D. Billardieri,** *Bridel;—Fl. N. Z* ii. 66. Stems stout, forming large tufts, 2–4 in. high. Leaves ¼ in. long, yellow-green, membranous, falcate, secund, ovate-lanceolate, acuminate, concave, not striated, serrulate towards the apex; nerve slender; perichætial longer. Fruitstalks longer than the perichætial leaves. Capsule subcylindrical, curved, strumose; operculum with a long beak.—Schwægr. Suppl. t. 121.

Var. β. *duriusculum*, Fl. Antarct. i. 129. Stems short, fastigiately branched; leaves more rigid, nerve broader. Fruitstalk longer.

Abundant in all the islands, and as far south as **Campbell's** Island, on old trees, etc. Var. β, **Lord Auckland's** group and **Campbell's** Island. (Australia, Tasmania, S. America.)

12. **D. setosum,** *Hook. f. and Wils. Fl. Antarct.* 129. *t.* 58. *f.* 5. *n.* 2; —*Fl. N. Z.* ii. 66. Stems densely tufted, short, brittle, 1–4 in. long, yellow, glossy, sparingly branched. Leaves ½ in. long, strict, suberect, crowded, fragile, very long lanceolate and setaceous, serrulate; nerve broad excurrent in a long seta, occupying the whole of the middle of the leaf; perichætial sheathing, short. Fruitstalk 1 in. long, stout. Capsule suberect, oblong, curved; operculum with the beak longer than the capsule. Calyptra pale-brown, red at the apex.

Var. β. *attenuatum*, Fl. Antarct. l. c. f. 5. n. 2. Stems longer and branched.

Northern and **Middle** Islands, *Colenso*, etc., and Port Preservation, *Lyall.* **Lord Auckland's** and **Campbell's** Islands: α and β, on the ground and on roots of trees, etc., common, *J. D. H.* (Tasmania, Chile.)

13. **D. Menziesii,** *Taylor in Phytologist,* ii. 1094;—*Fl. Antarct.* 128. *t.* 58. *f.* 4; *Fl. N. Z.* ii. 67. Stems tufted, rather slender, 1–2 in. long, covered with matted whitish fibrils. Leaves yellow-green, crowded, secund, rather rigid, straight when dry, subulate-lanceolate and serrulate at the base, narrowed into a very long straight setaceous point; nerve solid, very strong when dry; perichætial shorter. Fruitstalk very short, hardly longer than the

leaves. Capsule oblong, suberect, substrumose; operculum with a beak as long as the capsule. Male inflorescence nestling amongst the fibrils of the stem. —*D. brachypelma*, C. Muell. in Bot. Zeit. 1851. 550.

Var. *β. rigidum*, Fl. N. Z. Stem stouter; leaves less crowded, more rigid, spreading, subfalcate.

Common throughout the islands and in **Lord Auckland's** group. (Tasmania, Australia, Polynesia, Chile.)

Of *D. dichotomum*, Bridel, said by Montagne in Voy. au Pôle Sud, 297, to have been brought by Hombron from **Lord Auckland's** group, I know nothing. It is a Bourbon species.

12. DICRANODONTIUM, Brid.

Erect, tufted, usually terrestrial, diœcious mosses. Leaves imbricating all round the stems, with broad sheathing bases and a broad nerve. Fruitstalk terminal, curved when moist. Capsule pendulous or decurved, owing to the curvature of the fruitstalk, equal or unequal, striate or plicate; teeth 16, linear-lanceolate, remotely articulate, 2-fid to below the middle or to the base, divisions unequal, subulate; annulus narrow, persistent. Operculum oblique, subulate. Calyptra cucullate, smooth, not fringed at the base.

A small genus, chiefly insular and tropical, similar in habit and characters to *Campylopus*, but differing from that genus in wanting the lamellate leaves and fringed base to the calyptra, and from *Dicranum* in the teeth and pendulous capsule. Mitten refers the species to *Leptotrichum*, Hampe.

Leaf suddenly contracted from a broad base . . . . 1. *D. flexipes.*
Leaf linear-lanceolate . . . . 2. *D. lineare.*

1. **D. flexipes,** *Mitten, mss.*—*D. proscriptum*, Fl. N. Z. ii. 67, not of Hornschuch. Stems slender, simple, ⅛ in. long. Leaves falcate, secund, yellow-green, brown when old, base sheathing, then rigid setaceous or capillary, quite entire or serrulate about the middle; nerve stout. Fruitstalk curved when moist, straight when dry. Capsule elliptic, substriate when dry; annulus large. Inflorescence diœcious.

Var. *β.* Taller; leaves lax, squarrose, broader and more sheathing below.
Var. *γ.* Stem short; leaves crowded, falcate.

Northern Island: Bay of Islands, etc., *J. D. H.*, *Colenso*, etc. Mr. Mitten points out that this differs from the *D. proscriptum* of St. Helena in the gibbous capsule, small struma, and less suddenly subulate leaves.

2. **D. lineare,** *Mitten, mss.* Stems very short, tufted, less than ¼ in. high. Leaves pale-green, soft, spreading, linear-lanceolate; nerve continuous, keeled; margins serrulate; perichætial elliptic-ovate, convolute. Fruitstalk as long as the perichætial leaves. Capsule curved, ovoid, subequal, subapophysate, with 8 furrows when dry; teeth slender, yellow, broad; operculum conic-subulate, half as long as the capsule. Inflorescence monœcious.

Middle Island: Canterbury, *Travers*. Distinguished from all others by the cauline leaves not subulate from dilated bases, by the minutely serrulate leaves and defined nerve. (*Mitten*).

13. CAMPYLOPUS, Bridel.

Tufted, erect, pale-green or yellowish, often shining, diœcious mosses;

stems simple. Leaves in interrupted tufts, usually piliferous, with very broad nerves, lamellated at the back. Fruitstalk terminal, arcuate, often clustered. Capsule often turned down or completely round and up again by the curving of the fruitstalk, ovoid, equal or strumose, striated; annulus double, coiling off; teeth of *Dicranum*. Operculum obliquely beaked. Calyptra cucullate, fringed at the base.

A large genus, especially common in tropical and southern mountains and oceanic islands, chiefly distinguished from *Dicranum* by the often interrupted piliferous leaves, the curious coil in the fruitstalks, and fringed calyptra. The species are often very difficult of discrimination, and the New Zealand ones are probably far from satisfactorily determined, owing to the want of good specimens, and different views of authors.

I. Leaves with white hair-like points.
  α. Hair-like points reflexed . . . . 1. *C. introflexus.*
  β. Hair-like points straight.
    Leaves appressed when dry, gradually narrowed from an elliptic base . . . . 2. *C. appressifolius.*
    Leaves lanceolate, suddenly narrowed into short points . . 3. *C. clavatus.*
    Leaves lanceolate, suddenly narrowed into long points . . 4. *C. torquatus.*
II. Leaves without hair-like points . . . . 5. *C. bicolor.*

1. **C. introflexus,** *Hedw. Sp. Musc. t.* 29;—*Fl. N. Z.* ii. 69. Stems erect, branched, often proliferous and thickened at the ends. Leaves yellowish or yellow-brown or -green, densely imbricate, base broad, concave, lanceolate, sheathing, with a pellucid margin, suddenly contracted into a white, reflexed, toothed, hair-point; nerve broad, well defined below. Fruitstalks crowded, hidden amongst the perichætial leaves. Capsule obovate, unequal, striated; operculum conical.—*C. xanthophyllus*, Montagne; Fl. N. Z. ii. 68; *C. atro-virens*, De Notaris; Mont. in Voy. Pôle Sud; Bot. Crypt. 300?; *Dicranum leptocephalum*, C. Muell. in Bot. Zeit. 1851. 351.

Common throughout the islands and in **Lord Auckland's** group and **Campbell's** Island, *J. D. H.* (A common southern and northern species.)

2. **C. appressifolius,** *Mitten, mss.*—*C. clavatus*, Fl. N. Z. ii. 62 (not of Brown). Stem 1–2 in. long, slender, simple, thickened above. Leaves dull green, appressed when dry, spreading when moist, gradually dilated from an elliptic-lanceolate base to an acuminate, hair-like, serrulate, hyaline point; marginal cells below forming a subquadrate brown area on each side (*Mitten*).

Northern Island: Mount Eden, near Auckland, *Jupp* (*Herb. Mitten*). **Middle** Island, *Lyall*, barren.

3. **C. clavatus,** *Brown in Schwægrichen, Suppl. t.* 255 A;—*Fl. N. Z.* ii. 69, in part. Stems densely tufted, dichotomously branched; branches much thickened at the tips. Leaves golden-yellow, below lanceolate, suddenly acuminate into short, white, straight hair-points; nerve solid. Fruitstalk ¼ in. long, red. Capsule pendulous or erect, striate.

Northern Island: Taranaki hills, *Jupp* (*Herb. Mitten*). **Middle** Island: Nelson, *Herb. A. Richard;* Dusky Bay, *Lyall.* I am indebted to Mitten for distinguishing this from the preceding. (Tasmania, Australia, St. Paul's Island.)

4. **C. torquatus,** *Mitten in Fl. Tasman.* ii. 173.—*C. pallidus*, Hook. f. and Wils. Fl. N. Z. ii. 68. t. 84. f. 3, in part. Stems short, nearly simple. Leaves soft, pale green or whitish, of lax and spongy texture, crowded, sub-

erect; base ovate-lanceolate, narrowed into a long subulate, setaceous, white spinulous point; nerve very broad, spongy. Fruitstalks crowded, stout, spirally twisted. Capsule pyriform, pale, substance thick, grooved when dry; mouth purple; operculum small. Calyptra short.—*C. torfaceus*, Mitt. in Hook. Kew Journ. Bot. 1856. 257; *Dicranum flexuosum*, C. Muell. in Bot. Zeit. 1851. 551.

Northern and **Middle** Islands: East Coast, *Colenso*; Kiapara, *Mossman*; Auckland, *Sinclair*, *Knight*; Hokianga, *Sinclair*; Otago, *Hector and Buchanan*. **Campbell's** Island, barren, *J. D. H.* (Tasmania.)

5. **C. bicolor,** *Hornschuch;—Fl. N. Z.* ii. 69. Stems densely tufted. Leaves crowded, lower black, upper glossy green, stout, subulate-lanceolate, obtuse, concave at the apex, without hair-points; nerve broad.

Northern Island, *Colenso*; Kiapara, *Mossman*. (Tasmania, Australia.)

14. TREMATODON, Richard.

Usually short, tufted, annual, pale-green mosses, growing on earth. Leaves very slender. Fruitstalk terminal, flexuous or coiled. Capsule oblong, cernuous, with a long narrow neck (struma), annulate; teeth 16, in one row, subulate, transversely barred, 2-fid, often unequally, rough and granulated. Operculum long beaked. Calyptra cucullate, inflated.

A small genus, not rare in warm and temperate latitudes.

Capsule much shorter than its neck . . . . 1. *T. suberectus.*
Capsule as long as its neck.
  Capsule deflexed . . . . 2. *T. arcuatus.*
  Capsule suberect . . . . 3. *T. flexipes.*

1. **T. suberectus,** *Mitten, mss.*—*T. longicollis*, Fl. N. Z. ii. 69 (not of Richard). Stems short, slender, erect, sparingly branched. Leaves pale green, subulate, with setaceous denticulate points, flexuous when dry; nerve stout; perichætial very long, also denticulate at the apex. Fruitstalk $\frac{1}{2}$–$\frac{3}{4}$ in. long. Capsule elongate, nearly erect, much shorter than the somewhat curved neck; operculum conical, beaked, as long as the capsule. Inflorescence diœcious.

Northern and **Middle** Islands: Bay of Islands, on clay banks, *J. D. H.*; Wellington, *Lyall*. Mitten distinguishes this from *T. longicollis* by the shorter neck of the capsule and inflorescence.

2. **T. arcuatus,** *Mitten, mss.* Very similar to *T. suberectus*, but the capsule is as long as its neck, and deflexed.

Northern Island: Wellington, *Stephenson*.

3. **T. flexipes,** *Mitten in Fl. Tasman.* ii. 173. *t.* 172. *f.* 6. Stems very short. Leaves erecto-patent, lanceolate-subulate, quite entire; nerve broad, occupying the whole upper three-fourths of the leaf. Fruitstalk short, flexuose. Capsule suberect, red-brown, as long as its neck; teeth red, cleft; operculum with a curved beak.

Middle Island: Otago, *Hector and Buchanan*. (Tasmania.)

15. TRICHOSTOMUM, Bridel.

Usually tufted, short, green, slender, monœcious or diœcious mosses, growing on earth or rocks. Leaves narrow; cells minute above, large at the base; nerve strong, continuous or excurrent. Fruitstalk terminal. Capsule long, erect; annulus present or absent; teeth 32, in one row, approximated in pairs, filiform (very short in *T. mutabile*), tetragonous, granulate, closely barred, twisted inwards. Operculum elongate, obliquely beaked. Calyptra cucullate.

A large genus, both tropical and temperate, very closely allied to *Tortula*, and often hardly distinguishable. The teeth sometimes appear as 16, simple or perforated.

I. PILOPOGON.—Leaves with a hair-point . . . . 1. *T leptodum.*
II. Leaves linear, ligulate or lanceolate.
Leaves crowded, linear-oblong, erect when dry; margin reflexed; nerve vanishing . . . . 2. *T. lingulatum.*
Leaves oblong-lanceolate, acute, concave at the apex; margin flat 3. *T. phæum.*
Leaves linear-lanceolate; nerve excurrent; margin flat . . . 4. *T. mutabile.*
Leaves crowded, lanceolate, obtuse, concave at the tip; nerve excurrent; margins undulate . . . . 5. *T. rubripes.*
III. Leaves broad and sheathing below, above subulate, setaceous.
Leaves flexuous, lower ⅓ ovate. Capsule oblong . . . . . 6. *T. laxifolium.*
Leaves spreading, lower ¼ oblong. Capsule cylindric . . . . 7. *T. elongatum.*
Leaves strict, lower ⅓ ovate-lanceolate, keeled . . . . . . 8. *T. setosum.*
Leaves flexuous, lower ½ linear-oblong, upper part denticulate . 9. *T. australe.*

1. **T. (Pilopogon) leptodum,** *Mitten, mss.—Campylopus leptodus,* Montagne;—Fl. N. Z. ii. 68. Stems tufted, proliferous; branches fascicled. Leaves lanceolate-subulate, strict, with a short, white, toothed hair-point, upper recurved; nerve slender. Fruitstalks crowded, flexuous. Capsule oblong, straight, smooth; teeth divided nearly to the base, divisions very long and slender; operculum subulate.

Northern Island, *Colenso,* in various places; Auckland, *Knight.* (Habit of a *Campylopus.*)

2. **T. lingulatum,** *Hook. f. and Wils. Fl. N. Z.* ii. 71. *t.* 84. *f.* 4. Stems very short, ⅛ in. high. Leaves pale, spreading, crowded, rather flaccid, erect when dry, linear-oblong or ovate-lingulate, obtuse, keeled; margins slightly reflexed; nerve vanishing; cells rather large. Fruitstalks ¼–⅓ in. long, pale. Capsule erect, cylindric; annulus 0; operculum conico-subulate.

Var. *β.* Leaves longer; capsule subcylindric; operculum longer, beaked.
Northern Island: Bay of Islands, *J. D. H.* Var. *β. Colenso.*

3. **T. phæum,** *Hook. f. and Wils. Fl. N. Z.* ii. 72. *t.* 84. *f.* 5. Stems very short, tufted ⅓ in. high. Leaves purplish-brown, erecto-patent, rather dense and rigid, opaque, crisped when dry, oblong-lanceolate, keeled, acute, concave at the apex; margin flat, entire, opaque; nerve continuous, pellucid. Fruitstalk stout, ⅓ in. long. Capsule large, erect, oblong, dark red; teeth imperfect, connected by a basal membrane, oblique.

Northern Island: shores of Lake Waihau, *Colenso.* Specimens insufficient.

4. **T. mutabile,** *Bruch;—Fl. N. Z.* ii. 72. Stems tufted. Leaves

bright green, spreading, crisped when dry, linear-lanceolate, mucronate, somewhat keeled, margins nearly flat; nerve excurrent. Fruitstalk ½ in. long, yellowish. Capsule ovoid; annulus 0; teeth variable, very short, unequal; operculum with a long beak.—Wils. Bryol. Brit. 112. t. 41.

Northern Island, *Colenso; Canterbury, Travers.* (Europe.)

Mr. Wilson remarks that this moss, which is usually found near the seacoast in Britain, is rare in fruit; and the often imperfect peristome renders it difficult to distinguish from *Gymnostomum tortile.* Mitten says of Mr. Travers's specimens, that the leaves are only half as wide as in the European specimens, gradually tapering, and the operculum in a young state exceeds the capsule; it may be a different species.

5. **T. rubripes,** *Mitten, mss.* Stem short. Leaves dense, spreading, lanceolate, obtuse, concave at the apex; nerve pale, excurrent; margins flexuous, quite entire, incurved above. Fruitstalk red, flexuous. Capsule cylindric, erect; teeth oblique, red, nodular, free to the base; operculum conical, ⅓ as long as the capsule. Inflorescence diœcious.—Mitten.

New Zealand, *Kerr* (in Herb. Mitten). Almost a *Tortula,* and closely resembling *T. Knightii.*

6. **T. laxifolium,** *Hook. f. and Wils. Fl. N. Z.* ii. 72. Stems very short, ¼ in. long, simple. Leaves yellowish, distant, spreading, very long, setaceous-subulate, with an ovate sheathing base, flexuous, channelled, quite entire; nerve solid. Fruitstalk ½–1 in. long, slender, red. Capsule suberect, oblong, rather oblique, gibbous, narrowed towards the mouth; annulus distinct; operculum conical and subulate. Inflorescence monœcious.—*Dicranum flexifolium,* Hook. Musc. Exot. t. 144.

Northern Island: Bay of Islands, clay-hills, *J. D. H., Colenso;* Auckland, *Bolton.* (Tasmania, S. Africa, and S. America.)

7. **T. elongatum,** *Hook. f. and Wils. in Fl. Tasman.* ii. 176. *t.* 173, *f.* 1. Stem ¾–1 in. long, simple, rigid. Leaves lax, spreading, subsecund, rather flexuous and rigid, lower portion broad, elliptic-oblong, membranous; margin pellucid, suddenly contracted to a long subulate keeled point three times as long; margin roughish or subserrate. Fruitstalk 1 in. long, reddish. Capsule pale brown, cylindric, elongate; annulus large; operculum conic-subulate, half as long as the capsule. Inflorescence monœcious.

New Zealand (Fl. Tasman.). (Tasmania.)

8. **T. setosum,** *Hook. f. and Wils. Fl. N. Z.* ii. 73. *t.* 84. *f.* 6. Stems short, ¼–½ in. high, simple. Leaves pale yellow-green, lower ones reddish, strict, erect, crowded, setaceo-subulate from an ovate-lanceolate sheathing base, entire, sharply keeled to the apex; nerve broad, continuous. Fruitstalk very slender, ½–1 in. high, pale. Capsule oblong, oblique, mouth contracted; annulus distinct; operculum conical with a slender beak, half as long as the capsule. Inflorescence monœcious.—*Leptotrichum affine,* C. Mueller.

Northern and **Middle** Islands: Bay of Islands, etc., *J. D. H.; Colenso; Sinclair;* Auckland, *Knight;* Wellington, *Lyall.* (Australia, Chile.)

9. **T. australe,** *Mitten.—Leptotrichum,* Mitten in Linn. Soc. Journal iv. 66, Fl. Tasman. ii. 177; *T. longifolium,* Fl. N. Z. ii. 72. Stems elongate, branched, 1–4 in. high. Leaves lurid yellow-green, erecto-patent, strict,

lower half sheathing, linear-oblong, upper setaceous-subulate, denticulate; cells linear below, dense above; perichætial elongate, convolute. Fruitstalks ½ in. long, pale brown. Capsule ovoid, erect, brown; teeth short; operculum more than half as long as the capsule. Inflorescence monœcious, male terminal.—*Lophoidon longifolius* and *Didymodon longifolius*, var. 3, Fl. Antarct. 408. t. 59. f. 2; *Trichostomum longifolium* and *Distichium capillaceum*, Fl. N. Z. ii. 72 and 73.

Northern Island: mountainous districts, *Colenso;* Wairarapa Valley, *Knight.* **Middle** Island: Otago, *Lyall.* **Lord Auckland's** group and **Campbell's** Island, *J. D. H.* (Tasmania and Fuegia.)

Very imperfect specimens of a plant found with *T. lingulatum* at the Bay of Islands by myself, have been very doubtfully referred by Mr. Wilson to the European *T. strictum*, Bruch, (Fl. N. Z. ii. 72). The leaves are yellowish, strict, ovate-lanceolate, acuminate, mucronate by the thick excurrent nerve, erect or incurved when dry; the margin flat.

16. TORTULA, Schreber.

Tufted, green, yellowish or red-brown, usually short, annual or perennial monœcious or diœcious mosses, growing on the earth, sand, or stones, rarely on trees. Fruitstalk terminal. Capsule erect, oblong, annulate; teeth 32, very long, filiform, twisted to the left, uncoiling when moist, often united at the base into a long or short tube. Operculum conic, with an oblique, slender beak. Calyptra cucullate (mitriform in *Streptopogon*).

A very large genus, found in all parts of the world, and containing some of the most ubiquitous mosses.

I. *Calyptra cucullate. Nerve of the leaf produced into a white hair-like point.*

Leaves imbricate, concave; nerve proliferous . . . . 1. *T. chloronotos.*
Leaves spreading, concave; margin involute; nerve granular, green . 2. *T. papillosa.*
Leaves spreading; nerve red; margin revolute. . . . . 3. *T. Muelleri.*

II. *Calyptra cucullate. Nerve of the leaf excurrent, not white at the apex.*

a. *Perichætial leaves inconspicuous.*

† *Leaves serrulate or crenulate at the apex.*

Nerve smooth . . . . 4. *T. serrulata.*
Nerve rough, red . . . . 5. *T. rubra.*

†† *Leaves quite entire.*

Leaves lanceolate acuminate; margin revolute; nerve hardly excurrent 6. *T. torquata.*
Leaves lanceolate acuminate; margin revolute; nerve excurrent . . 7. *T. crispifolia.*
Leaves ligulate or lanceolate; nerve red; teeth scarcely twisted . . 8. *T. australasiæ.*
Leaves linear-lanceolate; nerve pellucid . . . . 9. *T. Knightii.*
β. *Perichætial leaves long and sheathing* . . . . 10. *T. calycina.*
III. STREPTOPOGON. *Calyptra mitriform. Leaves crisped, with pellucid margins* . . . . 11. *T. mnioides.*

1. **T. chloronotos,** *Bridel, Bryol. Univ. t.* 539;—*Fl. N. Z.* ii. 69. Stem very short, ½ in. high, nearly simple. Leaves imbricate, broadly ovate, concave, piliferous; margin entire, recurved; nerve filamentous, and bearing cellular buds; cells opaque. Fruitstalk slender, ⅓ in. long. Capsule elliptic-oblong.

Northern Island, *Colenso.* (Europe.)

2. **T. papillosa,** *Wils. Bryol. Brit.* 135. *t.* 44. Stems short, tufted,

sparingly branched. Leaves dull green, spreading, erect when dry, obovate, subacute, very concave, with the margin flat, shortly hair-pointed, strongly involute when dry, papillose at the back, and on the thick, spongy gemmiparous nerve; cells lax and succulent.

Northern Island. *Sinclair* (swamps only). (Europe and various other countries.) I have taken the description from the 'Bryologia Britannica.'

3. **T. Muelleri,** *Br. and Schimp.;—Fl. N. Z.* ii. 70. Stems elongated, covered with root-fibres. Leaves crowded, erecto-patent, straight, imbricate when dry, oblong-oval, obtuse, concave: margin reflexed; nerve purple-red, excurrent into a roughish hair-point. Fruitstalk long, purplish. Capsule cylindric, curved. Inflorescence monœcious.—*T. Antarctica,* Hampe; *Syntrichia princeps,* De Notaris.

Northern and **Middle** Islands: not uncommon, from the Bay of Islands, *J. D. H.* to Otago, *Lindsay.* Mitten refers to this common species *T. cuspidata* and *rubella,* Hook. f. and Wils. Fl. Tasman. 11, 175, 176. t. 172. f. 9 and 10. (Generally diffused.)

4. **T. serrulata,** *Hook. and Grev. in Brewst. Journ. Sc.* i. 229. *t.* 12;—*Fl. N. Z.* ii. 70. Stems elongate, somewhat branched, ½ in. long. Leaves orange-brown or rusty, subsquarrose when dry, lanceolate, acuminate, keeled; margin reflexed below, flat above, serrulate towards the apex; nerve stout, subexcurrent; cells minute. Fruitstalk stout, twisted, ½ in. long. Capsule inclined, cylindrical.

Northern Island: shores of Lake Waikau, *Colenso.* (Fuegia.)

T. rubra, *Mitten mss.—T. robusta,* β, Fl. Antarct. 409. Stems ½–2 in. high. Leaves at the apex of the stem, lower pale-red, upper greenish, crowded, spreading, oblong-lanceolate, broad at the base, gradually narrowed into an acute apex; margins recurved in the middle, crenate and serrulate at the apex; nerve red, scabrous at the back, excurrent. Fruitstalk red. Capsule cylindric.—Mitten.

Middle Island: Otago, *Hector and Buchanan.* (Falkland Islands, Australia). Distinguished from *T. Muelleri* by the serrulate leaves, and from *T. serrulata* by the scabrid nerve.

6. **T. torquata,** *Taylor;—Fl. N. Z.* ii. 70. Stem short, erect, tufted. Leaves crowded, erecto-patent, closely spirally twisted when dry, lanceolate, acuminate; margin quite entire, revolute; nerve stout, scarcely excurrent. Fruitstalk pale-red, slender, flexuose. Capsule erect, oblong-ovate.

Northern Island, *Colenso* (Australia, Tasmania). Closely allied to the European *T. unguiculata.*

7. **T. crispifolia,** *Mitten mss.* Stems 1 in. high. Leaves yellow-green, subcrisped when dry, spreading, lanceolate from a broad erect base, acuminate; margins recurved throughout; nerve concolorous, rather broad, excurrent into a pungent mucro. Fruitstalk red. Capsule cylindric-ovoid, oblique; teeth pale red and twisted; operculum half the length of the capsule. —Mitten.

Middle Island: Canterbury, *Sinclair and Haast.* Allied to the European *T. fallax,* but distinguished by the excurrent nerve.

8. **T. australasiæ,** *Hook. f. and Wils. Fl. N. Z.* ii. 70. Stem short,

simple, $\frac{1}{4}$–$\frac{1}{3}$ in. long. Leaves reddish, spreading, subrecurved, twisted and crisped when dry, subligulate linear or lanceolate, acute or obtuse, keeled; margin subreflexed; nerve red, running to the apex; cells minute. Fruitstalk slender, $\frac{1}{3}$ in. long, red. Capsule cylindrical-ovoid; teeth oblique, but hardly twisted; operculum with a short beak. Inflorescence diœcious.—*Trichostomum*, Hook. and Grev. in Brewst. Journ. Sc. i. t. 12; *T. fuscescens*, Hook. f. and Wils. in Fl. N. Z. ii. 73. t. 85. f. 1; *Tortula rufiseta*, Taylor.

Northern Island: Bay of Islands, etc., *Colenso, J. D. H.* Generically intermediate between *Tortula* and *Didymodon*, and perhaps a form of *T. cæspitosa*. Mitten identifies *Trichostomum fuscescens* with it. (Tasmania, Australia.)

9. **T. Knightii,** *Mitten in Fl. Tasman.* ii. 174. *t.* 172. *f.* 11.—*T. cæspitosa*, var., Fl. N. Z. ii. 70. Stems $\frac{1}{4}$–$\frac{1}{3}$ in. high. Leaves pale yellow-green, spreading, crisped when dry, linear-lanceolate from an oblong broader transparent base, very acuminate; margins quite entire, undulate throughout; nerve slender, excurrent. Fruitstalk slender, pale. Capsule cylindric.

Northern Island: Bay of Islands, *J. D. H., Colenso;* Auckland, *Knight.* Very near the European *T. Northiana.* (Tasmania.)

10. **T. calycina,** *Schwægr. Suppl. t.* 119;—*Fl. N. Z.* ii. 70. Stems very short, somewhat branched. Leaves yellow-green, subundulate, crisped when dry, narrow oblong-lanceolate, acuminate; margin flat; nerve stout; cells small; perichætial long, sheathing. Fruitstalk $\frac{1}{2}$–$1\frac{1}{2}$ in. long, red, yellow above. Capsule small, elliptic-oblong, suberect; operculum with a very long slender beak.

Northern and **Middle** Islands: Bay of Islands, *Logan, Colenso, J. D. H.;* Auckland, *Knight;* Nelson, *Jolliffe, Travers.* Apparently near *T. flexuosa* (Hook. Musc. Exot. t. 125) of South Africa, which has more erect leaves, less undulated and broader below. (Australia, Tasmania.)

11. **T. (Streptopogon) mnioides,** *Schwæg.:—Fl. N. Z.* ii. 71. Leaves yellowish, crisped, of firm texture, spreading, loosely imbricate, undulate, ovate-lanceolate, acuminate; margins pellucid, often gemmiparous at the apex.

Northern Island, *Colenso* (scrap only). Doubtfully referred to the European moss; it appears also to be Tasmanian.

17. DIDYMODON, Bruch and Schimper.

Short, tufted, monœcious or diœcious (rarely hermaphrodite) mosses, usually growing on earth. Leaves with minute cells above, larger at the base; nerve strong. Fruitstalk terminal, very slender, sometimes appearing lateral from the growth of shoots above it. Capsule oblong, erect; annulus persistent or unrolling; teeth 16, short, linear-subulate, excessively slender and fugacious, entire 2-fid or perforated along the middle line, not united by a membrane. Operculum conical, obtusely beaked. Calyptra cucullate.

A considerable genus, in both tropical and temperate regions, scarcely distinguishable from *Trichostomum*, except by the very tender fugacious peristome.

Leaves 3-farious, erect and subsecund when dry, quite entire . . . . 1. *D. papillatus.*
Leaves with acute, recurved, squarrose apices, serrulate towards the apex . . . . 2. *D. interruptus.*
Leaves twisted when dry, spreading, lanceolate, denticulate towards the apex . . . . 3. *D. erubescens.*

1. **D. papillatus,** *Hook. f. and Wils. Fl. N. Z.* ii. 73. *t.* 85. *f.* 2. Stems loosely tufted, 1 in. long, rigid, brittle, a little branched; branches filiform, erect. Leaves yellowish, 3-farious, spreading, recurved, erect when dry and subsecund, ovate-lanceolate, acuminate, keeled, entire, papillose on both surfaces; margin below recurved; nerve subsolid; perichætial convolute. Fruitstalk pale, ½–¾ in. long. Capsule erect, narrow-oblong, pale, contracted at the mouth; annulus small; teeth 16, irregular, cloven and anastomosing; operculum conical, subulate.—Mitten in Journ. Linn. Soc. iv. 70; *Zygodon tristichus,* C. Muell. in Bot. Zeit. 1852, 764.

Northern and **Middle** Islands: Bay of Islands, *J. D. H.*, *Colenso* (barren); Nelson, *Jolliffe, Travers;* Otago, *Hector and Buchanan.* (Tasmania and South Africa.)

2. **D. interruptus,** *Mitten. mss.* Stems loosely tufted, 1–½ in. high, erect, often interrupted by innovations. Leaves pale yellow-green, with an erect subquadrate base, above lanceolate, acute, recurved, squarrose, keeled; margin crenulate or serrulate at the apex; nerve vanishing, papillose at the back; cells above minutely papillose.

HAB. New Zealand, *Kerr* (in Herb Mitten), *Sinclair.*

3. **D. erubescens,** *Mitten mss.* Stems subsimple, tufted, 1 in. high and upwards, dull rusty-green. Leaves lax, twisted when dry, with a broad, erect, appressed base, then spreading, lanceolate, acute obtuse or with a recurved apiculus; margins recurved throughout their length, denticulate towards the apex; nerve obscure, percurrent, keeled. Fruitstalk ½ in. long, reddish. Capsule cylindric; teeth short, narrow, coherent at the base; annulus compound; operculum obliquely beaked, ⅓ as long as the capsule. Inflorescence monœcious. Antheridia amongst the upper longer leaves.—Mitten.

Middle Island: Otago, *Hector.*

18. DESMATODON, Bridel.

Small, tufted, perennial, green, monœcious mosses (male inflorescence axillary), usually growing on earth. Foliage of *Tortula.* Fruitstalk terminal. Capsule erect, cernuous or pendulous, with a short neck; annulus simple; teeth 16 (sometimes absent), in one row, granulated, remotely barred, 2-fid or 3-fid, divisions tetragonous, distantly articulate, joined by one or two bars, when moist erect, when dry incurved or convolute. Operculum obtusely beaked. Calyptra cucullate.

A small genus, intermediate between *Tortula* and *Trichostomum*, and perhaps eventually to be merged with these. I follow Wilson in keeping it distinct, though he says, that "when highly developed, the peristome is scarcely, if at all, distinguishable from *Tortula.*"

1. **D. nervosus,** *Bruch and Schimper;—Fl. N. Z.* ii. 71. Stems erect, ¼ in. high. Leaves yellow-green, spreading, convolute when dry, ovate-oblong, apiculate, concave; margin reflexed; nerve stout, thickened upwards, excurrent. Fruitstalk $\frac{1}{10}$–¼ in. long. Capsule ovoid, erect; teeth variable, unequal, sometimer twisted sometimes absent; annulus 0; operculum conical, hardly beaked.—Wils. Bryol. Brit. 103. t. 20; *Trichostomum convolutum,* Bridel.

Northern Island, *Colenso.* (Europe, S. America, Tasmania.)

19. DISTICHIUM, Br. and Schimp.

Stems tufted, erect, flexuose, beset with radicles, with shoots under each inflorescence. Leaves glossy, distichous or 3-ranked, with imbricate sheathing bases, setaceous, flexuose; nerve broad. Capsule erect or inclined, with a short tapering base; annulus of 2 rows of cells, spirally uncoiling; teeth 16, inserted below the orifice, free to the base, equidistant, linear or lanceolate, transversely barred, with the middle line entire perforated or cleft. Operculum conical, beaked. Calyptra cucullate.

A small genus of alpine European mosses.

1. **D. capillaceum,** *Br. and Sch.* Stems forming dense and wide-spreading tufts, 1–4 in. high. Leaves setaceous from a lanceolate sheathing base, quite entire. Capsule erect, ovoid-oblong or subcylindric; teeth narrow with distant articulations, irregularly 2-cleft.—Wils. Bryol. Brit. 104. t. 20 (not Fl. N. Z. ii. 73, which is *Trichostomum australe*).

Middle Island: Southern Alps, *Sinclair and Haast.* (Europe, Asia, N. Africa, N. America.)

20. CERATODON, Bridel.

Short, tufted, diœcious mosses, growing on earth, burnt wood, etc., foliage of *Trichostomum.* Fruitstalk terminal, very slender. Capsule suberect, narrow-oblong, curved, thick-walled, furrowed, angular when dry; teeth 16, lanceolate, split nearly to the base; joints prominent, distant in the upper part, close below, conniving when moist, spirally incurved when dry; annulus double, revolute. Operculum conical, beaked. Calyptra cucullate.

A small genus, some of the species of which are excessively widely distributed.

1. **C. purpureus,** *Bridel;—Fl. N. Z.* i. 178. Stems densely tufted, short, dichotomously branched. Leaves dull green, spreading, oblong-lanceolate, keeled, papillose on the back; margins recurved, somewhat twisted when dry; nerve strong, excurrent; perichætial larger, sheathing, acuminate. Fruitstalk purple. Capsule suberect, oblong, slightly incurved, striate, nearly horizontal when dry; teeth combined at the base, 2-fid, edges pale; operculum conic.—Wils. Bryol. Brit. 84. t. 20.

Abundant throughout the islands, as far south as **Campbell's** Island. (One of the commonest mosses in the world.)

21. ENCALYPTA, *Schreber.*

Short, densely tufted, green, monœcious or diœcious mosses, growing on stones, walls, and earth. Fruitstalk terminal. Capsule erect, cylindric; annulus single or 0; peristome 0 or simple or double; outer of 16 subulate teeth; inner a thin membrane produced into slender ciliæ. Operculum with a long beak. Calyptra large, campanulate, mitriform, much exceeding the capsule, tip subulate.

A considerable north temperate genus, rare in the tropics and southern hemispheres.

1. **E. australis,** *Mitten in Fl. Tasman.* ii. 182. Stems loosely tufted,

short. Leaves dull green, suberect, lower ligulate, upper oblong-spathulate, acute or rather obtuse; margin papillose; nerve percurrent, scabrous at the back below the tip. Fruitstalk $\frac{1}{8}$–$\frac{3}{4}$ in. long. Capsule cylindrical, smooth; peristome 0; operculum subulate, as long as the capsule. Calyptra smooth at the tip. Inflorescence monœcious.

Northern and **Middle** Islands, *Colenso;* Auckland, *Knight;* Banks's Peninsula, *Jolliffe;* Nelson, *Travers, Mantell.* Perhaps only a form of the European *E. vulgaris.* (Tasmania, Australia, Chiloe.)

22. HEDWIGIA, Ehrhart.

Loosely tufted, creeping or suberect, branched, monœcious mosses. Leaves imbricate, nerveless, concave, with diaphanous erose apices. Cells small, quadrate. Fruitstalk very short, terminal. Capsule incurved in the perichætium, globose; annulus and teeth 0. Operculum plano-convex. Calyptra small, conic, glabrous or hairy.

A small genus, native of both temperate hemispheres.

1. **H. ciliata,** *Ehr.*—Stems depressed, 1–2 in. long, forming loosely-tufted patches. Leaves dull, glaucous or yellowish, crowded, spreading, erect and imbricate when dry, ovate-lanceolate, with erose diaphanous points; margin recurved below, flattish above; perichætial larger, apices ciliate. Mouth of capsule dilated when dry.—Wils. Bryol. Brit. 146. t. 6.

Middle Island: Otago, *Hector and Buchanan.* (Europe, America, Tasmania.)

23. BRAUNIA, Schimper.

Loosely tufted, suberect, pinnately-branched, diœcious mosses. Leaves closely imbricated, concave, all but quite entire, with long flexuous hair-points. Cells small. Fruitstalk long, slender, on short lateral branchlets. Capsule globose, grooved; mouth large; annulus and teeth 0. Operculum convex, with a long slender beak. Calyptra small, cucullate, glabrous.

A small genus of chiefly southern mosses, found also in the mountains of the tropics.

1. **B. Humboldtii,** *Schimper.—Hedwigia,* Hook.;—Fl. N. Z. ii. 93.—Stems suberect, 2–6 in. high. Leaves obovate, somewhat contracted in the middle, acuminate, with long hair-points, coriaceous; perichætial larger, sheathing, reddish, glossy. Fruitstalk $\frac{1}{3}$–$\frac{1}{2}$ in. long. Male inflorescence axillary, abundant.—*Hedwigia,* Hook. Mus. Exot. t. 137; *Anæctangium,* Fl. Antarct. 135 and 415.

Middle Island, *Lyall;* Otago, *Hector and Buchanan.* **Lord Auckland's** group and **Campbell's** Island: abundant on moist rocks, *J. D. H.* (Australia, Tasmania, Andes, Fuegia.)

24. LEPTANGIUM, Montagne.

Minute tufted moss, growing in vegetable soil, having a succulent subterraneous stem, and very short branches. Leaves concave, nerveless; cells large, lax. Fruitstalk terminal, extremely short. Capsule almost sessile, globose, mouth dilated, peristome 0. Operculum conical. Calyptra minute, fugacious.

A very curious moss, also found in Australia, differing in its large lax cells from *Hedwigia* and *Anœctangium*.

1. **L. repens,** *Mitt. in Journ. Linn. Soc.* iv. 79.—*Anœctangium*, Hook. Musc. Exot. t. 106; Fl. N. Z. ii. 93.—Stems very minute, $\frac{1}{8}$–$\frac{1}{4}$ in. high; some barren, some fertile. Leaves on the barren stems spreading, nearly orbicular, apiculate; on the fertile larger, whitish, imbricating, ovate, narrowed into a slender acuminate point, quite entire. Capsule sunk amongst the perichætial leaves; mouth very wide, closed by a horizontal membrane; spores very large. Calyptra conical, covering only the tip of the operculum. Inflorescence monœcious; male usually axillary below the female; antheridia with filiform paraphyses.—*Anictangium*, Wilson in Lond. Journ. Bot. 1846, 143. t. 4 *a*.

Northern Island: Raukawa range, *Colenso*. (Australia, Tasmania.)

25. GRIMMIA, *Ehrhart*.

Tufted monœcious or diœcious, perennial mosses, often very dark coloured, and forming hemispherical tufts on rocks, etc., rarely on the ground, occasionally aquatic. Leaves often piliferous, white and transparent at the apex. Fruitstalk terminal, usually short. Capsule ovoid or oblong, annulate or not; teeth 16, in one row, lanceolate, barred, perforate, entire or 2-3-fid, reflexed when dry. Operculum conical, convex or nipple-shaped. Calyptra mitriform, oblique or lobed at the base, rarely cucullate.

A large genus, especially in temperate countries.

A. Schistidium. Capsule on a very short fruitstalk, hidden amongst the leaves. Operculum falling away with the columella attached. Calyptra minute, just covering the operculum . . . . 1. *G. apocarpa*.

B. Grimmia. Capsule on a short curved fruitstalk, exserted. Calyptra larger.

Leaves oblong-lanceolate, terminated abruptly by a hair; nerve vanishing 2. *G. pulvinata*.
Leaves linear-lanceolate, with a white hair-point; nerve vanishing . 3. *G. trichophylla*.
Leaves oblong-lanceolate, acuminate, hair-pointed; nerve percurrent . 4. *G. basaltica*.

1. **G. apocarpa,** *Hedwig.—Schistidium apocarpum*, Br. and Schimp.;—Fl. N. Z. i. 74. Nearly black; stems loosely tufted, $\frac{1}{2}$–3 in. long. Leaves erect, then spreading, ovate-lanceolate, apiculate or obtuse, margin reflexed; nerve disappearing below the transparent white tip; perichætial broader. Fruitstalk very short indeed. Capsule elliptic, sunk amongst the leaves, thick-walled; annulus 0; operculum very broad, falling away with the columella. Calyptra lobed, very small, capping the operculum only.—*Schistidium*, Wils. Bryol. Brit. 150. t. 13.

Northern and **Middle** Islands: stones in and close to water. Bay of Islands, banks of the Waitangi, *J. D. H.* Shores of Waikari Lake, *Colenso*; Nelson, *Travers*; Otago, *Hector and Buchanan*. (Australia, Tasmania, Kerguelen's Land, S. America, and throughout the northern hemisphere.)

2. **G. pulvinata,** *Smith;—Fl. N. Z.* ii. 75. Stems short, forming dense small hoary hemispherical tufts. Leaves spreading, oblong-lanceolate, keeled above, terminated by a white hair; margin entire, recurved; nerve vanishing. Fruitstalk short, decurved. Capsule turned downwards, ovoid,

8-furrowed; annulus large; operculum beaked. Calyptra mitriform. Inflorescence monœcious.—Wils. Bryol. Brit. 153. t. 13.

Var. β. *Africana*, Fl. N. Z. l. c.: smaller, capsule shorter, operculum conical mamillate, teeth very short.—*Fissidens pulvinatus*, β., Hedwig. Sp. Musc. t. 40. *G. cygnicolla*, Taylor.

Northern and **Middle** Islands: var. β. on rocks, etc., *Colenso*; Auckland, *Knight*; Nelson, *Travers*. (South Africa, Australia, and Tasmania.)

3. **G. trichophylla,** *Greville;—Fl. N. Z.* ii. 75. Stems short, forming small lax tufts. Leaves yellow-green, linear-lanceolate, narrowed into a white diaphanous hair-point, flexuous, crisped when dry; nerve vanishing. Fruitstalks short, curved or decurved. Capsule pendulous, ovoid, striated, angular when dry; annulus broad; teeth 2-fid; operculum conical, beaked. Inflorescence diœcious.—Wils. Bryol. Brit. 156. t. 32.

Middle Island: on stones. Ship Cove and Port Cooper, *Lyall*; Otago, *Lindsay, Hector and Buchanan*. (North temperate zone, South America, Tasmania.)

4. **G. basaltica,** *Mitten, mss.* Forming small hemispherical hoary tufts. Leaves spreading, oblong-lanceolate, acuminate with a hyaline hair-point, keeled by the percurrent nerve; perichætial with a much longer hair-point. Fruitstalk curved. Capsule ovoid-globose, plicate when dry; teeth red, perforated, subentire, reflexed when dry; operculum short, conic, acuminate. Inflorescence monœcious.

Middle Island: basalt rocks near Dunedin, *Lindsay* (*Herb. Mitten*). Closely resembling *G. orbicularis*, B. and S., of Europe, but capsule more plaited, less contracted at the mouth, and teeth reflexed when dry (*Mitten*).

26. RACOMITRIUM, *Bridel.*

Loosely tufted, often hoary, grey, diœcious, perennial mosses, growing in rocks or on the ground; stems usually elongate, branched. Leaves keeled, margins recurved, tips often white and transparent, nerve vanishing; cells long and sinuous at the base, square above. Fruitstalk terminal, usually short. Capsule elliptic or oblong, smooth, mouth narrow, annulate; teeth 16, 2–3-fid; divisions short, linear-subulate, unequal and irregularly combined, or very long filiform and free to the base; operculum conical, and subulate. Calyptra conical or mitriform, tip solid subulate, papillose; base membranous, fringed.

A genus of which most of the species abound in alpine regions, where some form large hoary patches on the ground and on rocks.

I. Dryptodon. Stems dichotomously branched, young shoots fastigiate, unbranched.

Leaves ovate-oblong. Fruitstalk very short. Operculum shorter than the capsule. Teeth 2-fid . . . . 1. *R. crispulum.*

Leaves ovate-lanceolate. Fruitstalk longer. Operculum acicular, as long as the capsule. Teeth entire . . . . 2. *R. rupestre.*

Leaves narrow-lanceolate. Operculum subulate, shorter than the capsule. Teeth cleft . . . . 3. *R. protensum.*

Leaves elliptic-lanceolate, 3-carinate, apiculate . . . . 4. *R. ptychophyllum.*

II. Racomitrium. Stems irregularly branched; branches with lateral branchlets; shoots not fastigiate.

Leaves deeply plaited, ovate-lanceolate, acute . . . . 5. *R. symphyodon.*

Leaves tortuous, lanceolate subulate, toothed, with diaphanous hair-points . . . . 6. *R. lanuginosum.*

1. **R. crispulum,** *Hook. f. and Wils.;—Fl. N. Z.* ii. 75.—*Dryptodon*, Fl. Antarct. 124. t. 57. f. 9. Stems forming loose tufts, 1–1½ in high, slender, branching. Leaves pale or dark lurid-green, erecto-patent and subrecurved, ovate-oblong, acuminate, keeled, margin entire, reflexed below; apex subdiaphanous; nerve percurrent; perichætial shorter, obtuse or acute. Fruitstalk very short, pale, lateral by the growth of side-shoots. Capsule elliptic-oblong, erect or inclined; annulus distinct; teeth red, 2-fid, reflexed when dry; operculum conical, subulate, not half as long as the capsule. Calyptra mitriform, base torn and inflexed.

α. Leaves gradually attenuated; cells of upper portion rounded.

β. Leaves with a suboval base, obtuse and apiculate, *R. chlorocarpum*, Mitten, mss.

Var. *α.* **Campbell's** Island, *J. D. H.* Var. *β.* **Northern** Island: on alpine rocks, *Colenso*. *Mitten* distinguishes the Northern Island form as a species (*R. chlorocarpum*). (Tasmania, Kerguelen's Land.)

2. **R. rupestre,** *Hook. f. and Wils.;—Fl. N. Z.* ii. 75.—*Dryptodon*, Fl. Antarct. 402. t. 152. f. 1. Stems densely tufted, slender, 1–2 in. long. Leaves lurid-green, spreading, densely subspirally imbricate when dry, ovate-lanceolate or ovate acuminate; lower subsquarrose, upper recurved, keeled, margin subrecurved, when dry more or less incurved and twisted; nerve scarcely continuous, red; perichætial broader, elliptic-oblong, obtuse. Fruitstalk very short, twisted when dry. Capsule small, erect, elliptic-oblong, red-brown; mouth contracted; teeth red, spreading when dry; operculum with a very slender beak, nearly as long as the capsule.—*? R. convolutum*, Montagne, Sylloge, 37.

Northern Island: moist rocks on the mountains, *Colenso;* also found in Tasmania, Fuegia, and Kerguelen's Land.

3. **R. protensum,** *Braun;—Fl. N. Z.* ii. 76. Stems loosely tufted. Leaves light green, spreading all round or subsecund, appressed when dry, rather rigid, elongate-lanceolate, not diaphanous at the summit; nerve strong, percurrent; perichætial somewhat sheathing. Fruitstalk pale, ½ in. long. Capsule subcylindric, pale brown, thin-walled; teeth long, irregularly split to the base; operculum usually as long as the capsule, beaked.—Wils. Bryol. Brit. 166. t. 45.

Northern Island: moist rocks in mountain regions, *Colenso*. (Europe, Fuegia.)

4. **R. ptychophyllum,** *Mitten, mss.* Stems tufted, 2 in. high; branches numerous, fastigiate. Lower leaves brown, upper yellow, spreading, imbricate when dry, and forming cuspidate tips to the branches, elliptic-lanceolate, subobtuse, apiculate, the tip diaphanous, keeled with the percurrent nerve, and with two deep folds on each side; margins quite entire, reflexed; cells narrow.

Middle Island: Otago (*Lindsay in Herb. Mitten*). Resembling *R. protensum*, var. 3 of Fl. Antarct., from Kerguelen's Land, but distinguished by the folds of the leaf (*Mitten*).

5. **R. symphiodon,** *Mitten in Fl. Tasman.* ii. 181. *t.* 173. *f.* 4.—*R. fasciculare*, Bridel, var.;—*Fl. N. Z.* ii. 76. Stems loosely tufted, slender, elongated; branches slender, subfascicled. Leaves spreading, subsecund, erect when dry, ovate-lanceolate, acuminate, with hyaline points, margin recurved below, quite entire; perichætial acute. Fruitstalk yellow, very short. Cap-

sule cylindric; mouth small; teeth short, with transverse connecting bars. —*R. fasciculare*, var. 2 and 3, Fl. Antarct. p. 402. *Grimmia emersa*, C. Muell. in Bot. Zeit. 1851, 562. *R. microcarpum ?*, Mont. in Voy. au Pôle Sud, Bot. Crypt. p. 284.

Northern and **Middle** Islands: common on rocks on the mountains. (Tasmania, Fuegia.)

6. **R. lanuginosum,** *Bridel;—Fl. Antarct.* i. 124;—*Fl. N. Z.* ii. 76. Stems elongate, loosely tufted, 2–12 in. long, branched. Leaves hoary, spreading, erecto-patent, lanceolate, acuminate, their apex hyaline toothed and granulate. Fruitstalk apparently lateral, short, erect, tubercled. Capsule minute, ovoid, pale brown; mouth small; walls thick; teeth 2-fid, divisions filiform; annulus large; operculum as long as the capsule, or longer. —Wils. Bryol. Brit. 169. t. 19.

Var. *pruinosum*, Fl. N. Z. iii. 76. Leaves more hoary, the hyaline point longer, spinulose-serrate; teeth erect.

Campbell's Island: on rocks, *J. D. H.*—Var. *pruinosum*, **Northern** and **Middle** Islands: rocks on the mountains, common, *Colenso, Travers, Haast, Hector and Buchanan* (not seen in fruit); the common form of this moss is abundant in the north and south temperate zones.

27. SCHLOTHEIMIA, *Bridel.*

Tufted, creeping, usually dark-coloured mosses, with erect short branches, growing in tufts on trees. Stems and branches red-brown or blackish below, yellow or greenish at the tips. Leaves imbricated all round, oblong; cells circular, opaque, punctiform. Fruitstalk terminal, long or short. Capsule erect, subcylindric, smooth or grooved, not annulate. Teeth in 2 rows; outer 16, in pairs, revolute; inner 16 or more, irregular, erect, conniving and orming a cone. Operculum conical, beaked. Calyptra conical or mitriform, rough at the tip, 4- or more lobed at the bottom; the lobes inflexed.

A considerable genus in the tropics and south temperate zone.

Leaves lingulate, shortly mucronate or not . . . . 1. *S. Brownii.*
Leaves oblong-lanceolate, with a long mucro . . . . 2. *S. Campbelliana.*

1. **S. Brownii,** *Schwægr. Suppl.* ii. *t.* 167;—*Fl. N. Z.* ii. 77. Stems loosely tufted, slender, ½ in. long; branches nearly as high. Leaves dusky red-brown, lingulate, obtuse or somewhat cuspidate, lax, erecto-patent; twisted when dry; nerve narrow, scarcely excurrent. Fruitstalk very slender, ½–¾ in. long. Capsule small, ovate-oblong, striated; inner peristome 32-parted; operculum very slender. Calyptra smooth.

Northern Island: Bay of Islands, *J. D. H.*; Port Nicholson, *Lyall* (not found in fruit). (Australia.)

2. **S. campbelliana,** *C. Muell. Synops. Musc.* i. 753.—*S. quadrifida*, Fl. Antarct. i. 126. t. 58. f. 1, not of Bridel. Stems loosely tufted; branches ½–1½ in. high, rather robust. Leaves red-brown, the upper green, spreading, twisted when dry, oblong-lanceolate, contracted in the middle, obtuse and emarginate or acute; nerve excurrent as a slender long mucro, concave, not striated. Fruitstalk stout, ¼ in. high. Capsule cylindrical,

erect, or inclined; operculum conical, with a slender beak, half as long as the capsule. Calyptra smooth.

Campbell's Island: on trunks of trees and on rocks, *J. D. H.*

28. MACROMITRIUM, *Bridel.*

Densely tufted, usually brown or blackish mosses, with creeping stems and erect branches, often forming large patches on trees and rocks. Stems and branches often dark below and yellowish-green at the tips. Leaves rather or very opaque; cells very minute, circular. Fruitstalk terminal. Capsule ovoid, narrowed below, erect, equal, thick-walled; mouth often when dry contracted by 8 folds; annulus 0; peristome 0, or single or double; outer 16 flat lanceolate teeth, inserted in pairs below the mouth; inner a lacerated membrane. Operculum with a straight beak. Calyptra mitriform, glabrous or pilose, longitudinally folded or grooved, laciniate at the base.

A very large tropical and subtropical genus, of which many of the species are either variable or difficult of discrimination. Of the New Zealand ones here described, I suspect some will have to be united hereafter.

I. *Calyptra glabrous.*

Peristome double . . . . 1. *M. sulcatum.*

Peristome single (0? *in* M. asperulum).

a. Leaves obtuse or acute, not mucronate (*see* 8, Hectori.)

Leaves linear-lanceolate, acuminate. Fruitstalk very short . . . 2. *M. longirostre.*

Leaves linear-lanceolate, acute. Fruitstalk very long . . . . . 3. *M. longipes.*

β. Leaves mucronate by the more or less excurrent nerve (*nerve vanishing in* M. Hectori).

Margin of leaves rough, with prominent papillæ . . . . 4. *M. asperulum.*

Stems slender. Leaves lanceolate-subulate, flexuous and twisted when dry . . . . 5. *M. gracile.*

Leaves subcirrhose when dry, linear-ligulate, obtuse . . . . . 6. *M. ligulare.*

Leaves straight, both when dry and moist . . . . 7. *M. orthophyllum.*

Leaves oblong-lanceolate, obtuse; nerve vanishing below the apex . 8. *M. Hectori.*

II. *Calyptra pilose. Peristome single.*

a. Leaves more or less lanceolate, acute or acuminate.

Leaves spreading and recurved, subobtuse. Capsule ovoid-oblong, striate . . . . 9. *M. recurvifolium.*

Leaves spreading and incurved, acuminate. Capsule urceolate, smooth . . . . 10. *M. Mauritianum?*

Leaves spreading and recurved. Capsule subpyriform, slightly grooved . . . . 11. *M. microphyllum.*

Leaves spreading, incurved, acute. Capsule ovoid, smooth . . . 12. *M. incurvifolium.*

β. Leaves more or less ligulate, obtuse, apiculate or mucronate.

Leaves spreading, subincurved; cells minute, dense. Capsule oblong; mouth plaited . . . . 13. *M. hemitrichodes.*

Leaves spreading, subincurved; cells lax. Capsule oblong; mouth plaited . . . . 14. *M. microstomum.*

Leaves erecto-patent; cells dense. Capsule ovoid . . . . . 15. *M. prorepens.*

Leaves spreading and incurved; margins erose; cells rather large . 16. *M. erosulum.*

γ. Leaves ligulate, retuse or 2-lobed . . . . 17. *M. retusum.*

III. *Calyptra unknown* . . . . 18. *M. aristatum.*

1. **M. sulcatum,** *Bridel;—Fl. N. Z.* ii. 77. Branches very short.

Leaves spreading, crisped when dry, linear-lanceolate, acuminate, undulate, quite entire; nerve percurrent. Fruitstalks ¼ in. long. Capsule ovoid, deeply grooved; outer teeth 16, linear, geminate; inner united into a membranous lacerate cone; operculum with a long straight beak. Calyptra glabrous, fimbriate to the middle.—*Schlotheimia*, Hook. Musc. Exot. t. 156.

Var. *β*. Leaves less acuminate, scarcely undulate; probably a different species.

Var. *β*. **Northern** Island, *Colenso*. A doubtful determination. The original plant is a native of Ceylon and Nepal.

2. **M. longirostre,** *Schwægr. Suppl. t.* 112;—*Fl. Antarct. t.* 126;—Fl. N. Z. ii. 78.—Branches erect, 1 in. high, crowded. Leaves bright yellow-red or bronze, lower black, subrigid, crowded, erecto-patent, twisted when dry, linear-lanceolate, very acute, with 2 parallel lines one on each side the solid stout nerve, margins recurved; nerve reddish. Fruitstalk stout, rather short, ⅛–¼ in. long, black. Capsule sulcate, elongate, narrow, almost linear-oblong, gradually narrowed into the stout fruitstalk; peristome single; operculum with a slender straight beak. Calyptra glabrous, fimbriate halfway up.—*Orthotrichum*, Hook. Musc. Exot. t. 25.

Var. *β*. *acutifolium*, Fl. N. Z. l. c. Leaves more acuminate, with a subexcurrent nerve. —*Orthotrichum*, Hook. and Grev. in Brewst. Journ. l. c. t. 5; Fl. Antarct. l. c. i. *M. Paccivanum*, De Notaris.

Middle Island: Dusky Bay, *Menzies*; Otago, *Hector and Buchanan*. **Lord Auckland's** group and **Campbell's** Island, *J. D. H.* **Chatham** Island, *Mr. Travers*.

Var. *β*. Port Preservation, *Lyall*. (Tasmania, Chiloe, a form.)

3. **M. longipes,** *Schwægr. Suppl. t.* 139;—*Fl. N. Z.* ii. 78. Branches elongate, erect, 1–1½ in. high. Leaves dull yellow-green, reddish below, crowded, erecto-patent, crisped and twisted when dry, linear- or subulate-lanceolate, subacute, margin recurved, nerve pale. Fruitstalk very long, rarely short, ½–1½ in., slender. Capsule elliptic; mouth contracted, plaited; peristome single; operculum very slender, straight. Calyptra glabrous.—*Orthotrichum*, Hook. Musc. Exot. t. 24.

Common throughout the **Northern** and **Middle** Islands, *Menzies*, etc. (Norfolk Island.)

4. **M. asperulum,** *Mitten in Fl. Tasman.* ii. 376.—*M. fimbriatum*, Fl. N. Z. ii. 77, not of Palisot. Stem slender, creeping; branches short, tufted. Leaves bright red-brown, dense, spreading, incurved when dry, ligulate from an oblong base, lower obtuse, upper acute, mucronate by the excurrent nerve, margins papillose and rough; perichætial shorter, acute. Fruitstalk ¼ in. long. Capsule ovoid, attenuate below; peristome 0; mouth darker, plaited; operculum conical, acuminate. Calyptra naked. *M. microstomum*, Fl. N. Z. ii. 79, in part.

Northern and **Middle** Islands; common on trunks of trees, etc. (Tasmania.)

5. **M. gracile,** *Schwægr. Suppl. t.* 112;—*Fl. N. Z.* ii. 78. Stems flexuous, slender, sparingly branched, suberect, 1–2 in. long. Leaves pale yellow, brown below, brittle, spreading and incurved, flexuous and twisted when dry, lanceolate-subulate, cuspidate by the excurrent nerve. Fruitstalk ⅛–¼ in. long. Capsule ovoid, striate when dry; peristome single. Calyptra glabrous.—*M. Mossmanianium*, C. Muell. in Bot. Zeit. 1851, 561. *Orthotrichum*, Hook. Musc. Exot. t. 27.

Northern and **Middle** Islands, probably common: Bay of Islands, *J. D. H.*; Dusky Bay, *Menzies*; Chalky Bay, *Lyall*; Nelson, *Travers*; Otago, *Hector and Buchanan*.

6. **M. ligulare,** *Mitten in Journ. Linn. Soc.* iv. 78;—Habit of *M. prorepens*. Leaves spreading, when dry crisped, twisted, and subcirrhose, linear-ligulate, obtuse, or shortly apiculate with the excurrent nerve, margin sub-erose; perichætial shorter, ovate, acute. Fruitstalk $\frac{1}{4}$ in. long, ovoid; mouth darker, plicate; teeth short; operculum conical, acuminate. Calyptra naked.

Northern Island: Waikehi, *Sinclair*; *Kerr* (*in Herb. Mitten*).

7. **M. orthophyllum,** *Mitten in Journ. Linn. Soc.* iv. 79. Habit, size, and colour of *M. longirostre*. Leaves spreading when moist, appressed when dry, striate in both cases, broadly lanceolate, keeled and shortly apiculate with the excurrent nerve; perichætial longer, broader, more erect, with longer points. Fruitstalk $\frac{1}{2}$ in. long. Capsule ovoid, attenuate at the base; mouth plicate; teeth short, free; operculum subulate, as long as the capsule. Calyptra glabrous.

Northern Island, *Kerr* (*in Herb. Mitten*); Auckland, *Knight*.

8. **M. Hectori,** *Mitten, mss.* Branches erect, divided. Leaves brown, almost shining, dense, spreading, scarcely twisted when dry, oblong-lanceolate, obtuse, apiculate, flat or waved, keeled; nerve vanishing below the apex; lower cells elongate, papillose; perichætial similar. Fruitstalk slender. Capsule fusiform, with a slender neck, plaited; teeth granular, cohering at the base; operculum long, subulate. Calyptra long, naked.

Middle Island: Otago, *Hector and Buchanan*.

9. **M. recurvifolium,** *Bridel*;—*Fl. N. Z.* ii. 78. Stems 1–2 in. long, creeping, rather slender. Leaves pale yellow-brown, widely spreading, or recurved when moist, spirally twisted when dry, subulate-lanceolate, rather obtuse, fragile; nerve pale, continuous; cells small, dense, opaque. Fruitstalk $\frac{1}{10}$–$\frac{1}{8}$ in. long, slender. Capsule oblong-ovoid, slightly grooved.—*Orthotrichum*, Hook. and Grev. in Brewst. Journ. Science, i. 120. t. 5.

Northern Island: Bay of Islands, *Logan*, *J. D. H.*, *Colenso*, *Kerr* (*in Herb. Mitten*). (Java.)

10. **M. Mauritianum**? *Schwægr. Suppl. t.* 189;—*Fl. N. Z.* ii. 79. Stems long, creeping; branches short or long. Leaves dense, yellow-green, spreading and incurved when moist, much crisped and twisted together when dry, lanceolate, acuminate; nerve stout, pale; cells dense, small. Fruitstalk short. Capsule urceolate, smooth. Calyptra slightly hairy.

Northern Island: Bay of Islands, *Logan*. A doubtful plant; the original is stated to be a native of Mauritius, Java, and Australia.

11. **M. microphyllum,** *Hook. and Grev.*;—*Fl. N. Z.* ii. 80. Stems and branches very slender, 1–2 in. long, $\frac{1}{20}$ in. diam. Leaves yellow-brown, loosely spreading and recurved when moist, striate and appressed when dry, ovate- or subulate-lanceolate, acute, keeled, with 2 striæ at the base, pellucid, more or less papillose above; nerve stout, vanishing below the tip; cells minute. Fruitstalk very slender, $\frac{1}{8}$–$\frac{1}{4}$ in. long. Capsule oval-oblong or sub-

pyriform, slightly grooved; peristome very short, undivided. Calyptra hairy. —*M. barbatum*, Mitten, mss.; *Orthotrichum*, Hook. and Grev. in Brewst. Journ. Sc. i. t. 6.

Northern and **Middle** Islands: East Coast, *Colenso;* Auckland, *Knight;* Port Nicholson, *Lyall;* Nelson, *Travers;* Otago, *Hector and Buchanan.* (Australia, Tasmania, and S. Africa.)

12. **M. incurvifolium,** *Schwægr. Suppl.;—Fl. N. Z.* ii. 79. Stem creeping. Leaves spreading and incurved when moist, crisped when dry, subulate-lanceolate, subacute, keeled, incurved at the tip. Capsule ovate, smooth; operculum acicular, straight. Calyptra laminate, pilose.—*Orthotrichum*, Hook. and Grev. in Brewst. Ed. Journ. i. 117. t. 4.

Middle Island: Dusky Bay, *Menzies.* (Doubtful.)

Mr. Wilson suspects some mistake in the habitat of this moss, which, though a native of Australia and the Pacific Islands, has not been found in New Zealand, except by Menzies.

13. **M. hemitrichodes,** *Schwægr. Suppl. t.* 193;—*Fl. N. Z.* ii. 79. Stems long, creeping, 3–4 in. long, rather stout; branches $\frac{1}{3}$ in. high. Leaves dark green, lower brown, spreading and subincurved when moist, crisp when dry, ligulate-lanceolate, obtuse, apiculate; margin recurved below; nerve stout, concolorous; cells very minute, dense, opaque. Fruitstalk $\frac{1}{6}$–$\frac{1}{4}$ in. long, slender. Capsule ovoid or subcylindric, smooth, plaited at the mouth. Calyptra somewhat hairy.

Northern Island, *Logan.* (Australia.)

14. **M. microstomum,** *Schwægr. Suppl., not of Fl. N. Z.* ii. 79. Stems and branches elongate, strict, branched at the apex. Leaves brownish-green, spreading and spirally twisted when moist, crisped when dry, ligulate-lanceolate, obtuse, apiculate, papillose; nerve reddish; cells lax, larger than in its allies, roundish. Capsule ovoid, smooth, plaited at the contracted mouth; peristome single; teeth narrow, geminate, trabeculate, white. Calyptra glabrous.—*Orthotrichum*, Hook. and Grev. in Brewst. Journ. Sc. i. t. 4.

Northern Island: Wellington, *Stephenson.* Nelson, *Travers.* (Tasmania.) Mitten refers the 'New Zealand Flora' plant of this name to *M. erosulum* and *asperulum.*

15. **M. prorepens,** *Schwægr. Suppl.* ii. *t.* 171;—*Fl. N. Z.* ii. 79. Stems 1–3 in. long; branches short, erect, simple. Leaves yellow-brown, erecto-patent when moist, scarcely crisped when dry, ligulate-lanceolate, obtuse, apiculate, keeled, scarcely papillose; cells small. Fruitstalk very slender, $\frac{1}{6}$–$\frac{1}{10}$ in. long. Capsule ovoid; mouth somewhat plaited. Calyptra pilose. Inflorescence monœcious.—*Orthotrichum*, Hook. Musc. Exot. t. 120.

Northern and **Middle** Islands: Waikehi, *Sinclair;* Dusky Bay, *Menzies;* Milford Sound, *Lyall;* Nelson, *Travers;* Otago, *Hector and Buchanan.*

16. **M. erosulum,** *Mitten in Journ. Linn. Soc.* iv. 78. Habit, size, and colour of *M. prorepens.* Leaves spreading when dry, compactly crisped and incurved, ligulate from an oblong base, keeled, mucronate; nerve yellow-brown, excurrent; lower cells narrow, upper papillose; perichætial broader, acute. Fruitstalk about $\frac{1}{4}$ in. long. Capsule ovoid, brown; mouth darker, plaited. Operculum subulate, as long as the capsule; teeth short. Calyptra pilose. *M. microstomum*, Fl. N. Z. ii. 79, in part.

Northern and **Middle** Islands: probably common.

17. **M. retusum,** *Hook. f. and Wils. Fl. N. Z.* ii. 79. *t.* 85. *f.* 6. Stems tufted, creeping; branches ½ in. long. Leaves pale yellow-green, crowded, erecto patent and recurved when moist, crisped when dry, ligulate-oblong, retuse and 2-fid at the apex, scarcely keeled; nerve pale; cells very minute, opaque. Fruit unknown.

Northern Island: habitat unknown, *Colenso.* Kiapara ?, *Ker* in Herb. Mitten.

18. **M. aristatum,** *Mitten, mss.* Small; branches short. Leaves greenish-brown, spreading, contracted when dry, broadly lanceolate, obtuse, with a subpiliferous apex, keeled; nerve vanishing below the apex; basal cells shortly oblong; upper scarcely papillose.—Mitten.

Northern Island: Auckland, *Knight;* specimen very small, and without fruit.

M. piliferum, Schwægr., a Sandwich Island plant, occurs in Mr. Menzies' collection, but has been found by no one else, and as the habitats of several of that emirent collector's plants were confounded, it is most probable that this is one such; it may be known by its spreading lanceolate-subulate leaves, with long hair-points.

M. Tongense, Sullivant, Mosses of the U. S. Expl. Exped. 7. t. 5 (*M. abbreviatum*, Mitten, mss.), is unknown to me.

29. ORTHOTRICHUM, Hedwig.

Densely tufted, bright or dark green, short-stemmed mosses, often forming small cushions on trees and rocks, rarely on earth, sometimes creeping. Fruitstalk terminal. Capsule erect, pyriform; neck more or less elongate, striate, ribbed when dry; annulus 0; peristome 0, single or double; outer 16 teeth mostly united in pairs and inserted below the mouth; inner 8–16 cilia. Operculum with a conical beak. Calyptra large, inflated, campanulate, plicate, often pilose; base crenate or lacerate.

A large European and north temperate genus, rare in the tropics.

Peristome double.

Leaves spreading, subacute. Capsule immersed . . . . 1. *O. calvum.*
Leaves spreading, subacute. Capsule exserted . . . . 2. *O. pumilum.*
Leaves spreading, erose. Capsule exserted . . . . 3. *O. luteum.*

Peristome single (BRACHYSTELEUM).

Leaves erecto-patent, oblong below, narrow above, obtuse . . . . 4. *O. crassifolium.*
Leaves suberect, subulate, acuminate . . . . 5. *O. angustifolium.*

1. **O. calvum,** *Hook. f. and Wils.;—Fl. N. Z.* ii. 8. *t.* 85. *f.* 7. Stems tufted, short, slender. Leaves dull green, spreading and rather recurved when moist, suberect when dry, oblong-lanceolate or ovate-oblong below, and more linear above, subacute; nerve solid; cells opaque. Fruitstalk $\frac{1}{12}$ in. long. Capsule about as long as the fruitstalk, minute, elliptic-oblong or clavate, striate; inner peristome of 8 cilia; operculum not seen. Calyptra yellow-brown, glabrous.

Northern Island: Manawata, on branches, *Colenso.*

2. **O. pumilum,** *Dickson;—Fl. N. Z.* ii. 80. Stems short, tufted, ¼ in. high. Leaves spreading when moist, appressed when dry, lanceolate, sub-acute, concave; margin revolute; nerve vanishing below the apex. Fruit-stalk very short. Capsule sunk amongst the leaves, ovoid, with a short neck, broadly striate; operculum conical, short. Calyptra naked.—Wils. Bryol. Brit. 178. t. 45; Schwægr. Suppl. t. 50.

Northern Island: Cliffs at Hawke's Bay, *Colenso*. A European plant, of which I have very imperfect New Zealand specimens.

3. **O. luteum,** *Mitten in Fl. Tasman.* ii. 184. Stems densely tufted, $\frac{1}{4}$ in. high. Leaves pale yellow, spreading when moist, crisped when dry, oblong and concave below, above linear-lanceolate, keeled; nerve yellow, vanishing below the apex; margins erose, recurved about the middle; perichætial longer, lanceolate. Fruitstalk $\frac{1}{8}$ in. long. Capsule cylindric-oblong, plicate; neck thickened; inner peristome of 8 cilia; operculum with a short beak. Calyptra pilose.—*Ulota lutea*, Mitten in Journ. Linn. Soc. iv. 77.

Northern Island: Kaipara Mossman, *Mitten*. (Tasmania.)

4. **O. crassifolium,** *Hook. f. and Wils. Fl. Antarct.* 125. *t.* 57. *f.* 8. Stems very short, tufted, $\frac{1}{4}$–$\frac{1}{3}$ in. high, Leaves dirty olive-green, erecto-patent when moist and dry (then horny), coriaceous, ovate or oblong and concave below, linear or ligulate above, obtuse; margin recurved; nerve stout, vanishing below the apex, upper leaves narrower above and broader at the base. Fruitstalk very short. Capsule exserted or immersed, pyriform, smooth; neck very short; peristome single, yellow; operculum convex with a short straight beak. Calyptra plaited, glabrous, 8-fid at the base.

Lord Auckland's group and **Campbell's** Island: on rocks and stones, near high-water mark, *J. D. H.* (Kerguelen's Land, Fuegia.)

5. **O. angustifolium,** *Hook. f. and Wils. Fl. Antarct.* 125. *t.* 57. *f.* 7. Stems very short, tufted, $\frac{1}{4}$–$\frac{1}{3}$ in. high. Leaves black-green, crowded, sub-erect both when moist and dry, strict, thick in texture, subulate-lanceolate, acuminate; nerve thick, continuous; perichætial longer, subsecund. Fruitstalk very short. Capsule immersed, small, elliptic-ovoid; mouth rather large; peristome single; operculum convex with a short beak. Calyptra not seen.

Campbell's Island: on rocks, on the hills, *J. D. H.*

30. ZYGODON, Hook.

Tufted, small, usually bright green, monœcious, diœcious, or hermaphrodite mosses, often creeping, growing on trees, rarely on the ground. Leaves lanceolate or oblong, keeled; margin plane; cells minute. Fruitstalk terminal. Capsule erect, pyriform or clavate, grooved or striate; annulus 0; peristome 0 or single or double; outer 16 flat teeth united in pairs, reflexed when dry; inner 8–16 linear hyaline cilia alternating with the outer, sometimes conniving and forming a campanulate cone. Operculum obliquely beaked. Calyptra oblique, cucullate.

* *Inner teeth horizontal, free at the base.*

a. Leaves with margins entire or nearly so.

† *Peristome double.*

Leaves erecto-patent, ligulate, obtuse . . . . 1. *Z. obtusifolius.*
Leaves spreading, squarrose, oblong, acute . . . . 2. *Z. Brownii.*

†† *Peristome single or* 0.

Leaves suberect, lingulate, subacute . . . . 3. *Z. intermedius.*
Leaves spreading, oblong-lanceolate . . . . 4. *Z. Reinwardtii.*

β. Leaves with the margins denticulate.

Leaves spreading, subincurved. Capsule immersed . . . . 5. *Z. cyathicarpus.*

** *Inner teeth conniving in a bell-shaped cone, united by a very short membrane at the base* (Codonoblepharum, *Schwægrichen*).

Leaves suberect, lingulate, apiculate . . . . 6. *Z. Menziesii.*

1. **Z. obtusifolius,** *Hook. Musc. Exot. t.* 159;—*Fl. N. Z.* ii. 80. Stems tufted, $\frac{1}{3}$ in. high. Leaves dull green, loosely erecto-patent, ligulate, obtuse. Fruitstalk $\frac{1}{16}$–$\frac{1}{10}$ in. long. Capsule deeply furrowed, 8-ribbed; peristome double; inner of 8 cilia; operculum with an inclined beak, $\frac{1}{4}$ as long as the capsule. Calyptra roughish above, subplicate below. Inflorescence monœcious.—Schwægr. Suppl. t. 136.

Northern Island: Bay of Islands, *J. D. H.*, *Colenso.* (East Indies.)

2. **Z. intermedius,** *Bruch and Schimper*;—*Fl. N. Z.* ii. 80. Stems loosely tufted, slender, $\frac{1}{2}$–1 in. high, branching. Leaves pale yellow-green, lax, erecto-patent, lanceolate-lingulate, subacute, or oblong-lanceolate; margins plane; nerve ceasing below the apex. Fruitstalk $\frac{1}{8}$–$\frac{1}{4}$ in. long. Capsule oblong, deeply grooved; more than half as long as the capsule; peristome single, outer absent, inner of 8 cilia; operculum with a slender inclined beak. Inflorescence diœcious.—*Z. conoideus*, *β*, Hook. and Grev. in Brewst. Journ. Sc. i. 132.

Middle Island: Dusky Bay, *Menzies.* (Tasmania, Australia, Chili.)

3. **Z. Brownii,** *Schwægr. Suppl. t.* 317 *b*;—*Fl. N. Z.* ii. 81. Stems tufted, branched, $\frac{1}{2}$–1 in. high. Leaves pale green, spreading, recurved and squarrose, oblong, acute; nerve continuous and excurrent. Fruitstalk very slender, $\frac{1}{4}$–$\frac{1}{2}$ in. long. Capsule pyriform, furrowed; peristome double; outer often rudimentary or irregular; inner of 8 cilia. Inflorescence diœcious.

Throughout the **Northern** and **Middle** Islands: common. (Australia, Tasmania.)

4. **Z. Reinwardtii,** *Braun*;—*Fl. N. Z.* ii. 81. Stems short, $\frac{1}{2}$–$\frac{1}{3}$ in. high. Leaves yellow-green, spreading, oblong-lanceolate; margins waved, minutely denticulate, keeled; nerve excurrent; perichætial shorter. Fruitstalk erect, flexuous, $\frac{1}{3}$–$\frac{2}{3}$ in. long. Capsule subpyriform, sulcate; peristome fugacious or absent (?) in the New Zealand form; operculum conic, beaked. Calyptra large, substriate, coriaceous. Inflorescence hermaphrodite.—Schwægr. Suppl. t. 312 *a*. *Zygodon denticulatus*, Tayl. in Lond. Journ. Bot. vi. 329. *Z. anomalus*, Dozy and Molkb.; Fl. Tasman. ii. 185.

Northern and **Middle** Islands: Port Nicholson, *Sinclair*; Ruahine mountains, *Colenso*; Southern Alps, *Haast*; Otago, *Hector and Buchanan.* (Tasmania, Java, S. America.) Inner peristome, in Java specimens, of 16 cilia; outer present and fugacious in the Tasmanian.

5. **Z. cyathicarpus,** *Montagne, Flor. Chili Crypt.* 132. *t.* 3;—*Fl. N. Z.* ii. 81. Stems $\frac{1}{2}$–1 in. high, tufted, rather stout. Leaves pale green, spreading, subrecurved, crisped when dry, linear-lanceolate, acute, keeled, remotely denticulate; perichætial overtopping the capsule. Fruitstalk very short. Capsule immersed, striate; peristome 0; operculum convex, with a short oblique beak. Inflorescence monœcious.—*Gymnostomum linearifolium*,

Tayl. in Lond. Journ. Bot. v. 42. *Didymodon cyathicarpus*, Mitten in Journ. Linn. Soc. iv. 70.

Northern Island: Makororo river, *Colenso*, with *Bartramia uncinata*. (North America, South Africa, Tasmania.) Mitten regards this as a Weissioid moss.

6. **Z. Menziesii,** *Mitten in Journ. Linn. Soc.* iv. 74. Stems tufted, densely fastigiately branched, ¼ in. high. Leaves crowded, suberect, yellow-brown when old, lingulate, apiculate, keeled; nerve nearly continuous, strong, reddish, vanishing below the apex. Fruitstalk $\frac{1}{10}$ in. long, slender. Capsule pyriform, pale, strongly grooved, contracted below the mouth; peristome double; inner of 16 cilia, conniving and forming a campanulate cone; operculum convex, obliquely beaked. Inflorescence diœcious.—*Codonoblepharum*, Schwægr. Suppl. ii. t. 137; Fl. Tasman. ii. 186.

Northern and **Middle** Islands: on bark of trees, *Colenso*. **Middle** Island: Dusky Bay, *Menzies*. (Tasmania, Australia.)

31. LEPTOSTOMUM, Br.

Densely-tufted, erect, bright green, soft, diœcious mosses, with matted, brown rootlets on the bases of the stems. Leaves piliferous, transparent; nerve stout; cells circular. Fruitstalk terminal, long. Capsule erect or pendulous, pyriform or clavate; annulus usually 0; peristome a membranous ring, more or less cleft, arising from the inner wall of the capsule. Operculum very short, conic. Calyptra cucullate, fugacious.

A genus of beautiful mosses, all natives of the southern temperate hemisphere, and growing on trunks of trees and rocks.

* *Capsule narrowed into the fruitstalk.*

Capsule erect or inclined; hair-point of leaf straight . . . . . . 1. *L. gracile.*
Capsule inclined; hair-point flexuous . . . . 2. *L. inclinans.*

** *Capsule erect, not narrowed below.*

Capsule large, hair-point branched . . . . 3. *L. macrocarpum.*

1. **L. gracile,** *Brown;—Fl. N. Z.* ii. 82. Stems densely tufted, 2–3 in. high. Leaves bright green, erect, crowded, closely imbricate when dry, lower ovate-lanceolate, upper oblong, all with a straight hair-point; margin reflexed; nerve stout. Fruitstalk 1–2 in. long. Capsule oblong, erect or inclined, subclavate; peristome a membranouns white ring, with 16 2-fid divisions; annulus obscure; operculum hemispherical.—Fl. Antarct. i. 122; Schwægr. Suppl. t. 104. *Gymnostomum*, Hook. Musc. Exot. t. 22.

Throughout the **Northern** and **Middle** Islands, *Menzies*, etc. **Campbell's** Island: barren, *J. D. H.*

2. **L. inclinans,** *Brown;—Fl. N. Z.* ii. 82. Stems densely tufted. Leaves bright green, erecto-patent, loosely imbricate, erect when dry, ovate-oblong, obtuse, with a flexuous long hair-point, which is toothed at the tip; margin recurved; nerve stout, pale. Fruitstalk slender. Capsule inclined, obvoid-clavate, pale; peristome as in *L. gracile;* operculum conical hemispheric.—Schwægr. Suppl. t. 213.—*Gymnostomum*, Hook. Musc. Exot. t. 168. *L. flexipile*, C. Mueller in Bot. Zeit. 1851, 547.

Northern Island: damp woods, Ruahamanga river, etc., *Colenso*.

3. **L. macrocarpum,** *Brown ;—Fl. N. Z.* ii. 82. Stems ½ in. high. Leaves bright green, elliptic- or obovate-oblong, obtuse, with a branched hair-point, concave; margin recurved; nerve slender. Fruitstalk very long. Capsule large, erect, ovate-oblong, not narrowed at the base; mouth small; peristome very small; operculum conical obtuse.—*Bryum,* Hedwig, Musc. Frond. iii. t. 10.

Abundant on rocks and trunks of trees throughout the **Northern** and **Middle** Islands. One of the handsomest New Zealand mosses.

32. AULACOMNIUM, Schwægr.

Tufted, perennial, erect mosses, usually growing in marshy places; stems tomentose with radicles. Leaves variable; cells roundish. Fruitstalk terminal, long, flexuous. Capsule erect or cernuous, furrowed when dry, with a slender struma; annulus compound; peristome double; outer teeth 16, equidistant, incurved when dry, closely trabeculate; inner a pellucid membrane, divided into 16 perforated processes, with interposed cilia. Operculum convex, obtusely beaked. Calyptra small, cucullate, smooth.

A small European genus.

1. **A. Gaudichaudii,** *Mitten in Hook. Kew Journ.* 1856, 262. Stems erect, branched. Leaves long, erecto-patent, oblong, cuspidate by the excurrent nerve, strongly toothed; perichætial longer and narrower. Capsule narrow, erect, cylindrical, plaited when dry; annulus large, adherent; outer teeth pale yellow, erect. Inflorescence diœcious.—*Leptotheca,* Schwægr. Suppl. t. 137.

Northern Island, Auckland, *Knight.* (Australia, Tasmania.) Mitten refers the *Brachymenum ovatum,* Fl. Antarct., of Falkland Islands, to this plant, but in that the leaves are quite entire.

33. ORTHODONTIUM, Schwægr.

Slender, tufted, pale green, erect mosses, usually growing on rocks. Leaves long, slender; nerve vanishing; cells lax large, oblong-square. Fruitstalk terminal, slender. Capsule erect or drooping, pyriform or clavate, exannulate, symmetrical, very thin-walled; cells lax; teeth in 2 rows; outer 16, lanceolate, inserted below the mouth, often inflexed when dry; inner 16, filiform, united at the base by a somewhat keeled membrane. Operculum short, conical, beaked. Calyptra small, cucullate, fugacious.

Delicate mosses, resembling *Brya.* There are but few species, chiefly tropical, one is British.

1. **O. sulcatum,** *Hook. and Wils. Ic. Plant. t.* 739 B;—*Fl. N. Z.* ii. 81. Stems tufted, nearly ½ in. long. Leaves pale, spreading and recurved, linear-lanceolate, nearly flat; nerve vanishing below the apex. Fruitstalk slender. Capsule elliptic-oblong, furrowed and inclined when dry; neck short. Operculum with a short oblique beak. Inflorescence monœcious or hermaphrodite.

Middle Island, Port William, *Lyall* (Australia.)

34. BRACHYMENIUM, Hook.

Small slender Bryoid mosses. Leaves closely imbricate; reticulations lax. Capsule suberect, oblong subclavate or pyriform, not furrowed, equal at the base; annulus simple; peristome double; outer of 16 linear-lanceolate teeth; inner a membrane irregularly split at the apex into 16 cilia. Operculum conical. Calyptra cucullate.

A small genus of tropical and southern mosses, very closely allied to *Bryum*.

1. **B. coarctatum,** *C. Muell. Synops.* i. 312 (*Bryum*). Stems short, $\frac{1}{4}$–$\frac{1}{2}$ in. high, tufted. Leaves minute, oblong- or ovate-lanceolate, acuminate, quite entire; nerve stout, ferruginous, exserted as a hyaline point; perichætial larger, longer, narrower. Fruitstalk slender, flexuous, from short basilar shoots. Capsule ovoid, erect; mouth contracted; operculum conic, acute.

Northern Island: Auckland, *Knight;* scraps only, picked out of other mosses by Mr. Mitten. (Java, Pacific islands.)

35. MIELICHOFERIA, Nees and Hornsch.

Characters and habit of *Bryum*, but the inflorescence is terminal, on short axillary branches, and appears lateral. Capsules pyriform or clavate, oblique or pendulous, with a tapering apophysis often of equal length; peristome simple, of 16 teeth, connate at the base; annulus large, unrolling spirally. Operculum short. Calyptra small, narrow, cucullate.

A genus of a few species of alpine mosses.

1. **M. longiseta,** *C. Muell. Synops.* i. 236. Stems short, erect, simple. Leaves yellowish, lanceolate, acuminate, with a slender excurrent nerve, slightly toothed at the apex; perichætial broader, nearly entire; nerve vanishing. Fruitstalk very long, slender, flexuose. Capsule pyriform, erect or inclined; apophysis of equal length.

Northern and **Middle** Islands; Auckland, *Knight;* Alps of Canterbury, *Sinclair and Haast.* (Andes of Columbia.)

36. BRYUM, Linn.

Usually erect, densely tufted perennial mosses, rarely creeping, growing in all situations, except aquatic. Stems proliferous at the tips. Leaves usually membranous; nerve distinct; cells rhomboidal. Fruitstalk terminal. Capsule pyriform or clavate, inclined or pendulous; neck short or long; annulus usually present; peristome double; outer 16 simple equidistant teeth, with a dark central line, transversely lamellate on the inner surface, incurved when dry; inner a membrane cleft into 16 keeled teeth, with often interposed cilia. Operculum short, convex, apiculate. Calyptra small, cucullate, fugacious.

One of the largest genera of mosses, found in all parts of the world, especially the cold and mountainous. Many of the New Zealand species are provisional only, and most of them require to be re-examined with more and better specimens. Their characters are often very obscure.

A. *Plants usually annual.* LEPTOBRYUM . . . . 1. *B. pyriforme.*

B. *Plants perennial.* BRYUM.

I. *Upper leaves larger, spreading (except in* B. campylothecium), *rosulate, more or less serrulate.*

Upper leaves spreading; margin thickened . . . . 2. *B. truncorum.*
Upper leaves erect, lanceolate; margin hardly thickened . . . 3. *B. campylothecium.*
Upper leaves spreading, ovate-oblong; margin not thickened . . 4. *B. Billardierii.*
Upper leaves spreading, spathulate; margin not thickened . . . 5. *B. rufescens.*

II. *Upper leaves not larger and spreading, distinctly serrulate (except in* B. eximium).

α. *Nerve vanishing below the apex of the leaf.*
Lower leaves ovate-acuminate . . . . 6. *B. Wahlenbergii.*
Lower leaves lanceolate . . . . 7. *B. crudum.*

β. *Nerve percurrent or excurrent.*
Leaves ½ in. long . . . . 8. *B. eximium.*
Leaves small, oblong-ovate, not crisped when dry. Inflorescence diœcious . . . . 9. *B. obconicum.*
Leaves small, oblong-ovate, crisped when dry. Inflorescence diœcious . . . . 10. *B. lævigatum.*
Leaves small, upper narrow-lanceolate. Inflorescence monœcious . . . . 11. *B. nutans.*

III. *Upper leaves not large and spreading, quite entire or very obscurely serrulate.*

α. *Leaves white and silvery* . . . . 12. *B. argenteum.*
β. *Leaves green, very obtuse* . . . . 13. *B. blandum.*
γ. *Leaves green, subulate-lanceolate* . . . . 14. *B. tenuifolium.*
γ. *Leaves ovate or ovate-lanceolate or oblong-lanceolate, acute or acuminate, not white or silvery, nerve excurrent or percurrent.*

† *Inflorescence hermaphrodite.*
Leaves ovate-lanceolate . . . . 15. *B. bimum.*
Leaves lanceolate, keeled, much crisped when dry . . . 16. *B. torquescens.*
Leaves oblong; cells large, soft . . . . 17. *B. mucronatum.*

†† *Inflorescence diœcious.*

§ *Capsule cylindric, clavate or long-pyriform.*
Leaves erecto-patent, oblong, cuspidate; nerve stout, excurrent . . . . 18. *B. curvicollum.*
Leaves suberect, lanceolate; nerve slender, excurrent in a hair-point . . . . 19. *B. creberrimum.*
Leaves appressed, ovate-oblong; nerve short . . . . 20. *B. crassum.*
Leaves erecto-patent, ovate-lanceolate. Operculum large, mamillate . . . . 21. *B. cæspiticium.*
Leaves erecto-patent, oblong-lanceolate. Operculum shortly conical, apiculate . . . . 22. *B. chrysoneuron.*

§§ *Capsule short, oblong or pyriform.*
Capsule pyriform . . . . 23. *B. annulatum.*
Capsule ventricose, depressed at base. Leaf-margin flat 24. *B. pachytheca.*
Capsule ovate-oblong, rounded at base. Leaf-margin recurved . . . . 25. *B. atro-purpureum.*

OF DUBIOUS SECTION.

B. flaccidum, B. varium, B. intermedium, B. incurvifolium.

1. **B. pyriforme,** *Hedwig.* Tufts annual or perennial, forming broad silky patches, bright green. Stems simple, slender, rooting at base only. Lower leaves lanceolate, entire, upper linear-setaceous, flexuous, slightly serrate; nerve extending almost to the apex; cells oblong. Fruitstalk slender, flexuous. Capsule inclined or pendulous, pyriform, thin in texture; mouth small; operculum convex.—*Leptobryum,* Wils. Bryol. Brit. 220. t. 28; Fl. Tasm. ii. 188.

Northern and **Middle** Islands: probably common, *Colenso, Haast, Hector*, etc. (N. and S. temperate regions). Messrs. Mitten and Spruce observe that this plant, which is annual in temperate regions, becomes perennial in the tropics.

2. **B. truncorum,** *Bory;—Fl. N. Z.* ii. 87. Stems 1–2 in. high. Leaves dark green, upper rosulate, spreading and recurved, crisped when dry, oblong-obovate, acuminate; margin thickened, recurved below, serrulate at the apex; nerve subexcurrent. Fruitstalk curved at the top, 1–3 in. long. Capsule narrowed into the fruitstalk, elongate, cylindric-pyriform, curved, drooping; operculum conical, apiculate.—Fl. Antarct. 134 and 415; Fl. Tasman. ii. 192; *B. leptothecium*, Taylor.

Northern and **Middle** Islands: common on stumps of trees, etc. **Campbell's** Island: barren, *J. D. H.* (Common in the southern hemisphere.) Mitten thinks the name *leptothecium* should be adopted for the New Zealand and Australian moss, whose identity with the original Mascarene one is not established.

3. **B. campylothecium,** *Taylor;—Fl. N. Z.* ii. 86. Stems 1–2 in. high, tomentose. Leaves pale yellowish, imbricate, coriaceous, erecto-patent, uppermost rosulate, but not spreading, appressed when dry, obovate-oblong, subacute, obscurely serrulate at the apex, not concave; margins thickened, reflexed; nerve excurrent as a long point. Fruitstalk slender. Capsule pendulous, curved, ovate-oblong or pyriform, with a narrow obconic apophysis; operculum conical, mamillate.

Northern and **Middle** Islands: probably common on trees and rocks, etc. **Campbell's** Island: mixed with *B. Billardieri*, J. D. H. (Australia and Tasmania.) Taylor's original description was, according to Wilson (Bryol. Brit. 242, in note), taken from two Swan River plants, of which one is the true *B. Billardieri*. Mitten considers this to be well distinguished from *B. truncorum* by the imbricate, not rosulate upper leaves.

4. **B. Billardierii,** *Schwægr. Suppl. t.* 76;—*Fl. N. Z.* ii. 86. Stems ½–1 in. high. Leaves pale green, interrupted, upper rosulate, spreading, crisped and waved when dry, ovate-oblong, subacute; margin not thickened, reflexed below; apex serrulate; nerve subexcurrent. Fruitstalk stout, curved at the top, 2 in. long. Capsule curved, clavate-pyriform, drooping, not narrowed into the fruitstalk; operculum subconical.—Fl. Antarct. p. 413; Wils. Bryol. Brit. 242. t. 4.

Northern Island: moist rocks and trunks, Bay of Islands, *Colenso, J. D. H.*; Auckland, *Sinclair*. When growing in the spray, this becomes of a rigid texture and lurid-green colour, the leaves spreading, but hardly rosulate. (Common in the southern hemisphere.)

5. **B. rufescens,** *Hook. f. and Wils.;—Fl. Tasm.* ii. 192. *t.* 174. *f.* 1. Very closely allied to *B. Billardieri*, but the stems are more slender. Leaves spathulate, softer, reddish-yellow, of looser texture at the base; nerve reddish; margin scarcely recurved. Capsule clavate, arcuate, rather pendulous; mouth purple.

Middle Island: Matama river, Otago, in boggy ground. *Hector and Buchanan*. Mere scraps, without fruit, identified by Mitten. The stems and leaves are quite black, except the tips, which are yellow-green. (Tasmania.)

6. **B. Wahlenbergii,** *Schwægr. Suppl. t.* 70;—*Fl. N. Z.* ii. 83. Stems ½–1 in. long, reddish, erect or ascending. Leaves pale glaucous-green, upper lanceolate, lower ovate-acuminate, concave, pellucid; nerve vanishing near

the serrulate apex. Fruitstalk 1–2 in. long. Capsule pendulous, shortly pyriform; annulus 0; teeth rather large; operculum convex or subconical, mamillate. Inflorescence diœcious.—Wils. Bryol. Brit. 227. t. 47; Fl. Antarct. 134, 414. *B. albescens*, auct.

Middle Island: Otago, *Hector and Buchanan*. **Stewart's** Island, *Lyall*. **Lord Auckland's** group (barren), *Lyall*. (Common in the temperate and cold parts of the N. and S. hemisphere.)

7. **B. crudum,** *Schreber*. Loosely cæspitose, glaucous-green, glossy and transparent. Lower leaves broadly ovate-lanceolate, entire, upper longer, ovate-lanceolate, uppermost linear-lanceolate, flexuose; nerve vanishing below the serrate apex. Fruitstalk reddish, flexuose. Capsule suberect, cernuous, oblong or obovate-pyriform; operculum convex, apiculate. Inflorescence in Europe hermaphrodite or diœcious.—Wils. Bryol. Brit. 224. t. 28.

Middle Island: Canterbury Alps, *Haast and Sinclair*. (Europe, N. America, India, Fuegia.) A beautiful moss.

8. **B. eximium,** *Mitten, mss*. Inflorescence lax. Stems simple, flaccid, 2–3 in. long. Leaves larger, $\frac{1}{4}$ in. long, uniform, spreading on all sides, bright green, lax, sometimes deflexed, ovate-oblong or linear-oblong, acute, concave; nerve percurrent; margin indistinctly bordered, obscurely serrate; cells small, oblong. Fruit unknown.

Northern Island, *Kerr* (*Herb. Mitten*); **Middle** Island, N. E. valley, Otago, *Hector and Buchanan*.

9. **B. obconicum,** *Hornsch.;—Fl. N. Z.* ii. 85. Stems with lateral shoots. Leaves clustered at the top of the stem, erecto-patent, not twisted when dry, oblong-ovate, acuminate, serrulate at the apex; margin slightly thickened, recurved; nerve opaque, keeled, excurrent, spinulose at the apex. Fruitstalk curved at the top. Capsule drooping, clavate, curved, with a long neck, which tapers into the fruitstalk; operculum convex, mamillate. Inflorescence diœcious.—Wils. Bryol. Brit. 239. t. 49.

Northern Island: Auckland, *Sinclair*. (Europe, Australia, S. Africa.)

10. **B. lævigatum,** *Hook. f. and Wils. Fl. Antarct.* 415. *t.* 154. *f.* 3. —*B. crassinerve*, Hook. f. and Wils.;—Fl. N. Z. ii. 83. Stems $\frac{1}{2}$–2 in. long, tomentose with purple root-fibres. Leaves pale yellow-green, erecto-patent, erect and crisped when dry, oblong or ovate-oblong, obtuse or acute, subserrulate towards the recurved apex, very concave, rather coriaceous; margin reflexed; nerve solid, continuous. Fruitstalk 1 in. long. Capsule pendulous, narrow-pyriform, not curved; operculum conical, obtuse. Inflorescence diœcious.

Northern and **Middle** Islands: Auckland, *Sinclair*; Canterbury, *Travers*; Otago, *Hector and Buchanan*. (Australia, Fuegia.)

11. **B. nutans,** *Schreb.;—Fl. Antarct.* 134. Stems $\frac{1}{2}$–2 in. high. Leaves spreading; upper narrow-lanceolate, serrate towards the apex; lower shorter, ovate-lanceolate, quite entire; nerve strong, continuous. Fruitstalk slender. Capsule nodding or pendulous, oblong-pyriform; operculum large, convex, papillose. Inflorescence monœcious; antheridia in pairs in the axils of the perichætial leaves.—Wils. Bryol. Brit. 225. t. 29.

Northern Island, *Kerr* (*in Herb. Mitten*). **Middle** Island: top of Kaikerai hill, Otago, *Lindsay*; Canterbury, *Haast*.

Lord Auckland's group: peaty soil on the hills (barren), *J. D. H.* (Europe, Tasmania.)

12. **B. argenteum,** *Linn.;—Fl. N. Z.* ii. 84. Stems very short, ¼–1 in. high. Leaves silvery or glaucous, very pale and glistening, imbricate, broadly ovate or oblong, concave, obtuse and apiculate or acuminate, quite entire; nerve vanishing beyond the middle; cells large and lax. Fruitstalk rather short, dark red. Capsule small, ovoid or oblong, pendulous, blood-red; operculum mamillate. Inflorescence diœcious.—Fl. Antarct. 413; Wils. Bryol. Brit. 247. t. 29.

Middle Island, *Lyall;* probably common throughout the islands. (Europe, Tasmania, America, India, Australia, Falkland Islands, in Cockburn Island, at the extreme southern limit of terrestrial vegetation, lat. 64° S.)

13. **B. blandum,** *Hook. f. and Wils. Fl. Antarct.* 134. *t.* 60. *f.* 1;—*Fl. N. Z.* ii. 83. Stems tufted, 1 in. high, sparingly branched, flaccid. Leaves upper pale yellow-green, shining, lower reddish, loosely imbricate, suberect, appressed when dry, oblong, obtuse, quite entire, very concave; margin not reflexed; nerve slender, reddish, not quite continuous to the tip; cells large, lax. Fruitstalk 1 in. long. Capsule clavate-pyriform, cernuous. Inflorescence diœcious.

Northern and **Middle** Islands: probably common, Makororo river, *Colenso;* Bay of Islands, *J. D. H.* (a lurid variety); Auckland, *Knight;* Otago, *Hector and Buchanan.*
Campbell's Island, in bogs (barren), *J. D. H.* A beautiful moss, also found in Tasmania.

14. **B. tenuifolium,** *Hook. f. and Wils. Fl. N. Z.* ii. 83. *t.* 85. *f.* 5. Stems short, simple, ½ in. high. Leaves pale green, erect or subsecund, subulate-lanceolate, acuminate; margin recurved, quite entire; nerve solid, excurrent; perichætial longer, slender. Fruitstalk pale. Capsule almost pendulous, oblong, straight or curved, obconic at the base; operculum conical; inner peristome inconstant. Inflorescence diœcious.

Northern Island, *Colenso;* Bay of Islands, on clay banks, *J. D. H.*

15. **B. bimum,** *Schreb.;—Fl. N. Z.* ii. 85. Stems ½–3 in. high, matted by purple radicles. Leaves yellowish-green, spreading, rather twisted when dry, ovate-lanceolate, keeled; margin recurved, quite entire; nerve excurrent. Fruitstalk 2 in. long. Capsule pendulous, obovate-pyriform or obconical, brown; operculum large, convex, mamillate. Inflorescence hermaphrodite. —Fl. Antarct. 413; Wils. Bryol. Brit. 238. t. 49.

Northern and **Middle** Islands: Hawke's Bay, in marshes, *Colenso;* Banks's Peninsula, *Jolliffe;* Otago, *Hector and Buchanan.* (Kerguelen's Land, Tasmania, Europe.)

16. **B. torquescens,** *Bruch and Schimp.;—Fl. Tasman.* ii. 190. Stems matted, covered with radicles. Leaves much crisped when dry; lower rather distant, narrow-lanceolate, keeled; margins reflexed; upper crowded, oblong-lanceolate, all concave, quite entire; nerve excurrent. Fruitstalk ½–2 in. high, curved at the top. Capsule drooping or pendulous, subincurved, narrow-obconic, with a tapering neck, red-brown; operculum convex, broad, mamillate, purple, shining. Inflorescence hermaphrodite.—Wils. Bryol. Brit. 239. t. 49.

Northern and **Middle** Islands: Auckland, *Knight;* Hokianga and Banks's Peninsula,

Jolliffe; Nelson, *Travers*. (Australia, Tasmania, Europe, S. Africa.) Apparently scarcely different from *B. bimum*.

17. **B. mucronatum,** *Mitten, mss.* Stems tufted, branching with innovations, 1 in. high. Leaves spreading, oblong, acuminate, with a subrecurved mucro; nerve brown, percurrent or excurrent; margins rather thickened, quite entire; cells large, soft. Fruitstalk elongate. Capsule elongate, pyriform; neck arched, horizontal or pendulous; peristome complete; operculum convex. Inflorescence hermaphrodite. (*Mitten*).

Middle Island: Canterbury Alps, *Haast*; Otago, in open ground, *Hector and Buchanan*. M. Mitten observes that this comes very near indeed to the European *B. uliginosum*, differing in the inflorescence and peristome.

18. **B. curvicollum,** *Mitten, mss.*—*B. clavatum*, Hook. f. and Wils.;—Fl. N. Z. ii. 84. t. 85. f. 3, not of Schimper. Stems ½ in. high, tomentose with brown radicles. Leaves erecto-patent, incurved, firm, rather opaque, ovate or oblong-lanceolate, very concave, cuspidate, quite entire; nerve stout, excurrent. Fruitstalk short, ⅓ in. long, red-brown. Capsule suberect or nodding or pendulous, curved, long, narrow, oblong-clavate; operculum subconical. Inflorescence diœcious.

Var. β. *extenuatum*, Fl. N. Z. l. c. Leaves spreading, hardly incurved. Fruitstalk longer. Capsule longer, less pendulous.

Var. γ. Capsule suberect; inner peristome adnate to the outer.

Northern and **Middle** Islands: Bay of Islands, *Logan*; Cape Turnagain, *Colenso*; Southern Alps, *Haast*; Otago, *Hector and Buchanan*.—Var. β. East Cape, *Sinclair*, *Colenso*.—Var. γ. Manawata river, *Colenso*. (Var. α. also found in Tasmania and Tristan d'Acunha.)

19. **B. creberrimum,** *Tayl.*;—*Fl. N. Z.* ii. 84. Stems 1 in. high, with long shoots. Leaves olive-green, lax, suberect, when dry erect, twisted and flexuous, lanceolate, acuminate, somewhat keeled; margin subrecurved, quite entire; nerve slender, excurrent as a very fine hair-point. Fruitstalk slender, flexuose; capsule nodding or pendulous, long, slender, subcylindric, narrowed into the slender apophysis; operculum convex, mamillate. Inflorescence diœcious.

Northern Island: Bay of Islands and Auckland, *Sinclair*, *Jolliffe*, etc. (Australia.)

20. **B. crassum,** *Hook. f. and Wils. Fl. N. Z.* ii. 86. *t.* 86. *f.* 1. Stems elongate, rigid, naked and tomentose, with fibrils at the base, ½ to 1 in. long. Leaves crowded at the tops of the stems, closely imbricate and appressed when dry; lower minute, scale-like; upper coriaceous, ovate-oblong, acute, concave; margin recurved, quite entire (serrulate at the tip?); nerve stout, scarcely excurrent. Fruitstalk ¾ in. high, arched at the top. Capsule pendulous, narrow-oblong, pyriform or subcylindric, scarcely narrowed into the fruitstalk; operculum hemispherical or subconical, apiculate. Inflorescence diœcious.

Northern Island: Manakau Bay, on scoriæ, *Colenso*. (Tasmania.)

21. **B. cæspiticium,** *Linn.* Tufts compact, dark green or yellowish, stems and branches covered with matted radicles. Leaves erecto-patent, straight when dry, terminal larger, ovate or ovate-lanceolate, acuminate; margin reflexed, entire or serrulate at the tip; nerve excurrent. Fruitstalk

elongate. Capsule pendulous, oblong-obovate, mouth contracted when dry; operculum large, mamillate. Inflorescence diœcious.—Wils. Bryol. Brit. 243. t. 29; Fl. Tasman. ii. 191.

Northern Island, *Colenso*. **Middle** Island: Otago, *Hector*. Specimens so named by Mitten. The capsule is cylindrical and much longer than in the usual European state. (Ubiquitous.)

22. **B. chrysoneuron,** *C. Mueller in Bot. Zeit.* 1851, 549.—*B. duriusculum*, Hook. f. and Wils.; Fl. N. Z. ii. 84. Stems short. Leaves spreading or erecto-patent, rather rigid, erect when dry, oblong-lanceolate, acuminate, concave; margins quite entire, reflexed; nerve stout, excurrent, or almost so. Fruitstalk very slender, flexuous. Capsule cernuous, oblong-pyriform or cylindric, narrow at the base; operculum shortly conical, apiculate. Inflorescence diœcious.

Var. *β*. Fl. N. Z. l. c. Leaves smaller, crowded, nerve scarcely excurrent. Fruitstalk and capsule shorter.

Var. *γ*. Fl. N. Z. l. c. Stems and fruitstalk longer. Leaves distant, spreading, lanceolate.

Common throughout the **Northern** and **Middle** islands on banks, etc. Very closely allied to and probably a form of the European *B. sanguineum*.

22. **B. annulatum,** *Hook. f. and Wils. Fl. Antarct.* 134. *t.* 60. *f.* 2;—*Fl. N. Z.* ii. 84. Stems very short, ⅛ in. high, tufted. Leaves dark green, crowded, rather rigid, spreading, not crisped when dry, ovate-lanceolate, acute, concave; margin quite entire; nerve stout, solid, continuous, hardly excurrent. Fruitstalk ¼–⅓ in. long, arched at the top. Capsule pendulous, shortly pyriform or obovoid, dark brown; annulus very large; teeth yellow, trabeculate; operculum convex, hardly conical. Inflorescence diœcious.

Northern and **Middle** Islands: Manawata river, on clay banks, *Colenso;* Otago, *Hector and Buchanan;* Canterbury, *Travers*. **Campbell's** Island: on the ground, *J. D. H.*

23. **B. pachytheca,** *C. Muell. Synops. Musc.* i. 307;—*Fl. Tasman.* ii. 191. Stems short, ¼–½ in. high. Leaves pale dull green, erect when dry, oblong-ovate, acuminate, concave; margin flat, quite entire; nerve stout, excurrent, red. Fruitstalk short, not arched above. Capsule pendulous from the top of the fruitstalk, ovoid, ventricose, rounded and somewhat lobed at the base, red-purple; operculum short, subconical. Inflorescence diœcious.

Middle Island: Canterbury, *Travers*. (Tasmania, Australia.)

24. **B. atro-purpureum,** *Weber and Mohr;—Fl. N. Z.* ii. 85. Stems very short, branched. Leaves imbricate, erecto-patent, erect when dry, ovate, acuminate, concave; margin recurved, quite entire; nerve stout, subexcurrent. Fruitstalk slender, curved at the top. Capsule pendulous, shortly ovate-oblong, abruptly rounded at the base; operculum broader than the contracted mouth, convex, mamillate. Inflorescence diœcious.—*B. dichotomum*, Hedwig; Wils. Bryol. Brit. 244. t. 50; Fl. N. Z. l. c. 85. *B. erythrocarpum*, Bridel.

Northern and **Middle** Islands: Bay of Islands, in clay banks, *J. D. H.* (Australia, South Africa, Europe, and elsewhere); Canterbury, *Travers*.

I have several other *Brya* from New Zealand, which are not in a good enough state to determine properly; amongst others are—

B. FLACCIDUM, *Bridel?, of Fl. N. Z.* ii. 85, which much resembles *B. cæspiticium*.

B. VARIUM, *Fl. N. Z.* ii. 85. *t.* 85. *f.* 4, a Dusky Bay plant without fruit, of which I find no specimens in Herb. Hook.

B. INTERMEDIUM, *Bridel?, Fl. N. Z.* ii. 87, of which a mere scrap was so named by my friend Mr. Wilson.

B. INCURVIFOLIUM, *C. Muell. in Bot. Zeit.* 1851, 549, a dubious plant, of which no fruit has ever been found, collected by Mossman at Kiapara.

37. MNIUM, Bruch and Schimp.

Erect, rarely creeping, loosely tufted, usually rather large perennial mosses; stems proliferous from the base, not from the tops. Leaves large, crowded and spreading at the tops of the stems; cells hexagonal. Fruitstalks terminal, often several together. Capsule pendulous, ovate-oblong, annulate; peristome double, as in *Bryum*. Operculum usually rostrate, rarely convex, and apiculate. Calyptra small, cucullate.

Usually handsome mosses, growing in woods, on moist rocks and trunks of trees, in many parts of the globe, differing in habit more than by tangible characters from *Bryum*; the New Zealand species may be known by the rostrate operculum.

Leaves ovate acuminate, nerve subcontinuous . . . . 1. *M. rostratum*.
Leaves longer, denser, nerve stronger . . . . 2. *M. rhynchophorum*.

1. **M. rostratum,** *Schwægr. Suppl. t.* 79;—*Fl. N. Z.* ii. 87. Stems decumbent, then ascending, ½–2 in. long, with creeping barren runners. Leaves large, crisped and undulate when dry, lower ovate acuminate, upper spreading, oblong, obtuse, apiculate; margin thickened, toothed; nerve subcontinuous. Fruitstalk crowded, 1–2 in. long. Capsules nodding or pendulous; operculum rostrate. Inflorescence hermaphrodite.—*Wils. Bryol. Brit.* 254. *t.* 31.

Common in the **Northern** and **Middle** Islands: *Colenso*, etc. (Europe, Asia, America.)

2. **M. rhynchophorum,** *Hook. and Harv. Ic. Pl. t.* 20. *f.* 3. Very similar to and perhaps a form of *M. rostratum*, but the leaves are longer, of denser substance, with minute cells, and the nerve stronger with a pellucid border.

Northern and **Middle** Islands: probably common in woods. Waikare lake, and Tuaranga, *Colenso*; Chalky Bay, *Lyall*. (Europe, Asia, America.)

38. MEESIA, Hedwig.

Large handsome mosses, with the habit of *Bryum*. Fruitstalk very long and slender. Capsule suberect, obovate or clavate, curved, gibbous at the back; mouth small and oblique; apophysis long, tapering. Peristome double; outer of 16 short entire or split teeth, more or less adherent to the inner peristome, which is longer and formed of a membrane split into 16 keeled processes; annulus small. Operculum small, conical. Calyptra submitriform, inflexed at the base, fugacious.

A small genus of more or less Alpine mosses.

1. **M. macrantha,** *Mitten in Hook. Kew Journ. Bot.* viii. 260. Stems tufted, short, branched. Leaves lanceolate, obtuse, margin recurved; nerve

vanishing below the apex; perichætial large. Fruitstalk long, slender. Capsule pyriform, curved; outer teeth short, obtuse; inner of long regular keeled processes; operculum conic, obtuse. Inflorescence monœcious; male on short branches, surrounded with broad perigonial leaves.

Middle Island: Otago, on open ground, *Hector and Buchanan.* (Alps of Victoria.)

39. CONOSTOMUM, Swartz.

Densely tufted, erect, bright green, diœcious or monœcious mosses, growing on the ground. Stems covered with radicles. Leaves closely imbricate, erect, narrow; nerve excurrent; cells minute, square. Fruitstalk terminal. Capsule inclined, globose, grooved; mouth small; annulus 0; teeth 16, narrow-subulate, conniving in a cone and joined by their tips. Operculum conic and beaked. Calyptra cucullate.

An Arctic and Antarctic genus, also found on the lofty mountains of temperate and tropical zones. It has been united with *Bartramia* § *Philonotis* by modern authors, but I follow Wilson in keeping it distinct.

Nerve broad . . . . 1. *C. australe.*
Nerve slender . . . . 2. *C. pusillum.*

1. **C. australe,** *Swartz;—Fl. N. Z.* ii. 87. Stems erect, branched, densely tufted, ½–1 in. high. Leaves erect, most densely imbricate, linear-lanceolate, acuminate, upper piliferous; nerve broad, excurrent. Fruitstalk 1–1½ in. high. Capsule inclined, subglobose; operculum obliquely rostrate. Inflorescence monœcious.—Schwægr. Suppl. t. 130; Fl. Antarct. 132 and 411. *Philonotis*, Mitten in Journ. Linn. Soc. iv. 81.

Northern and **Middle** Islands: Ruahine mountains, *Colenso;* Canterbury Alps, *Haast, Travers;* Otago, *Hector and Buchanan.* **Lord Auckland's** group and **Campbell's** Head: common on the hills, *J. D. H.* (Tasmania, Fuegia, Andes.)

2. **C. pusillum,** *Hook. f. and Wils. Fl. N. Z.* ii. 88. *t.* 86. *f.* 2. Stems ¼–½ in. high, loosely tufted. Leaves suberect, laxly imbricate, yellow-green, not appressed when dry, lanceolate, acuminate, piliferous; margin recurved, doubly serrate; nerve slender, excurrent; cells lax. Fruitstalk pale, ½–¾ in. long. Capsule suberect, pale, furrowed when dry; teeth slender, red; operculum not half as long as the capsule, beak inclined. Calyptra brown. Inflorescence monœcious.—*C. parvulum*, Hampe in Linnæa. *Philonotis*, Mitten, l. c.

Northern and **Middle** Islands: tops of the Ruahine mountains, *Colenso;* Otago, *Hector and Buchanan.* (An alpine, Tasmanian, and Australian moss.)

40. CRYPTOPODIUM, Bridel.

A rather large, loosely tufted, erect, branching, diœcious moss, growing on rocks and trees. Stems curved, covered with leaves throughout their length. Leaves long, erecto-patent, subsecund, sharply serrate; cells dense, square. Fruitstalks shorter than the capsules, often crowded, terminal on lateral shoots. Capsule ovoid, erect, smooth; annulus 0; mouth large; peristome double; outer of 16 lanceolate, acuminate, reflexed (when dry) teeth; inner a membrane split into 16 imperforate teeth, with often cilia between them. Operculum convex, apiculate. Calyptra cucullate. Inflorescence diœcious.

1. **C. bartramioides,** *Brid.*;—*Fl. N. Z.* ii. 88. Stems 2–6 in. high, long, curved, rigid, sparingly branched. Leaves linear-subulate, with broad sheathing bases, doubly spinulose-serrate; nerve solid. Capsules often 3 together, sunk amongst the leaves.—*Bryum bartramioides,* Hook. Musc. Exot. t. 18.

Northern and **Middle** Islands: Ruahine mountains, *Colenso*; Pelorus, *Jolliffe*; Dusky Bay, *Menzies*; Otago, *Hector and Buchanan.*—This fine moss has the habit and dense opaque foliage of a *Bartramia,* but the unstriated large-mouthed capsule of *Bryum.* Mitten considers it a congener of *Spiridens.*

41. BARTRAMIA, Hedwig.

Small or large, densely or loosely tufted, usually erect mosses, growing on trees, rocks, and earth. Branches fascicled, and stems often covered with matted radicles. Leaves narrow, rigid, serrate, papillose; nerve strong, continuous or excurrent. Cells small, square. Fruitstalk terminal. Capsule usually globose, furrowed when dry, erect, inclined or drooping; mouth contracted; annulus 0; peristome 0, or single or double; outer of 16 lanceolate red smooth teeth, marked with a central line, incurved when dry; inner 16 broad keeled teeth, with 2-fid pointed tips, with or without alternating cilia. Operculum short, flat. Calyptra small, cucullate.

A large genus of mosses, common in alpine and subalpine localities, rare or absent in hot ones. Some of the species here described are excessively near one another.

A. *Stems irregularly dichotomously branched. Leaves spreading, not striate. Inflorescence monœcious or diœcious.*

* *Capsule hidden amongst the leaves* . . . . 1. *B. Halleriana.*

** *Capsule exserted.*

Leaf-sheath oblong, blade subulate . . . . 2. *B. papillata.*
Leaf-sheath broad; blade linear . . . . 3. *B. patens.*
Leaf-sheath obovate-quadrate, blade subulate . . . . 4. *B. robusta.*
Leaf not sheathing; nerve very broad, excurrent . . . . . 5. *B. crassinervia.*

B. *Branches fascicled. Leaves not striate. Inflor. diœcious* (Philonotis).

* *Leaves serrulate, margins flat.*

Leaves lanceolate, distant, perichætial with long points . . . 6. *B. remotifolia.*
Leaves subulate-lanceolate, perichætial subulate . . . . . . 7. *B. australis.*

** *Leaves serrulate, margins recurved* . . . . 8. *B. tenuis.*

*** *Leaves quite entire* . . . . 9. *B. affinis.*

C. *Leaves sheathing, distinctly plaited. Inflor. diœcious* (Breutelia).

* *Branches densely fascicled.*

Capsule pendulous; upper cells of leaf short . . . . 10. *B. pendula.*
Capsule pendulous; upper cells of leaf elongated . . . . . . 11. *B. Sieberi.*
Capsule erect or incurved . . . . 12. *B. comosa.*

** *Branches sparingly fascicled.*

Sides of sheathing base of leaf rounded . . . . 13. *B. divaricata.*
Sides of sheathing base of leaf angled . . . . 14. *B. consimilis.*

*** *Stems dichotomously branched* . . . . 15. *B. ? elongata.*

1. **B. Halleriana,** *Hedwig*;—*Fl. N. Z.* ii. 88. Stems loosely tufted, tall, 1–2 in. high, subfastigiately branched. Leaves bright yellow-green,

spreading or secund, broad and sheathing at the base, suddenly contracted to a linear very long subulate blade, doubly serrulate; nerve continuous. Fruit-stalk very short, on side shoots, hence appearing lateral. Capsules immersed, often 2 together, much overlapped by the leaves; operculum conical. Inflorescence monœcious.—Wils. Bryol. Brit. 281. t. 23. *B. Mossmaniana*, C. Muell. in Bot. Zeit. 1851, 551.

Northern and **Middle** Islands: on rocks and trees, Ruahine range to Lake Waikare, *Colenso;* Nelson mountains, *Travers;* rocks and banks, Otago, *Hector and Buchanan.* (Europe, N. and S. America, Fuegia, Australia, and Tasmania, mountains of India and Africa.)

2. **B. papillata,** *Hook. f. and Wils. Fl. N. Z.* ii. 89. *t.* 86. *f.* 4. Stems 1 in. high, sparingly branched. Leaves yellow-green, crowded, patent or erecto-patent, striate, more or less crisped when dry, subulate, with an obovate sheathing base, minutely serrulate, papillose at the back; nerve solid. Fruitstalk ¼ in. long, appearing lateral from terminating short lateral shoots. Capsule inclined, subglobose; operculum small, convex. Inflorescence diœcious.—*B. acerosa,* Hampe in Linnæa. *B. fragilis,* Mitten in Linn. Soc. Journ. iv. 81.

Northern and **Middle** Islands: Bay of Islands, Falls of the Waitangi, *J. D. H.;* base of Tongariro and Ruahine range, *Colenso;* Nelson, *Travers;* Otago, on banks, etc., *Hector and Buchanan.* (Australia and Tasmania.)—Very near the European *B. ithyphylla.* M. Mitten informs me that his *B. fragilis* is a state with denser foliage.

3. **B. patens,** *Bridel;—Fl. Antarct.* 133. Stems erect, short, sparingly branched. Leaves spreading, light yellow-brown when dry, lax, rigid, linear or setaceous from a broader sheathing base, acuminate, serrulate; nerve broad, green. Fruitstalk ⅓–½ in. long. Capsule oblique, broadly oblong; operculum convex, shortly beaked.

Middle Island: Otago, *Hector and Buchanan;* **Campbell's** Island, rocks on the hills (barren), *J. D. H.*—Closely allied to *B. papillata.* (Fuegia.)

4. **B. robusta,** *Hook. f. and Wils. Fl. Antarct.* 133. *t.* 59. *f.* 4. Stems erect, loosely tufted, 1 in. high, very stout. Leaves yellow-green, dense, rigid, spreading, rather fragile; base short, oblong, obovate-quadrate, sheathing; blade long, subulate, acuminate, serrulate; nerve very broad and stout, covering the whole breadth of the upper part of leaf. Fruitstalk stout, red, ¾ in. long. Capsule erect, globose, brown; peristome not seen; operculum conical, rostrate, yellow. Inflorescence diœcious.

Lord Auckland's group and **Campbell's** Island: moist places in the hills, *J. D. H.*

5. **B. crassinervia,** *Mitten, mss.* Stems very short, tufted, branched. Leaves spreading, not sheathing at the base, which is subovate; blade lanceolate, acuminate; margins recurved, serrulate; nerve very broad, toothed at the back and apex; lower cells long, upper quadrate; perichætial broader. Fruitstalk short, red. Capsule subglobose, unequal, plaited; teeth narrow, red; operculum convex, apiculate. Inflorescence hermaphrodite.

Middle Island: Fagus forest, on the Hopkins, alt. 2500–3500 ft., *Haast.*—Allied to the European *B. gracilis,* but the nerve is thick, broad, and occupies one-fourth of the width of the leaf.

6. **B. remotifolia,** *Hook. f. and Wils. Fl. Tasman.* ii. 193. *t.* 174. *f.* 3. —*B. appressa,* Fl. N. Z. ii. 89. t. 86. f. 5. Stems short, slender, fascicu-

lately branched; branches slender, recurved. Leaves pale, glaucous, distant, spreading, rather crisped when dry, papillose, ovate or lanceolate, acuminate with long points, serrulate; margin plane; nerve slender, pellucid, excurrent; perichætial longer, erect or spreading, almost aristate at the point. Fruitstalk stout, 1 in. long, red. Capsule suberect. Inflorescence diœcious.—*Philonotis appressa*, Mitten in Linn. Soc. Journ. iv. 81. *B. exigua*, Sullivant in Mosses of U. S. Expl. Exped. 11. t. 8. *B. pusilla*, Sulliv. in Hook. Kew Journ. ii. 316. *Hypnum scabrifolium*, Fl. Antarct. 138. t. 60. f. 6.

Northern Island: Bay of Islands, Falls of the Waitangi, *J. D. H.*; Wairarapa valley, *Colenso*. **Middle** Island: Otago, *Hector*. **Lord Auckland's** group, *Lyall*. The name *appressa* was abandoned as being unsuited to the female plant, which has spreading leaves. (Tasmania, Fuegia.)

7. **B. australis,** *Mitten, mss.* Stems tufted, 1–2 in. high. Leaves spreading or subsecund, broad, lanceolate from a truncate base, gradually attenuated; margins serrulate; nerve excurrent, toothed at the back and apex; cells all elongate, papillose; perichætial ovate at the base. Fruitstalk long, slender. Capsule curved, reniform, plaited; mouth decurved; teeth red, inner yellow. Inflorescence diœcious.

Middle Island: Alps of Canterbury, *Travers*, and of Otago, *Hector and Buchanan*.—Allied to *B. calcarea*, but the leaves are gradually narrowed from the base upwards, the cells narrower, and margins not recurved, *Mitten*.

8. **B. tenuis,** *Taylor in Phytologist*, 1844, p. 1095.—*B. uncinata*, Schwæg.; *Marchica*, Brid.; and *radicalis*, Palisot,—Fl. N. Z. ii. 89. Stems short, slender; branches fascicled, very slender, often curved. Leaves crowded, subfalcate, secund, spreading, narrow, ovate-lanceolate, acuminate with a long point; margin flat or slightly recurved, serrulate; nerve excurrent; perichætial erect, with a setaceous point. Fruitstalk 1 in. long. Capsule cernuous; cilia of inner peristome connected at their apices (in Tasmanian specimens). Inflorescence diœcious; male capitulate.—Fl. Tasman. ii. 193. t. 174. f. 4; C. Muell. in Bot. Zeit. 1851, 552.

Northern and **Middle** Islands: abundant in wet places. **Kermadec** Islands, *Milne*. (Tasmania, Australia, Norfolk Island, Tristan d'Acunha.)

9. **B. affinis,** *Hook. Musc. Exot. t.* 176;—*Fl. N. Z.* ii. 90. Stems sparingly branched, 1–2 in. high; branches fascicled. Leaves crowded, erecto-patent, appressed when dry, strict, ovate-lanceolate, narrowed into a very slender point, margin much recurved, quite entire; nerve stout. Fruitstalk ½ in. high. Capsule ovoid, pendulous; operculum conical, acuminate. Inflorescence diœcious; male discoid.—Schwægr. Suppl. t. 137.

Northern and **Middle** Islands: Manakau Bay on scoriæ, *Colenso*; Auckland, *Knight*; Nelson, *Travers*; Otago, *Hector and Buchanan*. (Tasmania and Australia.)

10. **B. pendula,** *Hook. Musc. Exot. t.* 21;—*Fl. N. Z.* ii. 90. Stems tall, stout, 2–4 in. long, tomentose; branches fascicled, slender. Leaves spreading, ovate-lanceolate, acuminate, serrulate; nerve excurrent. Fruitstalk slender, 1 in. high. Capsule pendulous, oblong-clavate, grooved.—Schwægr. Suppl. t. 239. Fl. Antarct. 133 and 412.

Abundant throughout the islands, *Menzies*, etc., and in **Campbell's** Island, *J. D. H.* (Tasmania, Fuegia.)

11. **B. Sieberi,** *Mitten in Fl. Tasman.* ii. 194. *t.* 174. *f.* 6. Very similar to *B. pendula*, but stouter; branches more robust. Leaves yellow-green, not shining when dry; cells everywhere distinctly papillose, those of the upper portion of the leaf elongated. Capsule oblong, pendulous. Inflorescence diœcious.

Northern and **Middle** Islands: probably common; Bay of Islands, *Sinclair* (growing with *B. pendula*); Canterbury Alps, *Haast, Travers;* Otago, *Hector and Buchanan.* (Australia, Tasmania.)—Wilson does not distinguish this from *B. pendula.*

12. **B. comosa,** *Mitten in Fl. Tasman.* ii. 195. *t.* 174. *f.* 7. Stems tall, robust, very tomentose, 2–4 in. high, much fasciculately-branched above. Leaves divaricating, sheathing base short, erect, obovate; blade narrow, lanceolate, serrulate, plaited, minutely papillose; nerve excurrent in a slender point; perichætial smaller, ovate-lanceolate. Fruitstalk about 1½ in. high. Capsule suberect or inclined, ovoid, grooved; operculum conical. Inflorescence diœcious.

Northern Island: Auckland, *Knight;* Bay of Islands, *Cunningham, Jolliffe.* **Middle** Island: Milford Sound, *Lyall.*—Allied to *B. pendula,* but differing in the short erect base of the leaf; cells much shorter. (Tasmania.)

13. **B. divaricata,** *Mitten in Fl. Tasman.* ii. 195.—*B. gigantea,* Fl. N. Z. ii. 90, not Brid. Stems elongate, nearly simple, or with branches sparingly fascicled, tomentose below, 2–4 in. high. Leaves divaricating; base short, sheathing; blade lanceolate, plicate, papillose, serrulate; nerve excurrent; perichætial ovate, quite entire; nerve slender. Fruitstalk long. Capsule ovoid-oblong, horizontal, with a narrow pyriform neck; operculum conical. Inflorescence diœcious.

Northern and **Middle** Islands: Bay of Islands, *J. D. H., Sinclair;* Auckland, *Knight;* Milford Sound, *Lyall.* (Tasmania.)—Scarcely different from *B. comosa* and *pendula.*

14. **B. consimilis,** *C. Muell. Synops.* i. 492.—*Hypnum,* Fl. Antarct. 137. t. 60. f. 4. Stems 3–4 in. high, sparingly fasciculately branched, slender; branches erect. Leaves dark green, spreading, loosely imbricate, uppermost subsecund, gradually narrowed from a quadrate base whose sides are angular, minutely serrulate, plaited; nerve subexcurrent; cells rounded, punctiform. Fruit unknown.

Lord Auckland's group: marshy places on the hills, *J. D. H.*

15. **B.? elongata,** *Mitt. mss.*—*Hypnum,* Fl. Antarct. 137. t. 60. f. 3. Stems very robust, curved, ascending, 4–6 in. long, ⅓ in. diameter, very sparingly branched. Leaves closely imbricate, suberect and subsecund, pale yellow and shining when dry, ovate-lanceolate, acuminate, plaited, often waved, serrulate, papillose at the back; nerve slender, excurrent. Fruit unknown.

Middle Island: Otago, Waipori Creek, on the ground, *Hector and Buchanan.* **Lord Auckland's** group and **Campbell's** Island, *J. D. H.*—A most beautiful moss, doubtfully referred here.

42. FUNARIA, Schreber.

Short-stemmed, tufted, biennial, monœcious mosses, growing on earth and on burnt wood, soil, etc. Leaves very membranous, concave, reticulation lax; nerve strong. Fruitstalk terminal, often long and flexuous. Capsule inclined or pendulous, pyriform, curved, gibbous, smooth, oblique, often much contracted; annulus 0 or compound; peristome double; outer 16, oblique, closely-jointed teeth, united at their summits, and forming a reticulated disk, striated below; inner of 16 membranous, lanceolate, flat cilia, opposite the outer teeth, and adhering to their bases. Operculum very shortly conic. Calyptra large, cucullate, inflated, with a subulate tip.

Some of the plants of this genus are amongst the most widely diffused on the globe, *F. hygrometrica* especially appearing everywhere on burnt ground.

Leaves quite entire, nerve continuous. Operculum convex. Annulus compound . . . . *F. hygrometrica.*
Leaves serrulate at apex, nerve vanishing. Operculum flat. Annulus 0 *F. glabra.*
Leaves quite entire, nerve excurrent. Operculum flat. Annulus 0 . . *F. cuspidata.*

1. **F. hygrometrica,** *Hedwig;—Fl. N. Z.* ii. 90. Stems short. Leaves imbricate, crowded into an oblong head, broadly ovate-lanceolate, acuminate, concave, quite entire; nerve continuous; perichætial serrate at the apex. Fruitstalk arcuate. Capsule pendulous, incurved, pyriform; mouth wrinkled; border red; annulus compound; operculum convex.—Fl. Antarct. 135, 415; Wils. Bryol. Brit. 269. t. 20.

Var. *β. calvescens.* Stems more slender; upper leaves spreading, twisted when dry. Capsule suberect.—*F. calvescens.* Schwægr. Suppl. t. 65; Wilson, l. c.

Abundant throughout the islands, and as far south as **Campbell's** Island; also in the **Kermadec** Islands and **Chatham** Island. One of the commonest mosses in the world, almost sure to appear where wood has been burned.

2. **F. glabra,** *Taylor in Hook. Lond. Journ. Bot.* 1846, *p.* 57;—*Fl. N. Z.* ii. 91. Stems very short. Leaves erecto-patent, obovate, apiculate, serrulate at the apex; nerve vanishing below the apex. Fruitstalk erect, twisted to the left. Capsule clavate-pyriform, curved, gibbous; annulus 0; inner peristome imperfect; operculum nearly flat; margin not coloured.

Northern Island: Ahuriri and Raukawa mountains, *Colenso;* Auckland, *Knight.* (S.W. Australia, Tasmania.)

3. **F. cuspidata,** *Hook. f. and Wils. Fl. N. Z.* ii. 91. *t.* 86. *f.* 3. Stems $\frac{1}{8}$ in. high. Leaves spreading, nearly flat, ovate, acuminate, quite entire; nerve far excurrent. Fruitstalk pale, 1 in. long. Capsule erect, pyriform, small, quite symmetrical; annulus 0; inner peristome imperfect; operculum nearly flat.

Northern Island: Bay of Islands, *J. D. H.*

43. ENTOSTHODON, Schwægrichen.

Short, loosely tufted, biennial, monœcious mosses, growing in earth, rarely on stones. Stems short. Leaves membranous; cells large, lax.—Fruitstalk terminal. Capsule erect, pyriform; annulus 0; teeth 16, lanceolate, simple or in pairs, inserted below the mouth, trabeculate on the inner face, erect when dry. Operculum plano-convex. Calyptra cucullate, very membranous, inflated below.

A small genus, chiefly confined to southern temperate climates. One is British.

1. **E. gracilis,** *Hook. f. and Wils. Fl. N. Z.* ii. 91. *t.* 86. *f.* 7. Stems very short, $\frac{1}{8}$–$\frac{1}{4}$ in. high, unbranched. Leaves erect, imbricate, oblong, acute or acuminate, concave, quite entire; nerve vanishing below the apex. Fruitstalk very slender, 1 in. high, twisting to the left in drying. Capsule erect, apophysate; operculum nearly flat; annulus 0, contracted below the mouth when dry.

Northern Island, *Kerr;* Bay of Islands, *Sinclair.* (Tasmania.)

44. PHYSCOMITRIUM, Bridel.

Characters of *Entosthodon,* but teeth 0, and operculum conic.

A rather large genus of mosses, found in all temperate climates. The descriptions are so short that it is not necessary to add a key to the species.

1. **P. apophysatum,** *Taylor;—Fl. N. Z.* ii. 91. *t.* 86. *f.* 6. Stem very short, $\frac{1}{4}$ in. high. Leaves few, erecto-patent, ovate, acuminate or almost piliferous, subserrate, concave; nerve vanishing below the apex. Fruitstalk very short, stout. Capsule large, erect, clavate-pyriform, constricted below the mouth when dry.

Northern Island: Bay of Islands, on clay banks, *J. D. H.;* Hawke's Bay, *Colenso.* (Swan River, Tasmania.)

2. **P. pyriforme,** (*Bruch and Schimper,*) var. β, *Fl. N. Z.* ii. 92. Stems tufted. Leaves erecto-patent, subspathulate, subacute, concave, serrulate; nerve nearly continuous. Fruitstalk very short. Capsule suberect, turbinate or pyriform; mouth wide; annulus present; operculum conical, apiculate.—(?) *P. conicum,* Mitten in Fl. Tasman. ii. 197.

Northern Island: Bay of Islands, *Colenso, J. D. H.;* Auckland, *Knight.* (The *P. pyriforme* is a common European, American, and Australian plant.)

3. **P. pusillum,** *Hook. f. and Wils. Fl. N. Z.* ii. 92. *t.* 87. *f.* 1. Stems very short, $\frac{1}{8}$–$\frac{1}{4}$ in. high. Leaves spreading, spathulate, obovate- or ovate-oblong, acuminate, nearly entire; nerve nearly continuous. Fruitstalk extremely short. Capsule erect, immersed, subglobose; operculum conical. Calyptra small, conical-mitriform, covering the operculum only, torn at the base. Male inflorescence discoid.

Northern Island, *Sinclair.*

4. **P. Perottetii,** *Montagne, Syllog.* 30;—*Fl. N. Z.* ii. 92. Minute. Leaves imbricating and forming an ovoid capitulum, ovate acuminate, very concave, quite entire; nerve excurrent. Capsule pyriform; operculum flat.

Northern Island: Auckland, *Knight.*—Apparently the same as the Indian Moss of Montagne, but cells are larger and laxer, and nerve more continuous.

45. EREMODON, Bridel.

Tufted monœcious or diœcious mosses, growing on old wood, on earth, or on animal matter. Leaves membranous, acuminate, serrate; nerve vanishing; cells large and lax. Fruitstalk terminal. Capsule erect, upper part cylindric, lower forming a narrow apophysis; annulus 0; teeth 8 equidistant or 16 in pairs, inserted below the mouth, flat, reflexed when dry; spores compound, with 6–8 radiating lines. Operculum conic or convex. Calyptra mitriform or cucullate, glabrous or pilose, lacerate.

The *Eremodons* in the southern hemisphere are the representatives of the large genus *Splachnum* in the northern, but grow on decayed wood or moist ground, not on dung, as is usual with their Northern allies.

Leaves serrated . . . . 1. *E. robustus.*
Leaves quite entire.
 Nerve continuous or excurrent . . . . 2. *E. octoblepharis.*
 Nerve vanishing below the apex . . . . 3. *E. purpurascens.*

1. **E. robustus,** *Hook. f. and Wils. Fl. N. Z.* ii. 93. *t.* 87. *f.* 2. Stems 1–4 in. high, tomentose with radicles. Leaves pale, bright green, distant, lax, spreading, spathulate, lanceolate, with acuminate recurved apices, sharply serrated; nerve vanishing below the apex. Fruitstalk stout, ½ in. high. Capsule erect, oblong-clavate; teeth 8, incurved when dry, wide at the base, yellow; operculum subconic. Calyptra 4-partite at the base, rough at the apex.—*Dissodon callophyllus,* C. Muell. in Bot. Zeit. 1851, 546.

Northern Island: Bay of Islands, *Sinclair, Oldfield;* Auckland, *Knight;* Hawke's Bay, *Jolliffe.* (Tasmania.)

2. **E. octoblepharis,** *Hook. f. and Wils. Fl. N. Z.* ii. 94. Very similar to *C. robustus,* but smaller; stems 1 in. high. Leaves obovate, long acuminate, almost piliferous, quite entire; nerve continuous or excurrent. Capsule erect, clavate; teeth 8, double, perforated down the middle, reflexed when dry.—*Splachnum,* Hook. Musc. Exot. t. 167; Schwægr. Suppl. t. 129. *S. plagiopus,* Mont. Voy. au Pôle Sud, 285. *Dissodon plagiopus,* C. Muell.

Var. β. *pyriforme,* Fl. Antarct. 123. t. 57. f. 4. Leaves more erect, and crowded.
Var. γ. *major,* l. c. 123. t. 57. f. 4. Leaves larger, broader, lurid-green.
Throughout the islands, abundant. Var. β and γ, **Lord Auckland's** group and **Campbell's** Island, *J. D. H.* (Tasmania and Australia.)

3. **E. purpurascens,** *Hook. f. and Wils.*—*Splachnum,* Fl. Antarct. 123. t. 57. f. 6. Larger and laxer than *E. octoblepharis.* Leaves spreading, broadly obovate, acuminate, quite entire; nerve vanishing below the acuminate apex.

Northern Island: Auckland, etc., *Colenso* and *Botton, Knight.* **Lord Auckland's** group and **Campbell's** Island: in bogs, *J. D. H.* Not distinct, I suspect, from the *E. octoblepharis.*

46. POLYTRICHUM, Linn.

Erect, tufted, often large, rigid, dark green or brown, monœcious or diœcious mosses, growing on the ground or roots of trees, etc. Stems very rarely branched. Leaves usually long, thick, coriaceous, opaque; nerve very thick, with parallel grooves or plates on the upper surface; cells obscure, very minute. Fruitstalk terminal, stout. Capsule erect or slightly inclined, terete or 4–6-angled, rarely concave or flat on one side, and convex on the other, often contracted below the mouth; annulus 0; teeth 16, 32 or 64, very short, rigid or horny in texture, incurved, of several layers of superposed cells, with a circular membrane stretched loosely across their tips. Operculum flattish, often beaked. Calyptra small, cucullate, naked or densely clothed with a thatch of matted hairs.

A large and remarkable genus, of usually rigid, brown or lurid-green mosses, often large, sometimes gigantic; found in both temperate and tropical climates, but far most abundantly in the former, reaching the Arctic circle.

A. *Calyptra nearly glabrous.*

I. ATRICHUM, *Palisot.* Stem simple. Nerve of (toothed) leaf narrow. Capsule terete; operculum with a long beak . . 1. *P. angustatum.*
II. OLIGOTRICHUM, *DC.* Stem simple. Nerve of (entire) leaf broad. Capsule ovoid or oblong; operculum long . . . 2. *P. tenuirostre.*
III. PSILOPILUM, *Bridel.* Stem simple. Nerve of leaf grooved or lamellate. Capsule terete; operculum small.
Leaves spreading, toothed. Teeth 32 . . . . 3. *P. crispulum.*
Leaves suberect, quite entire. Teeth 16 . . . . 4. *P. australe.*
IV. POLYTRICHADELPHUS, *C. Muell.* (*Cyphoma,* Hook. f. and Wils.) Stem simple or dichotomous. Capsule horizontal, flat above, concave below . . . . 5. *P. Magellanicum.*
V. PHALACROMA, *Hook. f. and Wils.* Stem tall, fastigiously branched, dendroid. Capsule short, terete.
Stem 4–10 in. high. Leaves linear, from an ovate base . . 6. *P. dendroides.*
Stem 2–4 in. high, scaly. Leaves lanceolate-subulate . . . 7. *P. squamosum.*

B. *Calyptra clothed with matted hairs.*

VI. POGONATUM, *Palisot.* Stem simple or branched above. Capsule terete or nearly so. Apophysis 0.
Stem 1–2 in. Operculum shortly beaked . . . . 8. *P. tortile.*
Stem 2–4 in. Operculum with a long beak . . . . 9. *P. alpinum.*
VII. EUPOLYTRICHUM, *C. Muell.* Stem simple or dichotomous. Capsule angled, apophysate.
Capsule 4-angular. Leaves quite entire . . . . 10. *P. juniperinum.*
Capsule 4-angular. Leaves serrate; perichætial distinct . . 11. *P. commune.*
Capsule 4–6-angular. Leaves serrate; perichætial 0 . . . 12. *P. gracile.*

1. **P. angustatum,** *Hook. Musc. Exot. t.* 50;—*Fl. N. Z.* ii. 94. Stems slender, 2 in. high. Leaves spreading when dry, waved and crisped, narrow-lanceolate, doubly spinulose-serrate above; nerve slender, sparingly lamellate. Fruitstalks ½ in. long or more, 1–3 together. Capsule cylindric, slightly curved, brown; operculum with a long, slender, inclined beak. Calyptra slender, spinulose at the apex. Inflorescence diœcious.—*Atrichum ligulatum,* Mitten in Kew Journ. Bot. 1856, 262.

Northern and **Middle** Islands: shaded woods, Huiarau, *Colenso;* Otago, by creeks, *Hector and Buchanan.* (Europe and N. America, Tasmania.) Closely allied to the European *P. undulatum.* Mitten distinguishes the southern species as *Atrichum ligulatum,* distinguishing it by the leaves broader towards the apices, not involute when dry, and with wider thickened margins.

2. **P. tenuirostre,** *Menz.;—Hook. Musc. Exot. t.* 75;—*Fl. N. Z.* ii. 94. Stems rigid, short, ¼ in. high. Leaves spreading, incurved when dry, lower ovate-lanceolate, upper oblong-lanceolate, subacute, quite entire, concave; nerve slender, continuous; lamellæ indistinct. Fruitstalk 1–3 in. long, stiff, glossy. Capsule ovate-oblong, suberect, elliptic; operculum with a slender beak, as long or longer than the capsule. Calyptra not seen.

Middle Island: Dusky Bay, *Menzies, Lyall.*

3. **P. crispulum,** *Hook. f. and Wils. Fl. N. Z.* ii. 95. *t.* 87. *f.* 3. Stems 1 in. high. Leaves rather flaccid and spreading, crisped when dry, oblong-lanceolate, subacute, toothed, deeply grooved above the middle; nerve broad.

Fruitstalk stout, 1½ in. long. Capsule inclined, ovate-oblong, terete; mouth small; teeth 32; operculum with a slender beak. Calyptra hairy at the tip, ventricose. Inflorescence diœcious.

Northern Island: shaded woods, Huiarau, *Colenso*. (Tasmania.)

4. **P. australe,** *Hook. f. and Wils. Fl. N. Z.* ii. 95. *t.* 87. *f.* 6. Stems ½–1 in. high. Leaves crowded, erecto-patent, erect and incurved when dry, ovate below, gradually narrowed and oblong-subulate above, subacute, quite entire, densely lamellate; nerve indistinct. Fruitstalk stout, ½ in. long. Capsule inclined, ovoid, terete; mouth small; teeth 16; operculum with a short deflexed beak. Calyptra short, scabrid at the tip.

Northern and **Middle** Islands: Ruahine mountains, *Colenso;* mountains of Otago, alt. 5–6000 feet, *Hector and Buchanan.* (Tasmania.)

5. **P. Magellanicum,** *Hedwig, Sp. Musc. t.* 20;—*Fl. N. Z.* ii. 95. Stems stout, 1–4 in. high, dichotomously branched. Leaves erecto-patent or spreading and recurved, rigid, not altered when dry; base ovate, sheathing, thence subulate, serrate, obscurely lamellate. Fruitstalk stout. Capsule inclined or horizontal, oblong, semiterete; teeth 64; operculum conical, beaked. Calyptra bristly or rough at the apex.—Fl. Antarct. 132 and 411. t. 59. f. 3.—*Catharinea innovans,* C. Muell. in Bot. Zeit. 1851, 548.

Northern and **Middle** Islands: common on banks, *J. D. H.*, etc. **Lord Auckland's** group, *J. D. H.* (Tasmania and Fuegia.)

6. **P. dendroides,** *Conmerson;—Fl. N. Z.* ii. 96. Stems 6 in. to 1 foot high, stout, fasciculately much-branched above; the branches spreading like those of a tree. Leaves spreading, sheathing at the base, thence linear-subulate, sharply serrate, lamellate. Fruitstalks elongate, numerous. Capsule short, inclined, cylindrical; mouth large; teeth 64; operculum with a slender beak, longer than the capsule. Calyptra slightly hairy.—Schwægr. Suppl. ii. pt. 2. t. 151.

Northern Island, in subalpine and wooded districts, *Colenso*, etc., and throughout the **Middle** Island, *Lyall*, etc.—A magnificent moss. (Fuegia and Chili.)

7. **P. squamosum,** *Hook. f. and Wils. Fl. Antarct.* 411. *t.* 153. *f.* 8. Habit and appearance of *P. dendroideum*, but much smaller, 3–4 in. high; branches short, dense, curved. Leaves on the main stem scarious, those of the branches erecto-patent, subulate-lanceolate, strict, serrate, not half the length of those of *P. dendroideum.* Fruit unknown.

Northern Island: Taranaki hills, New Plymouth, *Jupp* (*Mitten*).—I have seen no New Zealand specimens of this species, which I discovered in Fuegia.

8. **P. tortile,** *Swartz;—Fl. N. Z.* ii. 96. Stems simple, 1–2 in. high. Leaves loosely spreading, incurved when dry, sheathing at the base, thence linear-lanceolate, flat, serrate. Fruitstalk 1 in. high. Capsule suberect, nearly terete, obscurely 6-lined; operculum convex, with a short beak. Calyptra clothed with matted hairs.—*P. subulatum*, Menzies in Linn. Trans. iv. 303. t. 6. f. 5.

Northern and **Middle** Islands: Bay of Islands, on clay banks, *J. D. H.;* Tehawera, *Colenso;* Hutt valley, *Sinclair;* Auckland, *Jolliffe;* Otago, *Lindsay.* (E. and W. Indies, S. America.)

A doubtful moss, very closely allied to the British *P. aloides*, but apparently identical with the tropical *P. tortile*.

9. **P. alpinum,** *Linn.* Stems 2–4 in. long, curved. Leaves spreading and recurved, long linear-lanceolate, sheathing at the base, suberect when dry; margin incurved, sharply serrate, spinulose at the back; nerve lamellar. Fruitstalk long, stout, yellowish. Capsule erect or inclined, ovoid or oblong, terete; teeth short, narrow; operculum with a long beak. Calyptra covered with red-brown matted hairs.—Wils. Bryol. Brit. 207. t. 11; Fl. Tasman. ii. 200.

Middle Island: Fagus forests of Otago, alt. 2–3000 feet, *Hector and Buchanan*. (Europe, N. America, Australia, Tasmania.)

10. **P. juniperinum,** *Hedwig;—Fl. N. Z.* ii. 96. Stems short, ½–1½ in. high, simple, stout. Leaves spreading and recurved, sheathing at the base, thence linear-lanceolate, acuminate and aristate; margins inflexed, quite entire; nerve not reaching the margin, lamellate. Fruitstalk 1–2 in. long. Capsule inclined or horizontal, 4-angled; teeth 64; operculum with a short beak. Calyptra covered with matted hairs.—Wils. Bryol. Brit. 213. t. 10.

Common throughout the **Northern** and **Middle** Islands; on banks, in bogs, etc. **Lord Auckland's** group, *Hombron and Jacquemont*. (Cosmopolitan.)

11. **P. commune,** *Linn.;—Fl. N. Z.* ii. 96. Stems simple, rigidly flexuous, 1–4 in. high, red-brown. Leaves spreading, recurved or erecto-patent, linear-subulate or lanceolate-subulate, serrate, acuminate, not aristate; nerve overlying the whole width, lamellate; perichætial membranous, sheathing, erect, piliferous. Fruitstalk 2–3 in. long. Capsule oblong, 4-angled, apophysate; teeth 64; operculum with a short beak. Calyptra covered with matted hairs.—Wils. Bryol. Brit. 211. t. 10.

Common throughout the **Northern** and **Middle** islands, *Colenso*, and **Chatham** Island, *W. Travers*. A very handsome moss, found all over the world, used for making brushes and hassocks in England.

12. **P. gracile,** *Menzies*. Stems densely tufted, 2–5 in. high. Leaves short, lanceolate from a sheathing base, with a broad, pellucid, serrate margin, flat; nerve lanceolate; perichætial 0. Fruitstalk 2–3 in. high. Capsule ovoid, obscurely 4–6-angled; teeth 32 or 64, irregular; operculum beaked. Calyptra shorter than the capsule, covered with pale matted hairs.—Wils. Bryol. Brit. 210. t. 46.

Middle Island: Canterbury, *Sinclair and Haast*. (Europe, N. America.)

47. DAWSONIA, *Br.*

Large, handsome, rigid, dark green mosses, with the habit and most of the characters of *Polytrichum*. Capsule inclined or horizontal, flat above, convex below; peristome a long brush of capillary cilia, unjointed, fringing the mouth, and also sometimes terminating the columella. Operculum subulate. Calyptra with a covering of matted hairs.

One of the handsomest and most curious genera of mosses, confined to New Zealand and Australia and Tasmania.

1. **D. superba,** *Grev.;—Fl. N. Z.* ii. 97. Stems 5–14 in. high, very rigid, naked below. Leaves rigid, dark brown, squarrose or suberect, very

narrow linear-subulate, with a large sheathing base, sometimes 1 in. long, spinulose-serrate. Fruitstalk short, stout, 2–4 in. long. Capsule generally partially hidden by the leaves; peristome of excessively numerous (above 500), filiform, cylindric cilia, in 8 or 9 concentric layers. Calyptra small, covered with smooth matted hairs.

Northern and **Middle** Islands; in dense forests, probably common; from Auckland southward, *Sinclair*; Massacre Bay, *Lyall*; Otago, *Lindsay*. One of the most magnificent known mosses. (Tasmania and Australia.)

48. ANŒCTANGIUM, Br. and Sch.

Tufted, perenial, diœcious mosses, growing on rocks and banks. Stems dichotomous; branches fastigiate. Leaves crowded, linear-lanceolate or subulate; cells small, roundish. Fruitstalk lateral, elongate, slender. Capsule erect, ovoid or obovoid, membranous, with a short inflated apophysis; annulus small; teeth 0. Operculum with a slender inclined beak.

A small European subalpine genus.

1. **A. compactum,** *Schwægr.* Tufts dense, bright yellow-green. Stems 1–4 in. high. Leaves 3-farious, spreading from an erect base, incurved, crisped when dry, lanceolate, acuminate, keeled; margin flat, slightly toothed towards the base; nerve pellucid, subexcurrent. Fruitstalk pale, $\frac{1}{12}$ in. long. Capsule small, oblong-ovoid, pale; operculum as long as the capsule.—Wils. Bryol. Brit. 311. t. 5. *Gymnostomum æstivum*, Hedwig.

Middle Island: open grounds, Otago, *Hector and Buchanan*. (Europe.)

49. AULACOPILUM, Wils.

A minute, creeping, sparingly-branched monœcious moss. Leaves distichous, nerveless, glaucous, papillose; cells most minute. Fruitstalk short, stout, lateral. Capsule erect, ovate-globose, truncate at the mouth, exannulate; teeth 0. Operculum conical, beaked. Calyptra large, including the capsule, and embracing the fruitstalk below it, grooved, split at the side.

1. **A. glaucum,** *Hook. f. and Wils. in Lond. Journ. Bot.* 1848, 90. *t.* iv. A;—*Fl. N. Z.* ii. 98. Stems creeping, $\frac{1}{4}$–$\frac{1}{2}$ in. long, sparingly branched. Leaves pale glaucous-green, spreading, appressed when dry, distichous, obliquely ovate, acuminate, papillose; perichætial erect, lanceolate. Fruitstalk $\frac{1}{12}$ in. high. Capsule about half as long as the fruitstalk.

Northern Island: Bay of Islands, on trees with *Fabronia*, Colenso. (South Africa, *Mitten*.)

50. FABRONIA, Bridel.

Minute, tufted, creeping, monœcious mosses, usually growing on trunks of trees. Leaves imbricate, usually ciliate-toothed or serrulate, membranous; cells large, lax. Fruitstalk lateral. Capsule erect, of thin texture, subglobose or pyriform; annulus 0; teeth 16, in pairs, inflexed, coriaceous. Operculum various. Calyptra cucullate.

Generally tropical or subtropical mosses; two are found in Europe.

1. **F. australis,** *Hook. Musc. Exot. t.* 160;—*Fl. N. Z.* ii. 98. Minute.

Stems very slender, creeping, with erect branches ¼ in. high, pale green. Leaves crowded, subsecund, turned upwards, rather concave, ovate or ovate-lanceolate, acuminate, serrulate; nerve reaching halfway; perichætial short, ovate. Capsule not ribbed, subglobose; operculum flattish.

Northern Island: Bay of Islands, *Colenso*, *J. D. H.* (Australia.)

51. LEUCODON, Bridel.

Creeping, branched, tufted loosely, rather stout, diœcious mosses, growing on rocks and trees; stems erect, often curved, terete; branches sometimes pendulous. Leaves densely imbricate, usually nerveless, grooved striated or smooth; cells long and narrow or dot-like. Fruitstalk lateral. Capsule erect, narrow-oblong; annulus 0 or partial; teeth 16, connate at the base, perforate, 2–3-partite. Operculum conical or rostellate. Calyptra cucullate.

A considerable genus, of which the species are not much allied to one another, and have been variously disposed of by Mitten and others.

Leaves not plaited or undulate, acuminate and piliferous; nerves 0 or 2 short . . . . 1. *L. Lagurus.*
Leaves plaited, mucronate, acuminate, nerveless . . . . 2. *L. implexus.*
Leaves not plaited, obtuse, nerveless . . . . 3. *L. nitidus.*

1. **L. Lagurus,** *Hook. Musc. Exot. t.* 126;—*Fl. Antarct.* i. 136. Stems tufted, creeping, with erect, short, stout branches, ½ in. high, covered with matted fibrils. Leaves bright green, glossy, closely imbricated, oblong-ovate, concave, acuminate and piliferous, not striated ribbed or plaited; nerves 0 or short, sometimes 2-nerved to the middle. Fruitstalk 1 in. long. Capsule subcylindrical; teeth united; operculum with a slender oblique beak, half as long as the capsule. Calyptra large. Male inflorescence of a few subglobular antheridia, on very slender stems with variable small leaves.—*Stereodon*, Mitten in Journ. Linn. Soc. vi. 88.

Campbell's Island (barren). (Fuegia and Tasmania.)—Wilson suggests (Bryol. Brit. 313) removing this from *Leucodon* under the name of *Lampurus*. The *L. Lagurus*, var. *β.* of 'Bryologia Britannica,' is a species of *Hypnum* (Mitten).

2. **L. implexus,** *Kunze.* Tufts dense, soft, bright green, shining, covered below with matted rootlets. Leaves imbricate, suberect, oblong-lanceolate, acuminate, piliferous, deeply plaited, quite entire or minutely toothed at the apex, nerveless. Fruitstalk short. Capsule erect, cylindric-oblong; teeth irregular; operculum conical, almost beaked.—*L. hexastichus*, Mont. *L. Kunzeanus* and *Neckera implexa*, C. Mueller.

Middle Island: Nelson Mountains, *Sinclair*; Otago, on rocks and trees, *Hector and Buchanan*. (Chili.) Determined by Mitten; the leaves have longer acuminate points than in Mueller's description of the Chili plant.

3. **L. nitidus,** *Hook. f. and Wils. Fl. N. Z.* ii. 99. *t.* 87. *f.* 4. Stem about 1 in. long, procumbent, branches obtuse. Leaves loosely imbricate, oblong, obtuse, concave, quite entire, not striate, nerveless, pale green, shining; margins subrecurved; perichætial longer, convulute. Fruitstalk ¼ in. long, red. Capsule erect, oblong, grooved when dry; teeth red, irregularly split halfway down, trabeculate; operculum long-beaked.—*Stereodon Lyallii*, Mitten in Journ. Linn. Soc. iv. 89.

Northern and **Middle** Islands: Bay of Islands, on trunks of trees, rare, *J. D. H.*; Nelson, *Travers*. Mr. Wilson remarks that this curious moss has the habit of *Pterogonium*, and may perhaps form a new genus, for which he proposes the name *Dichelodontium*.

52. LEPTODON, Mohr.

Creeping, much-branched, tufted, diœcious perennial mosses; branches pinnate, usually elastically curving inwards when dry. Leaves quite entire; nerve short; cells dot-like. Fruitstalk lateral. Capsule straight, erect or inclined; annulus 0; teeth 16, very short, linear-lanceolate, membranous. Operculum beaked. Calyptra cucullate, clothed below with very long hairs.

A genus of few species, some tropical, and one or two temperate.

1. **L. Smithii,** *Bridel;—Fl. N. Z.* ii. 99. Stems 1–3 in. long, slender; branches 2-pinnate. Leaves rounded-ovate, obtuse; nerve vanishing above the middle. Fruitstalk very short. Capsule suberect, oval-oblong.—Wils. Bryol. Brit. 317. t. 14.

Middle Island, *Lyall*, on bark; Otago, N.E. valley, on trunks of trees, with a north exposure, *Hector and Buchanan*. (Europe and South Africa.)

53. CLADOMNION, Hook. f. and Wils.

Loosely tufted, creeping mosses, with erect generally long and slender, nearly simple branches. Leaves imbricate all round, plaited and striate, nerveless. Fruitstalk very short, lateral. Capsule erect, straight, grooved; peristome double; outer of 16 lanceolate teeth; inner, a membrane deeply divided into 16 keeled cilia. Operculum beaked. Calyptra large, cucullate.

I follow Wilson in keeping this genus as defined in the 'Flora of New Zealand,' excluding *C. setosum*, which I refer back to *Cyrtopus* of Bridel. Mitten refers *C. ericoides* and *sciuroides* to the section *Achyrophyllum* of Hypnum, to which no doubt they are naturally much allied in foliage and in which the capsule is angled.

Leaves quite entire . . . . 1. *C. ericoides.*
Leaves serrated at the apex . . . . 2. *C. sciuroides.*

1. **C. ericoides,** *Hook. f. and Wils. Fl. N. Z.* ii. 99. Stems creeping; branches 2–8 in. long, suberect, nearly simple, stout. Leaves pale red or yellow-brown, shining, imbricate, ovate or oblong, acuminate, with recurved apices, concave, quite entire, plaited and striate; nerve 0. Fruitstalk $\frac{1}{3}$ in. long. Capsule erect, oblong, 8-grooved; outer teeth yellow, firm; inner of 16 keeled cilia, united halfway up; operculum with a slender beak; spores very large. Calyptra yellowish. Male inflorescence axillary, small, often clustered.—*Leskea*, Hook. Mus. Exot. t. 140. *Stereodon*, Mitten in Journ. Linn. Soc. iv. 89.

Northern and **Middle** Islands: common, south of Auckland, *Menzies*, etc.; Nelson, *Travers*; Otago, *Hector and Buchanan*.

2. **C. sciuroides,** *Hook. f. and Wils. Fl. N. Z.* ii. 100. Stems creeping; branches suberect, incurved, rather flattened. Leaves yellow-brown,

when dry rather shining, subsecund, erecto-patent, ovate, acuminate, serrate at the apex, plaited and striate; nerve 0. Fruitstalk ⅛ in. long. Capsule erect, oblong, 8-grooved; outer teeth pale; inner a membrane with rudimentary cilia; spores small; operculum conical-subulate, half as long as the capsule. Inflorescence diœcious.—*Leskea*, Hook. Musc. Exot. t. 175. *Stereodon*, Mitten in Journ. Linn. Soc. iv. 89. *Neckera glyptotheca*, Mueller (in Herb.).

Middle Island: Nelson, *Bidwill;* Otago, *Hector and Buchanan.* (Tasmania.)

54. METEORIUM, Bridel.

Flaccid, pale green, diœcious mosses. Stems creeping over trees and stones, then pendulous, often very long and flexuous; branches short, spreading. Leaves imbricate, usually very delicate, shining, concave, nerved or nerveless, smooth or striate; cells small. Fruitstalk lateral, very short. Capsule erect, straight, narrow, oblong or ovoid; annulus 0; peristome double; outer of 16 erect teeth; inner as many cilia, free, or united by a membrane at their bases. Operculum beaked. Calyptra cucullate.

Beautiful tropical and southern mosses; none are British.

Pendulous; stems 6–18 in. long.
Leaves oblong-spathulate, obtuse; nerve 0 . . . . 1. *M. molle.*
Leaves ovate-cordate, apiculate; nerve vanishing . . . . . . 2. *M. cuspidiferum.*
Leaves ovate-cordate, acuminate; nerve produced beyond the middle . . . . 3. *M. flexicaule.*
Pendulous; stems 1–4 in. long
Minute. Leaves ovate-lanceolate; nerve very short . . . . . 4. *M. pusillum.*
Leaves ovate-lanceolate; nerve produced to or beyond the middle 5. *M. nitens.*

1. **M. molle,** *Hook. f. and Wils. Fl. N. Z.* ii. 100. Stems very long, pendulous, slender, flexuous, flaccid, 1–8 in. long; branches nearly simple. Leaves pale straw-coloured, shining, imbricate, oblong-spathulate, obtuse, subcordate at the base, margins incurved; apex inflexed, concave, quite entire; nerve 0; perichætial twice as long, sheathing. Fruitstalk ⅛ in. long. Capsule ovoid; inner peristome divided to the middle into 16 keeled cilia; operculum with a slender beak as long as the capsule. Calyptra hairy, dimidiate.—*Leskea mollis*, Hedwig, Musc. Frond. 4. t. 40. *Neckera*, C. Muell. *Stereodon*, Mitten in Journ. Linn. Soc. iv. 88.

Throughout the **Northern** and **Middle** Islands: common in forests, etc. (Australia, Tasmania, Chili, Fuegia.)

2. **M. cuspidiferum,** *Tayl. mss.;—Fl. N. Z.* ii. 101. Stems as in *M. molle.* Leaves dark green or yellowish, loosely imbricate, erecto-patent, ovate-cordate, apiculate, semiamplexicaul and slightly toothed at the base, elsewhere quite entire, keeled, substriate; nerve vanishing; cells most minute. —*Trachypus Hornschuchii*, Mitten in Journ. Linn. Soc. iv. 91.

Var. *β. cerina;* leaves broader, waved when dry; margins reflexed at the base.—*T. cerinus*, Mitt. l. c.

Common in the **Northern** and **Middle** Islands: Saddle Hill, Otago, *Lindsay;* Bay of Islands, *Sinclair, J. D. H.* **Kermadec** Islands, *Milne.* *β.* **Middle** Island, *Lyall.* (Norfolk Island, Tasmania, East Indies.)

3. **M. flexicaule,** *Hook. f. and Wils. Fl. N. Z.* ii. 101. Stems pendulous, flexuous, very slender indeed, 4–10 in. long. Leaves dull yellow or brownish, loosely imbricate ovate-cordate or subspathulate, shortly acuminate, concave, not striated, margin quite entire except at the obscurely-toothed base; nerve produced to the middle.—*Trachypus flexicaulis,* Mitten, l. c. *Leskea flexicaulis,* Taylor.

Northern and **Middle** Islands, *Dr. Stanger;* Hawke's Bay, *Colenso;* Wellington, *Stephenson;* Otago, *Hector and Buchanan;* Nelson, *Travers.* (Tasmania, Andes.)

4. **M. pusillum,** *Hook. f. and Wils. Fl. N. Z.* ii. 101. *t.* 88. *f.* 1. Minute; stems very slender, weak, 1 in. long. Leaves dull green, loosely imbricate, suberect when dry, ovate-lanceolate, acuminate, margin quite entire, subrecurved; nerve very short; perichætial erect, lanceolate. Fruitstalk $\frac{1}{8}$ in. long. Capsule half as long as the fruitstalk, ovoid, 8-grooved, pale; operculum conical, very short.

Northern Island: Wairarapa valley, *Colenso.*

5. **M. nitens,** *Hook. f. and Wils. Fl. N. Z.* ii. 101. *t.* 87. *f.* 7. Stem 1–2 in. long, creeping, then pendulous, slender; branches numerous, short, ascending, $\frac{1}{10}$–$\frac{1}{4}$ in. long. Leaves shining, yellow-green, crowded, erect when dry, ovate-lanceolate, acuminate, margin subrecurved, minutely denticulate, somewhat striate; nerve vanishing about or above the middle; cells very narrow.

Northern Island, *Sinclair.*—A scrap without habitat or fruit.

55. CRYPHÆA, Mohr.

Slender, monœcious, creeping on trees or rocks; branches erect or pendulous, somewhat pinnately divided. Leaves imbricate, ovate, quite entire, nerved; cells very minute, like dots. Fruitstalk lateral, extremely short, hidden by the perichætial leaves. Capsule erect, narrow-oblong; annulus present; peristome double; outer 16 narrow erect teeth; inner 16 filiform free cilia. Operculum conical. Calyptra very small, mitriform, conic, glabrous, scarcely covering the operculum.

A genus of several tropical and temperate species.

Leaves ovate-lanceolate, quite entire; perichætial piliferous . . . . 1. *C. parvula.*
Leaves broadly ovate, long-acuminate, quite entire . . . . 2. *C. acuminata.*
Leaves very broadly ovate, obtuse, quite entire . . . . 3. *C. dilatata.*
Leaves orbicular-ovate, acute, minutely serrulate . . . . 4. *C. Tasmanica.*

1. **C. parvula,** *Mitten, mss.*—*C. consimilis,* Fl. N. Z. ii. 101, not of Montagne. Stems creeping, 1–2 in. long; branches distant, spreading, filiform. Leaves ovate-lanceolate, acuminate, margin quite entire, subrecurved; nerve nearly continuous; perichætial broadly obovate, retuse, suddenly produced into a piliferous acuminate point. Fruitstalk scarcely visible. Capsule ovoid; operculum conical, acute. Calyptra red-brown, rough at the apex. Inflorescence monœcious.

Northern Island: Wairarapa valley, *Colenso;* Otago, *Hector and Buchanan.* (Tasmania, Australia.)—Very near the Chilian *C. consimilis,* but differing in the perichætial leaves exceeding the operculum, and in the wider cauline leaves (Mitten).

2. **C. acuminata,** *Hook. f. and Wils. Fl. N. Z.* ii. 102. *t.* 88. *f.* 4. Stems very short, ¾ in. long, sparingly branched. Leaves dull yellow-green, erecto-patent, ovate, long-acuminate, quite entire; nerve vanishing above the middle; perichætial larger, setaceous at the apex. Capsules secund, sessile, hidden by the perichætial leaves; operculum conical, with a short beak. Inflorescence monœcious.

Northern Island: on trees, Hawke's Bay, etc., *Colenso*; Otago, *Hector and Buchanan.*

3. **C. dilatata,** *Hook. f. and Wils. Fl. N. Z.* ii. 102. *t.* 88. *f.* 2. Stems 2–6 in. long, slender, pendulous, branched; branches very short, spreading. Leaves dull green, spreading, erect when dry, loosely imbricate, broadly ovate, obtuse, quite entire, concave; nerve vanishing below the subcrenulate apex; perichætial lanceolate; nerve stronger. Capsules sunk amongst the perichætial leaves, often crowded, scattered or secund; operculum conic. Inflorescence monœcious.

Northern Island, *Colenso.*

4. **C. Tasmanica,** *Mitten in Fl. Tasman.* ii. 204. *t.* 175. *f.* 6. Stems elongate, naked below, above furnished with short close-set subsecund branches. Leaves spreading, orbicular-ovate, acute, margin flat, most minutely serrulate; nerve vanishing below the apex; perichætial ovate, then subulate. Capsule sunk in the perichætial leaves; operculum convex, acute.

Middle Island: Otago, on trees?, *Hector.* (Tasmania.)

56. CYRTOPUS, Bridel.

Stem erect, subpinnately branched. Leaves crowded, subsecund or suberect, serrate, with a stout nerve. Fruitstalk very short. Capsule oblong, erect; peristome of *Cladomnion.* Operculum beaked.

The genus *Cyrtopus* was established as a section of *Neckera,* by Bridel, upon this and a few other plants that have been placed in various genera by subsequent authors. I have thought it best in this work to retain it for this plant, which has been referred to five other genera, as its synonymy shows.

1. **C. setosus,** *Bridel, Bryol.* ii. 234.—*Cladomnion,* Fl. N. Z. ii. 100. Stems erect, flexuous, 6 in. long, nearly simple. Leaves dark greenish, crowded, subsecund, rigid, ovate at the base, then subulate with setaceous rigid apices, serrate; nerve solid. Fruitstalk very short. Capsule scarcely emerging from the leaves, erect, not grooved, oblong; teeth red; operculum with the beak shorter than the capsule. Calyptra yellow-brown.—*Neckera,* Hook. Musc. Exot. t. 8. *Spiridens,* Mitten in Journ. Linn. Soc. iv. 89. *Anœctangium,* Hedwig, Sp. Musc. 43. t. 5. f. 4, 6.—*Pilotrichum,* C. Muell.

Northern and **Middle** Islands: abundant from Bay of Islands, *Sinclair,* to Dusky Bay, *Menzies,* etc. (Tasmania, South America, Sandwich Islands.)

57. MESOTUS, Mitten.

A stout tufted moss; stem short, creeping, covered with matted rootlets; branches erect, stiff, simple or forked, crowded. Leaves close-set, spreading serrulate; nerve percurrent. Capsule ovoid-oblong, not furrowed, suberect, hidden amongst the leaves, lateral from the growth of innovations (but originally terminal according to Mitten); peristome single; teeth 16, red, equidistant, irregularly torn at the apex. Operculum conical, acuminate. Calyptra small, mitriform, lobed at the base.

A remarkable moss, of which Mitten says that it has the structure of leaf of *Symblepharis*, creeping stem of *Macromitrium*, and teeth of *Grimmia*. Dr. Schimper considers it to be truly pleurocarpous, and allied to *Esenbeckia*.

1. **M. celatus,** *Mitten, mss.* Branches 1–2½ in. high, robust. Leaves dull yellow-green, crowded, twisted when dry, erecto-patent when moist, ovate at the base, then lanceolate-subulate, keeled; margins flexuous, quite entire below, above serrulate; nerve percurrent; perichætial numerous, pale, convolute, broad below, suddenly contracted to a subulate point.

Middle Island: Otago, on dry banks, *Hector and Buchanan.*

58. PHYLLOGONIUM, Bridel.

Bright green shining or glistening mosses, creeping on the trunks of trees; stems short, pinnately branched; branches quite flat and frond-like. Leaves distichous, equitant, very closely imbricate, folded along the middle, sharply keeled, nerveless; cells long and narrow. Fruitstalk lateral, very short. Capsule suberect, straight; annulus 0; teeth 16, equidistant, flat. Operculum beaked. Calyptra large, nearly mitriform, rather hairy, torn at the base.

A tropical and subtropical genus of most beautiful mosses, remarkable for their golden-green hue and the almost metallic brilliancy of their lustre.

1. **P. elegans,** *Hook. f. and Wils. in Fl. N. Z.* ii. 102. *t.* 88. *f.* 6. Stems creeping, about 1 in. long; branches lanceolate, $\frac{1}{12}$ in. broad. Leaves oblong-rhomboid, obtuse, quite entire, so concave as not to be flattened without splitting; perichætial shorter, erect. Fruitstalk as long as or shorter than the capsule, which is turbinate, and when dry has a wide mouth and exserted columella.

Northern and **Middle** Islands: on smooth bark of trees, from the Bay of Islands, *J. D. H.* etc.; Auckland, *Jolliffe,* to Otago, *Hector and Buchanan.*

59. NECKERA, Hedwig.

Stems creeping on trees and rocks; branches erect, pinnately divided, flattened. Leaves complanate, sub-2-farious, 8-fariously inserted, oblique, often transversely waved, shining, transparent; cells very minute. Fruitstalk lateral, short or moderately long. Capsule erect, ovoid or oblong; annulus 0; peristome double; outer of 16 lanceolate teeth, trabeculate on the inner face; inner a membrane divided into 16 keeled teeth. Operculum obliquely beaked. Calyptra cucullate, glabrous.

A considerable genus, in both temperate and tropical climates, very rare or unknown in cold and frigid regions.

Leaves ovate-lanceolate, acuminate . . . . 1. *N. pennata.*
Leaves oblong, rounded at the apex . . . . 2. *N. lævigata.*

1. **N. pennata,** *Hedwig;—Fl. N. Z.* ii. 103. Stems pinnately branched. Leaves bright green, shining, undulate, ovate-lanceolate, acuminate, serrulate at the apex; nerve very short and inconspicuous; inner perichætial narrow-lanceolate, acuminate. Capsule hidden by the leaves; operculum with a short beak. Calyptra small, scarcely covering the operculum. Inflorescence monœcious.—*N. hymenodonta,* C. Muell.; Wils. Bryol. Brit. 414. t. 34.

Northern and **Middle** Islands: Wangera and Wawari, on trunks of trres, *Colenso;* Auckland, *Knight;* Otago, *Hector and Buchanan.* (Europe, Asia, and America.)

2. **N. lævigata,** *Hook. f. and Wils. Fl. N. Z.* ii. 103. *t.* 88. *f.* 3. Stems 1 in. and more high, pinnately branched, 2–4 in. long. Leaves bright green, loosely imbricate, not undulate, rather rigid, oblong, very obtuse, quite entire, convex; nerve short. Fruitstalk imbricating, convolute, lanceolate, sheathing the fruitstalk, which is $\frac{1}{10}$ in. long. Capsule exserted.

Middle Island: Banks's Peninsula, *Lyall;* Queensland Bush, Otago, *Martin (Hb. Lindsay), Hector and Buchanan.*

60. TRACHYLOMA, Bridel.

Stems stout, erect from a creeping rhizome, tree-like, flattened, pinnately branched above. Leaves distichous, not oblique, flattened, nerve indistinct; cells minute. Fruitstalk slender. Capsule erect, straight or slightly curved, oblong or cylindric; annulus 0; peristome double, outer 16, narrow, coriaceous, nodose teeth; inner 16, narrow, keeled, nodose cilia united by a membrane at the base; operculum with a long, straight beak. Calyptra cucullate.

A considerable genus in tropical and temperate regions.

1. **T. planifolium,** *Bridel;—Fl. N. Z.* ii. 103. Stems stout, erect, about 3 in. high, rising from a stout creeping rhizome. Leaves bright-green, shining, distichous, erecto-patent, ovate, serrate at the tip; nerve very faint, $\frac{1}{3}$ their length; perichætial long, narrow, serrate. Fruitstalk $\frac{1}{2}$ in. long and upwards, flexuous, often curved at the top. Capsule erect, subcylindric, equal or slightly unequal; outer teeth $\frac{1}{3}$ the length of the capsule, pale yellow, incurved when dry; inner almost white; operculum conic-subulate. —*Neckera planifolia,* Hook. Musc. Exot. t. 23 (scarcely of Hedwig).

Northern and **Middle** Islands: from Bay of Islands (*Wilkes's Exped.*) and Waikehi, *Sinclair,* to Dusky Bay, *Menzies,* and *Hector and Buchanan.*

61. ISOTHECIUM, Bridel.

Stout, stiff, erect, tree-like, diœcious mosses. Stem rising from a creeping rhizome; branched (often pinnately) above, naked below. Leaves inserted all round the branches, usually spreading or squarrose, nerveless or 1–2-nerved, often serrate; cells minute, dense. Fruitstalks lateral, often crowded towards the top of the stem, long or short, straight or curved at the top. Capsule erect or cernuous, straight or curved; annulate; peristome double; outer of 16 lanceolate teeth, each marked with a middle line, reflexed when dry, trabeculate in the inner surface; inner a membrane deeply divided into 16 cilia, with or without interposed cilia. Operculum conic or beaked. Calyptra cucullate.

A large tropical and temperate genus, differing from *Hypnum* in the erect dendroid habit and usually curved cernuous capsules. All the New Zealand species are diœcious, some are amongst the handsomest mosses in the islands. *Hypnum hispidum* (p. 472) should perhaps be transferred to near *I. pandum*. I have great difficulty in discriminating some of the species under sections A β and B β.

A. Stem erect, leafy, and more or less pinnately branched from the base.

a. Leaves more or less serrulate; nerve strong.

Stem 2–3 in. Nerve spinulose at the back. Capsule erect, grooved . 1. *I. sulcatum.*
Stem 3–5 in. Nerve not spinulose. Capsule terete . . . . 2. *I. pandum.*

β. Leaves entire or nearly so; nerve 0 *or* 2 *faint.*

Stem 2–3 in. Leaves spreading, broad-ovate, acute, entire, 2-nerved . 3. *I. Arbuscula.*
Stem 2–3 in. Leaves spreading, broad-ovate, acuminate, entire, 2-nerved. . . . . 4. *I. ramulosum.*
Stem 1–2 in. Leaves spreading, narrow ovate-oblong, serrulate, 2-nerved . . . . 5. *I. angustatum.*
Stem 1 in. Leaves erecto-patent, long acuminate, quite entire, nerveless . . . . 6. *I. pulvinatum.*
Stem 1 in., very slender. Leaves imbricate, oblong, obtuse . . . . 7. *I. gracile.*

B. HYPNODENDRON.—Stem naked below, rigid, fastigiately branched above.

a. Capsule terete; operculum conical.

Branches umbellate. Leaves ovate, serrate, cuspidate . . . . . 8. *I. Menziesii.*
Branches flattened. Nerve spinulose at the back . . . . 9. *I. Kerrii.*

β. Capsule grooved. Operculum with a slender beak.

* Branches not whorled, nearly simple, flattened. Leaves ovate, serrate . . . . 10. *I. spininervium.*
** Branches whorled, pinnate. Leaves oblong-lanceolate, serrate . 11. *I. marginatum.*
*** Branches stout, densely fascicled or whorled. Leaves rigid, serrate, subulate-lanceolate.
Branches whorled, suberect. Nerve terete, smooth . . . . . 12. *I. comosum.*
Branches fascicled, lateral, decurved. Nerve keeled. . . . . 13. *I. Sieberi.*
Branches fascicled, lateral, decurved. Nerve terete, spinulose at the point . . . . 14. *I. comatum.*

1. **I. sulcatum,** *Hook. f. and Wils. Fl. N. Z.* ii. 104. Stems suberect, 2–3 in. long, subovate, pinnately or 2-pinnately branched; branches divaricating, crowded. Leaves yellowish, crowded, imbricate, suberect, ovate, obtuse, mucronate, serrulate at the apex, concave; nerve solid, serrated at the back. Fruitstalk ¼ in. long, towards the top of the stem or on the branches. Capsule erect, cylindrical, 8-grooved, reddish; teeth yellow, outer incurved when dry; operculum subulate.—*Leskea*, Hook. Musc. Exot. t. 164. *Climacium*, Bridel.

Northern and **Middle** Islands: Bay of Islands, *J. D. H.*, etc.; Auckland, *Sinclair*; Canterbury, *Sinclair and Haast*. (Australia.)

2. **I. pandum,** *Hook. f. and Wils. Fl. N. Z.* ii. 105. *t.* 89. *f.* 1. Stems 3–5 in. high, naked below, arcuate or deflexed; branches pinnate. Leaves shining yellow-green, distichously imbricate, flattened, ovate or oblong, obtuse or apiculate, towards the apex serrulate; nerve stout; cells dot-like. Fruitstalk ½ in. long, arcuate at the top. Capsule horizontal, subcylindric, contracted at the base, nearly straight; operculum conical, beaked.

Northern and **Middle** Islands: not uncommon on rocks and stumps of trees, from the Bay of Islands, *J. D. H.*, to Otago, *Lyall.*

3. **I. Arbuscula,** *Hook. f. and Wils. Fl. N. Z.* ii. 104. Stems erect, 2–3 in. high, naked below, rigid, rising from a stout rhizome, 2-pinnately branched; branches sometimes decurved and rooting. Leaves pale straw-colour, imbricate, spreading, sometimes obscurely distichous, ovate, acute, quite entire, very concave; nerves 2, indistinct. Fruitstalk $\frac{1}{4}$ in. long, flexuous, stout, often curved at the top. Capsule suberect or cernuous, ovoid, straight or curved, not grooved; operculum conical.—*Hypnum,* Hook. Musc. Exot. t. 112. *Stereodon,* Mitten in Journ. Linn. Soc. iv. 88.

Var. *β. deflexum.* Apices of stem and branches elongated and deflexed, often rooting at the tips.—*Stereodon deflexus,* Mitten, l. c.

Throughout the islands, abundant on trees, **Lord Auckland's** group and **Campbell's** hold, *J. D. H.* (Australia, Tasmania, Aneiteum.) Mitten regards var. *deflexum* as a distinct species.

4. **I. ramulosum,** *Mitten, mss.* Very similar indeed to *I. Arbuscula,* but the leaves are not nearly so concave, and strongly acuminate.

Northern and **Middle** Islands: common on trunks of trees and from the Bay of Islands, *Jolliffe,* to Otago, *Hector and Buchanan.* **Chatham** Islands, *Travers.* (Victoria.)

5. **I. angustatum,** *Mitten (Stereodon), Journ. Linn. Soc.* iv. 88. Very similar to *I. Arbuscula,* but smaller in all its parts, 2–3-pinnately divided. Leaves spreading, subcompressed, narrow-oblong ovate, acute, concave, quite entire, those in the branches serrulate above; perichætial subulate, spreading. Capsule short, ovoid, horizontal.

Northern and **Middle** Islands: probably common, *Raoul;* Auckland, *Lyall, Knight;* Otago, Dunedin, *Hector and Buchanan;* Chain Hills, *Lindsay.*

6. **I. pulvinatum,** *Hook. f. and Wils. Fl. N. Z.* ii. 105. *t.* 88. *f.* 5. Small, stems 1 in. high, curved or decurved; branches pinnate, subincurved. Leaves bright green, erecto-patent, secund, ovate, long-acuminate, rather concave, quite entire; nerve 0; perichætial squarrose. Fruitstalk $\frac{1}{4}$ in. long, curved at the top. Capsule ovate, cernuous, terete; operculum conical.

Northern Island, *Colenso;* Bay of Islands, *J. D. H.;* Auckland, *Bolton.*

7. **I. gracile,** *Hook. f. and Wils.;—Fl. N. Z.* ii. 106.—*Hypnum,* Fl. Antarct. 141. t. 61. f. 3. Stems short, procumbent or suberect from a creeping rhizome, sparingly subpinnately branched; branches straight or recurved. Leaves pale green, opaque, imbricate, secund, oblong or ovate-oblong, hardly acute, concave, obscurely toothed; nerves 1–2, very short; perichætial squarrose. Fruitstalk $\frac{3}{4}$ in. long, quite smooth. Capsule cernuous, oblong; operculum conic; annulus large. Inflorescence diœcious.—*Hypnum gracilescens,* C. Muell. *Stereodon,* Mitten.

Middle Island: Otago, *Buchanan;* Nelson, *Jolliffe.* **Lord Auckland's** group, *J. D. H.* (Tasmania.) Very similar in habit to *Pterogonium filiforme.*

8. **I. Menziesii,** *Hook. f. and Wils. Fl. N. Z.* ii. 105. Stems stout, erect, 2–5 in. high; fastigiately branched above; branches subumbellate, spreading. Leaves yellow-green, shining, lower squarrose, ovate-cordate, with very slender acuminate apices; upper rather compressed, ovate, cuspidate, concave;

margin and strong nerve serrated at the back; nerve often vanishing below the apex; cells narrow. Fruitstalk 2–3 in. high. Capsule suberect or pendulous, smooth, large, oblong-cylindrical, not grooved; mouth not contracted; operculum short, conical.—*Hypnum*, Hook. Musc. Exot. t. 33.

Common throughout the islands, *Menzies*, etc.

9. **I. Kerrii,** *Mitten in Journ. Linn. Soc.* iv. 86 (*Trachyloma*). Habit and appearance of *I. Menziesii.* Leaves ovate, acute, toothed; nerve slender, excurrent, toothed at the back. Capsule unequal, horizontal, not grooved; operculum conic with a short beak.

Northern and **Middle** Islands: apparently common, Auckland, *Jolliffe;* Wellington, *Stephenson;* Waikehi, *Sinclair;* Canterbury, *Travers;* Otago, *Hector and Buchanan.*

10. **I. spininervium,** *Hook. f. and Wils. Fl. N. Z.* ii. 105. Stems robust, erect, 2–3 in. high, fastigiately branched at the top; branches simple, spreading. Leaves glossy, bright green, subdistichously imbricate, ovate, acute, serrated at the margin and back; nerve solid. Fruitstalk ¾–1½ in. high, erect or arched at the top. Capsule oblong, cylindric, grooved, cernuous; operculum with a slender beak, shorter than the capsule.—*Hypnum*, Hook. Musc. Exot. t. 29. *Rhacopilum*, C. Muell.

Var. *β. arcuatum.* Fruitstalk short, arcuate.—*Hypnum arcuatum*, Hedwig, Sp. Musc. t. 62. *Pterigophyllum*, Bridel, Bryol. ii. 348. *Trachyloma*, Mitten.

Abundant throughout the islands, *Menzies*, etc. (Tasmania, Java.)

11. **I. marginatum,** *Hook. f. and Wils. Fl. N. Z.* ii. 106. *t.* 89. *f.* 2. Stem robust, erect, red, 2–4 in. high; branched at the top; branches whorled, pinnately divided, decurved. Leaves dull green, crisped when dry, ovate-oblong or oblong-lanceolate, acute, rather concave; margin thickened, serrulate; nerve thick, spinulose at the back. Fruitstalk crowded, stout, 1–1½ in. high, arcuate at the top. Capsule cernuous, cylindrical, grooved; operculum with a slender beak.—*Hypnum limbatum*, Sullivant?

Northern and **Middle** Islands; common in moist forests. Mitten inclines to suppose that *Hypnum limbatum* is only a young state of *I. marginatum* (see under *Hypnum*).

12. **I. comosum,** *Hook. f. and Wils. Fl. N. Z.* ii. 106. Stems very stout, rigid, 1–3 in. high, densely covered with matted radicles; branches in one or more crowded whorls, stout, suberect or horizontal, somewhat pinnately divided. Leaves rigid, dusky-green, reddish when dry, spreading, subsecund, ovate-lanceolate, with a rigid subulate apex; margin serrate; nerve terete, not keeled, smooth at the excurrent point. Fruitstalks usually numerous, crowded at the top of the stem, stout. Capsule cernuous or pendulous, subcylindric, grooved; operculum nearly as long as the capsule, beak slender.—*Hypnum*, Labill. Fl .Nov. Holl. ii. t. 253, Schwægr. Suppl. t. 91. *Trachyloma*, Mitten.

Abundant throughout the islands, and in **Lord Auckland's** group and **Campbell's** Island. A stout handsome moss. (Tasmania, Australia, Java.)

13. **I. Sieberi,** *Hook. f. and Wils. Fl. Tasman.* ii. 296. Stem stout, erect, tomentose, 1–4 in. high, branched from the sides; branches not whorled, stout, deflexed, subcuspidate. Leaves green, not fulvous, crowded, erecto-patent, narrow-lanceolate, gradually acuminate, striate when dry;

margin thickened, coarsely serrate; nerve stout, keeled at the back, excurrent. Fruitstalks crowded, ¼–¾ in. high. Capsule horizontal, grooved; operculum with a long beak. Inflorescence diœcious.—*Hypnum Sieberi*, C. Muell. Synops. ii. 504.

Lord Auckland's group. (Tasmania.) Very similar to *I. comosum* but more robust, greenish, not fulvous, branches decurved, lateral, not whorled. Leaves longer, more crowded, more serrated and striated, cells larger, nerve not cylindrical, keeled at the back.

14. **I. comatum,** *C. Muell. Synops.* ii. 692. Habit and ramification of *I. Sieberi*, but branches fewer, not cuspidate. Leaves deep green, lax, patent, subsecund, rigid, cordate, lanceolate at the base, narrowed into a long subulate, setaceous, acutely-serrated apex; nerve excurrent, stout, punctulate at the back, serrated at the cylindric point; cells at the outer base large. Fruitstalks crowded, ¼–½ in. long. Capsule pendulous, elongate, cylindrical, 8-grooved, rather curved; operculum with a long beak.—*I. Colensoi*, Fl. Tasman. ii. 207. t. 176. f. 1.

Northern Islands: Taranaki hills, *Jupp, Kerr.* **Middle** Island: Nelson, *Travers;* Canterbury, *Haast;* Otago, *Hector and Buchanan.* (Tasmania.)

62. ENTODON, C. Muell.

Stem compressed, decumbent or procumbent, pinnate. Leaves ovate, imbricate all round the stem, shining, 2-nerved at the base, entire. Fruitstalk straight. Capsule erect, cylindric, regular. Peristome double; outer teeth 16, rigid, not hygrometric, inserted below the mouth of the capsule, split at the apex; inner of 16 narrow cilia; annulus present or 0; columella exserted. Operculum short. Calyptra dimidiate.

A small genus, native of Europe and other countries.

1. **E. truncorum,** *Mitten.* Tufts depressed; stems pinnately branched. Leaves spreading, compressed, imbricated so as to form cuspidate ends to the branches, concave, oblong, acuminate, quite entire; nerves very short; upper cells narrow, basal quadrate; inner perichætial erect, convolute, with lanceolate-subulate apices. Fruitstalk short. Capsule erect, cylindric, pale red; operculum subulate.

Middle Island: Otago, *Hector and Buchanan.* Resembles the northern *E. cladorhizans*, but leaves narrower, with more numerous distinct quadrate cells at the base. Peristome incomplete. (Mitten.)

63. HYPNUM, Linn.

Stems creeping, variously branched; branches erect prostrate or pendulous, terete, or compressed. Leaves distichous or imbricated all round, often secund; nerves 1, 2, or 0; cells and reticulation various. Fruitstalk lateral, slender, usually curved at the top and then smooth or scabrous. Capsule curved, horizontal, inclined or drooping, usually annulate; terete (grooved or angled in § *Acicularia*); peristome double, outer of 16 lanceolate teeth, marked with a middle line, inner surface trabeculate, inner a membrane cut to the middle into 16 keeled teeth with or without interposed cilia. Operculum various. Calyptra cucullate, glabrous (hairy in *H. pubescens*).

One of the largest genera of mosses, found in all parts of the globe. The species are extremely difficult of discrimination. Considerable experience and proficiency are required before a beginner can name any of the species correctly. Most of the New Zealand ones are referred to *Stereodon* by Mitten, but I have in this genus as in others followed Wilson's and my own generic nomenclature, as given in the Flora of New Zealand, being convinced that it is on the whole the best for a local flora; there are however no limits to the discrepancies of opinion as to the system and nomenclature of the pleurocarpous mosses, and as to which the highest authorities are completely at issue. The groups here adopted are pretty natural, but it is not always easy to perceive the characters upon which they are founded, especially those of the secund-leaved group (*Cupressiformia*), to which so many species belong.

KEY TO THE GROUPS AND SPECIES OF HYPNUM.

I. *Stem and branches covered with matted radicles amongst the leaves.*

A. TAMARISCINA. Leaves small. Nerve strong. Fruitstalk smooth.

II. *Stem and branches not covered with radicles.*

a. Leaves imbricate all round the stem, squarrose or erect or incurved, or secund or falcate, not 2-farious or distichous.

* *Leaves more or less falcate and secund, often circinate.*

B. ADUNCA. Stems flaccid. Operculum short, conic.
C. HISPIDA. Stems elongate, rigid. Nerve very strong. Operculum beaked.
D. CUPRESSIFORMIA. Stem flaccid. Nerve 0 or short and slender. Operculum beaked.

** *Leaves spreading, not squarrose nor large nor concave.*

E. PROLONGA. Fruitstalk scabrid. Operculum beaked.
F. CONFERTA. Fruitstalk smooth. Operculum beaked.
G. RUTABULA. Fruitstalk scabrid. Operculum conic.
H. SERPENTIA. Fruitstalk smooth. Operculum conic.

*** *Leaves squarrose and spreading, large and often concave.*

I. STELLATA. Capsule not grooved.
J. PTYCHOMNION. Capsule grooved.

**** *Leaves closely imbricate, not squarrose, large and concave.*

K. COCHLEARIFOLIA.

β. Stems compressed. Leaves imbricate, the lateral spreading 2-fariously.

L. DISTICHOPHYLLA.

A. **Tamariscina.**—*Stems pinnately or 2-pinnately branched, covered with matted green, fibrillous rootlets. Leaves imbricated all round the stems and branches, not 2-farious or distichous; nerve stout.*

* *Stems procumbent, straight or slightly arcuate.*

Stems pinnately branched; branches irregular in length. Leaves not plaited; perichætial with few cilia . . . . 1. *H. furfurosum.*
Stems pinnately branched; branches short, close-set, regular. Leaves plaited; perichætial with many cilia. . . . . 2. *H. fulvastrum.*
Stems 2-pinnately branched . . . . 3. *H. sparsum.*

** *Stems very arcuate, proliferous from the descending assurgent apices, 2-pinnate.*

Leaves of branches incurved; nerve prominent below the apex on the back . . . . 4. *H. lævivsculum.*
Leaves of branches with erect apices; nerve not prominent . . . 5. *H. denticulosum.*

B. **Adunca.**—*Leaves falcate, secund; nerve single, continuous or vanishing below the apex. Fruitstalk smooth. Operculum short, conical (unknown in* H. limbatum*). Often aquatic or marsh mosses.*

* *Nerve vanishing below the middle.*

Leaves circinate, striate. subserrulate . . . . 6. *H. uncinatum.*
Leaves not striate, deltoid at the base, quite entire . . . . 7. *H. Kneiffii.*
Leaves not striate, lanceolate, quite entire or subserrulate . . . . 8. *H. fluitans.*
Leaves circinate, lanceolate from an ovate base, quite entire . . . 9. *H. brachiatum.*

** *Nerve continuous.*

Leaves cordate at the base, quite entire . . . . 10. *H. filicinum.*
Leaves oblong- or linear-lanceolate, serrate . . . . 11. *H. limbatum.*

C. **Hispida.**—*Stems rigid, stiff, sparingly branched. Leaves falcate and secund, nerve very stout.*

Leaves ovate, rigid, nerve excurrent . . . . 12. *H. hispidum.*
Leaves ovate, rigid, nerve vanishing . . . . 13. *H. glauco-viride.*
Leaves ligulate from an ovate base . . . . 14. *H. umbrosum.*

D. Cupressiformia. Stems flattened. Leaves more or less secund, falcate or circinate, imbricate all round the stem or obscurely 2-farious; cells at the outer base often large; nerve 0 or 1–2, short. Fruitstalk smooth (rather rough in *H. cerviculatum*).—Species very difficult of discrimination.

* *Operculum as long as the capsule.*

† *Leaves strongly falcate or circinate.*

Leaves lanceolate, long acuminate; perichætial serrulate . . . . . 15. *H. cerviculatum.*
Leaves linear-lanceolate; perichætial similar. Branches long . . 16. *H. tenuirostre.*
Leaves linear-lanceolate; perichætial broader. Branches short . . 27. *H. amœnum.*

†† *Leaves slightly falcate.*

a. Leaves quite entire or obscurely serrulate.

Leaves lanceolate, subpiliferous. Capsule pendulous . . . . . . 18. *H. crassiusculum.*
Leaves oblong, acuminate; perichætial similar. Capsule horizontal 19. *H. Jolliffii.*
Leaves ovate or oblong, acuminate; perichætial narrower. Capsule suberect . . . . 20. *H. homomallum.*
Leaves lax, minute, lanceolate. Capsule pendulous . . . . . . 21. *H. molliculum.*

β. Leaves distinctly serrulate.

Leaves deltoid-ovate . . . . 22. *H. pubescens.*
Leaves linear-lanceolate . . . . 23. *H. leptorhynchum.*

** *Operculum shorter than the capsule.*

† *Leaves entire or nearly so, nerveless.*

a. Leaves strongly falcate, ovate-lanceolate.

Leaves acuminate; perichætial setaceous . . . . 24. *H. chrysogaster.*
Leaves acuminate; perichætial subpiliferous . . . . 25. *H. cupressiforme.*
Leaves subpiliferous. Capsule pendulous . . . . 26. *H. mundulum.*
Leaves subpiliferous. Capsule suberect . . . . 27. *H. limatum.*

β. Leaves obscurely falcate-secund.

Stems short, slender. Leaves narrow-lanceolate . . . . 28. *H. pulchellum.*
Stems elongate, slender. Leaves oblong-lanceolate . . . . 29. *H. acutifolium.*

†† *Leaves 2-nerved, serrulate, strongly falcate.*

Stems pinnately branched. Capsule pendulous . . . . 30. *H. sandwicense.*

E. Prælonga. Leaves spreading, imbricate all round the stem, not squarrose nor secund, serrulate in the N. Z. species; nerve reaching about halfway up. Fruitstalk scabrid; operculum beaked.—Small-leaved species, always creeping or prostrate. (Leaves subsecund in *H. austrinum.*)

Leaves lanceolate, subpiliferous . . . . 31. *H. muriculatum.*

Leaves ovate or ovate-lanceolate, acuminate . . . . 32. *H. austrinum*.
Leaves ovate-cordate acute . . . . 33. *H. remotifolium*.

F. CONFERTA. Leaves imbricating all round the stem, not squarrose nor secund, usually small and serrulate. Fruitstalk smooth (see *H. plumosum* in G.). Operculum beaked.

Leaves ovate, gradually acuminate; perichætial recurved . . . . 34. *H. tenuifolium*.
Leaves broadly ovate, suddenly acuminate; perichætial recurved. . 35. *H. elusum*.
Leaves oblong, piliferous; perichætial erect . . . . 36. *H. aristatum*.

G. RUTABULA. Leaves spreading, imbricate all round the stem, not squarrose, rarely subsecund; nerve reaching beyond the middle in the N. Z. species. Fruitstalk rough (obscurely so in *H. plumosum*). Operculum short, conical.

Leaves ovate-acuminate, serrulate. Operculum acuminate . . . 37. *H. rutabulum*.
Leaves ovate-lanceolate, acuminate, secund-falcate, serrulate. Operculum very obtuse . . . . 38. *H. paradoxum*.
Leaves subsecund, ovate-lanceolate. Operculum very acute . . . 39. *H. plumosum*.

H. SERPENTIA. Leaves imbricated all round the stem, not squarrose, rarely subsecund; nerve short or reaching beyond the middle. Fruitstalk smooth. Operculum short, conic (see *H. plumosum* in G.). Usually very minute mosses.

Leaves ovate-lanceolate . . . . 40. *H. serpens*.

I. STELLATA. Leaves imbricated all round the stem, small, spreading and squarrose. Fruitstalk smooth. Capsule not grooved.

Leaves ovate-lanceolate, acuminate; nerve reaching halfway . . . 41. *H. polygamum*.
Leaves ovate, acuminate; nerve solid, continuous . . . . . . 42. *H. relaxum*.
Leaves ovate, long-acuminate; nerve vanishing below the apex . . 43. *H. decussatum*.

J. ACICULARIA. Stems stout, suberect, sparingly branched. Leaves imbricated all round, squarrose, large, very concave, flaccid, serrate, nerveless. Capsule grooved or angled. Calyptra large, inflated.

Leaves squarrose and recurved, strongly serrate above . . . . . 44. *H. aciculare*.
Leaves refracted from the middle, minutely serrulate above . . . 45. *H. densifolium*.

K. COCHLEARIFOLIA. Stems stout or slender, prostrate, sometimes pendulous. Leaves imbricated all round, often appressed, not squarrose or secund, very concave and inflated, transparent, obtuse or apiculate; nerve 0 or obscure. Fruitstalk smooth. Capsule not grooved. Operculum conical.

* *Nerve very short or* 0.

Stems 2–4 in. long. Leaves hemispherical, very obtuse . . . . 46. *H. cochlearifolium*.
Stems 2–3 in. long. Leaves broadly oblong, obtuse, cordate at the base . . . . 47. *H. clandestinum*.
Stems 1–2 in. long. Leaves orbicular-quadrate, auricled at the base, round at the apex . . . . 48. *H. chlamydophyllum*.
Stems 1 in. high. Leaves large, oblong, very concave, acute or acuminate . . . . 49. *H. inflatum*.
Stems ascending. Leaves broadly ovate, shortly acuminate . . . . 50. *H. vagum*.

** *Nerve produced to the middle.*

Stems 2–5 in. long. Leaves broad oblong, subacute . . . . . 51. *H. divulsum*.

L. DISTICHOPHYLLA. Stems and branches compressed. Leaves inserted all round the stem, but distichous (the lateral only 2-farious in *H. polystictum*), flat or concave, nerveless or shortly 2-nerved. Fruitstalk and capsule smooth.

* *Leaves nerveless.*

Stems 2–4 in., pinnately branched. Leaves subserrulate; cells not papillose . . . . 52. *H. extenuatum*

Stem 1 in., vaguely branched. Leaves quite entire . . . . 53. *H. politum.*
Stem 1½ in., pinnately branched. Leaves serrulate; cells papillose . 54. *H. polystictum.*

** *Leaves 2-nerved at the base.*

Leaves entire or denticulate . . . . 55. *H. denticulatum.*

A. Tasmariscina.

1. **H. furfurosum,** *Hook. f. and Wils. Fl. N. Z.* ii. 107. *t.* 88. *f.* 7. Stem procumbent, slender, rigid, slightly arcuate, 1–2 in. long, pinnately branched, covered with matted fibrils; branches rather long, distant, filiform. Leaves green or yellowish, squarrose on the stem, suberect on the branches, incurved when dry, ovate-cordate, acuminate, somewhat keeled, scarcely plicate, quite entire or serrate towards the apex; perichætial larger, erect, long-acuminate, somewhat ciliate. Fruitstalk 1 in. long, smooth, red. Capsule pale-brown, cylindric-oblong, cernuous, slightly curved; operculum beaked, shorter than the capsule.—*H. unguiculatum*, Fl. Tasm. ii. 208. t. 176. f. 3. (?) *H. hastatum*, C. Muell. Synops. ii. 485.

Northern and **Middle** Islands: probably common on moist rocks and trunks of trees; Bay of Islands, *Cunningham*, etc.—A puzzling moss, difficult to discriminate from the two following. Mitten, who has paid much attention to the discrimination of the species of this section, informs me that the *H. hastatum*, C. Muell., may belong to this as well as *H. Stuartii*, for that is very variable. He has less doubt in referring *H. unguiculatum* to the same species. (Tasmania, Australia.)

2. **H. fulvastrum,** *Mitten in Journ. Linn. Soc.* iv. 92 (*Leskea*). Similar to *H. furfurosum*, but branches short and close-set, giving a linear outline to the frond. Cauline leaves secund, plaited, those of the branches and branchlets evenly disposed on every side; perichætial much ciliate.—Mitten, mss.

Northern and **Middle** Islands: probably common, Auckland, *Knight*, *Bolton*, *Kerr*, etc. (Tristan d'Acunha.)

3. **H. sparsum,** *Hook. f. and Wils. Fl. N. Z.* ii. 109. *t.* 89. *f.* 5. Stems very slender, matted, creeping, 1 in. long, 2-pinnately branched; branches short, very slender. Leaves dark green, very minute, spreading, incurved when dry, ovate or ovate-cordate, subobtuse, quite entire, but rough at the edges; nerve stout, pellucid, vanishing below the apex; perichætial much larger, long-acuminate, inner laciniate. Fruitstalk ½ in. long, smooth. Capsule inclined or cernuous, narrow-oblong. Inflorescence diœcious.

Northern Island, *Colenso*; Wangarei, *Bolton*; Waikehi, *Milne*. **Middle** Island, Nelson, *Travers*. This differs from *H. furfurosum* chiefly in the 2-pinnate stems. (Australian Alps.)

4. **H. læviusculum,** *Mitten in Fl. Tasm.* ii. 207 (*Leskea*). Stems arcuate, regularly 2-pinnate, proliferous from the descending assurgent apices. Leaves of *H. furfurosum*, those of the branchlets compressed, with incurved apices; nerve cristate at the back; cauline subcrenulate, with flexuous points; nerve not excurrent, vanishing below the filiform apex.—Mitten in Journ. Linn. Soc. iv. 92.

Northern and **Middle** Islands: probably common; Kiapara, *Mossman*; Wellington, *Stephenson*; Milford Sound, *Lyall*. (Tasmania.)

5. **H. denticulosum,** *Mitten, mss.* Very similar to *H. læviusculum*, but leaves of the branches patent, not compressed, their apices erect; nerve not protuberant on the back below the apex; cauline leaves not narrowed into a filiform apex, all the leaves with the margins minutely closely denticulate.—Mitten.

Northern Island: Auckland?, *Herb. Sinclair.*

B. Adunca.

6. **H. uncinatum,** *Hedwig;—Fl. N. Z.* ii. 107. Stems 1–2 in. high, slender, erect, pinnately branched. Leaves crowded, secund, falcate, circinate, subulate-lanceolate, narrowed into very slender points, striate, subserrulate; nerve vanishing below the apex; perichætial very long, striate. Fruitstalk 1 in. high, reddish. Capsule cernuous, oblong, smooth; operculum conical, apiculate. Inflorescence monœcious.—Hedwig, Musc. Frond. iv. t. 25; Wils. Bryol. Brit. 394. t. 26.

Northern and **Middle** Islands: in wet bogs, etc., Huiarau river, *Colenso;* Canterbury, *Sinclair and Haast;* Otago, *Hector and Buchanan.* (Almost all cold climates.)

7. **H. Kneiffii,** *Schimper;—Fl. N. Z.* ii. 107. Very similar to *H. uncinatum*, but leaves less crowded, not circinate, ovate-lanceolate, acuminate, deltoid at the base, concave, quite entire, not striated; nerve reaching halfway. Capsule oblong, cernuous; operculum conical. Inflorescence diœcious. —Wils. Bryol. Brit. 390. t. 58.

Northern and **Middle** Islands: East Cape, *Colenso;* Dusky Bay, *Lyall.*—Specimens all very imperfect and without fruit. (Europe, N. Asia, N. America.)

8. **H. fluitans,** *Linn.;—Fl. Antarct.* i. 141. Stems erect or floating, 1–3 in. long, subpinnately branched. Leaves yellow-brown, loosely imbricate, falcate, secund, lanceolate, acuminate, subserrulate or entire; nerve vanishing beyond the middle. Fruitstalk 2–3 in. long. Capsule oblong, cernuous; operculum conical.—Wils. Bryol. Brit. 387. t. 58.

Campbell's Island: swampy places, barren, *J. D. H.* (All temperate and cold latitudes.)

9. **H. brachiatum,** *Mitten, mss.* Stems matted, 4–5 in. long, pinnately branched, red-brown below, pale green and yellow above. Leaves falcate-secund, much incurved, concave, lanceolate from a broad ovate base, long-acuminate, quite entire; nerve vanishing below the apex; perichætial erect, elongate, lanceolate, convolute. Fruitstalk slender. Capsule shortly cylindric, arcuate; operculum conical, acuminate.

Northern and **Middle** Islands: Tobago Bay, *Knight;* Otago, in wet grounds, *Hector and Buchanan.*—Very nearly allied to the European *H. revolvens*, and scarcely distinguishable from it, except (according to Mitten) by the more ovate base of the cauline leaves, with more numerous cells at their angles.

10. **H. filicinum,** *Linn.;—Fl. Antarct.* i. 141. Stems slender, 2–3 in. high, compressed, pinnately branched, tomentose with purplish radicles. Leaves dull green, spreading, falcate or suberect, secund, cordate at the base, serrulate; nerve solid, continuous; perichætial striated. Fruitstalk 1 in.

long, smooth. Capsule cernuous, subcylindrical; operculum conical, acute. Inflorescence monœcious.—Wils. Bryol. Brit. 392. t. 26.

Lord Auckland's group: banks of streams, *Lyall*. A very slender form or variety, distinguished as var. β. in Fl. Antarct. (All cold and temperate latitudes.)

11. **H. limbatum,** *Sullivant in Proc. Am. Journ. Sc. and Art.* 183, *and in Musci of U. S. Expl. Exped.* 18. Stems 2–3 in. long, slender, flexuose, floating, sparingly branched. Leaves lax, spreading, oblong- or linear-lanceolate, acuminate, concave, keeled, serrate above; nerve stout, percurrent; apex serrated; margins thickened. Fruit unknown.

New Zealand: on stones in the bottom of streams (U. S. Expl. Exped.). Mitten is inclined to regard this as a state of *Isothecium marginatum*, p. 466

C. Hispida.

12. **H. hispidum,** *Hook. f. and Wils. Fl. Antarct.* i. 140. *t.* 61. *f.* 2. Stems prostrate or pendulous from a creeping rhizome, arched, subpinnately branched above, 3–10 in. long, rigid, subhispid; branches curved, often attenuate. Leaves dark green, rigid, secund, densely imbricate, ovate below, subulate above, very long-acuminate and setaceous, somewhat plaited, quite entire; nerve stout, excurrent; perichætial smaller, ovate; nerve long, exserted. Fruitstalk short, ½ in. long, red, smooth, stout, flexuous. Capsule ovoid, cernuous or horizontal; operculum with a slender curved beak as long as the capsule or shorter, Calyptra small, white. Inflorescence diœcious.—*Leskea*, Mitten. *H. aristatum*, Sullivant in Proc. Amer. Acad. Sc. and Art, 1854.

Abundant on rocks, in streams, and throughout the island, *J. D. H.* **Lord Auckland's** group: rocks on the hills, barren, *J. D. H.*—A puzzling moss. Wilson suggests that it may be better placed in *Isothecium*. I have followed Sullivant in placing it in section *Hispida*. (Tasmania, Australia, Norfolk Island.)

13. **H. glauco-viride,** *Mitten, mss.* Very closely resembling *H. hispidum*, but nerve vanishing below the apex.—Mitten.

Kermadec group: Sunday Island, *Milne*. (Norfolk Island.)

14. **H. umbrosum,** *Mitten in Journ. Linn. Soc.* iv. 92 (*Leskea*). Habit and colour of *H. hispidum*, but smaller and more slender. Stems 1–2 in. long, with a few irregular branches. Leaves spreading, subsecund, ligulate and acuminate from a subovate base, minutely serrulate above; nerve stout, vanishing in the elongated apex; perichætial ovate-subulate from a broad base, 1–2-toothed on each side. Fruitstalk ½ in. long, red. Capsule ovoid, horizontal; operculum rostrate, as long as the capsule. Inflorescence diœcious.

Northern Island, *Kerr* (*in Herb. Mitten*); Auckland, *Knight*.

D. Cupressiformia.

15. **H. cerviculatum,** *Hook. f. and Wils. Fl. N. Z.* ii. 113. *t.* 91. *f.* 2. Stems creeping, matted, ¼–1 in. long, subpinnately branched; branches nearly straight. Leaves yellow-green, shining, falcate and secund, lanceolate, acuminate, serrulate at the apex, margin scarcely recurved; nerve 0; perichætial erect, ovate-lanceolate, apiculate, serrulate. Fruitstalk short, ⅓

in. long, stout, roughish. Capsule oblong, horizontal, base substrumose; annulus 0; operculum with a long slender beak. Inflorescence diœcious. —*H. leptorhynchum*, *β*. Fl. Antarct. 141. *Stereodon*, Mitten in Journ. Linn. Soc. iv. 87.

Northern Island: Port Nicholson, *Sinclair;* woods at Tehawera, *Colenso.* **Lord Auckland's** group, *J. D. H.* (Tasmania.)

16. **H. tenuirostre,** *Hook. Musc. Exot. t.* 111;—*Fl. N. Z.* ii. 113. Stem creeping, subpinnately branched; branches elongate, erect, ¾–1 in. long. Leaves pale or dark green, shining, crowded, falcate, secund, ovate-lanceolate, acuminate, concave, quite entire or minutely serrulate towards the apex; nerve 0; perichætial similar. Fruitstalk long, slender, quite smooth. Capsule cernuous, ovoid-oblong; operculum with a long slender beak. Inflorescence monœcious.

Middle Island: Dusky Bay, *Menzies;* Bligh's Sound, Chalky Bay, and Otago, *Lyall, Hector and Buchanan.* (Tasmania.)

17. **H. amœnum,** *Hedw. Sp. Musc. t.* 77. *f.* 6–9. Stems prostrate, vaguely and pinnately branched, ½–1 in. long; branches short, their apices incurved. Leaves yellow-brown, shining, falcate and secund, lanceolate, long acuminate, concave, quite entire; cells large, pellucid at the marginal base; nerve 0; perichætial broader, entire or serrulate towards the apex. Fruitstalk ½–¾ in. long, very slender, red. Capsule horizontal, ovoid, urceolate when dry; operculum with a slender beak as long as the capsule.

Northern Island: Auckland, *Knight.* (Andes, Fuegia, Australia.)

18. **H. crassiusculum,** *Bridel;—Fl. N. Z.* ii. 113. Stem creeping, vaguely branched, 1–2 in. long, subpinnately branched; branches short, incurved, subcuspidate. Leaves pale green, loosely imbricate, subsecund, not falcate, lanceolate, acuminate, piliferous, quite entire, concave, margins scarcely recurved; nerve 0; cells inflated at the marginal base; perichætial longer, erect. Fruitstalk short, quite smooth. Capsule ovoid-oblong, subcernuous; operculum with a long slender beak. Inflorescence monœcious. —Schwægr. Suppl. t. 91. *H. contiguum*, Fl. Tasm. ii. 213. t. 177. f. 2.

Northern and **Middle** Islands: Auckland, *Sinclair;* Canterbury, *Sinclair and Haast;* Otago, *Hector and Buchanan.* (Tasmania, N. and S. America, Tristan d'Acunha.)

19. **H. Jolliffii,** *Mitten in Journ. Linn. Soc.* iv. 87 (*Stereodon*);—*Fl. Tasman.* ii. 213. *t.* 177. *f.* 1. Stems matted, 1–3 in. long, forming bright green shining patches; branches slender, with cuspidate apices. Leaves loosely imbricate, secund, not falcate, oblong or oblong-lanceolate, acuminate, quite entire or obscurely serrulate at the apex; cells at the marginal base larger and hyaline; nerve 0 or 2 obscure; perichætial similar, erecto-patent. Fruitstalk slender, 1 in. long. Capsule small, ovoid, inclined or horizontal; operculum slender, as long as the capsule.

Northern and **Middle** Islands, *Jolliffe;* Wangaroa, *Sinclair;* Otago, *Hector and Buchanan.* (Tasmania.)

20. **H. homomallum,** *C. Muell.;—Fl. Tasm.* ii. 213. Stems tufted, decumbent, prostrate, forming shining golden-yellow patches; branches short, flexuose. Leaves crowded, secund, slightly falcate, broadly ovate or oblong,

acuminate, very concave, quite entire, margin recurved; cells at the marginal base numerous, thick-walled, square; nerve 0; perichætial narrow, straight. Fruitstalk short, $\frac{1}{4}$–$\frac{1}{3}$ in., red. Capsule narrow-oblong, suberect; operculum with a slender beak as long as the capsule.—*H. Drummondii*, Tayl. *Leskea homomalla*, Hampe, Ic. Musc. t. 6.

Northern Island: Bay of Islands, *J. D. H.*; Auckland, *Jolliffe*. (Australia, Tasmania.)

21. **H. molliculum,** *Sullivant, Mosses of U. S. Expl. Exped.* 14. *t.* 11. A small, slender, flaccid, pale green plant, creeping on earth; branches weak, ascending. Leaves 2-farious, secund, hardly falcate, spreading, oblong-lanceolate, acuminate, concave; margins erect, quite entire; cells at the marginal base similar to the others; nerve 0 or very short; perichætial similar. Fruitstalk short, very slender. Capsule cernuous, small; operculum in our specimens slender, as long as the capsule.

Kermadec group: Raoul Island, *M'Gillivray and Milne*. (Sandwich Islands).—I have referred this plant doubtfully to Sullivants, the operculum being as long as the capsule and more slender than the figure of that author, who does not however describe this organ.

22. **H. pubescens,** *Hook. f. and Wils. Fl. N. Z.* ii. 113. *t.* 91. *f.* 3. Stems creeping loosely, subpinnately branched, 1 in. long; branches rather distant, nearly straight. Leaves yellow-green, shining, incurved, loosely imbricate, scarcely secund, deltoid-ovate, acuminate, serrulate, obscurely striate when dry; nerve obscure. Fruitstalk slightly rough, $\frac{2}{3}$ in. long. Capsule cernuous, ovoid; operculum conical, almost as long as the capsule. Calyptra hairy. Inflorescence diœcious.

Northern Island, *Colenso*; Auckland, *Sinclair*, *Knight*; Hick's Bay, *Jolliffe*. Very near the European *H. molluscum*.

23. **H. leptorhynchum,** *Bridel*;—*Fl. N. Z.* ii. 112. Stems densely tufted, creeping, pinnately branched, $\frac{1}{4}$–1 in. long; branches slender, matted. Leaves yellow-green, shining, falcate, secund, not circinate, twisted when dry, linear-lanceolate, acuminate, serrulate towards the apex; margin subrecurved; cells at the marginal base larger, inflated; nerve 0. Fruitstalk long, slender, smooth. Capsule oblong, nodding; operculum with a slender beak, as long as the capsule. Inflorescence monœcious.—Schwægr. Suppl. t. 93; Fl. Antarct. 141, excl. var. β; *Stereodon cyparoides*, Mitten in Journ. Linn. Soc. iv. 87 (not of Bridel.); *H. leucocytus*, C. Muell.

Abundant on trunks of trees throughout the **Northern** and **Middle** Islands, *Colenso*, etc., and in **Lord Auckland's** group, *J. D. H.* (Australia, Tasmania, S. America, S. Africa.)

24. **H. chrysogaster,** *C. Muell. in Linnæa*, 185.—*H. patale*, Fl. N. Z. ii. 112. t. 90. f. 6. Stems procumbent, 2–4 in. long, pectinately branched; branches crowded, spreading, flattened above, recurved. Leaves white or pale bright green, glossy, distichously spreading, strongly falcate, secund, almost circinate, ovate-lanceolate, long-acuminate, subserrulate at the apex; cells at the marginal base larger, collected into two yellow spots; nerve 0; perichætial erect, setaceous, serrulate. Fruitstalk slender, 1–2 in. long, smooth. Capsule oblong-ovoid, horizontal or cernuous; operculum convex, obtuse, apiculate. Inflorescence diœcious.—*Stereodon*, Mitten in Journ. Linn. Soc. iv. 87. *H. cupressiforme*, Fl. Antarct. 141.

Northern Island: common on trunks of trees, *J. D. H.*, etc.; Pelorus, *Jolliffe*. **Middle** Island: Chain-hills, Otago, *Lindsay*, *Hector and Buchanan*. **Campbell's** Island, *J. D. H.* **Auckland** Island, *J. D. H.*, *Bolton*. (Tasmania, Juan Fernandez.)

25. **H. cupressiforme,** *Linn.;—Fl. N. Z.* ii. 111. Stems creeping or prostrate, 1–3 in. long, vaguely pinnately branched; branches incurved. Leaves pale yellow, shining, circinately falcate, secund, incurved, ovate-lanceolate, acuminate, concave, quite entire or obscurely toothed; cells enlarged at the marginal base; nerve 0; perichætial erect, subpiliferous. Fruitstalk slender, ½–1 in. long. Capsule suberect, horizontal or cernuous, subcylindric; operculum short, conical, cuspidate. Inflorescence diœcious.—Wils. Bryol. Brit. 397. t. 27. *H. Mossmanianum*, C. Mueller in Bot. Zeit. 1851, 565.

Common throughout the islands, *Colenso*, etc. Very variable. Wilson considers the large cells at the marginal base of the leaf a good diagnostic character. (One of the most widely diffused mosses in the world.)

26. **H. mundulum,** *Hook. f. and Wils. Fl. N. Z.* ii. 112. *t.* 91. *f.* 1. Stem 1–1½ in. long, procumbent, slender, flexuous, pinnately branched; branches crowded, erecto-patent, subrecurved. Leaves very pale yellow, rather shining, falcate, secund, circinate, ovate-lanceolate below, narrowed to a long slender almost piliferous apex, quite entire; nerve 0; perichætial shorter, erect, acuminate, subserrulate. Fruitstalk slender, ½ in. long, red, quite smooth. Capsule ovoid, pendulous, dark red-purple; operculum shorter than the capsule, beaked. Inflorescence monœcious.

Middle Island, *Lyall.*

27. **H. limatum,** *Hook. f. and Wils. Fl. Antarct. Suppl.* 545. Stems pinnately branched, ¼ in. long; branches short, incurved. Leaves pale green or yellowish, shining, crowded, secund, falcate, subcircinate, membranous, ovate at the base, thence gradually narrowed into long subpiliferous points, quite entire; nerve 0; perichætial elongate, subsquarrose, subserrulate. Fruitstalk ½ in. long, quite smooth. Capsule suberect, cylindrical; operculum short, convex, hardly conical, apiculate, red. Inflorescence diœcious.—*H. terræ-Novæ*, Brid., *β. australe*, Fl. Antarct. 142. t. 61. f. 4.

Campbell's Island: on decaying wood, *J. D. H.* (Tasmania.)

28. **H. pulchellum,** *Dicks.?* Stems very slender, suberect, scarcely creeping, sparingly fastigiately branched. Leaves lax, more or less secund, slightly falcate, lanceolate, tapering from a broader base to an almost piliferous apex, quite entire; cells at the marginal base similar to the others; nerve 0; perichætial erect. Fruitstalk 1 in. high, slender, rising from the base of the stem. Capsule nearly erect, oblong, tapering into the fruitstalk, slightly curved; operculum conical, beaked.—Wils. Bryol. Brit. 404. t. 25. *H. nitidulum*, Wahlb.

Middle Island: Canterbury Alps, *Sinclair and Haast.*—I have queried Mitten's determination of this moss, which differs from the European one in laxer habit, paler colour, and longer, less falcate and rather narrower leaves.—Schimper remarks that it is nearer *H. Muellerianum*, but distinct. (*H. pulchellum* is a native of Temp. Europe and N. America.)

29. **H. acutifolium,** *Hook. f. and Wils. Fl. Antarct.* 138. *t.* 60. *f.* 5. Stems ascending, slender, ½–1 in. high, sparingly fastigiately branched; branches subcompressed. Leaves pale yellowish-brown, closely imbricate,

obscurely falcate-secund, or nearly erect, oblong-lanceolate, with a long acuminate apex, quite entire, concave, nerveless; cells at the outer base larger and darker coloured. Capsule unknown.

Campbell's Island: growing amongst *H. chlamydophyllum* on the hills, *J. D. H.*—Sullivant (Musci of U. S. Expl. Exped. 18) suspects this to be possibly a form of *H. crinitum*, from which, however, it widely differs.

30. **H. Sandwichense,** *Hook. and Arn. Fl. N. Z.* ii. 113. Stems procumbent, 1 in. long, pinnately branched; branches short, not crowded, distichously spreading. Leaves bright yellow-green, shining, spreading somewhat distichously, falcate and secund, ovate-lanceolate, acuminate, serrulate at the apex, margin scarcely recurved; cells at the outer base a little enlarged; nerves 0 or 2, very short; perichætial larger, broader, often serrulate. Fruitstalk ½ in. long, quite smooth. Capsule pendulous, ovoid; operculum convex, apiculate. Inflorescence monœcious.—*H. acinacifolium*, Hampe (fide Mitten).

Northern Island: Bay of Islands, on trunks of trees; rare, *J. D. H.*, *Colenso.* **Kermadec** group: Sunday Island, *Milne.* (Sandwich Islands.)

E. PRÆLONGA.

31. **H. muriculatum,** *Hook. f. and Wils. Fl. N. Z.* ii. 108. *t.* 89. *f.* 3. Stems small, slender, procumbent, subpinnately branched, 1–2 in. long; branches short, ¼ in. long, simple, subcompressed. Leaves spreading, ovate-lanceolate, acuminate, subserrulate, almost piliferous; nerve produced beyond the middle. Fruitstalk ⅓ in. long, rough. Capsule horizontal, oblong, curved at the base, where it joins the fruitstalk; operculum with a long beak. Inflorescence monœcious.

Var. *β.* branches flattened, leaves less acuminate, and rather twisted when dry.

Northern and **Middle** Islands: not uncommon in woods. Var. *β.* Bay of Islands, *J. D. H.*—Closely allied to several northern and other species. (Norfolk Island, Tasmania.)

32. **H. austrinum,** *Hook. f. and Wils. Fl. N. Z.* ii. 108. *t.* 89. *f.* 4. Stems 2–3 in. long, creeping, vaguely branched; branches 1 in. long, suberect, incurved. Leaves dark green, opaque, spreading, subsecund, ovate or ovate-cordate, acuminate, serrulate, concave; nerve faint, produced beyond the middle; perichætial erect. Fruitstalk ½ in. high, rough. Capsule ovoid, cernuous; operculum beaked. Inflorescence monœcious.

Var. *β.* Branches crowded, more slender; leaves smaller.

Northern and **Middle** islands: in woods from Auckland, *Knight*, to Akaroa, *Raoul.* Var. *β.* **Middle** Island, *Lyall.* (Tasmania, Australia.)

33. **H. remotifolium,** *Greville*;—*Fl. N. Z.* ii. 108. Stems creeping, rather slender, 1–2 in. long, subpinnately branched; branches ¼–½ in. long, spreading. Leaves pale, glossy, spreading, cordate-ovate, acuminate, almost piliferous, serrulate; nerve produced beyond the middle; perichætial ovate-oblong, piliferous. Fruitstalk rough, 1 in. high. Capsule cernuous, broadly oblong, curved; operculum with a slender beak, almost as long as the capsule. Inflorescence hermaphrodite.—Schwægr. Suppl. t. 200 (fruitstalk erroneously smooth). *H. asperipes*, Mitt. in Fl. Tasman. 209, t. 176, f. 4.

Northern Island: in various localities, *Sinclair and Colenso;* Auckland, *Knight.* **Middle** Island: Otago, *Hector and Buchanan.* (Tasmania, S. America?)

F. CONFERTA.

34. **H. tenuifolium,** *Hedw.—H. confertum,* Fl. N. Z. ii. 108, not of Smith. Stems matted, small, slender, arching, prostrate and ascending, pinnately branched, 1–2 in. long; branches compressed, long or short. Leaves dark green, subsecund, ovate-cordate, acuminate, serrulate, concave; nerve produced beyond the middle; cells linear; perichætial recurved, long-acuminate. Fruitstalk ½–1½ in. high, smooth. Capsule cernuous, ovoid-oblong, curved; operculum with a long slender beak. Inflorescence monœcious.—*H. collatum,* Hook. f. and Wils. Fl. Tasm. ii. 209. t. 176. f. 5.

Northern and **Middle** Islands: common on trunks of trees, from the Bay of Islands, *J. D. H.*, to Otago, *Lyall.* (Tasmania.)—I follow Mitten in referring this to the little-known *H. tenuifolium* (which is described as having quite entire leaves); the same author thinks it is probably a large form of the European *H. confertum.*

35. **H. elusum,** *Mitten, mss.* Stems depressed, subpinnately branched. Leaves spreading, somewhat compressed, broadly ovate, suddenly contracted to an acuminate apex, minutely serrulate above; nerve slender, produced to the middle; perichætial ovate-subulate, with reflexed apices, nerveless. Fruitstalk red, smooth. Capsule inclined, unequal; operculum subulate, beaked. Inflorescence monœcious.

Middle Island: N.E. valley, Otago, *Hector and Buchanan.*—Mitten describes this as having the appearance of *H. muriculatum,* from which its smooth fruitstalk distinguishes it; and as differing from *H. tenuifolium* in its smaller size and suddenly acuminate leaves.

36. **H. aristatum,** *Hook. f. and Wils. Fl. Tasm.* ii. 210. *t.* 176. *f.* 6. Stems small, tufted, 1–2 in. long, pinnately branched; branches slender, attenuate. Leaves dark green, not glossy, spreading, subcompressed, lax and flexuous when dry, oblong-acuminate and piliferous at the apex, serrulate, concave; margin not reflexed; nerve distinct, produced beyond the middle; perichætial erect, acuminate. Fruitstalk 1 in. long, smooth. Capsule cernuous, arcuate, oblong; annulus small, adhærent; operculum with a short beak. Inflorescence monœcious.

Northern and **Middle** Islands: Auckland, *Knight;* Saddle Hill, Otago, *Lindsay.* (Australia, Tasmania.)

G. RUTABULA.

37. **H. rutabulum,** *Linn.;—Fl. N. Z.* ii. 109. Stems rather stout, 1–3 in. long, suberect or prostrate, vaguely or subpinnately branched; branches erect, flexuous, very variable in length. Leaves pale, very shining, membranous, loosely imbricate, spreading, ovate, acuminate, concave, substriate, subserrulate; nerve produced beyond the middle; perichætial strongly recurved. Fruitstalk stout, rough. Capsule ovoid-oblong, cernuous; operculum short, conical. Inflorescence monœcious.—Wils. Bryol. Brit. 345. t. 26; Fl. Antarct. i. 138.

Northern and **Middle** Islands: in woods and on rocks, probably common, Bay of Islands, *J. D. H.;* Canterbury, *Haast;* Akaroa, *Raoul;* Otago, *Hector and Buchanan* **Campbell's** Island, *J. D. H.* (Tasmania, Europe, N. and S. America.)

38. **H. paradoxum,** *Hook. f. and Wils. Fl. Antarct.* 419. t. 155. f. 2; —*Fl. Tasm.* ii. 210. Stems 2–3 in. long, pinnately branched, creeping, loosely tufted; branches with curved tips. Leaves falcate and secund, shortly ovate-lanceolate, acuminate, strongly serrulate, striate; nerve stout, produced beyond the middle. Fruitstalk rough, brown, ½ in. long. Capsule ovoid-oblong, horizontal or cernuous; operculum conical, very obtuse.

Middle Island: Canterbury Alps, *Sinclair and Haast;* Otago, *Hector and Buchanan.* (Tasmania, Australia, Fuegia.)

39. **H. plumosum,** *Swartz;—Fl. N. Z.* ii. 109. Stems slender, 2–3 in. long, vaguely branched; branches suberect, rather incurved. Leaves erecto-patent, imbricate, subsecund, firm, ovate-lanceolate, concave, smooth, subserrulate at the apex; nerve produced beyond the middle; perichætial with long recurved points, nerveless. Fruitstalk ½–¾ in. long, roughish above the middle. Capsule ovoid, subcernuous; operculum small, conical, acute. Inflorescence monœcious.—Wils. Bryol. Brit. 340. t. 25.

Northern and **Middle** Islands: Bay of Islands, *Cunningham, J. D. H.;* Auckland, *Knight;* Bay of Plenty, *Jolliffe;* Bligh's Sound, *Lyall.* (Europe, Asia, America.)

H. Serpentia.

40. **H. serpens,** *Linn.;—Fl. N. Z.* ii. 109. Stems minute, matted, creeping, ¼–¾ in. long vaguely branched; branches filiform, simple, suberect. Leaves dark green, small, loosely imbricate, spreading, erect when dry, ovate-lanceolate, acuminate, quite entire; nerve produced beyond the middle, or much shorter; cells lax; perichætial larger, sheathing, membranous; nerve obsolete. Fruitstalk red, smooth. Capsule oblong, subcernuous; operculum conical, convex, acute.—Wils. Bryol. Brit. 362. t. 24.

Var. *β.* Fl. Antarct. 138. Nerve stronger and more solid.

Northern Island: Hawke's Bay, in bogs, *Colenso* (barren, and hence doubtful). Var. *β.* **Lord Auckland's** group, *J. D. H.* (Abundant in many temperate climates, but not hitherto in Tasmania.)

I. Stellata.

41. **H. polygamum,** *Bruch and Schimp.—H. nodiflorum,* Wils. in Fl. N. Z. ii. 109. Stems procumbent, vaguely subpinnately branched; branches short, slender, spreading. Leaves yellowish, firm, spreading, subsquarrose, ovate-lanceolate, acuminate, quite entire; cells narrow; nerve produced to the middle, or shorter; perichætial erect, striated, inner with subulate tips. Fruitstalk 1–2 in. long, smooth. Capsule cernuous, oblong, quite smooth; operculum conical. Inflorescence polygamous, male and female often crowded together.—Wils. Bryol. Brit. 365. t. 56.

Northern Island: bogs near Hawke's Bay, *Colenso;* Hick's Bay, *Jolliffe.* (Europe, N. America.)

42. **H. relaxum,** *Hook. f. and Wils. Fl. N. Z.* ii. 110. *t.* 90. *f.* 1. Stem procumbent, 2–4 in. long, vaguely subpinnately branched; branches long, erect, thickened at the top. Leaves yellowish, crowded, subsecund, spreading and recurved, soft, rather twisted when dry, ovate, acuminate, hardly serrulate; nerve stout, solid; cells oval. Inflorescence diœcious.

Northern and **Middle** Islands: probably common, *Colenso*; Wellington, *Lyall,* Akaroa, *Raoul;* Port Cooper, *Lyall.*

43. **H. decussatum,** *Hook. f. and Wils. Fl. N. Z.* ii. 110. *t.* 90. *f.* 2. Stems rather stout, 3–4 in. long, procumbent, pinnately branched; branches short, ⅓ in. long, simple, crowded, often recurved. Leaves yellowish or reddish, spreading, squarrose and recurved, not twisted when dry, ovate, long-acuminate, almost quite entire; nerve continuous nearly to the apex. Fruit unknown. Inflorescence diœcious.

Northern and **Middle** Island: East Cape, *Sinclair;* Cape Turnagain and Hawke's Bay, *Colenso;* Banks's Peninsula, *Jolliffe.* (Tasmania.)

J. ACICULARIA.

44. **H. aciculare,** *Labill.;—Fl. N. Z.* ii. 110. Stems large, stout, 2–10 in. long, vaguely branched; branches elongate, simple. Leaves crowded, spreading and squarrose, undulated, pale green, large, ovate, long-acuminate, serrate towards the apex, nerveless; perichætial small, oblong, abruptly piliferous. Fruitstalk 1–2 in. high, stout, purple. Capsule 8-angled, subcylindric; operculum longer than the capsule, with a slender beak. Calyptra large, chesnut-brown, coriaceous. Inflorescence diœcious; male axillary red.—Schwægr. Suppl. t. 92.

Abundant in woods throughout the **Northern** and **Middle** Islands, and in **Lord Auckland's** group and **Campbell's** Island. (Tasmania, Australia, temp. S. America, Sandwich Islands.)—One of the largest and commonest N. Z. mosses.

45. **H. densifolium,** *Brid. Sp. Musc.* ii. 204. Habit and appearance of *H. aciculare*, but leaves more crowded, all of their leaves strongly refracted from the middle, more shortly acuminate, and very finely serrulate.

Northern Island: Wellington, *Stephenson* (*Herb. Mitten.*) (Tristan d'Acunha.)

K. COCHLEARIFOLIA.

46. **H. cochlearifolium,** *Schwægr. Suppl. t.* 88;—*Fl. N. Z.* ii. 111. Stems long, robust, prostrate, almost creeping, 2–4 in. long, cylindrical, sparingly branched; branches short, 1 in. long, erect, stout. Leaves pale green, glossy, spreading, imbricate, orbicular, obtuse, quite entire, concave and almost hemispherical; nerve 0; cells narrow; perichætial sheathing. Fruitstalk ¾ in. long, smooth, arcuate at the top. Capsule subcernuous, ovoid; operculum conical, half as long as the capsule. Inflorescence diœcious. —Fl. Antarct. i. 139. *H. flexile*, Hook. Musc. Exot. t. 10 (not of Swartz).

Var. *β.* Stems more or less pendulous; branches shorter; fruitstalk very short, stout; capsule more rounded; operculum acuminate.

Throughout the **Northern** and **Middle** Islands from the Bay of Islands, *Cunningham*, to Dusky Bay, *Menzies*, most abundant. **Lord Auckland's** group and **Campbell's** Island, *J. D. H.* (Tasmania, Australia, Chiloe.)

47. **H. clandestinum,** *Hook. f. and Wils. Fl. N. Z.* ii. 111. *t.* 90. *f.* 3. Stem procumbent, 2–3 in. long, cylindric, vaguely sparingly branched; branches ¼–1 in. long, straight or incurved. Leaves pale greyish-green, glossy, spreading, imbricate, oblong-orbicular or broadly-oblong, very obtuse, cordate at the base, very concave like a boat, scarcely toothed; nerve 0; cells narrow; perichætial subsquarrose. Fruitstalk short, ¼–½ in. long, smooth. Capsule cernuous or horizontal, oblong-cylindric, smooth; operculum short,

conical. Inflorescence monœcious; male fl.: amongst or upon the leaves of the fertile stem.

Middle Island: moist forests, Port William, *Lyall;* Chain Hills, Otago, *Lindsay;* Nelson, *Travers;* Canterbury, *Haast.* (Tasmania, Australia.)

48. **H. chlamydophyllum,** *Hook. f. and Wils. Fl. Antarct. t.* 61. *f.* 1; —*Fl. N. Z.* ii. 111. Stems prostrate, 1–2 in. long, sparingly vaguely branched, cylindrical; branches long, erect, simple, cuspidate. Leaves pale green, shining, spreading, imbricate, orbicular-subquadrate, rounded at the apex, auricled at the base, quite entire, concave and inflated; nerves 1 or 2, very short; cells narrow-oblong; perichætial erect. Fruitstalk slender, 1½ in. long, smooth. Capsule cernuous or horizontal, narrow, ovoid-oblong; operculum short, beaked. Inflorescence monœcious.—*H. auriculatum,* Montagne.

Northern Island: abundant in moist woods in the interior, *Colenso.* **Middle** Island: Dusky Bay, *Menzies;* Nelson, *Travers;* Canterbury, *Haast;* Otago and Port William, *Lyall, Hector and Buchanan.* **Campbell's** Island: barren, *J. D. H.* (Tasmania, Fuegia.)

49. **H. inflatum,** *Hook. f. and Wils. Fl. N. Z.* iii. 111. *t.* 90. *f.* 5. Stems large, stout, tumid, erect, 2–3 in. high and ⅕ in. diameter, very sparingly branched. Leaves bright, shining, yellow, crowded, large and inflated, closely imbricate, suberect, oblong, apiculate or acuminate, quite entire, very concave, much undulated; apex recurved; nerve 0.

Northern Island: Manawaki, *Colenso.* **Middle** Island: Nelson, *Travers;* Murray Bay, *Jolliffe* (barren). A very remarkable moss, of which the genus cannot be determined till the fruit is discovered.

50. **H. vagum,** *Hornschuch.* Stems ascending, stout, much vaguely branched. Leaves broadly ovate, shortly acuminate, almost nerveless; perichætial long, sheathing, more or less reflexed. Fruitstalk elongate, erect, smooth, flexuous. Capsule oblong, nearly horizontal, purple; operculum conical, subulate, straight.

Northern Island, *Kerr* (*in Herb. Mitten*). I have seen no New Zealand specimens. (Australia.)

51. **H. divulsum,** *Hook. f. and Wils. Fl. N. Z.* ii. 111. *t.* 90. *f.* 4. Stems very long (2–5 in.), filiform, slender, cylindrical, vaguely sparingly branched; branches rigid, suberect, incurved or straight. Leaves grey-green, hardly glossy, spreading, loosely imbricate, rounded-obovate, obtuse or subacute, minutely serrulate; nerve reaching halfway; cells punctiform; perichætial squarrose. Fruitstalk ⅓–½ in. long. Capsule cylindric-ovoid, cernuous; operculum conical, acuminate. Inflorescence diœcious.—Fl. Tasman. ii. 211.

Northern Island: Auckland, *Knight.* **Middle** Island: Nelson, *Travers;* Canterbury, *Haast;* Port Cooper and Banks's Peninsula, *Jolliffe, Lyall;* Otago, *Hector and Buchanan.* Similar to *H. clandestinum,* but with nerved less-obtuse leaves and punctiform cells. (Tasmania, Australia.)

L. Distichophylla.

52. **H. extenuatum,** *Bridel.*—*H. crinitum,* Hook. f. and Wils. Fl. N. Z. ii. 114. t. 91. f. 4. Stems procumbent, 2–5 in. long, much pinnately branched; branches distant, ½ in. long, flattened, acute. Leaves pale green, shining, distichous, erecto-patent, subcompressed, concave, ovate-oblong, obtuse, with a piliferous point as long as themselves, subserrulate; nerve 0;

perichætial nearly similar. Fruitstalk 1–2 in. long, slender, smooth. Capsule ovoid-oblong, cernuous; operculum conical. Inflorescence diœcious.—Sullivant, Musci of U. S. Expl. Exped. 18.

Northern Island: not uncommon in woods, *A. Cunningham, Colenso*, etc. **Middle** Island: Pelorus harbour, *Jolliffe*; Nelson, *Travers*; Otago, *Hector and Buchanan*. **Lord Auckland's** Island, *United States Expl. Exped.* (Tasmania, Australia.) Our *H. crinitum* is hardly different from *H. extenuatum*, Bridel, though smaller, with rather larger points to the leaves (the figure in Fl. N. Z. represents it as too large and robust.) C. Mueller describes *H. extenuatum* as a native of N. Zealand, at Kiapara (Bot. Zeit. 1851, 566); there are a few barren scraps of it amongst my own Bay of Island collections, but I cannot distinguish the two species except as forms.

53. **H. politum,** *Hook. f. and Wils.; Fl. Antarct. t.* 154. *f.* 2.—*Fl. N. Z.* ii. 114. Stems suberect or prostrate, 1 in. long, sparingly branched, flattened; branches ¼ in., acute. Leaves bright green, shining, closely imbricate, distichous, spreading, oblong, compressed, keeled, very concave, acute, with subpiliferous apices, quite entire; nerve 0; cells very minute; perichætial much shorter, ovate, erect, long acuminate. Fruitstalk 1 in. high, smooth. Capsule inclined, slender, narrow oblong, with an attenuated apophysis; operculum conical, beaked, half as long as the capsule. Inflorescence diœcious.—*Phyllogonium callichroum*, Montagne, not of Bridel.

Northern Island, *Colenso* (scraps very barren). **Middle** Island: Otago, *Hector and Buchanan.* (Australia, Tasmania, Andes, Fuegia, Kerguelen Land, S. Africa.)

54. **H. polystictum,** *Mitten, mss.* Stem procumbent, flexuous, compressed, pinnately branched, 1–1½ in. long. Leaves pale yellow, shining, compressed, distichously spreading, complicate, broadly ovate, shortly acuminate, serrulate; nerve 0; cells elongate, very narrow, each with a row of about 5 papillæ.—Mitten.

Northern Island: On a Lichen (*Sticta*), *Knight.* Characterized by Mitten, who picked it from out of some Lichens sent by Dr. Knight. The leaves are hardly distichous, being imbricated all round, though the branches are compressed and the lateral leaves spread 2-fariously. The character of the cells is a very remarkable one, and compared by its author to that of the Indian *H. Nipalense* and S. American *H. planum*; it requires a high power of the microscope to detect it.

55. **H. denticulatum,** *Linn.* Stems prostrate, sparingly branched, 1 in. long; branches short, subfascicled. Leaves complanate, light green and glossy, inserted all round the stem but 2-farious, not undulate, concave, obliquely ovate, subacute or with a short tapering entire or serrulate apex; margin recurved below; nerves 2, short, basal. Fruitstalk basal, 1 in. long, stout, reddish. Capsule cylindric-oblong, inclined or horizontal; operculum conical, subacute. Inflorescence monœcious.—Wils. Bryol. Brit. 407. t. 24; *H. Donianum*, Smith.

Middle Island: Hopkin's River, in Beech forests, alt. 3–4000 ft., *Haast.* (Europe, N. America, Himalaya.) I follow Mitten in referring this plant to the European *H. denticulatum*, which is a very variable plant. The leaves of the New Zealand form are shorter, broader, less acute, and less very obscurely serrulate.

The European *H. riparum* is stated (Fl. N. Z. ii. 109) to be possibly a native of New Zealand, from very imperfect specimens collected at Hawke's Bay by Colenso. I do not now find the specimens, which were very small and incomplete, and I think that the identification is better suppressed.

64. OMALIA, Bridel.

Small, bright green, shining, creeping, stoloniferous, monœcious mosses; stems quite prostrate or suberect, pinnately branched; branches flat. Leaves distichously imbricated, often subsecund, but inserted 8-fariously, flattened, the lateral oblique, unsymmetrical, vertical, inflexed at the base on one side; nerve short or 0; cells very minute, narrow. Fruitstalk lateral, long, slender, curved at the tip. Capsule ovoid, drooping; annulus obscure; peristome of *Hypnum*. Operculum beaked. Calyptra cucullate.

A small genus, found in various temperate climates, on trees and rocks.

* *Leaves dark green, serrulate.*

Leaves obovate-oblong or spathulate . . . . 1. *O. pulchella.*
Leaves linear-oblong, undulate when dry . . . . 2. *O. oblongifolia.*

** *Leaves pale glossy green, quite entire.*

Leaves oblong, with a flat falcate point; nerve 0 . . . . 3. *O. falcifolia.*
Leaves rounded-obovate, nerved . . . . 4. *O. auriculata.*

1. **O. pulchella,** *Hook. f. and Wils. Fl. N. Z.* ii. 114. *t.* 91. *f.* 5. Stems prostrate or inclined from a creeping rhizome, rigid, much pinnately branched, ½–1 in. high; branches ¼ in. long. Leaves light green, not glossy, crowded, complanate, distichous, oblong or rounded-obovate or spathulate, apiculate, serrulate; nerve stout, reaching halfway; perichætial squarrose. Fruitstalk smooth, ⅓ in. high, arcuate at the top. Capsule nodding, broadly ovoid; operculum with a curved beak, as long as the capsule. Calyptra white. Inflorescence diœcious.—*Hookeria punctata*, nob. in Lond. Journ. Bot. 1844. 550.

Northern and **Middle** Islands: on trees, probably common; Bay of Islands, *J. D. H.*; Auckland, *Knight*; Port Nicholson and Milford Sound, *Lyall.* (Norfolk Island.)

2. **O. oblongifolia,** *Hook. f. and Wils. Fl. N. Z.* ii. 115. *t.* 91. *f.* 6. Similar in habit and most characters to *O. pulchella*, but the branches are narrower; leaves deeper green, undulate when dry, longer, linear-oblong, less serrulate, and the nerve is produced beyond the middle.—*Hookeria punctata*, var. β, nob. in Lond. Journ. Bot. 1844. 550.

Northern Island: Bay of Islands, with *O. pulchella*, *J. D. H.*; Auckland, *Knight* (drawing only), *Sinclair.*

3. **O. falcifolia,** *Hook. f. and Wils. Fl. N. Z.* ii. 115. *t.* 92 *f.* 1. Habit of *O. pulchella*, but more simple; branches much broader, ⅛ in. across. Leaves pale green, glossy, crowded, oblong, with a broad flat falcate obtuse apex, quite entire; nerve 0; perichætial long, erect, narrow. Fruitstalk very slender, ½–¾ in. long, arched at the top, smooth. Capsule oblong, horizontal or nodding; operculum short, conical. Inflorescence diœcious.—*Hypnum falcifolium*, nob. in Lond. Journ. Bot. 1844, 554.

Northern Island: Bay of Islands, *Cunningham*; Auckland, *Sinclair, Jolliffe*; Port Nicholson, *Lyall.* (Tasmania.)

4. **O. auriculata,** *Hook. f. and Wils. Fl. N. Z.* ii. 115. *t.* 92. *f.* 4. Stems creeping, broad, 2-pinnately branched, 1–2 in. long and broad; secondary branches broad, short. Leaves pale yellow-green, glossy, sub-

secund, obovate-rotundate, apiculate, concave, quite entire, auricled at the base; nerve produced to the middle, sometimes forked. Fruit unknown.

Northern Island: Auckland, on trees, *Knight, Sinclair.*

65. RHIZOGONIUM, Bridel.

Stems or branches erect, from a creeping rhizome, tufted, simple, rarely branched, rigid, often compressed. Leaves rigid, distichous or imbricate all round, simply or doubly spinulose, serrate; cells very lax, minute; nerve stout. Fruitstalk lateral, from the bases of the stems, rarely from the side, slender, long. Capsule inclined, suberect or nodding; peristome of *Hypnum*. Operculum usually beaked. Calyptra subulate, cucullate.

Tufted erect mosses, often very beautiful, common in tropical and subtropical and southern temperate forests.

A. *Leaves distichous.*

α. *Leaves simply toothed or serrate or quite entire.*

Leaves ovate-oblong, subobtuse; nerve vanishing below the serrulate apex . . . . 1. *R. distichum.*
Leaves ovate-oblong, denticulate; nerve excurrent . . . . . 2. *R. novæ-Hollandiæ.*
Leaves oblong-lanceolate, quite entire; nerve excurrent . . . 3. *R. pennatum.*

β. *Leaves doubly serrulate.*

Leaves ovate-lanceolate, acuminate; nerve continuous . . . . 4. *R. bifarium.*

B. *Leaves spreading all round the stem, not distichous.*

Leaves spreading, linear-subulate . . . . 5. *R. spiniforme.*
Leaves erecto-patent, lanceolate-subulate, twisted when dry . . 6. *R mnioides.*
Leaves spreading, linear-lanceolate . . . . 7. *R. subbasilare.*

1. **R. distichum,** *Bridel;—Fl. N. Z.* ii. 115. Stems ½ in. high. Leaves dark green, distichous, coriaceous, ovate-oblong, rather obtuse, toothed at the apex; margin not thickened; nerve stout at the base, vanishing below the apex. Fruitstalk stout, 1 in. high. Capsule horizontal, oblong; operculum conical, beaked. Inflorescence diœcious.—*R. Muelleri*, Hamp. in Linnæa; *Hypnum*, Schwæg. Suppl. t. 87.

Northern and **Middle** Islands: common on stumps of trees, etc., *Sinclair*, etc. (Tasmania and Australia.)

2. **R. novæ-Hollandiæ,** *Bridel;—Fl. N. Z.* ii. 116. Stems slender, ½–1 in. high. Leaves glossy, yellow-green, pellucid, distichous, often secund, rigid, oblong-ovate, acuminate, denticulate, margin scarcely thickened; nerve excurrent; cells rather large. Fruitstalk 1 in. high. Capsule horizontal, oblong, pale red; annulus large; operculum beaked, nearly as long as the capsule. Inflorescence diœcious.—*Leskea*, Schwægr. Suppl. t. 83; Fl. Antarct. i. 136.

Northern and **Middle** Islands: common on wood. **Lord Auckland's** group (barren), *J. D. H.* (Tasmania and Australia.)

3. **R. pennatum,** *Hook. f. and Wils. Fl. N. Z.* ii. 116. *t.* 92. *f.* 2. Stems 1 in. high, very slender and elegant. Leaves pale yellow-green, shining, pellucid, spreading, distichous, oblong-lanceolate, acuminate; margin thickened; nerve excurrent, stout. Fruitstalk 1 in. high. Capsule suberect,

small, almost turbinate, annulate; operculum conical, beaked, nearly as long as the capsule. Inflorescence diœcious.

Middle Island: on trees, Dusky Bay, *Menzies;* Port Preservation, *Lyall.*

4. **R. bifarium,** *Schimper in Bot. Zeit.* 1844, 125;—*Fl. N. Z.* ii. 116. Stems ½–1 in. high, slender, branched above; branches few, subfascicled, slender, curved, rachis flexuous. Leaves dull green, distant, spreading, distichous, rigid, ovate-lanceolate, acuminate; margin more or less thickened, doubly-serrate; nerve solid, continuous. Fruitstalk from the middle of the stem, curved at the top. Capsule small, horizontal, ovoid; operculum conical, beaked, shorter than the capsule.—*Hypnum,* Hook. Musc. Exot. t. 57, Fl. Antarct. i. 137. *Trachyloma,* Mitten.

Northern and **Middle** Islands: common in forests on stumps of trees, etc., *Menzies,* etc. **Chatham** Island, *W. Travers.* **Lord Auckland's** group, *J. D. H.* (Tasmania.)

5. **R. spiniforme,** *Bruch;—Fl. N. Z.* ii. 116. Stems rigid, 1–3 in. high, stout, unbranched. Leaves yellow-brown or reddish, loosely imbricate, spreading, rigid, linear-subulate; margin thickened, doubly spinulose-serrate on the edges and keel; nerve stout, excurrent. Fruitstalk basilar. Capsule horizontal, oblong, arcuate; operculum beaked.—*Hypnum,* Linn.; Hedwig Sp. Musc. iii. t. 25.

Var. β. Stems more slender; leaves shorter.—Fl. Antarct. i. 137. *R. Hookeri,* C. Muell. in Bot. Zeit. 1851, 547.

Northern Island: Bay of Islands, *Cunningham.* **Kermadec** group, *M'Gillivray,* Var. β. **Middle** Island: Queen Charlotte's Sound, *Jolliffe.* **Lord Auckland's** group and **Campbell's** Island: on rocks, barren, *J. D. H.* (Most tropical and temperate southern latitudes.)

6. **R. mnioides,** *Hook. f. and Wils. Fl. N. Z.* ii. 116. Stems ½–1 in. long. Leaves dull green, lurid, erecto-patent, twisted when dry, lanceolate-subulate, decurrent at the base; margin narrowly thickened, doubly serrate; nerve solid. Fruitstalk basilar, 1½ in. long. Capsule subcernuous, ovoid; operculum beaked.—*R. Mossmanianum,* C. Muell. in Bot. Zeit. 1851, 547. *Hypnum,* Hook. Musc. Exot. t. 77. *H. subbasilare,* Schwægr. Suppl. t. 256 (not of Hooker).

Northern Island: mountains of the interior, *Colenso;* Auckland, *Knight;* Port Nicholson, *Sinclair.* **Middle** Island: Dusky Bay, *Menzies, Lyall.* (Tasmania, Andes, Fuegia.)

7. **R. subbasilare,** *Schimper in Bot. Zeit.* 1844, 125. Branches erect, compressed, simple. Leaves subbifarious or subsecund but not distichous, erecto-patent, lax, linear-lanceolate, or linear-ligulate, acuminate, pellucid, serrulate; nerve vanishing below the apex; cells lax; perichætial minute, nerveless. Fruitstalk basilar, flexuous. Capsule cylindric-oblong, cernuous or horizontal; operculum conical. Inflorescence monœcious.—Fl. Tasman. ii. 216. *Hypnum,* Hook. Musc. Exot. t. 9. *H. mnioides,* Schwægr. Suppl. t. 257, not of Hook.

Middle Island: Canterbury, on shaded rocks, *Travers;* Otago, Lammermuir hills and Waipori Creeks; *Hector and Buchanan.* (Tasmania, Fuegia.)

66. HYMENODON, Hook. f. and Wils.

A small elegant diœcious moss, with the habit of *Rhizogonium.* Leaves distichous, oblong, piliferous, papillose; cells rounded, minute. Fruitstalk slender, lateral from near the base of the stem. Capsule suberect; peristome of 16 narrow, subulate, membranous equidistant fugacious teeth, cohering by their tips, and united by a narrow basal membrane. Operculum beaked. Calyptra split down one side. Inflorescence diœcious.

A small tropical genus, found in New Zealand, Java, and Brazil, on trunks of trees, etc.

1. **H. piliferus,** *Hook. f. and Wils. Fl. N. Z.* ii. 117. *t.* 92. *f.* 3. Stems quite simple, erect, ½ in. high. Leaves brittle, light glaucous-green, not shining, moistening slowly, laxly distichously imbricate, spreading, elliptic-oblong, flat, with long piliferous points, crenulate; nerve vanishing; perichætial lanceolate, acuminate, erect. Fruitstalk 1 in. long and slender, red. Capsule contracted at the mouth; teeth white, fugacious, faintly striate longitudinally and transversely. Calyptra small, white.—? *Hypnum Mougeotianum*, A. Rich.; Fl. N. Z. 57.

Northern and **Middle** Islands: Bay of Islands, on trunks of *Cyathea dealbata*, *J. D. H.*, etc.: Wangaroa and Auckland, *Jolliffe*; Otago, *Hector and Buchanan*. (Tasmania, Australia.)

67. HYPOPTERYGIUM, Bridel.

(*Including* LOPIDIUM *and* CATHAROMNION.)

Rhizome creeping on trees; stems erect, tree-like, 1–2–3-pinnately branched, flattened. Leaves 3-ranked, the lateral vertical and oblique; dorsal dissimilar, smaller, appressed to the stem, stipule-like; nerve produced to the apex, or vanishing; cells minute. Fruitstalk lateral, produced underneath the leaves, curved at the top. Capsule suberect, cernuous or pendulous, straight, annulate; peristome of *Hypnum* or *Leskea*, outer sometimes wanting. Operculum convex, beaked. Calyptra coriaceous, conical-subulate, usually dimidiate.

Tropical and south temperate mosses, abounding in damp woods, exceedingly pretty in habit; forming matted tufts in vegetable soil and on trees. Branches often radiating from the top of the stem.

I. HYPOPTERYGIUM.—Stem 2–3-pinnately branched above, there orbicular or deltoid in outline; branches radiating. Fruitstalk rather long. Capsule pendulous.

a. Leaves not mixed with bristles.

* *Branches 3-pinnate* . . . . 1. *H. filiculæforme.*

** *Branches 2-pinnate.*

Fruitstalk stout. Operculum conical, acuminate. Inflorescence monœcious . . . . 2. *H. viridulum.*
Fruitstalk slender. Operculum subulate, beaked, as long as the capsule. Inflorescence diœcious . . . . 3. *H. oceanicum.*
Leaves decurved; dorsal with a continuous nerve . . . . . 4. *H. novæ-Zelandiæ.*
Leaves ovate, acute; dorsal with the nerve vanishing beyond the middle . . . . 5. *H. discolor.*
Leaves glaucous; dorsal orbicular-ovate, acuminate; nerve produced towards the apex . . . . 6. *H. glaucum.*

β. Leaves with bristles intermixed.

Leaves spinulose-serrate . . . . 7. *H. tamariscinum.*
Leaves slightly toothed at the apex . . . . 8. *H. rotulatum.*

II. CATHAROMNION.—Stem elongate, pinnately branched above. Fruitstalk short. Peristome single (outer absent). Leaves intermixed with bristles.

Leaves ciliate . . . . 9. *H. ciliatum.*

III. LOPIDIUM.—Stem elongate, pinnately branched above. Fruitstalk short. Capsule inclined or cernuous. Peristome double.

Fruitstalk stout, smooth . . . . 10. *H. concinnum.*
Fruitstalk slender, rough . . . . 11. *H. Struthiopteris.*

1. **H. filiculæforme,** *Bridel;—Fl. N. Z.* ii. 117. Stems erect, from a creeping rhizome, naked below, 1–2 in. high, 3-pinnately radiatingly branched like a small fern above, deltoid or broadly ovate in outline; branches numerous. Leaves small, deep green, distichous, spreading, obliquely ovate-cordate, acuminate, subserrulate; nerve vanishing; dorsal orbicular, apiculate. Fruitstalk short, ½ in. long. Capsule ovoid, pendulous; operculum beaked, nearly as long as the capsule. Calyptra dimidiate. Inflorescence diœcious.—*Hypnum,* Schwægr. Suppl. t. 281 A.

Northern and **Middle** Islands: common on stumps of trees, etc., from the Bay of Islands, *J. D. H.*; to Otago, *Lyall.*

2. **H. viridulum,** *Mitten, mss.—H. rotulatum,* Fl. N. Z. ii. 118, not *Hypnum rotulatum,* Hedwig. Stems short, bright green, 2-pinnately branched; branches arranged in a deltoid form. Leaves on the main stem broadly ovate, acute, quite entire; nerve vanishing beyond the middle, on the branches similar, serrulate; dorsal suborbicular, apiculate, with the nerve continuous or ceasing below the apex; cells oblong-hexagonal; perichætial short, ovate, acuminate. Fruitstalk red. Capsule horizontal or pendulous; operculum conical, acuminate. Inflorescence monœcious.

Northern and **Middle** Islands: Wellington, *Stephenson;* Wangaroa and Akaroa, *Kerr* (*in Herb. Mitten*).—The above is as described by Mitten, who observes that this is a soft green moss with more slender branches than *H. novæ-Zelandiæ,* and leaves with larger cells that have acute ends. In appearance it comes nearest *H. discolor,* but it is smaller, with different inflorescence.

3. **H. oceanicum,** *Mitten, mss.* Stems short, $\frac{1}{6}$–$\frac{1}{3}$ in. high, expanding into a rounded-triangular green frond, ½ in. broad. Leaves on the primary branches ovate-oblong, apiculate, slightly serrulate, margined; nerve slender, vanishing beyond the middle; upper cells rounded, lower oval, pellucid; leaves of branchlets more oblong and denticulate; dorsal leaves small, orbicular, denticulate; nerve excurrent; perichætial erect, oblong, subulate-acuminate, quite entire, nerveless. Fruitstalk long. Capsule pendulous, oblong, rugose at the base; operculum with a subulate beak as long as the capsule. Inflorescence monœcious.—Mitten.

Kermadec group: Sunday Island, *Milne and M'Gillivray.*—Very similar to *H. viridulum,* but the fruitstalk is much longer and more slender. (Norfolk Island.)

4. **H. novæ-Zelandiæ,** *C. Muell. in Bot. Zeit.* 1851, p. 562.—*H. Smithianum,* Hook. f. and Wils. Fl. N. Z. ii. 118. Stems 1–2 in. high; branches arranged in a deltoid form. Leaves pale yellow-green, firm, more or less decurved, distichous, orbicular-ovate, apiculate, margin thickened, denticulate at the apex, rather concave; nerve stout, continuous; dorsal orbicular, acuminate, with a stout continuous nerve; perichætial lanceolate,

acuminate. Fruitstalk reddish, ½ in. long. Capsule pendulous; operculum beaked. Calyptra dimidiate. Inflorescence diœcious.—*Hookeria rotulata*, Smith in Linn. Trans. ix. 279 (not *Leskea rotulata* of Hedwig).

Abundant throughout the islands, at the roots of trees, etc., *Menzies*, etc. (Tasmania, Australia.)

5. **H. discolor,** *Mitten, mss.* Habit of *H. novæ-Zelandiæ*, but rather larger, not turning yellow when dry; branches recurved at the apex. Leaves on the stem broadly obliquely-ovate, subcordate, acute, serrulate; nerve vanishing at the middle, or the branches broader upwards, with longer nerve; dorsal leaves on the stem orbicular, acuminate with the short nerve, on the branches serrulate above with percurrent nerve; perichætial small, ovate, acuminate. Fruitstalk red. Capsule ovoid, horizontal; operculum with a subulate beak. Inflorescence diœcious.

Northern Island: Auckland, *Knight, Jupp;* Kiapara, *Mossmann.*

6. **H. glaucum,** *Sullivant, Musci of U. S. Expl. Exped.* 26.—*H. Smithianum*, β. Fl. N. Z. ii. 118. A very small species; stems tufted, ¼ in. high; branches forming a nearly orbicular crown, pale glaucous yellow-green, decurved. Leaves closely imbricate, very deflexed, broadly ovate-acuminate or apiculate, obscurely serrulate; nerve produced halfway or more; dorsal large compared with others, orbicular, abruptly acuminate, with the nerve continuous. Fruitstalk stout, arcuate. Capsule ovoid, cernuous; operculum with a slender beak, nearly as long as the capsule.

Northern Island: common at the roots of trees, from the Bay of Islands, *A. Cunningham*, etc., southwards.

7. **H. tamariscinum,** *Sullivant, Musci of U. S. Expl. Exped.* 26.—*H. setigerum*, Hook. f. and Wils. Fl. N. Z. ii. 118. Stems 2 in. high, branched as in *H. rotulatum*. Leaves yellow-green, rather glossy, mixed with small bristles, distichous, ovate-acuminate, spinulose-serrate; nerve vanishing; dorsal ovate, more acuminate, spinulose-ciliate. Fruitstalk very stout, ½ in. high; purplish, suddenly deflexed at the top. Capsule ovoid, pendulous, pale, abrupt and tubercled at the base; annulus distinct; operculum yellow, beaked, as long as the capsule. Calyptra not dimidiate. Inflorescence diœcious.—*Hypnum setigerum*, Palisot. *Leskea tamariscina*, Hedwig, Sp. Musc. t. 51 (not *Hypnum tamarisci* of Swartz); Fl. Antarct. i. 136.

Abundant throughout the islands, *Cunningham*, etc. **Lord Auckland's** group, *J. D. H.* (barren).—Sullivant restores the name of *tamariscinum* to this plant, it being identical with Hedwig's figure and description of *Leskea tamariscina*.

8. **H. rotulatum,** *Hedwig (Leskea), not of Fl. N. Z.* ii. 118. Stem short, naked below, 2-pinnately branched in a radiating manner above, there orbicular. Leaves mixed with a few setæ, dull green, not crowded, distichous, spreading, crisped or undulate when dry, orbicular-ovate; margin thickened, slightly toothed at the apex; nerve vanishing above halfway; dorsal orbicular, apiculate; nerve very short; perichætial ovate-lanceolate, concave, setulose on the back and apices, acuminate. Fruitstalk ½ in. long. Capsule pendulous; neck tubercled; operculum beaked. Inflorescence monœcious.—*Leskea*, Hedwig, Sp. Musc. t. 51.

Northern Island: Wangaroa, *Kerr* (*in Herb. Mitten*).—Mitten refers the *H. rotulatum* of Fl. N. Z. to *H. viridulum*.

9. **H. ciliatum,** *Bridel.—Catharomnion,* Fl. N. Z. ii. 119. Stems 1 in. high, naked below, and covered with matted radicles, pinnately branched above; branches crowded, alternate. Leaves pale green, hoary (from the long cilia), mixed with setæ, distichous, orbicular-ovate, acuminate, long-ciliate; nerve vanishing; dorsal much smaller, ovate-lanceolate, acuminate, ciliate. Fruitstalk very short, ¼ in. long, purplish, stout. Capsule nearly erect, oblong, somewhat unequal, reddish-brown; peristome large, single; operculum beaked. Calyptra mitriform, inflexed, and laciniate at the base. Inflorescence diœcious.—*Pterigynandrum,* Hedw. Sp. Musc. t. 17.

Northern and **Middle** Islands: common in forests, on the ground, from the Bay of Islands, *J. D. H.*, to Nelson, *Travers*. (Tasmania.)

10. **H. concinnum,** *Bridel.—Lopidium,* Fl. N. Z. ii. 119. Stems 1–3 in. high, naked below, pinnately branched above, then erect; ramification ovate-oblong in outline. Leaves yellowish or yellow-green, distichous, crowded, oblong-ligulate, acuminate, margin thickened, serrulate at the apex; nerve subcontinuous; dorsal more dilated at the base; cells very small, roundish; perichætial nearly entire; nerve nearly continuous. Fruitstalk ⅙ in. long, stout, thickened upwards, scarcely longer than the perichætial leaves, smooth. Capsule suberect; operculum conical-subulate, nearly as long as the capsule. Calyptra conical-subulate. Inflorescence diœcious.—*Leskea,* Hook. Musc. Exot. t. 34; Fl. Antarct. i. 136.

Northern and **Middle** Islands: Bay of Islands, *Wilkes's Expedition;* Nelson, *Travers;* Dusky Bay, *Menzies;* Otago, *Hector and Buchanan.* **Lord Auckland's** group: on rocks, *J. D. H.*

11. **H. Struthiopteris,** *Bridel.—Lopidium pallens,* Fl. N. Z. ii. 119. Habit and ramification of *H. concinnum*. Leaves pale, crisped when dry, ovate-oblong, shortly acuminate, toothed at the apex, margin thickened; nerve stout, excurrent; dorsal cordate, acuminate; perichætial with longer points. Fruitstalk slender, ¼ in. long, rough. Capsule horizontal, annulate; operculum beaked, as long as the capsule. Calyptra cleft on one side. Inflorescence monœcious.—Mitten in Journ. Linn. Soc. iv. 96. *Leskea concinna,* Schwægr. Suppl. t. 269 (not of Hooker).

Northern and **Middle** Islands: common in damp forests, etc. (Australia, Tasmania, Chili, Java.)

68. CYATHOPHORUM, Palisot.

Bright green, large, simple, diœcious mosses. Rhizome creeping on trees; stems erect, simple. Leaves 3-ranked; lateral vertical, spreading, oblique; dorsal smaller, stipule-like, rounded, appressed to the stem; nerve short, often 2-fid; cells large, lax. Fruitstalk lateral, very short, inserted in a cup-shaped tumid sheath on the under-surface of the stems. Capsule oblong, straight, annulate; peristome of *Hypnum*. Operculum convex, acuminate. Calyptra small, mitriform.

A most beautiful genus of mosses, confined to India, Australia, and N. Zealand.

1. **C. pennatum,** *Bridel;—Fl. N. Z.* ii. 120. Stems erect, 2 in. to 1 ft. long, simple, rarely 2-fid, naked below, very flat, bright green and translucent. Leaves distichous; lateral $\frac{1}{6}$–$\frac{1}{4}$ in. long, ovate-oblong, oblique, serrate; nerve half as long, often forked; dorsal leaves rounded, apiculate; perichætial small, slender. Fruitstalk very short, stout, curved; sheath large, conspicuous. Capsules large, ovoid, pale brown; spores small; operculum conical, acuminate, half as long as the capsule. Calyptra very small, cellular, brown, covering the upper half of the operculum. Antheridia mixed with club-shaped paraphyses.—*Hookeria pennata*, Hook. Musc. Exot. t. 163.

Var. *β. minus.* Leaves more distant, acute, Fl. Antarct. i. 143. t. 62. f. 3.

Var. *γ. apiculatum.* Stems short; leaves shorter, apiculate, more sharply serrate.

Abundant throughout the islands from the Bay of Islands to Lord Auckland's Island. Var. *β.* Lord Auckland's Island, *J. D. H.*; Totara, *Colenso.* Var. *γ.* Middle Island, *Lyall.* (Tasmania and S.E. Australia.)

69. CALOMNION, Hook. f. and Wils.

A minute, densely tufted, diœcious, bright green, shining moss, growing on trunks of trees. Stems simple, flattened, lanceolate, erect from a creeping rhizome. Leaves 3-farious, of 2 forms; lateral distichous, lanceolate; dorsal appressed, broader, both quite entire, with stout nerves; cells minute, rounded. Fruitstalk terminal, stout, erect. Capsule erect, cylindric-oblong, annulate; mouth contracted; teeth 0. Operculum with a long beak. Calyptra cucullate.

A very curious moss, of doubtful affinity.

1. **C. lætum,** *Hook. f. and Wils. Fl. N. Z.* ii. 97. *t.* 87. *f.* 5. Stems $\frac{1}{4}$ in. high, persistent, the old ones bristle-like. Lateral leaves vertical, but not as to insertion, oblong-lanceolate, acute; nerve reaching the tip; dorsal leaves appressed, nearly orbicular, apiculate; perichætial erect, linear, rather obtuse. Fruitstalk about as long as the stem; operculum nearly as long as the capsule. Male flowers gemmiform, with closely imbricate scales; paraphyses few or 0.—*Eucladon complanatum*, Hook. f. and Wils. in Lond. Journ. Bot. iii. 538.

Northern and **Middle** Islands: Bay of Islands, *Sinclair;* on dead trees and tree-fern trunks, *J. D. H.;* Waikehi, *Sinclair;* Auckland, *Knight;* N.E. valley, Otago, *Hector and Buchanan.*

70. RACOPILUM, Palisot.

Rigid dark green mosses. Stems creeping, on wood, rock, and the ground, more or less pinnately branched, often tomentose with radicles. Leaves more or less distichous, or imbricate all round, usually of two forms; lateral spreading, subdistichous, subpapillose; dorsal smaller; nerve excurrent, piliferous; cells minute, rounded. Fruitstalk lateral, stout or slender, erect, usually curved at the top. Capsule unequal, elongate, curved, drooping, erect, or pendulous, grooved, annulate; peristome of *Hypnum.* Operculum beaked. Calyptra mitriform, more or less hairy below, rarely glabrous, sometimes dimidiate, base inflexed, fringed.

A rather large tropical and subtropical genus.

α. Leaves of two forms; dorsal smaller.

Leaves ovate-oblong, acuminate. Fruitstalk stout, ½ in. long . . . 1. *R. strumiferum.*
Leaves oblong, obtuse. Fruitstalk very slender . . . . 2. *R. cristatum.*
Leaves ovate-oblong, acuminate. Fruitstalk stout, 1–1½ in. long . . 3. *R. lætum.*

β. Leaves uniform.

Leaves obliquely ovate-cordate, acuminate . . . . 4. *R. robustum.*

1. **R. strumiferum,** *C. Muell. Bot. Zeit.* 1851, 563.—*R. australe,* Fl. N. Z. ii. 121. t. 92. f. 7. Stem ½–6 in. long, tomentose with brown radicles, vaguely subpinnately branched; branches short, straight or curved. Leaves yellow or green, crisped when dry, lateral distichous, spreading, ovate or oblong-ovate, acuminate and almost piliferous by the long, stout, pellucid excurrent nerve, serrulate at the apex, dorsal leaves smaller. Fruitstalk ½ in. long, stout, 3-angular when dry. Capsule yellow, afterwards red, horizontal, decurved, strumose, narrow, cylindrical, grooved; operculum with a short beak. Calyptra hairy, inflated below, hardly dimidiate. Inflorescence monœcious or diœcious.

Northern and **Middle** Islands: common in forests, on the roots of trees, *Cunningham*, etc. (Tasmania.)

2. **R. cristatum,** *Hook. f. and Wils. Fl. N. Z.* ii. 121. *t.* 92. *f.* 7. Stem 1–3 in. long, slender, vaguely branched. Leaves pale green, much smaller than in the others, lateral distichous, spreading, crisped when dry, oblong, obtuse, serrulate at the apex; nerve excurrent in a hair-point; dorsal much smaller, ovate-cordate, acuminate. Fruitstalk ¾ in. long, very slender, terete when dry. Capsule horizontal, curved, grooved; teeth scabrid at the back; operculum with a short beak. Calyptra sparingly pilose, dimidiate.—? *R. tomentosum*, C. Muell. in Bot. Zeit. 1851, 563.

Northern Island: forests at Tehawera, *Colenso.* **Kermadec** group, *M'Gillivray.* (Norfolk Island, Tasmania, Australia, India, Japan, Pacific.)

3. **R. lætum,** *Mitten in Journ. Linn. Soc.* iv. 93. Stem slender, 1–2 in. long. Leaves pale yellow-green, lax, crisped when dry, lateral, subdistichous, oblong-lanceolate, acuminate, subserrulate; nerve concolorous, excurrent; dorsal broader, shorter, ovate, acute. Fruitstalk 1–1½ in. long, twisted when dry. Capsule horizontal, slightly curved; operculum with a subulate beak, ½ as long as the capsule. Calyptra sparingly pilose.

Northern Island, *Kerr; Auckland, Sinclair.* **Middle** Island, *Nelson, Jolliffe, Mantell.* Very similar to *R. strumiferum*, but the fruitstalk is much longer.

4. **R. robustum,** *Hook. f. and Wils. Fl. N. Z.* ii. 121. *t.* 92. *f.* 5. Stems stout, ½–1½ in. long, sparingly branched; branches ascending. Leaves all uniform, pale yellow-green, crowded, spreading, complicate and involute when dry, obliquely ovate-cordate, acuminate, subserrate at the apex, piliferous apex long. Fruitstalk stout, ½ in. long, twisted when dry. Capsule much curved, long cylindric, lower part erect, upper horizontal, grooved; operculum shortly beaked. Calyptra mitriform, hairy, shortly split at the base. Inflorescence monœcious?

Northern Island: forests at Tehau-totara, etc., *Colenso, Jolliffe.* **Chatham** Island, *W. Travers.*

71. HOOKERIA, Smith.

Monœcious or diœcious mosses. Stems prostrate or creeping, rarely sub-erect, variously branched, often pinnately; branches usually compressed or quite flat. Leaves usually membranous, imbricate all round or distichous; lateral often dissimilar, stipule-like and oblique; nerve 0 or 1-2; cells lax, rhomboid or hexagonal. Fruitstalk lateral, slender, often recurved at the top. Capsule ovoid, cernuous; annulus obscure or 0; peristome double; outer of 16 lanceolate-subulate incurved teeth, ribbed on the outer face, trabeculate on the inner; inner a membrane divided into 16 keeled teeth, with rarely interposed cilia; operculum beaked. Calyptra mitriform, naked or fringed at the base.

A large genus of tropical and temperate mosses, of which few species occur in the north temperate zone, but many in the south.

§ I. SAULOMA, *Hook f. and Wils.* Leaves subsecund, not margined; cells large, lax. Outer teeth remotely barred at the back; inner keeled, without interposed cilia. Calyptra naked, inflexed at the base, not fimbriate.

Leaves oblong-ovate, acute, quite entire . . . . 1. *H. tenella.*

§ II. MNIADELPHUS, *C. Muell.* Leaves margined or not; nerve 0 or vanishing. Inner teeth of peristome without interposed cilia. Calyptra fimbriate at the base (*Distichophyllum*, Dozy and Molkb.).

a. Leaves with thickened margins.

* *Leaves serrulate; nerve* 0 *or reaching halfway.*

Leaves orbicular-obovate; cells large, lax; nerve 0 . . . . . . 2. *H. apiculata.*
Leaves orbicular-ovate; cells large, lax; nerve reaching halfway . 3. *H. rotundifolia.*

** *Leaves quite entire; nerve reaching halfway.*

Leaves green, obovate or oblong; cells small; perichætial acuminate . . . . 4. *H. crispula.*
Leaves white, obovate, apiculate; cells small; perichætial obtuse . 5. *H. pulchella.*
Leaves yellowish, obovate, apiculate, crisped when dry; perichætial obtuse . . . . 6. *H. sinuosa.*
Leaves yellow-green, obovate, not apiculate; margin reflexed; perichætial obtuse . . . . 7. *H. amblyophylla.*
Leaves yellow, oblong, acute, filmy; perichætial acute . . . . 8. *H. adnata.*
Leaves obovate-spathulate, apiculate, opaque; perichætial ovate, acute . . . . 9. *H. flexuosa.*

β. Leaves without thickened margins.

Leaves obovate, obtuse . . . . 10. *H. microcarpa,*

§ III. PTERYGOPHYLLUM, *Bridel.* Leaves flattened, not margined, toothed; nerve short, 2-fid; cells large, lax. Fruitstalk glabrous. Outer teeth remotely barred. Calyptra not fimbriate at the base, often laciniate. (The species of this section are probably all forms of one.)

Stems 2–10 in. long. Leaves pale green, obovate or rhomboid . 11. *H. quadrifaria.*
Stems ½–1 in. long. Leaves blackish, oblong-ovate or obovate . 12. *H. denticulata.*
Stems 1–4 in. long. Leaves pale green, ovate-oblong . . . . 13. *H. robusta.*
Stems ½–1 in. long. Leaves brown or blackish and spathulate . 13. *H. nigella.*

§ IV. ERIOPUS, *Bridel.* Leaves flattened, scarcely margined, serrate; nerves 2, short; cells large, lax. Fruitstalk crested above with long hairs. Outer teeth trabeculate; inner with interposed cilia. Calyptra fimbriate at the base; papillose at the apex.

Stems 2–4 in. long. Leaves obovate . . . . 14. *H. cristata.*
Stems 1 in. high. Leaves orbicular-oblong . . . . 15. *H. flexicollis.*

§ I. SAULOMA.

1. **H. tenella,** *Hook. f. and Wils. Fl. N. Z.* ii. 122. *t.* 92. *f.* 8. Stems short, stout, sparingly branched, ⅓ in. high. Leaves almost white, glossy, crowded, imbricate, secund, plicate, narrow ovate-oblong or lanceolate, acute; margin quite entire, not thickened, recurved; nerve 0 or very faint; cells lax; perichætial small, ovate-lanceolate, erect. Fruitstalk ½ in. high, smooth. Capsule erect or subcernuous, small, ovoid, contracted below the mouth when dry; operculum short, conical, hardly beaked. Calyptra naked, even at the base. Inflorescence diœcious.

Northern and **Middle** Islands: on dead wood, *Colenso;* Banks's Peninsula, *Lyall.* (Tasmania.)

§ II. MNIADELPHUS.

2. **H. apiculata,** *Hook. f. and Wils. Fl. Antarct.* 421, 155. *f.* 6;—*Fl. N. Z.* ii. 122. Stems tufted, 1 in. high, stout, erect, sparingly branched, below covered with black radicles; branches compressed, fertile, procumbent. Leaves loosely imbricate, undulate when dry, lateral spreading; dorsal appressed, ovate-orbicular, apiculate; margin thickened, subserrate at the apex; nerve 0; cells large, hexagonal; perichætial erect, ovate-lanceolate, acute. Fruitstalk flexuous, rough. Capsule horizontal or cernuous, ovoid, subapophysate; operculum shorter than the capsule, with a straight beak. Calyptra small, hairy, white, fimbriate at the base. Inflorescence diœcious.

Northern Island: creeping amongst *Racopilum* (barren), *Colenso.* Mitten observes that this is truly an *Eriopus,* though wanting the technical character of the hairs on the fruitstalk. (Tasmania, Fuegia.)

3. **H. rotundifolia,** *Hook. f. and Wils. in Lond. Journ. Bot.* 1844, *p.* 551;—*Fl. N. Z.* ii. 122. *t.* 93. *f.* 1. Small; stems prostrate, ¼–½ in. long, sparingly branched; branches compressed, slender. Leaves dark green, lax, spreading, crisped when dry, orbicular-ovate, apiculate; margin thickened, toothed; nerve reaching halfway; cells large, lax; perichætial toothed. Fruitstalk slender, ¼ in. long, smooth. Capsule horizontal, narrow-oblong, pale; operculum with a slender straight beak, nearly as long as the capsule. Calyptra fimbriate at the base. Inflorescence diœcious.

Northern and **Middle** Islands: not uncommon on fallen trees, etc., from the Bay of Islands, *J. D. H.,* to Milford Sound, *Lyall.*

4. **H. crispula,** *Hook. f. and Wils. Fl. N. Z.* ii. 122. *t.* 93. *f.* 2. Stems short, green, rather broad, prostrate, ¼ in. long, very sparingly branched; branches flattened. Leaves bright green, opaque, not glossy, spreading, distichous, crisped when dry, obovate-oblong or broadly oblong, apiculate; margin thickened, quite entire; nerve reaching halfway; cells very minute; perichætial ovate, acuminate. Fruitstalk slender, smooth. Capsule, etc., as in *H. rotundifolia.*

Northern Island: Bay of Islands, on clay banks, *J. D. H.* **Lord Auckland's** group, *Hombron* (fide *Montagne*). (Tasmania.)

5. **H. pulchella,** *Hook. f. and Wils. Fl. Antarct.* 142. *t.* 62. *f.* 1;—*Fl. N. Z.* ii. 123. Stems prostrate, ½ in. long, soft, flattened, sparingly branched. Leaves white, pellucid, flaccid, crowded, distichous, broadly obovate, obtuse or apiculate, with a narrow, thickened, quite entire margin; nerve reaching halfway; cells small, rounded; perichætial erect, smaller, ovate, obtuse; tips reflexed. Fruitstalk ½ in. high, rough and thickened at the top. Capsule suberect or cernuous, small, oblong, with a narrow apophysis; operculum shorter than the capsule; beak straight. Calyptra fimbriate. Inflorescence diœcious.

Northern Island: Tararua, *Colenso.* **Middle** Island: common, Bligh's Sound, *Lyall;* Canterbury, *Travers;* on the ground, Otago, *Hector and Buchanan.* **Lord Auckland's** group: on twigs of trees, *J. D. H.* (Tasmania.)

6. **H. sinuosa,** *Hook. f. and Wils. Fl. Tasm.* ii. 219. *t.* 177. *f.* 3. Stems ½ in. long, compressed, sparingly branched. Leaves yellowish, crowded, firm, subdistichous, crisped and undulate when dry, obovate, obtuse; margin broadly thickened, quite entire; nerve reaching halfway; cells small; perichætial small, obtuse. Fruitstalk 1 in. high, glossy, smooth, red. Capsule imperfect. Calyptra fimbriate. Inflorescence diœcious.

Northern Island: probably from near Auckland, *Sinclair.* (Tasmania.)

7. **H. amblyophylla,** *Hook. f. and Wils. Fl. N. Z.* ii. 123. *t.* 93. *f.* 3. Stems large, stout, prostrate, 2–3 in. long, sparingly branched, compressed. Leaves yellow-green, lurid when old, crowded, distichous, erecto-patent, appressed when dry, obovate, rounded at the apex; margin thickened, quite entire, and often recurved at the apex; nerve reaching halfway; cells minute; perichætial obtuse. Fruitstalk 1 in. high, slender, smooth. Capsule small, narrow, horizontal or cernuous. Calyptra fimbriate at the base. Inflorescence diœcious.

Northern Island; Port Nicholson, *Sinclair;* Auckland, *Knight.* **Middle** Island: bogs, Mataura river, *Hector and Buchanan.* (Tasmania.)

8. **H. adnata,** *Hook. f. and Wils. Fl. N. Z.* ii. 123. *t.* 93. *f.* 4. Stems short, ¼ in. long, reddish, sparingly branched; branches broad, compressed. Leaves pale yellow, delicately membranous, distichous, undulate when dry, oblong or subspathulate, acute; margin very narrowly thickened, quite entire; nerve reaching halfway; cells minute; perichætial ovate, acute. Fruitstalk very slender, ⅓ in. long, smooth, red. Capsule very small, cernuous, ovoid; operculum as long as the capsule; beak slender. Calyptra fimbriate at the base. Inflorescence monœcious.

Northern Island: common on leaves of trees and ferns, especially on *Trichomanes elongatum, Cunningham,* etc.

9. **H. flexuosa,** *Mitten.* A small species, intermediate between *H. adnata* and *crispula.* Leaves spreading, obovate-spathulate obovate-oblong or broadly oblong, shortly apiculate; margin thin, flexuous, cartilaginous; nerve vanishing beyond the middle; cells very minute, rounded; perichætial small, ovate, acute. Fruitstalk curved and subscabrid at the apex. Capsule ovoid, horizontal; operculum subulate.—Mitten, mss.

Middle Island: Banks's Peninsula, *Herb. Knight.*—I have seen no specimens of this species, which Mitten describes from two stems sent by Dr. Knight; he observes that the very minute cells ($\frac{1}{3000}$ in. diam.) give a more obscure appearance to this moss than to any of its allies.

10. **H. microcarpa,** *Hook. f. and Wils. Fl. N. Z.* ii. 123. Stems 1–5 in. long, vaguely branched; branches broad, compressed. Leaves large, pale green, whitish in the middle, distichous, imbricate, appressed when dry, obovate, obtuse; margin not thickened, quite entire; nerve 0; cells large and pellucid near the centre, smaller towards the margin; perichætial minute, ovate-lanceolate. Fruitstalk $\frac{1}{2}$ in. long. Capsule small, erect or cernuous, ovoid; operculum short, beaked. Calyptra fimbriate at the base. Inflorescence diœcious.—*Hypnum*, Hedwig, Sp. Musc. t. 59. *Pterigophyllum*, Bridel.

Abundant throughout the **Northern** and **Middle** Islands, *Cunningham*, etc. (Tasmania.)

§ III. PTERIGOPHYLLUM.

11. **H. quadrifaria,** *Smith;—Fl. N. Z.* ii. 124. Very large. Stems suberect, 2–10 in. long, sparingly branched; branches compressed, $\frac{1}{3}$ in. broad. Leaves pale green, large, whitish when old, loosely imbricate, distichous, vertical, obliquely broadly obovate or subrhomboid, obtuse; margin not thickened, denticulate (quite entire on barren shoots); dorsal suborbicular; nerve forked, vanishing; cells large, lax, hexagonal; perichætial small, ovate. Fruitstalk 1 in. long. Capsule pendulous, oblong, tubercled; operculum conical, with a nearly straight subulate beak. Calyptra small; base not fimbriate, lacerate. Inflorescence diœcious.—Hook. Musc. Exot. t. 109; Schwægr. Suppl. t. 162.

Throughout the **Northern** and **Middle** Islands: abundant, *Menzies*, etc.; in dark woods, on the ground.

12. **H. denticulata,** *Hook. f. and Wils. Fl. Antarct.* i. *t.* 62. *f.* 2. Stem suberect, sparingly branched, 1 in. high; branches compressed, subincurved, rigid when dry. Leaves shining, pale green when fresh, opaque black crisped and fragile when dry, loosely sub-4-fariously imbricate, lateral distichous, oblong, ovate or obovate, obtuse or subacute; margin toothed, not thickened; dorsal orbicular; nerve short, vanishing, 2-fid; cells large, hexagonal; perichætial small, ovate. Fruitstalk $\frac{1}{2}$ in. long, stout, red. Capsule cernuous or pendulous, oblong, narrowed at the base; operculum short, beaked. Calyptra glabrous, coriaceous, not fimbriate at the base. Inflorescence diœcious.

Lord Auckland's group and **Campbell's** Island: in wet places, *J. D. H.* (Fuegia and Falkland Islands.)

13. **H. robusta,** *Hook. f. and Wils. Fl. N. Z.* ii. 124. *t.* 93. *f.* 5. Very large; stems 1-4 in. high, suberect, rigid, sparingly branched; branches flattened. Leaves pale yellow-green, subglaucous, slightly undulate when dry, distichous, vertical, coriaceous, ovate-oblong, obtuse; margin not thickened, toothed; nerve purplish, reaching to the middle; dorsal erect, appressed, obovate. Fruitstalk $\frac{3}{4}$ in. long, smooth. Capsule pendulous, ovoid; operculum very short. Calyptra glabrous. Inflorescence diœcious.

Northern Island: moist woods, in the interior, *Colenso*, etc.

14. **H. nigella,** *Hook. f. and Wils. Fl. N. Z.* ii. 124. *t.* 93. *f.* 6. Stems suberect, ¼–1 in. high; branches and foliage as in *H. denticulata*, but lurid-green, spathulate-obovate, blackish when dry; perichætial forming a conspicuous bulb, and operculum shorter than the capsule.

Throughout the **Northern** and **Middle** Islands: in dark woods. **Kermadec** group, *Milne*. **Lord Auckland's** Island, U. S. Expl. Exped. (Tasmania, Australia.)

§ IV. Eriopus.

15. **H. cristata,** *Arnott;—Fl. N. Z.* ii. 125. Large, robust. Stems rigid, suberect, 2–4 in. high, sparingly branched; branches flattened, ¼–⅓ in. broad. Leaves green, pale, crisped when dry, laxly imbricate, distichous, broadly obovate, subacute; margin slightly thickened, sharply serrate; dorsal orbicular, apiculate; nerves 2, basal; perichætial broadly ovate, acuminate, piliferous. Fruitstalk short, ⅓–¾ in., arched at the top, then crested with long white hairs. Capsule subpendulous, pyriform; operculum beaked. Calyptra short, ciliate, fimbriate at the base, papillose at the top. Inflorescence diœcious.—*Leskea*, Hedwig, Sp. Musc. t. 49. *Chætophora*, Bridel.

Common in damp places throughout the **Northern** and **Middle** Islands: on trunks of trees, etc.

16. **H. flexicollis,** *Mitten, mss.* Stems ½–1 in. high, loosely tufted, ascending. Leaves pale yellow-green, lateral spreading, orbicular-oblong, apiculate; margin cartilaginous, serrulate above; nerve very short; cells small, oblong; dorsal smaller; perichætial ovate-subulate. Fruitstalk pale, crested with cilia, flexuous. Capsule small, short, oval, horizontal; operculum subulate. Calyptra pale, with appressed cilia.

Middle Island: Canterbury Alps, *Sinclair and Haast;* Otago, wet rocks, Dunedin, *Hector and Buchanan*. Closely allied to *H. apiculata*, but the fruitstalk is crested with bristles.

72. DALTONIA, Hook. and Tayl.

Small, erect, tufted, monœcious or hermaphrodite mosses. Leaves imbricated all round the stem; margin thickened; nerve strong; cells rhomboid. Fruitstalk lateral, erect, rather rough. Capsule erect or suberect, ovoid, obscurely apophysate; annulus 0; peristome double, outer of 16 narrow teeth, incurved when dry, recurved when moist, with a middle line, slightly barred; inner a membrane divided to the base into 16 keeled cilia. Operculum with a straight beak. Calyptra conico-mitriform, ciliate at the base.

A small genus of tropical and temperate mosses, often growing on small twigs of trees and bushes.

1. **D. nervosa,** *Hook. f. and Wils. in Wils. Bryol. Brit.* 419 *in note.*—*Hookeria*, Fl. Antarct. 142. t. 61. f. 5. Stems crowded, suberect, ½–1½ in. high, sparingly fastigiately branched. Leaves pale green, closely imbricate, erecto-patent, appressed when dry, ovate-lanceolate, acuminate, keeled; margin thickened, quite entire; nerve strong, continuous, subexcurrent; cells rounded; perichætial shorter, erect, oblong, subacute. Fruitstalk ¼ in. long, red, rough near the top. Capsule small, suberect, cylindric-ovoid, apophysate; operculum conical, with a short beak. Calyptra fimbriate at the base.—*D. Novæ-Zelandiæ*, Mitten in Journ. Linn. Soc. iv. 95.

Northern Island: Wellington, *Stephenson*. **Lord Auckland's** group: on twigs of bushes, etc., *J. D. H.* Much resembling the British *D. splachnoides*, in which the nerve ceases below the apex of the leaf.

ORDER V. HEPATICÆ.

(*Including* JUNGERMANNIEÆ *and* MARCHANTIEÆ.)

Small plants, rarely more than 1–2 in. high, loosely cellular, without woody fibres or vascular tissue. Fronds either furnished with stem and leaves, or forming flat, more or less divided foliaceous expansions. Leaves usually distichous, sometimes tristichous, those on one surface being smaller (stipules). Reproductive organs of three kinds: *capsules* containing minute spores, produced (as in *Musci*) from archegonia; *antheridia*, which are minute membranous sacs, by means of whose contents the archegonia are fertilized, and *gemmæ*. 1. CAPSULE lateral or terminal, either sunk in the frond or more often raised on a white cellular stalk; sometimes hidden in cavities of a stalked green receptacle, usually bursting into 4 valves placed crosswise, and containing microscopic spores mixed with spiral fibres. At an early stage the capsule is covered with a *calyptra*, which is tipped with a brown styliform process and sheathed at the base by a tubular or flattened *perianth*, which again is usually surrounded by altered leaves which form the *involucre*. 2. ANTHERIDIA oblong or globose, membranous sacs, as in *Musci*. 3. GEMMÆ or cellular buds, found chiefly in the frondose division (MARCHANTIEÆ), in variously-shaped receptacles on the surface of the frond

A large Order of plants, found in all parts of the world, but most abundant in the humid, warm, and temperate climates.

The *Hepaticæ* may be divided into two primary groups, one of them *foliaceous*, which are supplied with distinct leaves, and the other *frondose*, in which the green axis is continuous and simple, or forked or lobed at the edge; two New Zealand genera, *Noteroclada* and *Fossombronia*, are intermediate between these groups, and may be regarded either as frondose, with the frond pinnatifid to the midrib, or as foliaceous, with the leaves united more or less at the base.

To understand the structure of a foliaceous *Hepatica*, search any copse or forest for green or brown moss-like plants, bearing slender white cellular fruitstalks surmounted by 4 little brown radiating arms (valves) placed cross-wise. These differ conspicuously from all mosses in the white cellular fruitstalk, and capsule divided into the above 4 radiating valves; they further differ from most mosses in the leaves being invariably 2-ranked or sometimes 3-ranked, the third rank being different in form and called stipules; also, unlike so many mosses, they rarely grow upright, and the rootlets all descend from one side only of the stem and branches. Now let the specimen be held up with the side from which the roots descend away from the observer; this surface is called the *ventral*, that towards the observer the *dorsal*. The third rank of leaves (stipules), if present, is invariably dorsal, and the rootlets often grow from the base of the stipules. Next observe the direction in which the leaves are set in the stem; they are rarely horizontally set on, usually obliquely. If the obliquity is such that the lower margin is next the observer (dorsal), the insertion of the leaves is called *succubous*, in this case (still looking at the dorsal side of the stem) each leaf overlaps the one below it, or would do so if they were close together. If, on the other hand, the upper margin is next the observer (dorsal), the insertion is called *incubous*, and in this case each leaf overlaps the one above it. This must be clearly understood, for no genus has both incubous and succubous leaves. In some genera one margin of the leaf, always the ventral, unites with the stipule; if the genus is succubous, the union is between the leaf and the stipule above it (*Lepidozia*); if incubous, between the leaf and the stipule below it (*Mastigobryum*).

The fruiting organs appear at the tips or sides of the stems or branches, or sometimes of short lateral proper branchlets (*i. e.* of branchlets formed for the purpose) and consist of, first, an *involucre* formed of the two (or few) last pairs of leaves and the corresponding stipules, which become erect, often elongate, and are more lobed than usual; within this is second, the *perianth*, an erect tubular or inflated sheath, often compressed, winged or angled, with a contracted or dilated entire lobed or 2–3-lipped mouth; within this again is, third, the calyptra, an oblong cylindric body, split at the side or top; and fourth, inserted within the calyptra, is the capsule, which, when ripe, is carried up on a long or short white cellular fruitstalk.

The capsule splits into 4 valves placed crosswise, and contains spores mixed with extremely delicate long cells, which contain usually 2 coiled fibres.

Of all these organs the calyptra alone is never absent: there may be no involucre, or the involucral leaves may be adnate to the perianth, or reduced to scales growing on the perianth. The calyptra is sometimes confluent with the perianth; in *Gottschea* it is sunk in the hollowed tip of the stem; in *Saccogyne* and *Gymnogyne*, it is enclosed in a curious fleshy tubular sac that descends from the branch into the ground, or swings from the tip of the branch.

So much for the plant when in fruit. The *female inflorescence* consists of one or more *archegonia*, analogous to those of mosses; these are slender flagon-shaped bodies, with a central cavity communicating with the air by the tubular neck. In the cavity of the archegonium is a solitary loose cell; to this the antherozoid (from the antheridia or male inflorescence) gains access and fertilizes it. After fertilization the archegonium swells (often carrying up on its surface the other archegonia, which, not being fertilized, do not change) and its aperture closes; meanwhile the enclosed cell is becoming rapidly developed into a capsule, with its fruitstalk, which eventually bursts through the side of the enlarged archegonium (thereafter called calyptra), rises into the air, splits into 4 valves, and discharges the spores. Thus, one great difference between *Musci* and *Hepaticæ* is, that in the former the archegonium after being developed into a calyptra, is ruptured at its base and carried up on the top of the capsule, whilst in *Hepaticæ* it remains attached to the stem and the capsule bursts through it.

The *male inflorescence* consists of minute pedicelled sacs (*antheridia*), usually solitary in the axils of modified leaves (*perigonial*), which sometimes occupy proper branchlets. The antheridia are on the same or different plants from the archegonia, and contain cells with an enclosed spiral filament (antherozoids), which are supposed to gain access to and fertilize the pistillidia.

Of the frondose *Hepaticæ* many have the same kind of fruit as has been described; but in *Marchantieæ* the involucres perianth and capsule, instead of rising erect from the frond, are placed on the under side, and are consequently pendulous from a stalked peltate hemispherical receptacle, and the antheridia are in cups or are imbedded in the substance of the frond; in these plants, too, are often found reproductive, green, structureless, globose or oblong bodies, called *gemmæ*, which grow in special cups. In *Anthoceros* the capsule is erect, as in the foliaceous species, but is very long, linear, and splits down one side. Finally, in *Riccia*, the spherical calyptra is sunk in the substance of the frond, and does not emerge from its substance. It firmly coheres with the enclosed capsule, and both burst as one body, discharging the contained spores.

Of the *Hepaticæ* (about 212) here enumerated, the greater majority were discovered by Mr. Colenso and myself, and were new to science on the return of the Antarctic Expedition to England. They were placed, at the late Dr. Taylor's request, in his hands for immediate publication, and the descriptions of many appeared, in 1844, in Hooker's 'London Journal of Botany,' and again in the 'Botany of the Antarctic Voyage.' At that time the valuable 'Synopsis Hepaticarum' of Gottsche, Lindenberg, and Nees ab Esenbeck, was not published (it was not completed till 1847), and the difficulty of ascertaining and defining the genera and species was very great indeed. Owing to this cause and to Dr. Taylor's precipitancy in publication, much confusion crept into his work; the same plant appearing under several generic names, and some descriptions answering to subsequently cancelled species, having been also published as good species. It is still more unfortunate that Dr. Taylor did not return to the Hookerian herbarian specimens of all the species which he received for publication, and his herbarium having been sold since his death, there is now little chance of some of his New Zealand species ever being recognized. Whilst feeling it my duty to make these facts clearly known, I must add, that those who study this extensive Order of plants by the aid of modern works, can have no conception of the difficulties which Dr. Taylor had to overcome, twenty-two years ago, in classifying and naming the 300 or 400 *Hepaticæ* with which I supplied him, and of which not fifty had been published; imperfect and hasty as his work was, it showed great skill, no little sagacity, and indomitable perseverance as a microscopic investigator.

In the present work I have followed in Mr. Mitten's footsteps, he having drawn up all the descriptions for the 'Flora Novæ-Zelandiæ,' with great care; and to him, too, I am indebted for naming the species of this Order received since the publication of that work; he has also detected many of Dr. Taylor's errors, and rediscovered, in later collections, many of his lost species.

It only remains to append the same caution to the student of this as of every other Order of Cryptogams, viz. that his advance must be slow to be sure, and that it will take some months' careful collecting and study with the microscope to arrive at any clear general idea of the genera of this difficult Order. The most useful books are Gottsche, Lindenberg, and Nees's 'Synopsis Hepaticarum,' Hamburg, 1844 (but not completed till 1847), an 8vo volume of 834 pages, without plates, this I have quoted under the abbreviated title Syn. Hep.; and for plates, the 'Flora Novæ-Zelandiæ,' 'Flora Antarctica,' and Hooker's 'Musci Exotici,' 8vo, all contain figures of many species.

The following sketch of the genera and species is, no doubt, very incomplete, and probably also far from satisfactory in various points; but there really are not materials for a satisfactory and sufficient examination of a great many of the species, and long descriptions of imperfect or small specimens are sure to mislead. It can be regarded as a mere outline only, representing the condition of our knowledge at the present time. I have found it impossible to construct satisfactory keys to the species, but the descriptions are so short that these will scarcely be missed.

KEY TO THE GENERA OF THE NEW ZEALAND HEPATICÆ.

I. FOLIOSE.—*Plants with distinct leaves and often stipules. Capsule solitary, on an erect fruitstalk, 4-valved. Elaters with 2 spiral fibres.*

A. *Leaves succubous.*

* *Perianth* 0.

1. GYMNOMITRIUM. Leaves entire or 2-fid. Stipules minute or 0.

** *Perianth of the same substance as the leaves, not produced downwards into a fleshy pendent tube.*

† *Perianth terminal* (*lateral in some* Plagiochilæ).

2. JUNGERMANNIA. Perianth tubular; mouth toothed. Stipules 0 or present.
3. PLAGIOCHILA. Perianth compressed at right angles to insertion of the leaves, obconic, 2-lipped. Stipules 0.
4. LEIOSCYPHUS. Perianth obconic, compressed, 2-lipped. Stipules 2–4-fid.
5. LOPHOCOLEA. Perianth 3-gonous, 3-lobed; lobes toothed. Stipules 2–4-cleft.
6. SCAPANIA. Perianth compressed parallel to the insertion of the leaves. Leaves 2-lobed. Stipules 0.
7. GOTTSCHEA. Perianth sunk in the tumid top of the stem. Leaves large, complicate, imbriate or crisped.

†† *Perianth lateral, or on extremely short lateral branches.*

8. CHILOSCYPHUS. Perianth obovate or campanulate, 2–3-lobed. Stipules present.
9. PSILOCLADA. Perianth tubular with falcate laciniæ. Leaves and stipules 4-cleft.
10. ADELANTHUS. Perianth tubular; mouth connivent, toothed. Stipules 0.

*** *Perianth or involucre a pendulous fleshy tube.*

11. GYMNANTHE. Perianth terminal.
12. SACCOGYNA. Perianth lateral.

B. *Leaves incubous.*

* *Perianth free. Leaves without an inflexed basal lobule.*

† *Perianth lateral, or on very short lateral branchlets.*

13. LEPIDOZIA. Leaves and stipules very minute, 4-cleft. Perianth 3-plicate; mouth toothed.
14. MASTIGOBRYUM. Leaves entire or 3-dentate. Perianth 3-gonous.
15. MICROPTERYGIUM. Leaves concave, 2-fid. Perianth long, membranous.

†† *Fruit terminal.*

16. ISOTACHYS. Leaves and stipules nearly equal. Perianth tubular; mouth contracted, toothed.

** *Perianth 0 or clothed with the adnate involucral leaves.*

† *Leaves without an inflexed basal lobule.*

17. TRICHOCOLEA. Leaves and stipules multifid; segments hair-like. Perianth 0.
18. SENDTNERA. Leaves and stipules deeply cleft. Perianth with a many-cleft mouth.

†† *Leaves with an inflexed basal lobule.*

19. POLYOTUS. Leaves closely imbricate. Stipules often 4-fid. Perianth 0.

*** *Perianth free, lateral or axillary on the upper branches. Leaves with an inflexed or adnate basal lobule.*

† *Lobule membranous, incurved or appressed.*

20. RADULA. Perianth terete or flat. Stipules 0.
21. MADOTHECA. Perianth compressed; mouth contracted, 2-lipped. Stipules large.
22. LEJEUNIA. Perianth various; mouth contracted. Stipules entire or toothed, rarely 0. Leaves usually pale green.

†† *Lobule clavate, lunate, cylindric or half orbicular.*

23. FRULLANIA. Perianth keeled. Leaves and stipules purple or black, rarely green.

II. FRONDOSE.—*Plants with indistinct leaves, or consisting of a continuous scale- or leaf-like, simple or divided frond. (The 2 first genera are almost foliose.)*

A. *Capsule solitary, globose or oblong, usually splitting into 4 valves (as in* I. Foliose). *Elaters with 2 or 3 spiral fibres.*

* *Perianth perfect.*

† *Frond pinnate or pinnatifid. Perianth with adnate involucral scales.*

24. FOSSOMBRONIA. Perianth dorsal. Lobes of frond angular.
25. NOTEROCLADA. Perianth subterminal. Lobes of frond orbicular.

†† *Frond continuous, simpled, forked or divided.*

26. PETALOPHYLLUM. Frond crisped. Perianth rising from the midrib underneath, confluent with the perianth.
27. ZOOPSIS. Frond minute, slender, of bladdery cells, the marginal with cilia. Perianth lateral, large, pedicelled, laciniate.
28. PODOMITRION. Frond erect, stalked, oblong. Perianth dorsal from the midrib, tubular.
29. STEETZIA. Fronds linear, forked. Perianth subterminal from the midrib, tubular; mouth toothed.

** *Perianth* 0.

30. SYMPHYOGYNA. Frond linear, midrib indistinct. Involucre a toothed scale, dorsal.
31. METZGERIA. Frond linear; midrib distinct. Involucre a ventricose 2-lipped scale on the midrib beneath.
32. ANEURA. Frond linear, opaque, with no apparent midrib. Involucre cup-shaped, torn, marginal.

B. (MARCHANTIEÆ).—*Capsules several, pendent from a stalked peltate cup-like receptacle.*

33. PLAGIOCHASMA. Receptacle minute, 3–4-lobed; lobes concealed by the ascending involucres.

34. MARCHANTIA. Receptacle large, 8–10-rayed; involucres 2-valved, with several capsules.

35. DUMORTIERA. Receptacle convex, hairy; involucres horizontal, opening by a lateral slit, with one capsule.

36. REBOULIA. Receptacle 1–6-lobed to the middle; involucres continuous with the margin of the lobe, with one capsule.

37. FIMBRIARIA. Receptacle conical, tubercled; margins forming 4 campanulate involucres, with one capsule.

C. (TARGIONIEÆ).—*Capsule solitary towards the tip of the frond, erect, shortly stalked, bursting irregularly, enclosed in a sessile 2-valved involucre.*

38. TARGIONIA. Involucre of 2 longitudinal valves.

D. (ANTHOCEROTEÆ).—*Capsule solitary on a long fruitstalk, erect, linear, very long, 2-valved, with a central filiform columella, bursting by* 1 *or* 2 *linear valves.*

39. ANTHOCEROS. Involucre tubular. Capsule with 2 linear valves; elaters 0.

E. (RICCIÆ).—*Capsules imbedded in the substance of the frond, valveless.*

40. RICCIA. Involucre 0. Calyptra cohering with the capsule.

1. GYMNOMITRIUM, Corda.

Stems erect or creeping, usually very slender or capillary. Leaves succubous, distichous, flat or concave, entire or 2-lobed Stipules 0 or very inconspicuous. Fruit terminal. Involucral leaves 2–4, convolute, emarginate. Perianth 0. Calyptra short. Capsule globose on a slender fruitstalk; elaters with 2 spiral fibres. Antheridia axillary, obovate.

The genus has not been found in Tasmania; it passes into *ungermannia.*

1. **G. stygium,** *Hook. f. and Tayl.;—Fl. Antarct.* 144. *t.* 62. *f.* 4. Minute; stems erect, capillary, $\frac{1}{4}$–$\frac{1}{2}$ in. high, vaguely branched, olive-brown or purplish. Leaves erect, subsecund, loosely imbricate, broadly obovate, rounded or retuse at the apex; perichætial orbicular, hyaline at the margins.

Campbell's Island: creeping amongst mosses, etc., on rocks on the hills, *J. D. H.*—The southern representative of the northern *G. concinnatum,* which has 2-fid leaves.

2. **G. acinacifolium,** *Hook. f. and Tayl. Fl. Antarct.* 144. *t.* 62. *f.* 5. Stems dark purple, erect, tufted, sparingly branched, 1 in. high. Leaves secund, closely imbricate, opaque, obliquely oblong or acinaciform, rounded at the apex, posterior margin recurved, anterior convex.

Campbell's Island: creeping amongst mosses and on rocks on the hills, *J. D. H.*

3. **G. ochrophyllum,** *Hook. f. and Tayl. Fl. Antarct.* 145. *t.* 62. *f.* 6. Stems pale green, filiform, sparingly branched, 1–1$\frac{1}{2}$ in. high. Leaves distichous, loosely imbricate, erecto-patent, concave, obovate or subquadrate, shortly 2-fid; segments obtuse, cells lax. Stipules very minute or absent.—*G. minuta,* Fl. Antarct. 152, not of Crantz.

Lord Auckland's group: amongst mosses on the hilltops, *J. D. H.*

4. **G. concinnatum,** *Corda.—Jungermannia,* Fl. N. Z. ii. 128. Stem erect, short; branches straight, thickened, obtuse and compressed at the apex.

Leaves most densely imbricate, ovate, 2-fid, with a narrow membranous border. Stipules 0.—Hook. Brit. Jung. t. 3.

Northern Islands: top of the Ruahine Mountains, *Colenso.* (Europe, Fuegia.)

2. JUNGERMANNIA, *Linn.*

Stems prostrate, creeping. Leaves succubous, distichous, entire or lobed. Stipules usually present. Fruit terminal on the main stem or on a lateral branch. Involucral leaves free. Perianth tubular, angular; mouth toothed or laciniate. Calyptra usually included. Capsule ovoid or globose, on a slender fruitstalk; elaters with 2 spiral fibres. Antheridia in the bases of inflated perigonial leaves.

One of the largest genera of *Hepaticæ*, found all over the globe.

§ 1. *Stipules* 0. *Leaves entire or obscurely* 1-*toothed.*

1. **J. monodon,** *Hook. f. and Tayl.;—Fl. N. Z.* ii. 128. *t.* 94. *f.* 2. Stems 1–2 in., purple or brown, flexuous. Leaves closely imbricate, ovate-lanceolate, acuminate, oblique, sometimes a tooth on one side; involucral 2-multifid, toothed. Perianth oblong, 4-plicate, mouth ciliated.—Syn. Hep. p. 664.

Northern and **Middle** Islands: common, *J. D. H.* (Tasmania.)

2. **J. inundata,** *Hook. f. and Tayl.;—Fl. N. Z.* ii. 129. *t.* 94. *f.* 3. Stems $\frac{1}{4}$–$\frac{1}{2}$ in., dirty green, procumbent. Leaves imbricate, almost vertical, orbicular, quite entire, dorsal margin decurrent, opaque; involucral larger, spreading. Stipules 0. Perianth turbinate, 4–5-plicate and laciniate. Capsule globose.—Syn. Hepat. 669.

Northern Island: inundated places, *J. D. H.;* Cape Kidnapper, *Colenso.*

3. **J. colorata,** *Lehmann;—Fl. N. Z.* ii. 128. Stems 1–2 in. high, purple or dusky green, sending off long thread-like flagella covered with scales. Leaves closely imbricate, nearly vertical, orbicular, quite entire; involucral incised. Stipules 0. Perianth ovoid, 8–10-plicate; mouth contracted, toothed.—Syn. Hep. 36 and 673.

Northern Island: top of the Ruahine range in bogs, *Colenso,* and probably on all the Alps. **Lord Auckland's** group and **Campbell's** Island, *J. D. H.* (Australia, Tasmania, S. Africa, Juan Fernandez, Chili, Fuegia, etc.)

4. **J. flexicaulis,** *Nees;—Fl. N. Z.* ii. 128. Closely resembling *J. colorata,* but the leaves are subcordate, the involucral quite entire.—Syn. Hep. 87.

Northern and **Middle** Islands: Tararua Mountains, *Colenso;* Canterbury, *Travers.* (Java, Sandwich Islands.)

13. **J. perigonalis,** *Hook. f. and Tayl.;—Fl. Antarct.* 145. *t.* 62. *f.* 7 Minute; stems slender, $\frac{1}{2}$ in. high. Leaves dark brown or purplish, distant, suberect, secund, concave, semiamplexicaul, ovate-orbicular or obovate, retuse or unequally 2-fid at the apex, quite entire; involucral acutely 2-dentate. Perianth tubular, elongate; mouth plicate, contracted, toothed.—Fl. Tasm. ii. 222. f. 178. f. 1.

Lord Auckland's group: rocks on the hilltops, *J. D. H.* (Tasmania.)

§ 2. *Stipules* 0. *Cauline leaves lobed or cut.*

5. **J. punicea,** *Nees.* Stems short, curved; branches rigid, narrowed at the tips, giving off stout flagella. Leaves subvertical, imbricate, nearly orbicular, 2-fid, teeth conniving, acute; involucral spreading, toothed, connate. Stipules 0. Perianth 3-gonous, plaited; mouth toothed.—Syn. Hep. 97 and 676; Mont. Voy. au Pôle Sud, i. 261. t. 17. f. 3.

Lord Auckland's group, *Hombron.*

6. **J. schismoides,** *Mont.*;—*Fl. N. Z.* ii. 129. Stems purple, 1–2 in. high, erect, rather stout, apices curved. Leaves closely imbricate, very concave or complicate, obliquely 2-lobed; lobes conniving, unequal, acute, cuspidate, entire or serrulate. Stipules 0.—Mont. Voy. au Pôle Sud, 258. t. 17. f. 1; Fl. Antarct. t. 161. f. 9.

Southern Island, *Lyall.* **Lord Auckland's** group and **Campbell's** Island: common, *Hombron, J. D. H.* (Fuegia.)—The leaves are entire in N. Zealand specimens, serrulate in Fuegian.

6. **J. multicuspidata,** *Hook. f. and Tayl.*;—*Fl. Antarct.* 150. Stems loosely tufted, prostrate, sparingly branched. Leaves distant, whitish, suberect, membranous, pellucid, obovate, 2–4-fid to the middle, sinus obtuse, segments acute. Perianth lateral or terminal, long, cylindric, split at the side, mouth 3–4-toothed.—Syn. Hep. 686.

Campbell's Island: in pools near the sea, *J. D. H.*—Taylor compares this with *J. bicuspidata.*

§ 3. *Stipules present* (*sometimes absent in* J. rotata). *Leaves quite entire.*

8. **J. rotata,** *Hook. f. and Tayl.*;—*Fl. N. Z.* ii. 129. *t.* 24. *f.* 4. Stems procumbent, tufted, branched, recurved, flexuous, 1–3 in. high. Leaves imbricate, dark green, secund, erecto-patent, almost vertical, orbicular, concave, quite entire, ventral margin subdecurrent; involucral similar. Stipules appressed, lanceolate, or 0. Perianth obovate, above obtusely 4-gonous, with 4 inflexed laminæ.—Syn. Hep. 672.

Northern Island: watery place near Taupo, *Colenso.* **Lord Auckland's** group: mixed with *Plag. fuscella,* J. D. H.

§ 4. *Stipules present. Leaves more or less toothed.*

9. **J. dentata,** *Raddi*;—*Fl. N. Z.* ii. 128. Stems creeping, with erect branches, swelling at the tips. Leaves rather remote, broader than the stem, subvertical, complicate-concave, orbicular-ovate, 2-fid to the middle, toothed; involucral numerous, imbricate, deeply 2-fid. Stipules subulate. Perianth narrow, membranous, 4-plicate, mouth denticulate.—Syn. Hep. 143.

Northern Island: Bay of Islands, *J. D. H.*; Te Aute, *Colenso.* (Europe, Tasmania.)

10. **J. squarrosa,** *Hook. Musc. Exot. t.* 78;—*Fl. N. Z.* ii. 127. Stem elongate, simple or dichotomous. Leaves patent, squarrose, most densely imbricate, quadrate, undulate, 2-cuspidate; lobes quite entire or 1-toothed, deflexed. Stipules large, similar, but the lobes have flexuous incurved teeth. Perianth ovate, plicate, mouth lacerate-ciliate.—Syn. Hep. 130.

Northern Island: Auckland, D'Urville, *Sinclair*; Ruahine Mountains, *Colenso.* **Middle** Island: Dusky Bay, *Menzies*; Nelson, *Mantell.*—The undulations of the leaves are omitted in the 'Musci Exotici' plate.

11. **J. pulchella,** *Hook. Musc. Exot. t.* 94;—*Fl. N. Z.* ii. 128. Stems tall, erect, nearly simple. Leaves half-vertical, subquadrate, 4-fid, very membranous. Stipules broad, 5–6-fid; divisions obtuse, ciliated. Perianth tubular, 3-gonous above; mouth truncate, fringed with long cilia that point in all directions.

Northern and **Middle** Islands: Auckland, *Sinclair;* Dusky Bay, *Menzies;* Port Nicholson and Port Preservation, *Lyall.* (Tasmania.)

12. **J. quadrifida,** *Mitten in Fl. N. Z.* ii. 128. *t.* 94. *f.* 1. Stems ½ in. long, creeping, ascending, stoloniferous below, dirty brown, tips pink or yellowish. Leaves vertical, spreading, 4-fid to the middle, sinus obtuse, segments lanceolate, entire or toothed. Stipules similar, and involucral sharply toothed. Perianth elongate, 3-gonous above, mouth truncate-lacerate, laminæ erect or spreading.

Northern Island: Patea village, on the ground, *Colenso.*

3. PLAGIOCHILA, Nees and Montagne.

Stems from a creeping rhizome, erect, ascending, or creeping, often large and rigid. Leaves distichous, succubous, dorsal margin decurrent and reflexed, often obliquely. Stipules 0 in the New Zealand species. Fruit terminal or lateral. Involucral leaves 2, larger than the cauline. Perianth compressed at right angles to the insertion of the leaves; mouth truncate, entire or toothed. Calyptra membranous. Capsule on a long or short fruitstalk, ovoid; elaters with 2 spiral fibres. Antheridia covered by small imbricated ventricose perigonial leaves.

One of the largest genera of *Hepaticæ*, found in all parts of the world. Of the following sections, § 1 is well marked, but the others are often difficult of recognition. The species are very puzzling to identify.

§ 1. *Leaves opposite, vertical, connate at the base.*

1. **P. conjugata,** *Lindb.*—*Fl. N.Z.* ii. 130. Stems creeping; branches erect, proliferously divided, stoloniferous. Leaves obliquely orbicular-reniform, denticulate and ciliate. Perianth terminal, obconic; mouth truncate, denticulate and ciliate.—Hook. Musc. Exot. t. 91; Syn. Hep. 52.

Middle Island: Dusky Bay, *Menzies.*

2. **P. connexa,** *Hook. f. and Tayl. Fl. N. Z.* ii. 130. Stems sparingly-branched, olive-brown or yellow, 1 in. high; apices incurved. Leaves orbicular, lower entire, upper subdenticulate.—Syn. Hep. 648.

Northern Islands: Bay of Islands, *A. Cunningham.*—There are no specimens of this in Herb. Hook.

3. **P. prolifera,** *Mitten in Fl. N. Z.* ii. 131. *t.* 94. *f.* 5. Stems 2–3 in. long, creeping. Leaves rather remote, orbicular, toothed; involucral spinulose. Perianth obovate, compressed; mouth toothed. Male spikes fascicled, flagelliform, with attenuated apices.

Northern Island: Bay of Islands, *J. D. H.*

§ 2. *Stems sparingly branched. Leaves alternate, concave, more horizontal than vertical. Lower margin not much decurrent nor recurved.*

4. **P. circinalis,** *Lehm.;*—*Fl. Antarct.* i. 348. Stems tufted, erect,

branching, stout, 1–3 in. high, olive-brown. Leaves closely imbricate, erecto-patent, obliquely cordate, concave, quite entire or minutely toothed; margins recurved, dorsal gibbous; involucral similar, but large. Perianth obconic, compressed; mouth entire or slightly toothed.—Syn. Hep. 53, 652. *P. hemicardia*, Fl. Antarct. 148. t. 63. f. 2; Syn. Hep. 627.

Lord Auckland's group and **Campbell's** Island: on rocks on the hills, *J. D. H.* (Australia.)

5. **P. pleurota,** *Hook. f. and Tayl. Fl. Antarct.* 149. *t.* 63. *f.* 4;—*Fl. N. Z.* ii. 135. Stems tufted, erect, rigid, 1 in. high, sparingly branched, yellow-green. Leaves lax, erecto-patent, obovate, rather obtuse, sparingly irregularly toothed; ventral margin slightly decurrent and recurved. Perianth exserted, narrow obovate, almost winged; mouth contracted, toothed.—Syn. Hep. 633. *P. cognata*, l. c. 14. t. 53. f. 3; Syn. Hep. 625.

Middle Island: Port William, *Lyall.* **Lord Auckland's** group on moist stones, etc., *J. D. H.*—Nearly allied to *P. Dicksoni*, but not dendroid, and leaves more obovate.

6. **P. fuscella,** *Hook. f. and Tayl. Fl. Antarct.* 149. *t.* 63. *f.* 5. Stems loosely tufted, patent, ascending, vaguely branched, brown or blackish, 1–1½ in. high; apices incurved. Leaves crisped when dry, erecto-patent, obliquely orbicular-oblong; dorsal margin subrecurved, ventral subserrulate recurved. Perianth obovate, immersed, truncate, deflexed.—Syn. Hep. 648.

Lord Auckland's group: in boggy places, *J. D. H.*

§ 3. *Leaves more or less vertical, alternate, the margins recurved, dorsal decurrent.*

† *Stems erect from a creeping rhizome, tall, very much branched in a tree-like or fascicled manner. Leaves toothed and usually spinulose.*

7. **P. Stephensoniana,** *Mitten in Fl. N. Z.* ii. 133. *t.* 95. Stems 4–10 in. high, 2–3-pinnately branched. Leaves olive-green, ovate or obliquely cordate, subquadrate at the apex; dorsal margin quite entire, ventral spinulose; involucral similar, more toothed. Perianth on short proper branches, ovate, compressed; mouth toothed.

Northern and **Middle** Islands: in ravines, common, *Colenso, Lyall, Stephenson,* etc. —A magnificent plant, the noblest of all *Hepaticæ*.

8. **P. gigantea,** *Lindb.;—Fl. N. Z.* ii. 133. Stems ascending, fastigiately divided in a fan-shaped manner, recurved. Leaves spreading, orbicular-quadrate, sharply denticulate and ciliate, nearly entire at the base. Perianth oblong, compressed; mouth ciliate, dilated.—Hook. Musc. Exot. t. 93; Syn. Hep. 51.

Abundant throughout the **Northern** and **Middle** Islands, *Menzies*, etc.

9. **P. ramosissima,** *Lindb.;—Fl. N. Z.* ii. 133. Stems erect, excessively fastigiately branched. Leaves subopposite, erecto-patent, orbicular-ovate; dorsal margin quite entire, ventral spinulose dentate at the apex. Perianth urceolate; mouth truncate, compressed, ciliate.—Hook. Musc. Exot. t. 92; Syn. Hep. 58.

Middle Island: Dusky Bay, *Menzies;* Port Preservation, *Lyall*

10. **P. Arbuscula,** *Lehm. and Lindb.;—Fl. N. Z.* ii. 132. Stems erect,

much branched, dendroid. Leaves spreading, imbricate, dimidiate-ovate, acuminate; ventral margin and apex toothed, with a terminal spine. Perianth terminal and on the forks, long exserted, oblong; mouth compressed, 2-labiate, lips acute, ciliate-toothed.—Syn. Hep. 27.

Abundant throughout the **Northern** and **Middle** Islands, *Cunningham*, etc. (Java.)

11. **P. fasciculata,** *Lindl.;—Fl. N. Z.* ii. 132. Stems ascending, dichotomous below, fastigiately branched above. Leaves obliquely orbicular-oblong, convex; ventral margin and apex unequally toothed. Perianth lateral and on the forks, long exserted, obovate; mouth compressed, obliquely truncate, ciliate.—Syn. Hep. 27; Fl. Tasm. ii. 224. *P. Colensoi?* Taylor in Lond. Journ. Bot. 1846, 269.

Northern and **Middle** Islands: common on trees, *J. D. H.*, etc. **Lord Auckland's** group, *J. D. H.* (Tasmania.)

12. **P. Dicksoni,** *Hook. f. and Tayl. Fl. N. Z.* ii. 134. *t.* 96. *f.* 3. Stems erect, flexuous, 2–3 in. high, sparingly branched in a dendroid manner. Leaves spreading, ovate-oblong, truncate and 2–3-toothed at the apex; dorsal margin quite entire, ventral sparingly toothed; involucral broader, spinulose. Perianth elongate-obovate, compressed; lips rounded, toothed.—Syn. Hep. 637.

Northern and **Middle** Islands, *Menzies*, etc.—This is sometimes as little branched as the following, or indeed less so, and both of them may be referred to either section of these two.

†† *Stems not dendroid nor fasciculately branched, either simple or dichotomous or sparingly divided.*

a. Leaves entire or very slightly toothed, or 1–2-toothed at the apex.

13. **P. ansata,** *Hook. f. and Tayl. Fl. Antarct.* 425. *t.* 156. *f.* 6;—*Fl. N. Z.* ii. 131. Stems suberect, tufted, nearly simple, 1–1½ in. high. Leaves red-brown, closely imbricate, suberect, appressed, flat, obliquely-orbicular, quite entire; involucral truncate, the dorsal margin toothed at the base. Perianth obconic, compressed; mouth denticulate.—Syn. Hep. 649.

Middle Island: Port William, *Lyall.* (Falkland Islands.)

14. **P. microdictyum,** *Mitten in Fl. N. Z.* ii. 131. *t.* 94. *f.* 6. Stems 2–3 in. high, slender, sparingly branched; branches with incurved tips. Leaves green, spreading, closely imbricate, deltoid-ovate, the angles rounded; dorsal margin quite entire, ventral sometimes slightly toothed; involucral a little toothed; perianth oblong, compressed; lips ciliate-toothed.

Northern and **Middle** Islands: Bay of Islands, *J. D. H.*, with *P. prolifera* and *Sendtnera attenuata*; Otago, *Hector and Buchanan.*

15. **P. radiculosa,** *Mitten in Fl. N. Z.* ii. 133. *t.* 96 *f.* 1. Stems 2–3 in. long, prostrate, rooting; branches few, ascending. Leaves green, imbricate, diverging, obliquely oblong-cordate, 1–2-toothed at the apex or quite entire, dorsally very decurrent; involucral broader, toothed and ciliate. Perianth immature, compressed, broadly obconic; mouth truncate, toothed.

Northern Island: on *Weinmannia* bark, Tarawera, *Colenso.*

β. Leaves much toothed and spinulose.

16. **P. deltoidea,** *Lindb.;—Fl. N. Z.* ii. 131. Variable in size and

form; stems erect; branches somewhat fascicled, 2–4 in. high. Leaves vertical, closely imbricate, subsecund, rhomboid-obovate; dorsal margin quite entire; ventral arched and the apex toothed. Perianth terminal, ovate, compressed; mouth toothed and ciliate.—Syn. Hep. 55. *P. gregaria*, Hook. f. and Tayl. in Lond. Journ. Bot. 1844, 564; Syn. Hep. 654. *P. læta*, Mitten.

Northern and **Middle** Islands: common, on stumps of trees, etc., from the Bay of Islands, *J. D. H.*, to Thomson Sound, *Lyall.* (Tasmania.)

17. **P. annotina,** *Lindb.;—Fl. N. Z.* ii. 131. Stems stout, erect, 3–5 in. high, dichotomous. Leaves very closely imbricate, dimidiate-ovate, convex, obtuse, toothed and ciliate; ventral bases conniving and forming a crest. Perianth terminal and lateral, oblong, with a narrow toothed wing; mouth truncate, compressed, fimbriate.—Syn. Hep. 41, 643. *J. adiantoides*, Hook. Musc. Exot. t. 90.

Northern and **Middle** Islands: on trees; Tarawera, *Colenso;* Dusky Bay, *Menzies;* Otago, *Hector and Buchanan;* Southern Island, *Lyall.* (Tasmania.)

18. **P. incurvicolla,** *Hook. f. and Tayl.;—Fl. N. Z.* ii. 132. *t.* 96. *f.* 2. Stems suberect, 2 in. high, sparingly branched, flexuous; tips decurved. Leaves close-set, brown-green, rigid, spreading, ovate or obovate, coarsely spinous-toothed all round. Perianth narrow, oblong-clavate, compressed; tips rounded, toothed.—Syn. Hep. 651.

Northern and **Middle** Islands: Auckland, *Sinclair;* Wairarapa valley, *Colenso;* Milford Sound, etc., *Lyall;* Otago, *Hector and Buchanan.* Allied to *P. fasciculata*, but not branched like that plant.

19. **P. Lyallii,** *Mitten in Fl. N. Z.* ii. 132. Stems erect, dichotomously and fastigiately branched, slender, flexuous, 2–3 in. high. Leaves brown-green, rather remote and rigid, broadly ovate or semicordate; dorsal margin quite entire; ventral and apex spinous; involucral similar, more toothed. Perianth as in *P. incurvicolla*, from which it differs in the entire dorsal margin of the leaf.

Northern and **Middle** Islands: Auckland, *Sinclair;* Port Preservation, *Lyall.* The description and figure of the perianth in the Flora of N. Z. represent it as too short and broad. This again most closely resembles *P. fasciculata.* (Tasmania.)

20. **P. Sinclairii,** *Mitten in Fl. N. Z.* ii. 132. *t.* 96. *f.* 5. Stems 4–5 in. long, erect, strict, sparingly branched. Leaves loosely imbricate, olive-brown, divergent, obliquely ovate-oblong, obtuse; dorsal margin reflexed, quite entire; ventral and apex spinulose, lower orbicular. Perianth $\frac{1}{6}$ in. long, terminal, oblong-ovate, compressed; mouth truncate, toothed.

Northern Island: Auckland, *Sinclair;* Tehawera, *Colenso;* Port Nicholson, *Mantell.* A noble species.

4. LEIOSCYPHUS, Mitten.

Stems prostrate, creeping or ascending; branches spreading. Leaves succubous, spinulose, distichous, close set, entire, rarely 2-fid. Stipules small, 2–4-fid. Fruit terminal. Involucral leaves like the cauline. Perianth dilated upwards, compressed at right angles to the direction of the leaves; mouth entire or toothed. Capsule on a slender fruitstalk. Antheridia as in *Chiloscyphus.*

1. **L. repens,** *Mitten in Fl. N. Z.* ii. 134. *t.* 97. *f.* 1. Stems creeping, 1 in. long, sparingly branched. Leaves pale green, spreading, close set, ovate, 2-toothed; dorsal margin straight, ventral arched; cells rounded. Stipules 2-fid; teeth subulate, 1-toothed, decurrent and united to the leaf below. Perianth elongate, obovate, much compressed; mouth toothed.

Northern Island: Bay of Islands, creeping over *Lepidozia Lindenbergii*, *J. D. H.* Similar to *Lophocolea bidentata*, but differing remarkably in the perianth.

2. **L. chiloscyphoideus,** *Mitten in Fl. Tasman.* ii. 225. Stem creeping, sparingly branched. Leaves subimbricate, semi-vertical, convex, orbicular; dorsal margin reflexed. Stipules small, free, distant, ovate, 2-partite; segments subulate, flexuous; margins 1-toothed. Perianth elongate, obliquely obovate; lips dilated, incurved.—*Chiloscyphus amphibolius*, Nees, and *retusatus*, Hook. f. and Tayl. Fl. Antarct. 441. t. 161. f. 3. *Plagiochila*, Syn. Hep. 647.

Lord Auckland's group: on bark of trees, etc., *J. D. H.* (Tasmania, Fuegia, Falkland Islands.)

3. **L. turgescens,** *Mitten.—Jungermannia*, Hook. f. and Tayl. Fl. Antarct. i. 150. t. 64. f. 2. Stems procumbent, sparingly branched, ½–1 in. high, curved. Leaves pale olive-green, closely imbricate, suberect, secund, very concave, orbicular-reniform, quite entire, pellucid; ventral margin decurrent. Stipules obovate or orbicular, entire or 2-fid. Perianth oblong, truncate, compressed; mouth entire or obscurely lobed.—Syn. Hep. 671.

Lord Auckland's group: on the ground near the hilltops, *J. D. H.* Perhaps a *Chiloscyphus*, like the *C. notophylla.*

4. **L. strongylophylla,** *Mitten.—Jungermannia*, Hook. f. and Tayl. Fl. Antarct. i. 146. t. 62. f. 9. Minute, stems tufted, slender, curved, sparingly branched. Leaves very minute, greenish-brown, loosely imbricate, suberect, orbicular, concave, quite entire, pellucid, truncate, 2-fid; involucral oblong-obovate, longer than the perianth. Stipules small. Perianth oblong, subcompressed; mouth rounded, 2-lipped.

Lord Auckland's group and **Campbell's** Island, amongst grass, roots of trees, etc., *J. D. H.*

5. LOPHOCOLEA, Nees.

Stems prostrate, creeping. Leaves succubous, distichous, flaccid, decurrent at the base, 2-multifid. Stipules 2–4-cleft. Fruit terminal. Involucre of 2–4 large leaves. Perianth tubular below, 3-gonous above; lobes toothed and crested. Calyptra short, membranous, rupturing transversely at the base or irregularly at the apex. Capsule on a slender fruitstalk, oblong; elaters and antheridia as in *Chiloscyphus.*

A large genus, widely distributed.

* *Stipules connate by the decurrent bases on one or both sides, with the leaves below them.*

1. **L. pallida,** *Mitten;—Fl. N. Z.* ii. 135. Stems 1 in. long, prostrate, sparingly branched or simple. Leaves almost opposite, pale green, ovate, obtuse retuse or minutely 2–3-toothed; involucral similar or toothed on the ventral margin; cells large. Stipules appressed, 4-toothed; involucral ovate, 2-fid, entire or toothed. Perianth 3-winged; wings and mouth

toothed.—*L. multipenna*, in part, of Fl. Antarct. 155. ?*L. connata*, Swartz, of Montagne in Voy. au Pôle Sud, 255.

Northern Island: Auckland, *Sinclair*. **Lord Auckland's** group, *J. D. H.*, mixed with *L. decipiens* and *Leioscyphus chiloscyphoideus*. The description in Fl. Antarct. is very inaccurate, owing to the intermixture of these species. Odour sweet.

2. **L. heterophylloides,** *Nees;—Fl. N. Z.* ii. 135. Stems ½ in. long, procumbent, nearly simple. Leaves yellow-green, imbricate, horizontal or semivertical, flat, orbicular-ovate, rather retuse. Stipules 2-fid, ciliate, toothed at the base, decurrent and connate on one side to the leaf below. Perianth triquetrous; mouth 3-lipped, toothed.—Syn. Hep. 157. *Chiloscyphus canaliculatus*, Hook. f. and Tayl. Lond. Journ. Bot. 1844, 563 (perianth inaccurately described); Syn. Hep. 710.

Northern and **Middle** Islands: from the Bay of Islands, *J. D. H.*, to Canterbury, *Lyall*. (Tasmania.) Often fragrant.

3. **L. biciliata,** *Mitten in Fl. N. Z.* ii. 137. *t.* 97. *f.* 4. Stem 1 in. long, procumbent, nearly simple. Leaves broad, spreading, brownish-green, loosely imbricate, deltoid-ovate, truncate and 2-toothed, teeth slender. Stipules small, 2-partite; segments 2-toothed on one side.—*Chiloscyphus*, Hook. f. and Tayl. in Lond. Journ. Bot. 1845, 84; Syn. Hep. 707.

Northern Island, *Colenso*. (Specimens imperfect, and affinity doubtful.)

4. **L. Colensoi,** *Mitten in Fl. N. Z.* ii. 138. *t.* 97. *f.* 6. Stems 2–3 in. long, creeping, branched. Leaves brownish-green, imbricate, ovate-oblong, truncate, shortly 2-dentate; teeth sometimes obsolete. Stipules 2-partite; segments 1-toothed on each side.

Northern Island: East Coast, on wood, *Colenso*. The largest known species of the genus.

5. **L. subporosa,** *Mitten in Fl. N. Z.* ii. 137. *t.* 97. *f.* 3. Stems 1 in. high, creeping, branched. Leaves pale green, imbricate, rather rigid; ovate, 2-dentate; cells unequal, involucral spinulose. Stipules 4-toothed, narrowly decurrent on one or both sides. Perianth narrow-oblong, 3-gonous, dorsal angle with a toothed wing; laminæ subobtuse, entire or toothed.

Northern Island: Auckland, *Sinclair*; Wellington, *Stephenson*.

5. **L. triacantha,** *Hook. f and Tayl.;—Fl. N. Z.* ii. 138. Stems 2 in. long, creeping, branched. Leaves spreading, flat, fuscous-green, close-set; ovate-quadrate, truncate and 3-cuspidate. Stipules palmately 4-fid; segments setaceous, one side decurrent. Perianth unknown.

Northern and **Middle** Islands: hills at Pahiatua, *Colenso*; Akaroa, *Raoul*; Port Nicholson, *Mantell*.

6. **L. leucophylla,** *Tayl.;—Fl. N. Z.* ii. 138. Stems 2–3 in. long, slender, creeping, branched. Leaves pale greenish or whitish-brown, closely imbricate, rather horizontal than vertical, convex, membranous, cellular, pellucid, triangular-ovate; margins strongly recurved, broadly connate, with the stipules, sharply toothed all round. Perianth terminal; laciniæ entire, obtuse.—*Chiloscyphus*, Fl. Antarct. 157. t. 65. f. 4; Syn. Hep. 181 and 706.

Northern Island: top of the Ruahine mountains, *Colenso*.

Lord Auckland's group: on the ground, *J. D. H.*

A beautiful and very peculiar species. The character of the original *L. leucophylla*, Tayl. (published in Syn. Hep. 155, from Herb. Greville), differs totally from this, but as the authors of that work cite this as synonymous with the present *Chiloscyphus leucophyllus* (Lond. Journ. Bot. 1844, 348), it is better to retain this name, and abolish the *L. leucophylla*, whatever it may have been.

** *Stipules free.*

+ *Leaves entire or nearly so.*

7. **L. novæ-Zelandiæ,** *Nees;—Fl. N. Z.* ii. 135. Stems procumbent, branched. Leaves yellowish-green, cellular, imbricate, spreading, orbicular-quadrate, quite entire or slightly emarginate; involucral oblique, 2-toothed. Stipules free, ovate, reflexed, 2-fid. Perianth terminal, obovate, 3-gonous, wingless, 3-fid; segments unequally toothed.—*L. subviridis*, Fl. Antarct. 438. t. 159. f. 4. *L. rivalis*, 437. t. 158. f. 7. *L. sabuletorum*, 437. t. 158. f. 8. *J. subintegra*, 443. t. 160. f. 5.

Common throughout the **Northern** and **Middle** Islands, and very variable indeed, *Menzies*, etc. **Lord Auckland's** group, *J. D. H.* Often fragrant. (Tasmania, Falkland Islands.)

8. **L. australis,** *Mitten;—Fl. Tasman.* ii. 226 *in note.* Stems 1 in. long, suberect or procumbent and rooting, branched. Leaves dark brown, upper closely imbricate, appressed, erect, orbicular, concave, quite entire or obscurely lobed; cells large, opaque. Stipules 2-partite, variously toothed. Perianth curved, oblong, cylindric below, 3-gonous above, angles undulate; lips crenulate.—*Chiloscyphus*, Fl. Antarct. 156. t. 65. f. 3 (perianth erroneously described); Syn. Hep. 189. 709.

Lord Auckland's group and **Campbell's** Island: moist banks, etc., common, *J. D. H.*

10. **L. planiuscula,** *Hook. f. and Tayl. Fl. Antarct.* 156. *t.* 65. *f.* 2. Stems 2 in. long, procumbent, flaccid, branched. Leaves purplish, variable in size, very membranous, pellucid, distichous, orbicular-ovate, quite entire, base broad, decurrent; tips recurved, margins sometimes waved. Stipules 2-fid, entire or 1-toothed on each side.—Syn. Hep. 165.

Lord Auckland's group: wet rocks near the sea, *J. D. H.*

†† *Leaves 2-fid or 2-dentate at the apex (not serrate).*

11. **L. bidentata,** *Nees;—Fl. N. Z.* ii. 136. Stems variable in length, sparingly branched, prostrate. Leaves divaricating, flat, pale green, triangular ovate, flaccid, 2-dentate, with a shallow sinus. Stipules small, distant, 2-partite; segments 2-fid, entire or toothed. Perianth subsessile, angles not or slightly winged; mouth laciniate and toothed.—Syn. Hep. 159. *J. recurvifolia*, Hook. f. and Tayl. in Lond. Journ. Bot. 1844, 562. *J. leptantha*, *J. divaricata*, and *J. alternifolia*, Fl. Antarct. t. 159. f. 6, and 161. f. 2 and 8. *J. textilis*, Fl. Antarct. 435. t. 158. f. 9; Fl. N. Z. ii. 137.

Northern Island: Bay of Islands, *J. D. H.* A very common and variable plant, found in many parts of the world and throughout the temperate regions, to which I suspect some of the following species of this section may be referable.—*L. textilis* differs only in the larger cells of the foliage.

12. **L. bispinosa,** *Hook. f. and Tayl. Fl. Antarct.* 153. *t.* 64. *f.* 7. Stems

$\frac{1}{2}$–$\frac{1}{3}$ in. long, creeping and rooting, slender, tips ascending. Leaves straw-coloured, loosely set, erecto-patent or suberect, oblong-ovate or subquadrate, 2-fid; segments spreading, acuminate. Stipules minute, 2-partite; segments subulate, entire or 1-toothed at the base, rarely multifid.—Syn. Hep. 162. *J. perpusilla*, Fl. Antarct. 154. t. 64. f. 9; Syn. Hep. 163.

Campbell's Island: moist trunks of trees, etc., *J. D. H.*—Closely allied to *L. bidentata*, but pale straw-coloured; leaves more deeply 2-fid and cells minute.

13. **L. lenta,** *Hook. f. and Tayl. Fl. Antarct.* 154;—*Fl. N. Z.* ii. 136. *t.* 97. *f.* 2. Habit and foliage of *L. bispinosa*, but more erect. Leaves spreading and greener. Perianth with spinulose wings and tips.—Syn. Hep. 692. *J. diademata*, Hook. f. and Tayl. in Lond. Journ. Bot. 1844, 560; Syn. Hep. 692. *J. secundifolia*, Fl. Antarct. 438. t. 159. f. 2; Syn. Hep. 693.

Northern and **Middle** Islands: Bay of Islands, *J. D. H.*; Wairarapa valley, etc., *Colenso*; Otago, *Hector and Buchanan*. **Lord Auckland's** group: on trunks of trees, *J. D. H.* (Tasmania, Fuegia, etc.)—Scarcely different from *L. bidentata.*

14. **L. spinifera,** *Hook. f. and Tayl. Fl. Antarct.* 155. *t.* 65. *f.* 1;—*Fl. N. Z.* ii. 137. Stems $\frac{1}{2}$ in. long, prostrate, subpinnately branched. Leaves yellow or whitish-green, closely imbricate, 2-fid, sinus obtuse, margin recurved, dorsal very decurrent; cells large, lax. Stipules large, broad, reniform, 6-fid, divisions subulate; involucral toothed all round; teeth recurved. Perianth ovate, angles winged and toothed above, tips toothed and lacerate. —Syn. Hep. 163 and 693.

Northern Island: Auckland, *Sinclair*; Wellington, *Stephenson*. **Lord Auckland's** group: on the hills, *J. D. H.*—The large stipules best characterize this.

15. **L. allodonta,** *Hook. f. and Tayl. Fl. Antarct.* 155; *Fl. N. Z.* ii. 137. *t.* 97. *f.* 5. Stems 2 in. long, prostrate, rooting. Leaves dull green, widely spreading, distichous, horizontal, imbricating, membranous, alternate, ovate-oblong, obtuse or truncate, 2-dentate or with 2 spinous processes separated by a shallow sinus. Stipules minute, 2-partite.—Syn. Hep. 163 and 693.

Northern Island: Tarawera, on *Weinmannia* trees, *Colenso*. **Lord Auckland's** group: on bark, *J. D. H.*—Closely allied to *L. biciliata*, etc. Odour strong, aromatic.

††† *Leaves toothed.*

16. **L. muricata,** *Nees*;—*Fl. N. Z.* ii. 138. Stems minute, $\frac{1}{6}$–$\frac{1}{4}$ in. long, procumbent, branched. Leaves close-set, pale, subhorizontal, and stipules subquadrate-ovate, acutely 2-dentate, spinulose ciliate, muricate with short hairs above. Perianth terminal, ovate, plaited; mouth 5–6-fid.—Syn. Hep. 169, 703.

Northern and **Middle** Islands: amongst mosses, etc., probably common. Ruahine range and Tehawera, *Colenso*; Wellington, *Stephenson*; Port William, *Lyall*. (Tasmania, S. Africa.)

6. SCAPANIA, Lindenberg.

Stems erect or ascending. Leaves succubous, distichous, concave or complicate, 2-lobed. Stipules 0. Fruit terminal. Involucral leaves 2, larger than the cauline. Perianth compressed parallel to the direction of the leaves; mouth entire or ciliated. Capsule on a slender fruitstalk, ovoid; elaters with 2 spiral fibres. Antheridia in the forks of 2-lobed perigonial leaves.

A small genus, native of the north and south temperate zones, rare in the tropical.

1. **S. densifolia,** *Nees.—S. vertebralis,* Gottsch.;—Fl. Antarct. 153. Stems 3–4 in. high, erect, rarely branched. Leaves olive-green or brown, distichous, crowded, vertical, spreading, 2-fariously imbricate, 2-fid, ciliated; lobes incurved, twisted.—Syn. Hep. 73; Hook. Musc. Exot. t. 36.

Lord Auckland's group: on trees, rare, *J. D. H.* (Tasmania.)
A most beautiful plant, originally described as from Staten Land, but probably erroneously, and brought from Dusky Bay by Menzies, who confounded the habitats of several of his plants.

7. GOTTSCHEA, Nees.

Stems erect or suberect from a creeping rhizome, simple, rarely branched, very stout, fleshy and cellular. Leaves succubous, very large, distichous, cellular, and fleshy, 2-lobed, complicate, serrate or pinnatifid; lobes sometimes lamellate. Stipules rarely absent. Fruit terminal in the hollowed apex of the stem. Involucre tubular, with a lacerate mouth, or formed of imbricating stipular leaves, free or connate with the tumid apex of the stem, together forming the perianth. Calyptra ovate. Capsule on a stout fruitstalk, oblong or ovoid; elaters with 2 spiral filaments. Antheridia in ventricose imbricate perigonial leaves similar to the cauline.

A noble genus, almost confined to the southern temperate hemisphere, and most abundant in New Zealand.

* *Leaves stipulate.*

† *Leaves with toothed ridges or crests on the blade.*

1. **G. Lehmaniana,** *Lindb.;—Fl. N. Z.* ii. 151. Stem 2–3 in. high. Leaves ovate-oblong, crested with short lamellæ, serrate, ventral lobe ovate-lanceolate, dorsal as long. Stipules ovate, 2–4-fid; segments ciliate, furnished at the base with pinnatifid leaflets. Perianth terminal; involucral leaves connate, subpinnatifid at the apex, acute, ciliate-serrate.—Syn. Hep. 20; Mont. Voy. au Pôle Sud, 276. t. 16. f. 1. *G. Hombroniana,* Mont.

New Zealand, *Banks and Solander.* **Middle** Island: Chalky Bay, *Lyall.* **Lord Auckland's** group, *Hombron, J. D. H.* (Australia, Tasmania.)

2. **G. glaucescens,** *Nees;—Fl. N. Z.* ii. 151. Stems 3–4 in. high, covered with hairs and leafy scales. Leaves glaucous, ovate-oblong, pubescent and with ciliate erect lamellæ, unequally 2-fid to the middle; lobes obtuse; dorsal ovate, one-third shorter than the oblong ventral, ciliate. Stipules orbicular, 4-fid, pubescent and ciliate-serrate. Perianth terminal, cylindric-oblong; involucral leaves elongate, margins reflexed, closely ciliate; mouth obliquely truncate toothed.—Syn. Hep. 20; Hook. Musc. Exot. t. 39.

Northern and **Middle** Islands: Wairarapa valley, *Colenso;* Otago, *Hector and Buchanan;* Port William and Port Preservation, *Lyall.*

3. **G. Balfouriana,** *Hook. f. and Tayl. Fl. N. Z.* ii. 151. *t.* 101. *f.* 2. Stems 1–3 in. high, covered with ciliated scales. Leaves pale green,

crowded, oblong-lanceolate, obtuse, toothed and ciliate; ventral lobe oblong, falcate, with 3 short-toothed lamellæ; dorsal $\frac{1}{2}$-ovate. Stipules suborbicular, 4-fid; segments obtuse, toothed and ciliate.—Syn. Hep. 622.

Northern and **Middle** Islands, *Colenso, Stanger;* Chalky Bay, *Lyall.* **Lord Auckland's** group, *J. D. H.*

4. **G. repleta,** *Hook. f. and Tayl. Fl. N. Z.* ii. 153. *t.* 101. *f.* 3. Stems 1–2 in. high, covered below with purple radicles, glabrous above. Leaves yellow-green, erecto-patent, imbricate, amplexicaul, ovate-oblong, margins sharply toothed; ventral lobe broadly ovate-lanceolate, subacute, with 3–4-toothed lamellæ; dorsal much shorter, obliquely orbicular-ovate. Stipules imbricate, oblong, 2-fid; segments 2-partite, toothed and ciliate.—Syn. Hep. 622.

Northern and **Middle** Islands, *Colenso;* Port William, *Lyall.*

5. **G. unguicularis,** *Hook. f. and Tayl. Fl. N. Z.* ii. 151. *t.* 102. *f.* 1. Stems 1 in. long, covered below with purple radicles, above with scattered 2-fid or toothed scales. Leaves pale green, imbricate, spreading, toothed; ventral lobe oblong-obtuse, with few short lamellæ; dorsal broadly ovate, much smaller. Stipules 4-fid; segments toothed and ciliate. Involucre elongate, loosely clothed with leaves which do not sheath.—Syn. Hep. 622.

Northern and **Middle** Islands, *Colenso;* Auckland, *Sinclair and Bolton;* Nelson, *Mantell.*

†† *Leaves without lamellæ.*

6. **G. appendiculata,** *Nees;—Fl. N. Z.* ii. 150. Stems 4–6 in. high. Leaves very large, yellow-green, oblong; ventral lobe ovate-oblong, pinnatifid at the apex; laminæ serrulate; dorsal $\frac{1}{4}$ shorter, dimidiate-ovate, acute, sharply toothed. Stipules 2-partite; laminæ inciso-serrate. Involucre formed of the uppermost leaves and stipule, cup-shaped; mouth cut into 8–9 equal toothed segments.—Syn. Hep. 14; Hook. Musc. Exot. t. 15.

Northern and **Middle** Islands: Auckland, *Sinclair*; Hokianga, *Sinclair*; Dusky Bay, *Menzies;* Port Cooper, *Lyall;* Otago, *Hector and Buchanan.*

7. **G. nobilis,** *Nees;—Fl. N. Z.* ii. 151. Stems 3–8 in. high, stout, almost woody, with tufted leaflet-like scales at the bases of the leaves. Leaves yellow-green, 2-fid to the middle; lobes ovate, acute, spinulose-servate; dorsal $\frac{1}{4}$ shorter. Stipules suborbicular, 2-dentate, spinulose-serrate. Perianth terminal, ovoid, subplicate; mouth with serrate laciniæ.—Syn. Hep. 21; Hook. Musc. Exot. t. 11.

Northern and **Middle** Islands: Tararua, *Colenso;* Dusky Bay, *Menzies;* South Land, *Lyall;* Otago, *Hector and Buchanan.*

8. **G. ciliata,** *Mitten in Fl. N. Z.* ii. 151 *t.* 101. *f.* 4. Stems 2–3 in. high, glabrous. Leaves purplish-green, closely imbricate, very broadly ovate, ciliated all round with hair-like cilia; lobes nearly equal. Stipules 2–4-fid; lobes rounded, long, ciliated.

Northern Island: creeping on fern fronds, Ruahine range, *Colenso.* (Tasmania.)

9. **G. splachnophylla,** *Hook. f. and Tayl. Fl. Antarct.* 424. *t.* 156. *f.* 2;—*Fl. N. Z.* ii. 150. Stem procumbent, 2 in. high, very succulent and

fragile, covered with rootlets. Leaves dull olive-green or whitish, very thick, fleshy, brittle; ventral lobe ovate-oblong, truncate, entire; dorsal shorter, broadly ovate or semicordate, also truncate, both often crenate at the apex. Stipules suborbicular or oblong, retuse or 2-fid.—? *G. pachyphylla*, Nees, Syn. Hep. 621; Fl. Antarct. 147.

Northern Island: summit of the Ruahine mountains, *Colenso.*

? **Lord Auckland's** group: on the hills, *J. D. H.*—A very difficult plant to dissect after being dried. Of the Auckland Island plant referred to *G. pachyphylla* (a Tristan d'Acunha species), there were but a few scraps, which, after examination by Dr. Taylor, were not returned to Kew. Mr. Mitten suspects that they are referable to this. (Fuegia.)

** *Stipules* 0.

10. **G. pinnatifolia,** *Nees;—Fl. N. Z.* ii. 149. Stems 2–3 in. high. Leaves dull green, closely imbricate, obliquely ovate-acuminate or ovate-lanceolate, irregularly toothed and ciliate; dorsal lobe nearly as long as but narrower than the ventral; involucral toothed. Stipule 0. Perianth cylindric-oblong; mouth 5-lobed; lobes toothed.—Syn. Hep. 22; Hook. Musc. Exot. t. 114; Fl. Antarct. 147. t. 63. f. 1. *G. ciliigera*, Hook. f. and Tayl. in Lond. Journ. Bot. iii. 376.

Northern and **Middle** Islands: Tararua range, *Colenso;* Dusky Bay, *Menzies;* Port William, *Lyall;* Otago, *Hector and Buchanan.* **Lord Auckland's** group, *J. D. H.* (Tasmania.)

11. **G. tuloides,** *Hook. f. and Tayl. Fl. N. Z.* ii. 150. *t.* 101. *f.* 1. Stems short, stout, 1 in. long, broad. Leaves dark green, imbricate, spreading, with minute asperities on the surface towards the apex, toothed; ventral lobe ovate-lanceolate, acute, concave; margin flat; dorsal shorter, ovate, truncate, anterior margin recurved; involucral free, sheathing. Stipule 0.—Syn. Hep. 626.

Northern and **Middle** Islands: forests on the east coast, *Colenso;* Port Preservation, *Lyall.*

8. CHILOSCYPHUS, Corda.

Stems prostrate, creeping, rooting from the bases of the stipules. Leaves succubous, distichous, decurrent. Stipules often decurrent and connate with the leaves below them. Fruit terminal on very short lateral branches. Involucre of 2–6 leaves. Perianth 2–3-partite. Calyptra herbaceous, globose or clavate, often longer than the perianth, bursting irregularly at the apex. Capsule ovoid, on a slender fruitstalk; elaters with 2 spiral fibres. Antheridia in the saccate dorsal bases of perigonial leaves which resemble the cauline.

Similar in habit to *Lophocolea*, with which some of the species are probably confounded. A large genus found in all temperate and many tropical parts of the world. In *C. Billardieri*, the commonest species, the character of leaves connate at the base or free, and entire at the apex or toothed, breaks down. I suspect that a good many of the following species will be found very unstable.

§ I. *Leaves opposite. Stipules united to both the leaves below them by their decurrent margins.*

* *Leaves united by their dorsal bases* (*sometimes free in* J. Billardieri).

† *Leaves entire or nearly so at the apex.*

C. Menziesii, *Mitten in Fl. N. Z.* ii. 139. *t.* 98. *f.* 1. Stem 1 in. long, creeping, flexuous, branched. Leaves brownish-green, opposite; dorsal bases connate, broadly ovate or ½-orbicular, obtuse, quite entire, thick, succulent; involucral concave denticulate. Stipules ovate, 2-dentate; margins obtusely 3-denticulate, broadly connate with the leaves. Perianth campanulate; teeth of the mouth short, incurved.

Middle Island: Dusky Bay, *Menzies*, on Lichens; Port Preservation, on bark of trees, *Lyall*. **Lord Auckland's** group, *J. D. H.*

2. **C. Billardieri,** *Nees;—Fl. N. Z.* ii. 139. Stems 3–5 in. long, prostrate, dichotomously branched. Leaves green or brown, opposite, ovate, subacute, obscurely 2-dentate; ventral margin arched; dorsal straight, 6–9-toothed at the base, free or connate with that of opposite leaf. Stipules imbricate, transverse, convex, 2-toothed, connate with the leaves below. Perianth campanulate; mouth laciniate, fimbriate.—Hook. Musc. Exot. t. 61; Syn. Hep. 175.

Northern and **Middle** Islands: common in woods, etc., *Menzies*, etc. **Lord Auckland's** group and **Campbell's** Island: *J. D. H.* (Australia, Tasmania.) A very variable plant in size, colour, toothing of the leaves, which are free or connate at their bases.

3. **C. sinuosus,** *Nees;—Fl. N. Z.* ii. 141. Stems 3–4 in. long, procumbent. Leaves crowded, brownish-green, opposite, ovate-oblong, obtuse, obscurely 2-toothed at the apex, connate or almost so by their dorsal bases; ventral margin undulate; dorsal very decurrent; involucral incised. Stipules distant, spreading, 5-partite; segments toothed, connate with the leaves below. Perianth ovate; mouth laciniate, inflexed.—Syn. Hep. 175; Hook. Musc. Exot. t. 113. *C. oblongifolius*, Hook. f. and Tayl. in Lond. Journ. Bot. 1845, 563; Syn. Hep. 705.

Northern and **Middle** Islands: abundant from the Bay of Islands, *J. D. H.*, to Dusky Bay, *Menzies*. **Lord Auckland's** group, *J. D. H.*

† *Leaves strongly* 2–5-*dentate at the apex.*

4. **C. fissistipus,** *Hook. f. and Tayl. Fl. Antarct.* 157. Stems ½–1 in. long, procumbent, branched. Leaves ovate, united by their dorsal bases, 2–4-toothed at the apex; ventral margin undulate, 1–2-toothed; dorsal straight; involucral 2–3-fid; margins toothed and laciniate. Stipules imbricate, reniform, spinulose, connate with the leaves below them. Perianth campanulate; mouth open, laciniate.—Syn. Hep. 175 and 704.

Northern Island, *Colenso;* Auckland, *Bolton, Sinclair*. **Lord Auckland's** group, *J. D. H.* (Tasmania.)—A handsome, strongly scented species.

5. **C. aculeatus,** *Mitten in Fl. N. Z.* ii. 140. *t.* 98. *f.* 4. Stem 1 in. long, creeping. Leaves green, opposite, imbricate, connate by their dorsal bases, ovate; apex 2-toothed, with an obtuse sinus; dorsal margin with 1 spine, ventral with 2 or 3. Stipules transversely oblong, 4-spinous, broadly united with the leaves below them.

Northern Island?, creeping over *Hypopterygium concinnum (Herb. Mitten).*

** *Dorsal bases of the leaves not connate.*

† *Leaves nearly entire at the apex.*

6. **C. supinus,** *Hook. f. and Tayl. Fl. N. Z.* ii. 142. *t.* 99. *f.* 2. Stems

creeping, 1½ in. long. Leaves brownish-green, imbricate, opposite, membranous, ovate, obtuse or truncate, rarely obscurely toothed; dorsal margin arched, rather recurved; ventral straight. Stipules rather large, suborbicular, 2-dentate with a rounded sinus, spinous-toothed, connate with the leaves below. Perianth campanulate; mouth laciniate, inflexed.—Syn. Hep. 708.

Northern Island: Bay of Islands, *Sinclair;* Ruahine range, *Colenso.*—Abnormal leaves are 2-toothed.

7. **C. polycladus,** *Mitten in Fl. N. Z.* ii. 142. *t.* 99. *f.* 3. Very similar in size and appearance to *C. supinus,* but the ventral margin of the leaves is toothed towards the base, and the stipule deeply divided into about 6 spreading laciniæ.—*Lophocolea,* Hook. f. and Tayl. in Lond. Journ. Bot. 1846, 367; Syn. Hep. 697.

Middle Island: Akaroa, *Raoul.*

8. **C. decipiens,** *Gottsche;—Fl. N. Z.* ii. 139. Stems procumbent, branched. Leaves deltoid-ovate; ventral margin arched, inflexed; dorsal bases decurrent, straight, toothed towards the base and free; apex entire; involucral denticulate. Stipules reniform; margins involute, connate with the leaves below. Perianth ovate-oblong, plaited, tips rounded, toothed.—Syn. Hep. 176.

Northern and **Middle** Islands: Tararua, *Colenso;* Dusky Bay, *Menzies;* Port Preservation, *Lyall.*

†† *Leaves 2–4-toothed at the apex.*

9. **C. coalitus,** *Nees;—Fl. N. Z.* ii. 141. Stem 1 in. long, creeping. Leaves opposite, ovate-quadrate, truncate, 2-dentate; teeth subulate; involucral small, 2-fid. Stipules 4–6-toothed, connate with the leaves below them. Perianth obovate; mouth 4-toothed.—Hook. Musc. Exot. t. 123; Syn. Hep. 180 and 706.

Northern and **Middle** Islands: from the Bay of Islands, *J. D. H.*, to Dusky Bay, *Menzies.* **Lord Auckland's** group, *J. D. H.* (Tasmania, Java.)

10. **C. Lyallii,** *Mitten in Fl. N. Z.* ii. 140. *t.* 98. *f.* 3. Stems 2–3 in. long, creeping. Leaves green, opposite, imbricate, trapezoid, truncate and 2-dentate at the apex, membranous; dorsal margin entire; ventral spinulose-toothed. Stipules 2-partite; margins toothed, connate with the leaves below them.

Middle Island: Port Preservation, *Lyall,* with *C. coalitus.*—A very handsome species.

11. **C. trispinosus,** *Mitten in Fl. N. Z.* ii. 140. *t.* 98. *f.* 5. Stems creeping, 1–1½ in. long. Leaves pale green, pellucid, nearly opposite, ovate-trapezoid, truncate, 2-toothed, cells large; dorsal margin entire; ventral 1-spinose towards the middle. Stipules short, 2-partite; segments 2- or 3-spinose, narrowly united to the leaves below them.

Middle Island: Bligh's Sound, creeping over *C. coalitus, Lyall.*—Odour heavy, aromatic.

12. **C. odoratus,** *Mitten in Fl. N. Z.* ii. 140. *t.* 98. *f.* 6. Closely allied in size, habit, and other characters to *C. trispinosus,* but the leaves are

of an olive or brownish-green colour, more deeply 2-toothed at the apex, and stipules 4-spinose. Perianth shortly campanulate; mouth shortly toothed.

Northern and **Middle** Islands: Auckland, *Sinclair*, etc., Port Preservation, *Lyall*.—Odour pungent, peppery.

13. **C. physanthus,** *Mitten in Fl. N. Z.* ii. 141. *t.* 98. *f.* 7. Stem 1 in. long, creeping. Leaves lurid or dirty green, subopposite, imbricate, ovate, 2-dentate; sinus obtuse; teeth diverging; involucral 4-toothed. Stipules spinous, 4-toothed, narrowly connate with the leaves below them. Perianth campanulate, rather plaited above; mouth laciniate, inflexed.—Syn. Hep. 700.

Northern Island: Bay of Islands, *J. D. H.*; Auckland, *Sinclair*.

§ II. *Leaves alternate. Stipules united by one decurrent margin to a leaf below them.*

14. **C. laxus,** *Mitten in Fl. N. Z.* ii. 142. *t.* 99. *f.* 1. Stem 2 in. long, brittle. Leaves bright green, subpellucid, alternate, rather remote, ovate-oblong, unequally 3–4-toothed; cells large, lax. Stipules small, 2-partite; segments 1-toothed.

Northern Island: Auckland, *Sinclair*. **Middle** Island, *Lyall*. (Tasmania.)

15. **C. tridentatus,** *Mitten in Fl. Tasman.* ii. 228. *t.* 179. *f.* 1.—*C. combinatus*, Fl. N. Z. ii. 141. Stems procumbent, slender, 1 in. long. Leaves brownish-green, convex, rigid, shortly oblong or subquadrate, truncate, 2- or 3-toothed at the apex; margins recurved; stipules small, 4-toothed, connate on one side with a leaf below.

Northern Island: Bay of Islands, *Cunningham*; Cape Turnagain, *Colenso*; Wellington, *Stephenson*. (Tasmania.)

§ III. *Leaves opposite or alternate. Stipules free.*

16. **C. piperitus,** *Mitten in Fl. N. Z.* ii. 141. *t.* 98. *f.* 8. Stem ½ in. long, creeping. Leaves green or greenish-brown, subopposite, oblong or ovate-oblong, truncate, 2-toothed; sinus shallow. Stipules small, 2-dentate; teeth lanceolate, diverging.

New Zealand, *Herb. Mitten*.—A little-known and imperfectly-characterized species. Odour of black pepper.

17. **C. chlorophyllus,** *Mitten in Fl. N. Z.* ii. 139. *t.* 98. *f.* 2. Stem ½ in., creeping. Leaves alternate, pale green or brownish, imbricate, ovate, 2-cuspidate; dorsal margin entire; ventral 1–2-toothed near the apex; involucral similar, denticulate. Stipules free, ovate, 2-dentate, toothed on the sides. Perianth campanulate; mouth open, lobed, and spinous-laciniate.—*Lophocolea*, Hook. f. and Tayl. in Lond. Journ. Bot. 1844, 562; Syn. Hep. 698.

Northern Island: Bay of Islands, *J. D. H.*; Auckland, *Sinclair*.

18. **C. echinellus,** *Mitten in Fl. N. Z.* ii. 141. Stems small, procumbent. Leaves subimbricate, horizontal, subrotund, spinulose-toothed. Stipules 2-partite; segments lanceolate-subulate, toothed.—*Lophocolea*, Syn. Hep. 703.

Middle Island: Dusky Bay, said to have been picked out of specimens of *Plagiochila ramosissima* from Dusky Bay, sent by Dr. Taylor to the authors of the species (Lindenberg and Gottsche), but this is more than doubtful, as Dr. Taylor had no Dusky Bay mosses.

19. **C. cymbaliferus,** *Hook. f. and Tayl.—Jungermannia,* Fl. Antarct. 151. t. 64. f. 5. Stems tufted, 1 in. high, suberect or procumbent, curved. Leaves pale green or yellowish, flaccid, pellucid, densely and closely imbricate, erecto-patent, very concave, subsecund, obliquely orbicular, serrulate; dorsal margin produced into an obovate inflated auricle. Stipules free, very broad, transverse, tumid, closely imbricating, obscurely 3-lobed or quite entire. Perianth lateral, ovate-oblong, tumid below, compressed below the plaited toothed mouth.—Syn. Hep. 711.

Lord Auckland's group: roots of old trees in woods, rare, *J. D. H.*—A very peculiar plant. (Tasmania.)

9. PSILOCLADA, Mitten.

Hirsute; stems capillary, creeping, subpinnately branched. Leaves succubous, distant, minute, quadrate, 4-cleft. Stipules similar. Fruit lateral, on very short branches. Involucral leaves 2-fid, large, falcate. Perianth subcylindric, smooth; mouth laciniate; laciniæ falcate, secund. Calyptra, etc., unknown.

1. **P. clandestina,** *Mitten in Fl. N. Z.* ii. 143. *t.* 99. *f.* 4. Stems 1 in. long, capillary, procumbent, sparingly branched. Leaves bright green, microscopic, scarcely broader than the stem, remote, square in outline, consisting of a short horizontal blade and 4 erect subulate lobes at right angles to it; cells large, inflated; involucral, imbricate, large, falcate-secund, more cut. Stipules similar, smaller.

Northern Island: Wellington, creeping on mosses, *Stephenson.* (Tasmania.)
A very imperfectly known plant, of which but a few scraps have been found. The only species of the genus.

10. ADELANTHUS, Mitten.

Stems erect from a creeping rhizome, branched. Leaves succubous, distichous, vertical, the dorsal margin decurrent. Stipules 0. Fruit terminal on short ventral branchlets, concealed at the bases of the branches. Involucral leaves 3-farious. Perianth tubular, subtrigonous; mouth connivent, toothed. Capsule on a slender fruitstalk. Elaters with 2 spiral fibres. Antheridia in small ventral spikes.

A small genus, consisting of 4 species, found in various parts of the globe. The fructification is very brittle, and hence difficult to examine in a dry state. Dr. Taylor confounded that of an *Aneura* (growing amongst it) with it.

1. **A. falcatus,** *Mitten in Journ. Linn. Soc.* vii. 243.—*Plagiochila,* Fl. N. Z. ii. 131. Stems much branched, 1–3 in. high; branches falcate, incurved, thickened upwards. Leaves dingy-green or brown or black, imbricate, vertical, erect, subopposite, orbicular; dorsal margin subinflexed, quite entire, decurrent; ventral toothed.—*J. falcata,* Hook. Musc. Exot. t. 89. *Plagiochila,* Syn. Hep. 649. *Aulicularia occlusa,* Fl. Antarct. 146. t. 62. f. 8; Syn. Hep. 619.

Northern and **Middle** Islands: in bogs, summit of the Tararua and Oparapara mountains, *Colenso;* Dusky Bay, *Menzies;* Port Preservation, *Lyall.* **Campbell's** Island: in bogs, *J. D. H.* (Tasmania.)

11. GYMNANTHE, Taylor.

Stems prostrate or ascending, vaguely branched. Leaves succubous, distichous. Stipules present or absent. Fruit terminal. Involucre a cylindrical pendulous tube, often fleshy and buried in the ground. Perianth 0 or adnate with the involucre. Capsule on a slender fruitstalk; elaters with 2 spiral fibres. Antheridia free in the axils of the leaves.

A small genus, with one species in the northern temperate zone and several in the southern.

* *Stipules present, very small.*

1. **G. unguiculata,** *Mitten in Fl. N. Z.* ii. 144. *t.* 99. *f.* 6. Stems club-shaped, creeping, with numerous white rootlets, ascending, 1 in. long. Leaves pale green or tinged brown or purple, distichous, densely imbricate, connivent and appressed, suberect, suborbicular or obovate, upper large, obscurely 12-lobed, deeply spinulose-dentate, with 8–10 large teeth; dorsal lobe smaller. Stipules variable, 2–3-fid or palmate.

Northern and **Middle** Islands: sulphur springs at Waimata, on a *Dicranum*, *Colenso*. A curious little plant; the stems appear clavate from the closely imbricate leaves being progressively larger upwards. Fruit unknown, but plant very similar to *G. Wilsoni* of Britain.

2. **G. diplophylla,** *Mitten in Fl. Tasm.* ii. 230. *t.* 179. *f.* 5. Stems short, ½–1 in. long, creeping. Leaves white, pellucid, membranous, cellular, closely imbricate, divaricating, conduplicate, unequally 2-lobed, upper lobe smaller, stipuliform, toothed and ciliate. Stipules 2-fid; segments toothed and ciliate. Perianth terminal, short, cylindric, purple.—*J. diplophylla*, Fl. Antarct. 152. t. 64. f. 4. *Gottschea*, Nees, Syn. Hep. 624.

Northern Island, *Sinclair*, *Kerr* (*in Herb. Mitten*). **Lord Auckland's** group: on tree-ferns, *J. D. H.*

** *Stipules* 0.

† *Leaves quite entire.*

3. **G. Drummondii,** *Mitten in Fl. N. Z.* ii. 144. *t.* 99. *f.* 8. Stems ½ in. long, stout, creeping, rooting at the tips. Leaves greenish-brown, imbricate, spreading, flat, upper larger, ovate, obtuse, quite entire; dorsal margin subrecurved. Involucre as long as the plant, clavate, with scattered rootlets. —*Riccia squamata*, Tayl. in Drummond's Swan River Mosses. *Podanthe squamata*, Tayl. in Hook. Lond. Journ. Bot. 1846, p. 413; Syn. Hep. 789.

Northern Island: forests of Titiokura, *Colenso*. (Australia and Tasmania.)

†† *Leaves 2-cuspidate, otherwise quite entire.*

4. **G. lophocoleoides,** *Mitten in Fl. N. Z.* ii. 144. *t.* 99. *f.* 7. Stem 1½ in. long, creeping, covered with rootlets below. Leaves pale yellow-green, distichous, spreading, flat, obovate-quadrate or cuneate, 2-lobed at the apex; lobes subulate. Involucre unknown.

Northern Island: Tararua mountains, creeping amongst *Plagiochila falcata*, *Colenso*. This quite resembles a *Lophocolea* in habit, and its genus must be very doubtful in the absence of fructification.

††† *Leaves more or less toothed.*

5. **G. setulosa,** *Mitten in Fl. N. Z.* ii. 144. *t.* 99. *f.* 5. Stem 1 in.

long, erect from a creeping rhizome, densely setose. Leaves bright yellow-green, spreading, 2-farious, closely imbricate, concave, ovate or obovate, unequally 2-lobed; margins waved and spinulose-toothed all round. Involucre unknown.

Northern Island: Tararua mountains, with *Plagiochila falcata*, *Colenso*.—A very curious plant, but in the absence of fruit a doubtful *Gymnanthe*.

6. **G. saccata,** *Tayl. in Fl. Antarct.* 153. Stems erect from a creeping rhizome, 2–3 in. long, flexuous, fertile incurved. Leaves distant, 3-farious, distichous, ovate-quadrate, flat, decurrent, truncate, emarginate; ventral margin and apex toothed; dorsal subreflexed, entire. Involucre terminal, fleshy.—Hook. Musc. Exot. t. 16. *G. Urvilleana*, Tayl. in Fl. Antarct. 153; Syn. Hep. 193. *G. tenella*, Hook. f. and Tayl. Fl. N. Z. ii. 143; Syn. Hep. 192 and 712; Fl. Tasman. ii. 229. t. 179. f. 3. *Plagiochila Urvilleana*, Mont. in Voy. au Pôle Sud, t. 16. *J. abbreviata*, Hook. f. and Tayl. in Lond. Journ. Bot. 1843, 374; Syn. Hep. 647 (*Plagiochila*).

Throughout the **Northern** and **Middle** Islands: abundant, from the Bay of Islands, *J. D. H.*, to Dusky Bay, *Menzies*. **Lord Auckland's** group, *J. D. H.* (Tasmania, Fuegia.)—A common and very variable plant, *G. tenella* seems to be a small state of it.

12. SACCOGYNA, Dumont.

Stems procumbent and rooting, vaguely branched. Leaves succubous, distichous, subhorizontal, entire. Stipules entire or toothed. Fruit lateral. Involucre fleshy, pendulous from the under side of the stem, giving off radicles, cylindric. Perianth 0. Calyptra adnate with the involucre, or with the apex free. Capsule on a slender fruitstalk; elaters with 2 spiral fibres. Antheridia in the axils of imbricating minute perigonial leaves on proper branches rising from the axils of the stipules.

This and a single European species, with the habit of *Lophocolea*, are the only known species.

1. **S. australis,** *Mitten in Fl. N. Z.* ii. 145. *t.* 100. *f.* 1. Stems 2–3 in. long, creeping. Leaves brownish-green or yellowish, distichous, horizontal, imbricating, obliquely ovate-oblong, 2-toothed at the apex or quite entire. Stipules small, 2-toothed, connate with the leaves below them. Involucre oblong, fleshy, with a few rootlets, crowned with a few small toothed leaves.

Northern Island: forests of Tararua, *Colenso*.

13. LEPIDOZIA, Nees.

Stems creeping, often very minute and slender, throwing out rootlets from the ventral surface. Leaves incubous, obscurely distichous, 4-toothed or 4-cleft. Stipules present. Fruit terminal (and antheridia) on proper branches from the under side of the stem. Involucral leaves numerous, short, broad, toothed. Perianth elongate, 3-angled, the faces hollow; mouth toothed. Calyptra membranous. Capsule on slender fruitstalks, globose; elaters with 2 spiral fibres. Antheridia solitary in the bases of conduplicate 2–3-cleft perigonial leaves.

A large genus, of very minute-leaved plants, often themselves minute and almost microscopic. The species are in many cases very badly defined, and I suspect that not a few of the following may be united hereafter. Very few species indeed inhabit the N. temperate zone. The larger resemble *Sendtnera* in general appearance.

§ 1. *Leaves 3–5-fid, but not toothed nor serrate.*

* *Stems flexuose, wiry, 1–4 in. long, much pinnately or 2-pinnately branched; branches decurved, attenuated.*

1. **L. microphylla,** *Lindb.;—Fl. N. Z.* ii. 145. Stems 1–2 in. long, pinnately branched; branches pendulous with capillary tips. Leaves minute, distant, appressed, palmately 4-partite; involucre oblong-ovate, 2–4-fid. Stipules quadrate, flat, deeply 4-fid. Perianth cylindric, attenuate, incurved, obsoletely toothed.—Hook. Musc. Exot. t. 80; Syn. Hep. 202.

Middle Island: Dusky Bay, *Menzies;* Otago, *Hector;* Port Preservation, *Lyall;* Otago, *Hector and Buchanan.* (Tasmania.)

2. **L. capilligera,** *Lindb.;—Fl. N. Z.* ii. 145. Stems 1 in. long, erect, simply pinnate; branches with capillary tips. Leaves olive-brown, subimbricate, ½ vertical or horizontal, and stipules obovate-quadrate or cuneate, 4-fid to the middle; lobes divaricate, subulate. Fruit unknown.—*L. tetrapila*, Tayl. in Lond. Journ. Bot. 1846, 370; Syn. Hep. 716.

Northern and **Middle** Islands: Tararua, with *Saccogyna australis, Colenso;* Otago, *Hector and Buchanan.* (Tasmania.)

3. **L. prænitens,** *Lehm. and Lindb.;—Fl. N. Z.* ii. 145. Stems procumbent, alternately 2-pinnate. Leaves subimbricate, ½-vertical, obovate-cuneate, 4-fid; segments lanceolate. Stipules patent, transversely quadrate, 4-partite; lobes divaricate. Perianth subsessile, curved; mouth sub-4-toothed.—Syn. Hep. 206.

Northern and **Middle** Islands: Auckland, *Sinclair;* Wellington, *Stephenson;* Tararua, *Colenso;* Dusky Bay, *Menzies.* **Lord Auckland's** group, *J. D. H.* (Tasmania.)

4. **L. Gottscheana,** *Lindb.;—Fl. N. Z.* ii. 145. Stems procumbent, irregularly subpinnately branched. Leaves approximate, subhorizontal, flat, obcuneate or quadrate, 4-fid; lobes subulate. Stipules remote, 4-partite. Perianth as in *L. prænitens.*—Syn. Hep. 206.

Middle Island: Dusky Bay, *Menzies.* (Tasmania.)—This and the two preceding seem not essentially different.

5. **L. lævifolia,** *Hook. f. and Tayl. Fl. Antarct.* 157;—*Fl. N. Z.* ii. 146. Stems 1–2 in. long, excessively pinnately branched; branches deflexed. Leaves yellow or olive-brown, subvertical, rather close-set, subimbricate on the branches, ovate-cordate, 3–5-fid; lobes flat or incurved; involucre small. Stipules distant, spreading, ovate, cordate or orbicular, 4-fid. Perianth as in *L. prænitens.*—Syn. Hep. 208.

Northern Island: probably common, Wellington, *Stephenson;* Port Nicholson, *Lyall.* Tehawera, *Colenso.* **Lord Auckland's** group and **Campbell's** Island, *J. D. H.* (Tasmania.)

6. **L. pendulina,** *Lindb.;—Fl. N. Z.* ii. 146. Stem 2–3 in. long, erect, 2-pinnate; branches pendulous, fascicled; tips capillary. Leaves imbricate,

subvertical, obliquely secund, and stipules orbicular-ovate, convex, deeply 4-fid; segments lanceolate, conniving, somewhat waved.—Syn. Hep. 208.

Northern and **Middle** Islands: top of the Ruahine range, *Colenso;* Dusky Bay, *Menzies;* Canterbury, *Haast;* Otago, *Hector and Buchanan.* (Tasmania.)

7. **L. spinosissima,** *Mitten in Fl. N. Z.* ii. 146. *t.* 100. *f.* 2. Stems 3 in. long, 2-pinnate; branches fastigiate, capillary at the ends. Leaves pale-brown, erecto-patent, distant, and stipules flat, cuneate or oblong-quadrate, 4-fid below the middle; segments subulate, rigid.—*Sendtnera,* Hook. f. and Tayl. in Lond. Journ. Bot. 1846, 373; Syn. Hep. 723.

Northern Island: common, *Edgerley, Colenso,* etc.

8. **L. filamentosa,** *Lindb.;—Syn. Hep.* 207. Stems suberect, pinnate; branches with capillary tips. Leaves remote, ½-vertical, decurrent, convex, orbicular, 3–4-fid; lobes broadly lanceolate, with incurved tips; involucre oblong, incised and toothed. Stipules ovate-quadrate, 3–4-fid. Perianth cylindric-pyriform; mouth toothed.—Mont. Voy. au Pôle Sud, 246.

Lord Auckland's group, *D'Urville.* A N.W. American and Fuegian plant, of which I have seen no N. Z. specimens.

** *Stems usually less than* 1 *in. long, capillary, vaguely branched.*

9. **L. dispar,** *Mont. Voy. au Pôle Sud,* 248;—*Fl. Antarct.* 158. Stem capillary, minute, ½ in. long, tufted; branches very long, capillary, spreading. Leaves vertical, distant, spreading, incurved, obovate-cuneate, narrow at the base, 3-fid to the middle; segments subulate subacute. Stipules similar, smaller. Perianth terminal, capitate, large.—Syn. Hep. 203.

Lord Auckland's group and **Campbell's** Island, *Hombron, J. D. H.*

10. **L. patentissima,** *Hook. f. and Tayl. Fl. Antarct.* i. 158. *t.* 65. *f.* 5. Stem minute, short, slender, ½ in. long, creeping, tufted, subpinnately branched; branches short. Leaves olive-green, imbricate, rather close-set, obliquely spreading, cellular, obovate-quadrate, narrowed at the base, 3–5-fid; segments shortly ovate-subulate, straight or incurved. Stipules similar, minute.—Syn. Hep. 204.

Lord Auckland's group: creeping on trunks of trees, etc., *J. D. H.*

11. **L. Lindenbergii,** *Gottsche;—Fl. N. Z.* ii. 146. Stem creeping, 1–2-pinnately branched; branches crowded, spreading. Leaves pale yellow-green, distant, distichous, subvertical, spreading, obovate-quadrate; segments capillary, articulate, straight and incurved; involucre unequally divided. Stipules orbicular-ovate, 3–5-parted. Perianth cylindric, elongate; mouth contracted, ciliate.—Syn. Hep. 213.

Northern and **Middle** Islands: probably common, from the Bay of Islands, *J. D. H.*, to Dusky Bay, *Menzies.*—This is the plant alluded to in Fl. Antarct. 158, under *L. tetradactyla,* as from New Zealand. (Tasmania.)

12. **L. capillaris,** *Lindb.;—Fl. N. Z.* ii. 146. Stems ½ in. long, creeping, capillary, pinnately branched or decompound; branches diverging. Leaves vertical, subimbricate, obovate-quadrate, and minute stipules 3–4-partite; segments lanceolate, subulate, incurved, obtuse; involucral shortly incised, ciliate; margins toothed. Perianth with the mouth ciliate.—Syn.

Hep. 212. *L. hippuroides*, Hook. f. and Tayl. Fl. Antarct. 159. t. 65. f. 7, and *I. nemoides*, Tayl. in Hook. Lond. Journ. Bot. 1845, p. 84.

Northern Island, Bay of Islands, *Sinclair*. **Lord Auckland's** group: on bark, etc., *J. D. H.*—A common tropical and southern plant, first described from West Indian specimens. (Jamaica, S. Africa, Tasmania, etc.)

13. **L. plumulosa,** *Lehm. and Lindb.*—*L. tetradactyla*, Hook. f. and Tayl. Fl. Antarct. 158. Stems 1 in. long, procumbent, pinnately decompound; branches subequal, narrowing towards their tips. Leaves close-set, subvertical, quadrate-obovate, deeply 3–6-fid; segments subulate. Stipules quadrate, spreading at the base, acutely 4-fid. Perianth cylindric; mouth contracted, toothed.—Syn. Hep. 211.

Lord Auckland's group: on bark, etc. *J. D. H.* (Fuegia.)—This was confounded in Fl. Antarct. with the *L. Lindenbergii*, which has a ciliate mouth to the perianth.

§ 2. *Leaves* 3–5-*fid and also toothed.*

14. **L. tenax,** *Lindb.*;—*Fl. Antarct.* 158. Stems pinnately compound or decompound; branches incurved; tips convolute. Leaves imbricate, vertical, ovate, concave, cauline appressed, 8-partite, sides lacerate and ciliate, those of the branches spreading, 3–4-fid; segments subulate. Stipules ovate, flat, 4–5-parted, ciliate and lacerate at the base.—Greville in Annals of New York Lyceum, i. 277. t. 23; Syn. Hep. 212.

Lord Auckland's group: on the ground and on trunks of trees, *J. D. H.* (Australia.)

15. **L. albula,** *Hook. f. and Tayl. Fl. Antarct.* 159. *t.* 65. *f.* 6. Stems ½–1 in. long, procumbent, subpinnately branched; branches decurved, narrowed to the tips. Leaves yellow or greenish-white, pellucid, densely imbricate, very broad, amplexicaul, spreading, very concave, cellular, obliquely-oblong, 4-fid and deeply incised all round; dorsal margin dilated; laciniæ entire or 2-fid, incurved. Stipules large, orbicular, concave, irregularly deeply 6–8-toothed.—Syn. Hep. 211.

Lord Auckland's group: creeping over other *Hepaticæ*, *J. D. H.*—Very much more beautiful and stouter than the other species, of a different habit and texture, but, owing to the absence of fruit, doubtful as to genus.

14. MASTIGOBRYUM, Nees.

Stems creeping and rooting or ascending, large, sparingly branched, giving off numerous filiform leafless shoots. Leaves distichous, incubous, usually 3-fid at the apex. Stipules toothed, often connate with the leaves above them. Fruit (and antheridia) terminal, on short proper branches, arising from the bracts of the stipules. Involucral leaves small, narrow, incised at the apex. Perianth elongate, 3-angular; mouth 3-toothed. Calyptra membranous. Capsule globose, on a slender fruitstalk; elaters with 2 spiral fibres. Antheridia 2 in the axil of each perigonial leaf.

A tropical and subtropical genus, rare in Europe and N. America, most abundant in Australia and New Zealand. The species are often broad and flat, some resembling *Lophocoleæ* in general habit, but the stipules are connate with the leaves above (not below) them.

§ 1. *Stipules quite free from the leaves.*

* *Leaves quite entire.*

1. **M. convexum,** *Lindb.;—Fl. N. Z.* i. 147. Small; stems procumbent, flexuous, subdichotomous; branches narrowed to the tips. Leaves imbricate at their bases, ½-vertical, ovate, retuse, entire or obscurely 3-denticulate. Stipules remote, free, ovate, 4-fid; segments acute, incurved. Perianth ovate, incurved; mouth plaited, toothed.—Syn. Hep. 215.

Northern Island: Tararua mountains, *Colenso.* (S. Africa, Australia, Peru, Mauritius.)

2. **M. tenacifolium,** *Hook. f. and Tayl. Fl. Antarct.* 152. *t.* 64. *f.* 6. Stems 1–1½ in. long, suberect, slender, stiff, rigid, flexuous. Leaves purple-brown or black, opaque, distant, spreading, rigid, elliptical-oblong, obtuse, quite entire, concave. Stipules smaller, entire or 2-fid.—Syn. Hep. 687.

Lord Auckland's group: on rocks near the hilltops —Mitten (Fl. N. Z. ii. 147) confidently refers this to the genus *Mastigobryum;* the fruit is unknown. (Fuegia.)

** *Leaves 2-dentate or 2-fid.*

3. **M. anisostomum,** *Lehm. and Lindb.;—Fl. N. Z.* ii. 146. Stems 1–2 in. long, suberect or creeping, dichotomous; branches decurved. Leaves olive-brown, opaque, horizontal or ½-vertical, lax or closely imbricate, concave, deflexed; those of the branches subsecund, triangular-ovate, unequally 2-dentate. Stipules minute, orbicular, 3-fid. Perianth cylindric; mouth 4-toothed.—Syn. Hep. 219. *M. atro-virens,* Hook. f. and Tayl. Fl. Antarct. 160; Syn. Hep. 218.

Middle Island: Dusky Bay, *Menzies;* Port Preservation, *Lyall.* **Lord Auckland's** group: creeping amongst mosses, *J. D. H.*

4. **M. Colensoanum,** *Mitten in Fl. N. Z.* ii. 147. *t.* 100. *f.* 3. Small; stem 1 in. long, procumbent, dichotomous, stoloniferous. Leaves pale green, membranous, spreading, flat, imbricate, oblong, 2-dentate, smaller tooth on the ventral side; sinus acute; dorsal margin arched, ventral straight. Stipules appressed, minute, 3-toothed.

Northern Island: Tararua, with *Saccogyna australis, Colenso.* (Australia, Tasmania.)

*** *Leaves 3-dentate.*

5. **M. Taylorianum,** *Mitten;—Fl. N. Z.* ii. 147. *t.* 100. *f.* 5. Small; stems ½–1 in. long, creeping, dichotomous. Leaves pale green, laxly imbricate, spreading, flat, obliquely oblong-quadrate, 3-dentate; dorsal margin arched, ventral straight, with a band of large translucent cells. Stipules small, orbicular, 4-toothed.

Northern Island: forests of Tehawera, *Colenso.*

6. **M. monilinerve,** *Nees;—Fl. N. Z.* ii. 148. Stem procumbent, dichotomous. Leaves approximate, ½-vertical, spreading, convex, obliquely oblong, 3-dentate; ventral margin with a band of large translucent cells. Stipules close-set, orbicular-ovate, crenulate. Perianth plicate above; mouth toothed.—Syn. Hep. 223.

Stewart's Island: on *Plagiochila annotina, Lyall.* (Australia, Tasmania.)

§ 2. *Stipules connate on both sides with the leaves above them.*

7. **M. novæ-Hollandiæ,** *Nees;—Fl. N. Z.* ii. 148. Stems 2–3 in. long, procumbent, dichotomous; branches equal, often recurved. Leaves imbricate, divergent, flat or convex, ovate-oblong, subfalcate, unequally serrulate, erose or dentate at the apex; involucral appressed, inciso-serrate. Stipules close-set, orbicular-quadrate, dentate or multifid, usually connate with the leaves above them. Perianth cylindric-ovate, narrowed upwards and plaited; mouth dentate.—Syn. Hep. 221. *M. adnexum,* Lehm. and Lindb.; Mont. Voy. au Pôle Sud, 243.

Common throughout the **Northern** and **Middle** Islands, and in **Lord Auckland's** group.—A common and very variable plant in the Southern Hemisphere.

8. **M. novæ-Zelandiæ,** *Mitten in Fl. N. Z.* ii. 148. *t.* 100. *f.* 6. Large; stems 2–3 in. long, procumbent, dichotomous, with long stout radicles below. Leaves olive-brown or -green, subopposite, distichous, spreading and deflexed, imbricate, obliquely ovate, 3-dentate; dorsal margin arched, ventral sinuate; stipules quadrate, spreading, crenulate, united to the leaves above them.

Northern and **Middle** Islands: forests of Tehawera and Tararua, *Colenso;* Auckland, *Sinclair;* Canterbury, *Haast.*—A noble species.

9. **M. involutum,** *Lindb.;—Fl. N. Z.* ii. 148. Large; stems 2–3 in. long, forked; branches dense. Leaves densely imbricate, diverging and deflexed, obliquely oblong, concave; apex 3-dentate, incurved or involute. Stipules suborbicular, repand; apex crenate, reflexed, connate with the leaves above them.—Syn. Hep. 221; Montagne in Voy. au Pôle Sud, t. 18. f. 2. *Herpetium,* Montagne, Cent. iv. n. 30.

Middle Island: Milford Sound, *Lyall;* Otago, *Hector and Buchanan.* **Lord Auckland's** group, *Hombron, J. D. H.* (Tasmania.)

10. **M. affine,** *Mitten in Fl. N. Z.* ii. 147. *t.* 100. *f.* 4. Stem 1 in. long, dichotomously branched, creeping, sending down long rootlets. Leaves dirty green, imbricate, spreading, flat, obliquely ovate, obliquely truncate, 3-crenate or -toothed; dorsal margin arched, ventral straight. Stipules transversely oblong, toothed and spinescent, connate with the leaves above.

Northern Island: forests of Tararua and Tehawera, *Colenso.*

11. **M. decrescens,** *Lehm. and Lindb.* Stem ½–1 in. long, creeping, branched, narrowed at the apex. Leaves imbricate, ½-ovate; dorsal margin subundulate, ventral margin decurrent, connate with the stipule; apex truncate, 3-toothed; teeth crenulate. Stipule reniform, spreading, quite entire. Perianth cylindric-ovate; apex narrow, plaited; mouth toothed.—Syn. Hep. 219; Mont. Voy. au Pôle Sud, 243. t. 19. f. 4 (*Herpetium*).

Lord Auckland's group, *Hombron.*—I have seen no New Zealand specimens of this, the description of which is taken from the 'Synopsis Hepaticarum.' (Mauritius.)

15. MICROPTERYGIUM, Lindenberg.

Stems creeping, with ascending flat branches, incurved or involute apices, and flagelli from the under surface. Leaves incubous, imbricate, complicate, kceled in the American but not in the only N. Z. species. Stipules entire or 2–4-fid. Fruits terminal on short lateral branches. Involucral leaves elongate, ciliate. Perianth long, membranous, triquetrous; mouth laciniate. Calyptra slender, membranous, included. Capsule on a slender fruitstalk; elaters with 2 spiral fibres. Antheridia as in *Mastigobryum?*

A small genus of temperate and tropical American species.

1. **M. nutans,** *Mitten in Fl. N. Z.* ii. 148. Stem 1 in. high, stout, tufted, erect from a creeping rhizome, thickened upwards, succulent, nearly simple; tips incurved or nodding. Leaves pale green or whitish, cellular, spreading, densely imbricate, concave, broadly ovate, unequally 2-fid, subentire; segments incurved, lower smaller. Stipules erect, broadly ovate, 2-fid or -toothed. Perianth lateral, large, lanceolate, trigonous above.—*Mastigobryum,* Fl. Antarct. 160. t. 65. f. 8; Syn. Hep. 219 and 717.

Northern Island: forests of Titiokura, *Colenso.* **Lord Auckland's** group: wet places, at the roots of trees, etc., *J. D. H.*—Mitten observes that the 2-fid leaves and absence of a keel are against its association with the American species, but its habit, etc., is similar, and widely different from that of *Mastigobryum.*

16. ISOTACHIS, Mitten.

Stem erect, branching with innovations, almost trifarious. Leaves incubous, imbricating, conduplicate, serrulate. Stipules nearly as large. Fruit terminal. Involucral leaves, inner minute, outer like the cauline. Perianth erect, tubular, fleshy, rigid; mouth contracted, toothed.

A southern genus, of which its author says it may be recognized by its evenly arranged leaves and stipules, the latter so closely resembling leaves that the leaves may almost be called trifarious. It resembles *Sendtnera* in habit, but differs in the form of the perianth and free calyptra.

1. **I. Lyallii,** *Mitten in Fl. N. Z.* ii. 149. *t.* 100. *f.* 7. Stem 2–4 in. high, erect; branches decurved at the tips. Leaves pale brown and purple, imbricate, subquadrate or ½-cordate, truncate; dorsal margin arched, toothed at the apex; ventral spinulose; segments toothed, recurved. Stipules obovate, 3–4-fid. Perianth subcylindric, obtuse, papillose.

Northern and **Middle** Islands: top of the Ruahine range, *Colenso;* Port Preservation, *Lyall.*—A tall and handsome plant.

2. **I. subtrifida,** *Hook. f. and Tayl. in Lond. Journ. Bot.* 1844, 579;—*Fl. N. Z.* ii. 149. Stem 1 in. high, slender, simple; tips decurved. Leaves pale brown, upper rosy, loosely imbricate, secund, erecto-patent, ovate, conduplicate, 2–3-fid; teeth and dorsal margin entire; ventral entire or toothed. Stipules similar, smaller.—Syn. Hep. 681; Fl. Tasman. t. 179. f. 7.

Northern Island: Bay of Islands, *J. D. H.* (Tasmania.)

3. **I. intortifolia,** *Hook. f. and Tayl. Fl. Antarct.* 150. *t.* 64. *f.* 1. Stems 1½–2 in. high, erect, flaccid, sparingly branched. Leaves purplish, spreading, imbricate, flaccid, cellular, very concave, ventricose, amplexicaul, orbicular oblong, unequally 2–4-fid, rarely entire; segments acuminate, incurved, toothed. Stipules imbricate, large, orbicular, very concave, emarginate 2-fid or toothed, membranous.—Fl. Tasman. ii. 233.

Campbell's Island: bogs on the hills, *J. D. H.* (Tasmania.)

17. TRICHOCOLEA, Nees.

Stems erect or inclined, tufted, much branched, very soft, white and woolly to appearance. Leaves incubous and distichous, but clothing the stem, deeply palmately divided, the lobes laciniated. Stipules present. Fruit in the forks of branches. Involucral leaves many, connate into a hairy tube, which is adnate with the calyptra, coriaceous; mouth truncate. Perianth 0. Capsule oblong, on a slender fruitstalk; elaters with 2 spiral fibres. Antheridia in the axils of leaves on the upper side of the stem.

A small genus of very beautiful pale *Hepaticæ*, with the leaves so much ciliate and cut as to give the whole stem a woolly appearance.

1. **T. tomentella,** *Nees;—Fl. N. Z.* ii. 153. Stems 2–5 in. long, 3-pinnate. Leaves white, 2-partite; segments capillary, multifid; ventral lobe smaller, inclined forwards. Stipules subquadrate, 4-partite, capillaceo-multifid.—Syn. Hep. 237. *T. mollissima,* Fl. Antarct. 161; Syn. Hep. 237. *T. tomentella, γ javanica,* Syn. Hep. 721.

Throughout the **Northern** and **Middle** Islands, abundant in damp woods, also in **Lord Auckland's** group and **Campbell's** Island. (Tasmania, India, Europe, and America.)

2. **T. lanata,** *Nees;—Fl. N. Z.* ii. 153. Stem 2–3 in. long, distantly simply pinnate. Leaves with the dorsal lobe incised, ciliate and laciniate. Stipules 4-fid to the middle only; segments capillaceo-multifid.—Syn. Hep. 238; Hook. Musc. Exot. t. 116.

Abundant throughout the **Northern** and **Middle** Islands, *Menzies,* etc.

3. **T. polyacantha,** *Hook. f. and Tayl. Fl. Antarct.* i. 161. *t.* 65. *f.* 9. Minute; stems ½ in. long, very slender, capillary, sparingly branched. Leaves very minute, brittle, olive-green, loosely imbricate, spreading, and 4-fid; segments multifid, setaceous, articulate, cellular. Perianth terminal, large, erect, subclavate, hispid, 4-toothed. Fruitstalk very long.—Syn. Hep. 239.

Lord Auckland's group: in woods near the sea, *J. D. H.*—A very doubtful *Trichocolea,* perhaps a *Lepidozia.*

18. SENDTNERA, Endlicher.

Stems erect or inclined, tufted, pinnately branched; branches often recurved, attenuated. Leaves obscurely distichous, incubous, closely imbricate, 2–5-cleft, segments entire. Stipules 2- or many-cleft. Fruit terminal on long branches, or lateral, subsessile. Involucral leaves numerous, incised, connate with the perianth. Perianth tubular, deeply cleft, membranous at the base. Calyptra chartaceous. Capsule globose, on a short fruitstalk; elaters free, with 2 spiral fibres. Antheridia in the axils of tumid perigonial leaves, on proper branches.

A small genus, scattered over various parts of the world.

I. Schisma, *Dumont.—Fruit terminal on long branches.*

1. **S. juniperina,** *Nees;—Fl. N. Z.* ii. 153. Stems 3–5 in. long, suberect, slender. Leaves pale brown, and stipules oblong, 2-fid, with an

obtuse sinus; segments lanceolate, acuminate, straight or slightly diverging. —Syn. Hep. 239; Hook. Brit. Jungerm. t. 4 et Suppl. t. 1.

Northern Island: top of the Ruahine range with *J. colorata, Colenso.* (Tasmania, India, Europe, etc.)

2. **S. attenuata,** *Mitten in Fl. N. Z.* ii. 153. *t.* 102. *f.* 2. Stems 3 in. long, pinnately branched; branches decurved, attenuated, flagelliform. Leaves pale yellow-brown, ovate, 2-fid; segments 2-partite, with subulate lobes, ventral margin entire, dorsal spinulose-toothed. Stipules similar, but deeper cut, toothed at the base. Fruit terminal. Involucre ovate, covered with leafy scales.

Northern Island: Bay of Islands, *J. D. H.*

3. **S. ochroleuca,** *Nees;—Fl. N. Z.* ii. 153. Stems variable in size, slender, creeping or ascending. Leaves and stipules ochreous-yellow, 3–5-fid, ciliate at the base; segments of the stipules channelled or flat. Perianth campanulate, covered with foliaceous scales.—Syn. Hep. 240. *J. hirsuta,* Nees; Fl. Antarct. 160 and 443.

Middle Island: Milford Sound and Port Preservation, *Lyall.* **Campbell's** Island, *J. D. H.; Otago, Hector and Buchanan.* (Fuegia, India, S. Africa, Mexico.)

II. MASTIGOPHORA.—*Fruit on short lateral branches or almost sessile.*

4. **S. scolopendra,** *Nees;—Fl. N. Z.* ii. 153. Stem 3–5 in. long, erect, pinnate; branches deflexed, attenuated to the tips, flagelliform, naked, often rooting. Leaves and stipules closely imbricate, scarious, rigid, oblong, 2-fid; segments diverging, again 2-fid, acuminate, diaphanous. Perianth in the axils of lateral branches, obovate, 4-fid, covered with imbricating leaves. —Hook. Musc. Exot. t. 40; Fl. Antarct. p. 160. *Schisma,* Nees, etc.

Middle Island: Dusky Bay, *Menzies.* **Lord Auckland's** group and **Campbell's** Island: abundant on the hills, *J. D. H.* (Tasmania.)

5. **S. flagellifera,** *Nees;—Fl. N. Z.* ii. 153. Stems 2–3 in. long, erect, pinnately branched; branches attenuate, flagelliform. Leaves greenish-brown, 2-farious, horizontal, unequally 2-fid; lobes conduplicate; dorsal larger, acute, often slightly toothed; ventral more lanceolate, entire or 2-fid. Stipules ovate, 2-partite, rarely 4-fid, toothed at the base.—Syn. Hep. 242; Hook. Musc. Exot. t. 59.

Northern and **Middle** Islands: common, from Bay of Islands, *Cunningham,* etc., to Dusky Bay, *Menzies.* (Tasmania.)

19. POLYOTUS, Gottsche.

Stems prostrate, pinnately branched. Leaves incubous, closely imbricate, auricled, the auricle often spinous, with a lamina of various shape between it and the blade of the leaf. Stipules usually 4-fid, the middle lobes clavate. Fruit lateral or axillary. Involucre of many confluent leaves. Perianth 0. Calyptra confluent with the involucre, its apex free, bearing sterile pistils near the apex. Capsule oblong, on short fruitstalk; elaters with 2 spiral fibres. Antheridia solitary in the axils of terminal perigonial leaves.

A small Australian, New Zealand, and Antarctic genus, not found in the northern hemispheres.

1. **P. claviger,** *Gottsche;—Fl. N. Z.* ii. 152. Stem creeping, 3–4 in. long, 2-pinnately branched. Leaves yellow-brown, closely imbricate, plane or convex, ovate-cordate, acute or apiculate, quite entire or more or less toothed; auricles clavate, with a large triangular lamina. Stipules 4-partite; segments with revolute margins, entire or toothed, 2 intermediate often saccate. Involucre cylindric-ovoid, rough with the adnate involucral leaves. —Syn. Hep. 215; Hook. Musc. Exot. t. 70.

Var. α. Leaves quite entire.
Var. β. *Stangeri.* Leaves of the branches toothed.
Var. γ. *Taylori.* Cauline leaves toothed and spinous on the ventral margin.—*P. Taylori,* Syn. Hep. 246.

Northern and **Middle** Islands: common from Auckland, *Sinclair,* to Dusky Bay, *Menzies.* **Campbell's** Island: on alpine rocks, *J. D. H.* (Tasmania.)—A very variable plant.

2. **P. palpebrifolius,** *Gottsche;—Fl. N. Z.* ii. 246. Stems 4–6 in. long, creeping, flexuose, 1–2-pinnate. Leaves brown, closely imbricate, ovate, spinous-ciliate; auricle club-shaped, split, with a spine above the fork, and interposed lamina ciliate and variable in form and lobing. Stipules 4-partite, ciliate, intermediate lobes on the branches saccate. Involucre axillary, with ciliated leaflets.—Syn. Hep. 246; Hook. Musc. Exot. t. 71.

Middle Island: Dusky Bay, *Menzies;* Thomson's Sound, *Lyall.* (Fuegia.)

3. **P. brachycladus,** *Gottsche;—Fl. N. Z.* ii. 152. Stem 3–4 in. long, 2-pinnate; branches very short. Leaves dark purple-brown, orbicular-cordate or broadly ovate, ciliate; auricles club-shaped, with no spine, saccate, purple, rarely flat and ciliate. Cauline stipules 4-partite, segments entire toothed or cut, those of the branches with 2 intermediate auricles. Involucre conical, coriaceous, leafy; leaflets spinous-ciliate.—Syn. Hep. 247; Fl. Tasm. ii. 234. t. 180. f. 2.

Northern Island: top of the Ruahine mountains, *Colenso.* (Tasmania.)

4. **P. Magellanicus,** *Gottsche;—Fl. N. Z.* ii. 153. Stem creeping, 3-pinnate. Leaves brown-purple, imbricate, ovate-orbicular, spinous-ciliate; cauline auricles variable, clavate, saccate, setigerous or plane, with laciniate segments; laminæ larger, 3-angular, sometimes subcucullate. Cauline stipules orbicular-ovate, 2-fid, entire or ciliate, those on the branches 2-fid lacerate or clavate. Involucre very large; leaves convolute, ciliate.—Syn. Hep. 248; Hook. Musc. Exot. t. 115.

Northern and **Middle** Islands: forests, Wairarapa river, *Colenso;* Chalky Bay, *Lyall.* **Campbell's** Island: on alpine rocks, *J. D. H.* (Fuegia, Tasmania, Australia.)

20. RADULA, Nees.

Stems erect or creeping, pinnately branched. Leaves distichous, incubous, 2-lobed; ventral lobe small, inflexed, rooting at the base. Stipules 0. Inflorescence monœcious. Fruit in the fork or apex of short branches. Involucral leaves 2, 2-lobed. Perianth terete or compressed; mouth dilated. Calyptra pyriform, persistent, bursting below the apex. Capsule on a short fruitstalk, ovoid; elaters with 2 spiral fibres; spores large. Antheridia in the inflated bases of perigonial leaves.

A considerable genus, found in all parts of the world.

1. **R. buccinifera,** *Hook. f. and Tayl. Fl. N. Z.* ii. 154. Stem 1 in. long, prostrate, pinnate. Leaves yellow-green or brown, subimbricate, spreading, rounded at the apex, upper lobe obovate-orbicular, convex, its apex incurved, lower minute, trapezoid, appressed. Perianth at length axillary, very long, subcylindric, compressêd above; mouth dilated, quite entire.—Syn. Hep. 261.

Northern and **Middle** Islands: common in forests on the mountains, *Colenso*, etc.; Otago, *Hector and Buchanan.* (Tasmania.)

2. **R. physoloba,** *Mont.*;—*Fl. N. Z.* ii. 154. Stem procumbent, rigid, flexuous, much pinnately branched. Leaves yellow-green or brown, subvertical, orbicular, convex, quite entire; apex inflexed; lobule large, inflated below the apex, retuse or emarginate. Perianth terminal or axillary, long, cochleariform, truncate.—Syn. Hep. 254; Mont. Voy. au Pôle Sud, 256. t. 17. f. 4. *L. aquilegia*, Hook. f. and Tayl. in Lond. Journ. Bot. 1846, 291.

Northern Island: forests of the Manawata, *Colenso.* **Lord Auckland's** group, *Hombron.* (Tasmania.)

3. **R. uvifera,** *Hook. f. and Tayl. Fl. Antarct.* 162;—*Fl. N. Z.* ii. 154. Stem 1–2 in. long, stout, much branched. Leaves olive-brown or purplish, imbricate, convex, tumid, obovate-orbicular, quite entire; upper lobe larger, convex, broadly oblong-orbicular; apex recurved; lower ovate-oblong, obtuse, tumid below, its apex appressed to the upper. Perianth obovate-oblong, much compressed, subcochleariform, 4–5-costate below; margins subulate; mouth truncate, entire.—Syn. Hep. 258, 729. *R. multicarinata*, Syn. Hep. 258.

Northern Island: in mountainous districts, common, *Colenso*, etc. **Lord Auckland's** group: abundant on bark of trees, *J. D. H.*

4. **R. plicata,** *Mitten in Fl. N. Z.* ii. 154. Stem 1 in. long, slender, procumbent, pinnate. Leaves imbricate, orbicular-obovate, as in *R. buccinifera.* Perianth linear-elongate, flattened, with 3 keels or ribs on the ventral side, and 2 on the dorsal; mouth truncate, quite entire.

Northern Island: Auckland, with *R. buccinifera*, *Sinclair.* Closely allied to *R. buccinifera*, but larger and with a very different perianth.

5. **R. complanata,** *Dumortier*;—*Fl. N. Z.* ii. 155. Stem 1–2 in. long, flattened, creeping, subpinnate. Leaves pale greenish-brown, distichous, closely imbricate; upper lobe orbicular, almost flat, quite entire; lower lobe $\frac{1}{4}$ smaller, appressed with a rounded angle. Perianth flattened, oblong, truncate.—Hook. Brit. Jungerm. t. 81; Syn. Hep. 257.

Northern Island: Raukawa, on trunks of trees, *Colenso.* **Campbell's** Island, *J. D. H.* —The Raukawa plant is a small form, without fruit, bearing buds on the edges of the leaves, as is common in Europe. (A very common plant in Europe, America, S. Africa, and India.)

6. **R. marginata,** *Hook. f. and Tayl. Fl. N. Z.* ii. 154. Stems flattened,

1–3 in. long, vaguely branched. Leaves dirty yellow or olive-green, subimbricate, distichous, spreading; upper lobe oblong-orbicular, with a thickened quite entire margin; lower lobes obliquely ovate-lanceolate, obtuse, suberect, appressed. Perianth terminal, elongate, flattened, narrowed below; margins thickened; lips truncate, quite entire.—Syn. Hep. 261.

Northern and **Middle** Islands: Bay of Islands, *J. D. H.*; Auckland, *Sinclair*; Nelson, *Mantell.*—The thickened margins of the leaf at once distinguish this curious species.

7. **R. dentata,** *Mitten, mss.*—*Lejeunia dentata*, Mitten in Fl. N. Z. ii. 159. Leaves orbicular, unequally 2-dentate, with an obtuse sinus, dorsal margin 3-dentate; ventral quite entire; lobule large, ovate-oblong, obtuse, quite entire.

Northern Island, *Herb. Mitten*; Auckland, *Sinclair.*

21. MADOTHECA, Dumortier.

Stems creeping or pendulous, often large, pinnately branched. Leaves distichous, incubous, with an inflexed membranous lobule at the ventral base. Stipules large, decurrent. Inflorescence diœcious. Fruit lateral, nearly sessile. Involucral leaves, 2 or 4, 2-lobed. Perianth obovoid, 2-convex; mouth 2-lipped, entire or lacerate. Calyptra globose, persistent, bursting below the apex. Capsules on short fruitstalks, globose; elaters free, with 2 spiral fibres; spores large, angular. Antheridia as in *Frullania.*

A considerable and widely distributed genus, both tropical and temperate.

1. **M. Stangeri,** *Gottsche*;—*Fl. N. Z.* ii. 155. Stems 3–12 in. long, erect; branches spreading. Leaves yellow-green, closely imbricate, very convex, decurved, suborbicular; apex incurved, quite entire or with the ventral margin waved; lobule triangular-ovate, outer side somewhat reflexed; involucral unequal, toothed. Stipules ovate, obtuse, gibbous in the centre; margins subrecurved, quite entire. Perianth lateral, sessile, 2-lipped; lips toothed.—Syn. Hep. 281. *M. partita*, Hook. f. and Tayl. in Lond. Journ. Bot. 1844, 392; Syn. Hep. 279. *M. elegantula*, Mont. Voy. au Pôle Sud, 232. t. 18. f. 3.

Throughout the **Northern** and **Middle** Islands: abundant, *Menzies*, etc. **Lord Auckland's** group: on trunks of trees, *J. D. H.*—A very fine but variable species, the southern representative of the European *M. platyphylla*, Dum.

22. LEJEUNIA, Libert.

Stems prostrate or creeping. Leaves distichous, incubous. Stipules usually present. Inflorescence diœcious. Fruit lateral or terminal on proper branches. Involucral leaves 2, 2-lobed. Perianth ovoid or obovoid, terete or angled; angles winged or crested; mouth 3–4-lobed. Calyptra obovoid, persistent, bursting below the apex. Capsule on short fruitstalk, globose, pale, 4-cleft halfway; elaters adhering to the valves, erect, upper end truncate, with 1 spiral fibre. Antheridia as in *Frullania*, on proper branches.

A very large tropical and temperate genus, of which 236 species were published in the 'Synopsis Hepaticarum' in 1844, and many have been added since.

§ 1. *Stipules* 0.

1. **L. lævigata,** *Mitten in Fl. N. Z.* ii. 157. Stem 1 in. long, creeping, flattened and closely appressed, pinnate. Leaves imbricate, suborbicular, obtuse; lobule oblong or obovate, 3-dentate, one-quarter smaller than the leaf; involucral narrower, with obtuse entire lobes. Perianth obconic or turbinate, obtuse; dorsal face flat; ventral with 2 appressed folds.

Northern Islands: on fronds of ferns, leaves, etc.; Auckland, *Sinclair;* Tehawera, *Colenso.*

2. **L. pulchella,** *Mitten in Fl. N. Z.* ii. 157. *t.* 103. *f.* 2. Stem very minute, $\frac{1}{12}$–$\frac{1}{8}$ in. long. Leaves white, membranous, remote, incurved recurved or flexuous, oblong, obtuse, quite entire; lobule small, subquadrate, 2–3-dentate; involucral smaller, narrower. Perianth lateral, flattened, obconic or obcordate, retuse; margins quite entire.

Northern Island: Manawata river, *Colenso.*

§ 2. *Stipules large, suborbicular, quite entire or very inconspicuously toothed.*

3. **L. olivacea,** *Hook. f. and Tayl. Fl. N. Z.* ii. 157. Stems 1 in. long, procumbent, subpinnate. Leaves brownish, wrinkled when dry, imbricate, spreading, oblong-ovate, rounded, incurved and often white at the apex. lobule small, oblong, acute, tumid, quite entire; involucral narrower, acute. Stipules large, orbicular-obovate, obscurely emarginate. Perianth oblong-obovate, flattened, 3-ribbed on either face.—Syn. Hep. 334. *Phragmicoma olivacea*, Mitten, mss.

Northern Island: Wellington, *Stephenson, Colenso;* Auckland, *Sinclair.*

4. **L. scutellata,** *Mitten in Fl. N. Z.* ii. 155. *t.* 102. *f.* 4. Stem 1–2 in. long, tufted; branches pinnate, ascending. Leaves olive-green or brown, imbricate, spreading, convex, ovate-oblong, quite entire; apex acute, recurved, sinuate at the base; lobule small, quite entire, orbicular, involute; involucral immersed, ovate-lanceolate, nearly entire; lobules lanceolate. Stipules orbicular-obovate, with lanceolate lobules; involucral 2-fid. Perianth axillary, oblong, with entire angles.—*Thysananthus scutellatus*, Hook. f. and Tayl. in Lond. Journ. Bot. 1846, 383; Syn. Hep. 739.

Northern and **Middle** Islands: Bay of Islands, *Cunningham*, etc.; Nelson, *Mantell;* Otago, *Hector and Buchanan.*

5. **L. anguiformis,** *Hook. f. and Tayl. Fl. N. Z.* ii. 156. *t.* 102. *f.* 5. Stem 1–1½ in. high. Leaves olive-green or brownish, imbricate, spreading, oblong-ovate, quite entire; apex obtuse or acute and ventral margin incurved; lobule half as long, narrow, elongate, rounded at the apex, quite entire; involucral broader, with an acute lobule. Stipules larger, orbicular, broader than long, quite entire. Perianth obovate, compressed; margins and apex toothed; faces obscurely keeled.—*Thysananthus anguiformis*, Hook. f. and Tayl. in Lond. Journ. Bot. 1844, 567; Syn. Hep. 289.

Northern Island: Tararua, on trees, *Colenso.*

Two species, of which descriptions were published by Dr. Taylor, but of which the specimens were not returned to the Hookerian herbarium, are referred with probability, by Mr. Mitten, to this, viz. *L. mollis*, Mitten, Fl. N. Z. ii. 157 (*Ptychanthus*, Hook. f. and Tayl. l. c. 334), with leaves sinuate, plicate at the base, and stipules smaller (Bay of Islands,

J. D. H.); and *Z. ophiocephala*, Mitten, Fl. N. Z. ii. 157 (*Thysananthus*, Hook, f. Tayl. l. c. 384), with an acute lobule and stipules with recurved tips (Bay of Islands, *Sinclair*). To these may be added *L. plicatiscypha*, Gottsche, Syn. Hep. 748; Fl. N. Z. ii. 156 (*Phragmicoma*, Hook. f. and Tayl. l. c.).

§ 3. *Stipules large, many-toothed.*

6. **L. Stephensoniana,** *Mitten in Fl. N. Z.* ii. 155. *t.* 102. *f.* 3. Stems 1½ in. long, dichotomously branched. Leaves olive-brown, imbricate, oblong-ovate, acute, quite entire, flat or with the apex incurved or involute; ventral margin incurved; lobule small, toothed. Stipules obovate, subtruncate, toothed; margins slightly recurved.

Northern Island: Wellington, *Stephenson.*—Described from a fragment picked from off mosses sent to Mr. Mitten.

§ 4. *Stipules 2-dentate or 2-fid.*

a. Leaves ciliolate or papillose.

7. **L. papillata,** *Mitten in Fl. N. Z.* ii. 158. *t.* 103. *f.* 5. Stems ⅛–⅓ in. long, creeping, branched. Leaves pale green, rather remote, spreading, deflexed, falcate, ovate-acinaciform, rounded at the incurved apex, ciliolate, with papillæ all round; dorsal margin convex; ventral rather concave; lobule small, inflated; involucral smaller, with an ovate lobule. Stipules not ciliate, orbicular, 2-fid; segments 1–2-toothed. Perianth obovate-quadrate, retuse, compressed, 5-gonous; angles papillose.

Northern Island: creeping on fern fronds and mosses, *Sinclair*, *Colenso*; Wellington, *Stephenson.*

β. Leaves not ciliate, very obtuse, rarely acute in L. rufescens. *Stipules minute.*

8. **L. rufescens,** *Lindenberg.*—*L. implexicaulis*, Fl. N. Z. ii. 158. Stems ¼–¾ in. long, creeping, branched. Leaves olive-green, cellular, imbricate, spreading or suberect, oblong-obovate, orbicular-ovate or ovate, convex; apex incurved, obtuse subacute or truncate; lobule small, involute. Stipules usually narrower than the branch, more or less orbicular, acutely 2-fid. Perianth lateral, sessile, compressed, with 1 keel on the dorsal and 2 on the ventral side.—Syn. Hep. 366. *L. implexicaulis*, *mimosa*, *albo-virens*, and *primordialis*, Hook. f. and Tayl. in Lond. Journ. Bot. 1844, 397, 398, and 1845, 92; Fl. Antarct. 166. t. 66. f. 4.

Northern and **Middle** Islands, probably common, creeping amongst mosses, etc.; Manawata river, *Colenso*; Wellington, *Stephenson*; Canterbury, *Travers*. **Lord Auckland's** group and **Campbell's** Island, *J. D. H.* (Fuegia, Tasmania.)—Probably a very common southern species (of which *L. primordialis* is an exceedingly small variety), and the representative of the European *L. serpyllifolia.*

9. **L. cucullata,** *Nees.* Stems very minute, capillary, almost microscopic, creeping, branched. Leaves pale green, lax, erecto-patent, concave, orbicular-quadrate, subtruncate, quite entire; lobule as large as the leaf, incurved, appressed; summit angled; base tumid. Stipules most minute, 2-fid; segments linear, subincurved.—Syn. Hep. 389. *L. plicatiloba*, Hook. f. and Tayl. Fl. Antarct. 166; Syn. Hep. 369; Mont. Voy. au Pôle Sud, 218.

Lord Auckland's group, *Hombron*; creeping on *Parmelia*, *J. D. H.*

γ. Leaves not ciliate, sometimes papillose at the edge, subacute or acute. Stipules minute.

10. **L. latitans,** *Hook. f. and Tayl. Fl. N. Z.* ii. 159. Stems minute, creeping, slender, $\frac{1}{4}$–$\frac{1}{3}$ in. long. Leaves pale green, distant, erecto-patent, ovate, acute or acuminate; margins papillose; lobule ovate, involute, half the size of the leaf; involucral united with the obovate involucral stipule. Stipules minute, 2-fid; segments lanceolate, obtuse.—Syn. Hep. 345; Fl. Antarct. 166.

Northern Island: on moist banks, *Colenso*. **Lord Auckland's** group: creeping on mosses, etc., *J. D. H.*

11. **L. comitans,** *Hook. f. and Tayl. Fl. N. Z.* ii. 159. Stems minute, $\frac{1}{10}$–$\frac{1}{8}$ in. long, procumbent, with thickened shoots. Leaves pale olive-green, closely imbricate, spreading, minutely cellular, tumid, oblong, subacute, quite entire; apex recurved; lobule ovate, inflexed, subacute. Stipules minute, distant, orbicular, 2-fid. Perianth axillary, obcordate, flattened dorsally, with 1 broad rib on the ventral face.—Syn. Hep. 760.

Northern Island: Bay of Islands, with *L. mollis, J. D. H.*

δ. Leaves not ciliate, obtuse. Stipules rather large, conspicuous.

12. **L. nudipes,** *Hook. f. and Tayl. Fl. N. Z.* ii. 158. *t.* 103. *f.* 4. Stems minute, creeping, branched, $\frac{1}{4}$ in. long. Leaves pale green, distant, spreading, concave, opaque, obovate, narrowed at the base, rounded at the apex, subcrenulate; lobule small, narrow, subacute; involucral similar. Stipules ovate, acutely 2-fid, involucral obovate with ovate segments. Perianth clavate, narrow below and substipitate, 5-gonous, retuse, with crenulate angles.—Syn. Hep. 372.

Northern Island: Bay of Islands, creeping over mosses, *J. D. H.*

14. **L. thymifolia,** *Nees;—Fl. N. Z.* 158. Stem creeping, branched. Leaves yellow, imbricate, oval, narrowed at the obtuse apex, quite entire; lobule orbicular, truncate; involucral ovate-lanceolate, obtuse, with a tongue-shaped lobule. Stipules large, imbricate, ovate-cordate, 2-fid. Perianth terminal and lateral, obovate, 5-gonous.—Syn. Hep. 373.

Northern Island: Bay of Islands, on bark, *Sinclair;* Cape Turnagain, *Colenso.* (Java, S. America, Madeira.)

15. **L. tumida,** *Mitten in Fl. N. Z.* ii. 157. *t.* 103. *f.* 3. Stems very slender, $\frac{1}{4}$–$\frac{1}{2}$ in. long, much branched. Leaves bright green, loosely imbricate, cells pellucid, very convex, spreading, rather horizontal, obliquely ovate-oblong, rounded at the apex, quite entire; lobule small, inflexed, inflated; involucral with a longer obtuse lobule. Stipules orbicular, acutely 2-fid. Perianth obovoid, tumid, retuse, inflated and obscurely 5-angled at the summit.

Northern Island: probably common on trees, *Sinclair, Colenso, Stephenson,* etc. **Campbell's** Island, *J. D. H.* (Tasmania, Australia.)

24. FRULLANIA, Raddi.

Stems prostrate or creeping, flattened, usually purplish-brown. Leaves distichous, incubous, convex, quite entire, with a simple or rarely double lobule at or near the base, which is erect or appressed, club-shaped lunate trumpet-shaped or inflated. Stipules sometimes rooting at the base, usually 2-fid. Inflorescence diœcious. Fruit terminal on proper branches. Involucral leaves 2 or 3, not auricled. Perianth ovoid or obovoid, terete or 3–4-angled; mouth contracted, tubular. Calyptra pyriform, persistent, bursting below the apex. Capsule on a very short fruitstalk, globose, 4-cleft halfway; elaters adhering to the valves, truncate, with 1 spiral fibre; spores large, irregular. Antheridia in the saccate bases of closely imbricate 2-lobed perigonial leaves. Pistillidia 2 or 4 in each perianth.

A very large genus, growing on rocks, bark of trees, etc.; a few species are European, but most are southern.

§ 1. *Lobule lunate or kidney-shaped, rarely hemispherical, usually longer than broad, more or less parallel to the curvature of the leaf.*

* *Lobules 2. Stipules quite entire.*

1. **F. cornigera,** *Mitten in Fl. N. Z.* ii. 163. *t.* 104. *f.* 8. Stems 1–2 in. long, pinnate. Leaves olive-brown, opaque, suborbicular, with hyaline margins; lobules 2, brown, tubular, decurved. Stipules transversely oblong, quite entire.

Northern and **Middle** Islands: Bay of Islands, *J. D. H.*; Port Cooper, *Lyall.*

** *Lobule solitary. Base of the leaf inflexed. Stipules 2-fid.*

2. **F. patula,** *Mitten in Fl. N. Z.* ii. 159. *t.* 104. *f.* 1. Stem 1 in. long, pinnate. Leaves olive-brown, loosely imbricate, flattish, orbicular; ventral base broadly inflexed; lobule small, brown, reniform; involucral elongate, ovate, obtuse, ventricose below and sheathing the perianth. Stipules orbicular-cordate, very large, shortly 2-fid. Perianth small, quite smooth.

Northern Island: Tarawera, on dead bark, Manawata river, on fences, etc., *Colenso.*

3. **F. squarrosula,** *Hook. f. and Tayl. Fl. N. Z.* ii. 160. *t.* 103. *f.* 6. Stems 1 in. long, 2-pinnate. Leaves olive-green or almost black, patent and recurved, squarrose, ovate, rounded at the apex; dorsal margin often white; ventral sinuate in the middle; lobule small, on the incurved margin of the leaf kidney-shaped or hooded, brown; involucral with an ovate acute lobule. Stipules orbicular-ovate, 2-fid. Perianth oblong-ovate, obtuse, smooth at the back, convex and obscurely 2-keeled above; ventral face with one keel.—Syn. Hep. 412.

Northern Island: probably common, Bay of Islands, on Lichen and rocks, *J. D. H.*; Port Nicholson and Titiokura, etc., *Colenso*; Wellington, *Stephenson.*

4. **F. pycnantha,** *Hook. f. and Tayl. Fl. N. Z.* ii. 160. Very similar to *F. squarrosula*, but the perianth is covered with minute foliaceous scales, the stipules keeled down the middle, with subrecurved margins, the involucral stipule and leaves more cut.—Syn. Hep. 411.

Northern and **Middle** Islands: probably common, on Licheus, etc., from the Bay of Islands, *J. D. H.*, to Thomson's Sound, *Lyall.*

5. **F. scandens,** *Mont. Voy. au Pôle Sud*, 227. *t.* 19. *f.* 2;—*Fl. Antarct.* 165. Stem 1½ in. long; branches pinnate, very short. Leaves pale

green, closely imbricate, obliquely ovate-orbicular, quite entire, obtuse, concave, lower margin incurved?; lobule lunate; involucral not seen; stipule orbicular, 2-dentate.

Lord Auckland's group: on branches of trees, *Hombron*.—I have seen no specimens, and am uncertain, from the figure, whether the base of the leaf is incurved or not.

*** *Lobule solitary. Base of leaf not inflexed. Stipules 2-fid* (*toothed in* F. Hampeana).

6. **F. falciloba,** *Hook. f. and Tayl. Fl. N. Z.* ii. 160. Stems 1–2 in. long, branched. Leaves brown, ovate or orbicular-oblong, slightly incurved; lobule large, elongate, falcate; involucral 2-lobed; lobule much cut. Stipules orbicular-oblong, 2-dentate; involucral large, 2-fid, 1-dentate on each side. Perianth convex on the dorsal, and having a broad keel on the ventral face.—Syn. Hep. 423.

Northern Island: Wairarapa valley, on bark, *Colenso*. (Tasmania.)

7. **F. cranialis,** *Tayl.* Stems 1–2 in. long, tufted, procumbent, subpinnately branched. Leaves olive-brown, loosely imbricate, spreading, rounded, concave, quite entire; lobule helmet-shaped, ⅓ the size of the upper lobe. Stipules ovate, 2-fid, with one tooth on each side.—Tayl. in Hook. Lond. Journ. Bot. 1845, 86.

Middle Island: Canterbury, *Haast*. (Australia.)

8. **F. Hampeana,** *Nees;—Fl. N. Z.* ii. 160. Stems pinnate. Leaves pale green, imbricate, ½-vertical, suborbicular; lobule arched, deflexed, acuminate. Stipules suborbicular, 6–8-toothed and 2-fid.—Syn. Hep. 426.

Northern Island: trunks of trees, Wairarapa valley, *Colenso*. (Australia, Tasmania.)—Readily recognized by the many-toothed stipules.

9. **F. spinifera,** *Hook. f. and Tayl. Fl. N. Z.* ii. 161. *t.* 104. *f.* 2. Stems 1–1½ in. long, pinnate. Leaves brownish-green, loosely imbricate, diverging, broadly oblong; lobule small, arched, with acuminate ends; involucral lanceolate, acute, with a lanceolate subulate 1-dentate lobule. Stipules orbicular, shortly 2-fid; involucral broadly 2-fid, with lanceolate lobes. Perianth elongate, subcylindric, tumid, smooth; retuse with a mucronate apex.—Syn. Hep. 776.

Northern Island: Auckland, *Sinclair*; Tehawera, on trees in the forest, *Colenso*.

10. **F. deplanata,** *Mitten in Fl. N. Z.* ii. 161. *t.* 104. *f.* 3. Stem 1 in. long, pinnate. Leaves red-brown, diverging, imbricate, orbicular-ovate or oblong; apex rounded, incurved; lobule large, falcate, recurved, acuminate; involucral ovate, acute, entire or toothed, with a lanceolate toothed lobule. Stipules suborbicular, shortly 2-fid; involucral elongate, laciniate. Perianth obcordate, elongate, flattened at the top, retuse, mucronate, smooth.

Northern Island: east coast and interior, on bark, *Colenso*; Wellington, on Lichens, *Stephenson*.

§ 2. *Lobule vertically elongate, clavate tubular trumpet-shaped or ½-orbicular.*

* *Lobule on the surface of the leaf, ½-orbicular.*

11. **F. reptans,** *Mitten in Fl. N. Z.* ii. 161. *t.* 104. *f.* 4. Stem small, slender, ½ in. long, creeping, pinnate. Leaves dark olive-brown or red or

blackish, loosely imbricate, spreading, orbicular-obovate; lobule very large, occupying the centre of the leaf, and $\frac{1}{3}$ smaller, compressed, semiorbicular, black; involucral with a lanceolate 1-dentate lobule. Stipules small, cuneate, 4-toothed; involucral 2-fid, 1-dentate on each side. Perianth oblong-obovate, mucronate; dorsal face flat, 2-keeled; ventral convex, 4-keeled; keels toothed above.

Northern Island: East coast, on bark of *Edwardsia*, *Colenso;* Auckland, *Sinclair.* (Tasmania.)

12. **F. pentapleura,** *Hook. f. and Tayl. Fl. N. Z.* ii. 162. *t.* 104. *f.* 7. Stems small, slender, $\frac{1}{6}$–$\frac{1}{8}$ in. long, pinnate. Leaves red-brown or blackish or olive-green, spreading, orbicular; lobule on the face of the leaf, compressed, $\frac{1}{2}$-orbicular; involucral ovate, with an ovate 1-dentate lobule. Stipules small, suborbicular, 2-fid, entire or dentate on each side. Perianth oblong-obovate, mucronate, compressed, 2-keeled on each face; keels and margins waved.—Syn. Hep. 775.

Northern Island: Tarawera, in woods, *Colenso;* Auckland, *Sinclair.*

** *Lobule on the margin of the leaf, cylindric, with a dilated trumpet-shaped mouth.*

13. **F. ptychantha,** *Montagne;—Fl. N. Z.* ii. 163. Stems 2 in. long, creeping, pinnate, thickened upwards. Leaves from olive-green to red-purple, loosely imbricate, very tender; cells minute, obovate-cuneate; lobule small, pedicelled, oblong, inflated, opening downwards; involucral acuminate, quite entire, united with the stipular. Stipules ovate-oblong, $\frac{1}{2}$ as large as the leaf, deeply acutely 2-fid. Perianth exserted, pyriform, 9-plicate.—Mont. in Voy. Pôle Sud, 225. t. 19. f. 3; Syn. Hep. 442. *F. myosota*, Hook. f. and Tayl. in Lond. Journ. Bot. 1844, 393.

Northern and **Middle** Islands: Wellington, *Stephenson, Lyall.* **Lord Auckland's** group and **Campbell's** Island, *Hombron, J. D. H.*

14. **F. hypoleuca,** *Nees;—Fl. N. Z.* ii. 163. General characters of *F. ptychantha*, but the leaves are closely imbricate, broader; the lobule distant from the stem, with an interposed triangular lamina, the involucral leaves subserrate, and the perianth concave on the dorsal face, 1-keeled on the ventral. —Syn. Hep. 443.

Northern Island: Wellington, on Lichens, *Stephenson* (*in Herb. Mitten*). (Oahu.)

15. **F. fugax,** *Hook. f. and Tayl. Fl. N. Z.* ii. 161. *t.* 104. *f.* 5. Stem small, slender, creeping, pinnate, $\frac{1}{2}$ in. long. Leaves red-brown, imbricate, diverging, orbicular; lobule very large, cylindric-oblong, with a dilated mouth below; involucral elongate, obtuse, with a subulate-lanceolate lobule. Stipules small, 2-fid, and 1–2-toothed; involucral more deeply 2-fid; segments entire or toothed. Perianth sessile, oblong; dorsal face flat, 2-keeled; ventral with 1 broad keel.—Syn. Hep. 445.

Northern Island: probably common on Lichens, Bay of Islands, *J. D. H.;* east coast, etc., *Colenso;* Wellington, *Stephenson.*—The perianths here described are not fully formed.

16. **F. incumbens,** *Mitten in Fl. N. Z.* ii. 162. *t.* 104. *f.* 6. Very similar in size and structure to *F. fugax*, but paler. Leaves more incurved; lobule very much smaller. Stipules not toothed, and with the obovate perianth keeled sharply on its ventral face.

Northern Island: on bark, Wairarapa valley, east coast and interior, *Colenso.*

17. **F. rostrata,** *Hook. f. and Tayl. Fl. Antarct.* 163. Stems minute, slender, subpinnate. Leaves red-brown, imbricate, spreading, orbicular, subapiculate; lobule oblong, with 2 short processes on its inner margin; involucral orbicular-oblong, incurved, apiculate. Stipule minute, orbicular. Perianth obovate from a slender narrow base, much broader than the stem.—Syn. Hep. 445.

Lord Auckland's group: on *Parmelia enteromorpha, J. D. H.*

18. **F. magellanica,** *Spreng.;—Fl. Antarct.* 162. Stems 2–3 in. long, much 2-pinnately branched. Leaves red and olive-brown or blackish, imbricate, orbicular-ovate, sometimes obscurely denticulate; lobule obovate, distant from the stem; involucral acute, entire, rarely with the lobule toothed. Stipule minute, appressed, 2-fid. Perianth obovate, impressed on the dorsal face, obtusely keeled on the ventral.—Syn. Hep. 446.

Campbell's Island: on alpine rocks, *J. D. H.* (Fuegia.)

*** *Lobule on the margin of the leaf, erect or diverging, clavate or pedicelled.*

19. **F. aterrima,** *Hook. f. and Tayl. Fl. N. Z.* ii. 162. Stems ½–1 in. high, small, slender, sparingly branched. Leaves black, shining, upper purplish, imbricate, spreading, concave, oblong or orbicular; tips recurved; cells large at the ventral side; lobule erect, pyriform, inflated; involucral longer, narrower, acute. Stipules minute, 2-partite, and 1–2-toothed. Perianth obovate, 3-quetrous.—Fl. Antarct. t. 66. f. 3; Syn. Hep. 450.

Northern Island: Wairarapa valley and Ruahine range, Beech-trees, etc., *Colenso.* **Lord Auckland's** group: rocks on the hills, *J. D. H.*

20. **F. congesta,** *Hook. f. and Tayl. Fl. N. Z.* ii. 162. Very similar to *F. aterrima*, of a red-brown colour, with rounder stipules, which are never toothed, the leaves have not the cluster of enlarged cells at the base, and the involucral are white at the tips.—Fl. Antarct. 164; Syn. Hep. 451.

Northern Island: probably common on Lichens, bark of trees, rocks, etc., Auckland, *Sinclair;* Titiokura, Tehawera, etc., *Colenso;* Wellington, *Stephenson.* **Lord Auckland's** group: on rocks and bark, *J. D. H.*

21. **F. allophylla,** *Hook. f. and Tayl. Fl. Antarct.* 163. *t.* 66. *f.* 1. Stems minute, slender, flexuous, $\frac{1}{6}$–$\frac{1}{8}$ in. long, sparingly branched. Leaves red-brown, black when dry, lax, distant, erecto-patent or spreading, convex, subpetiolate, obliquely ovate, acute or obtuse, sometimes 2–3-toothed; cells large; lobule oblong or clavate, pedicelled, diverging at the base, then incurved or decurved. Stipules minute, 2-fid; segments lanceolate, but sometimes clavate like the lobules.

Campbell's Island: on hills, intermixed with other *Hepaticæ, J. D. H.*—A minute, very curious and beautiful little plant.

22. **F. reticulata,** *Hook. f. and Tayl. Fl. Antarct.* 163. *t.* 66. *f.* 2. Stems 1–2 in. long, pinnate, flattish. Leaves black, young red-purple, distichous, imbricate, very tender, convex, lower orbicular, entire, upper ovate, coarsely toothed; cells large, pellucid; lobules clavate, sometimes absent in the lower leaves, sometimes 2-fid or double. Stipules 4-partite; segments subulate, or some or all clavate.

Lord Auckland's group: rocks on the mountains, *J. D. H.*—A larger plant than *F. rostrata*, with larger leaf-cells and 4-partite stipules.

23. **F. gracilis,** *Nees.* Stem creeping, pinnately and 2-pinnately branched. Leaves imbricate, orbicular, obtuse, quite entire; lobule auricled, oblique, exposed; involucral inciso-serrate. Stipules imbricate, flat, subrotund, 2-fid, quite entire; involucral inciso-serrate. Perianth 3-gonous, prismatic, smooth; dorsal surface convex, ventral with 1 keel.—Syn. Hep. 452; Mont. in Voy. au Pôle Sud, 223.

Lord Auckland's group, *Hombron.*—I have seen no Auckland Island specimens; the species is also found in Java and the Malayan peninsula.

24. FOSSOMBRONIA, Raddi.

Frond prostrate, creeping, lobed pinnately; lobes succubous, distichous, quadrate, 3–5-lobed, flaccid. Stipules 0. Fruit terminal, or lateral by the growth of the stem. Involucre of subulate leaves, adnate to the perianth, which is campanulate; mouth large, lobed. Calyptra pyriform. Capsule on a short fruitstalk, globose, irregularly 4-valved; elaters short, with 2–3 spiral fibres. Antheridia naked on the under surface.

A small genus, found in various parts of the world, intermediate between the frondose and leafy *Hepaticæ*; the stem may be regarded as a frond pinnatifid to the midrib, or as furnished with adnate leaves.

1. **F. pusilla,** *Nees;—Fl. N. Z.* ii. 163. Minute; stems $\frac{1}{4}$ in. long, simple or forked. Leaves obliquely spreading, lower waved and lobed, upper crisped, 3–4-lobed or -angled. Perianth obconic, toothed.—Syn. Hep. 467.

Northern Island: east coast and Hawke's Bay, on banks, palings, etc., *Colenso.* (America, Canary Islands, S. Africa, Tasmania, Kerguelen's Land.)

2. **F. intestinalis,** *Tayl. Fl. N. Z.* ii. 163. Minute; stems $\frac{1}{8}$–$\frac{1}{4}$ in. long, with purple rootlets. Leaves ascending, tumid, plaited, convolute, incised and lobed; lobes acute. Perianth campanulate, crenate.—Syn. Hep. 469.

Northern Island: Bay of Islands, *J. D. H.* (Tasmania.)

25. NOTEROCLADA, Taylor.

Frond appressed to the ground, creeping, rooting, stout, forked, pinnately lobed; lobes broad, sessile, decurrent, soft, quite entire. Stipules 0. Fruit lateral or terminal. Involucre confluent with the perianth, erect; cylindrical or compressed; mouth somewhat 2- or 4-lobed. Capsule subglobose, on a slender fruitstalk, 4-valved or bursting irregularly; elaters short, with 2? spiral fibres. Antheridia immersed in the back of the stem.

A small Brazilian and Australian genus, closely allied to *Fossombronia.*

1. **N. porphyrorhiza,** *Mitten in Fl. N. Z.* ii. 163. Stem simple, $\frac{1}{2}$ in. long, flexuous, with the leaves $\frac{1}{4}$ in. broad, green or purple. Leaves distichous, oblique, approximate, ovate or suborbicular; margins papillose. Perianth small, obovate; lobes crenate-dentate.—*Androcryphia*, Nees; Syn. Hep. 470. *Jungermannia*, Nees; Montagne in Ann. Sc. Nat. 1839, t. 1. f. 1.

N. confluens, Tayl. in Lond. Journ. Bot. 1844, 478; Fl. Antarct. 446. t. 161. f. 7.

Northern and **Middle** Islands: Titiokura, in watercourses, *Colenso;* Auckland, *Sinclair;* Otago, lagoons, *Lindsay*. (Brazil, Fuegia, Falkland Islands, Kerguelen's Land.)

26. PETALOPHYLLUM, Gottsche.

Fronds simple or 2-fid, rooting below from the midrib, lamellate above. Inflorescence diœcious. Fruit from the back of the midrib. Involucre confluent with the perianth, quadrate, campanulate; mouth broad, open, undulate, toothed. Calyptra large, globose, as long as the perianth or shorter. Capsule large, globose, on a slender fruitstalk, bursting irregularly; elaters filiform, with 2 or 3 spiral fibres; spores large, reticulate. Antheridia dorsal at the forks, crowded, globose.

A small British and Australian genus.

1. **P. Preissii,** *Gottsche;—Fl. N. Z.* ii. 164. Frond $\frac{1}{4}$ in. long, obovate or cuneate; ventral surface rugulose, striate; dorsal with radiating lamellæ, which are connate, forming little pockets at the costa.—Syn. Hep. 472.

Northern Island: Hawke's Bay, on clay soil, *Colenso*. (S.W. Australia.)

27. ZOOPSIS, Hook. f. and Tayl.

Frond very slender, tufted, creeping, rigid, silvery-green, almost capillary, sparingly branched, of large tumid hexagonal cells; midrib stout; margins waved. Fruit lateral. Involucre of a few lanceolate scales. Perianth large, pedicelled, obovate-oblong, deeply laciniate.

A very curious little plant, which I discovered in Lord Auckland's group, and has since been found in Tasmania and New Zealand.

1. **Z. argentea,** *Hook. f. and Tayl. Fl. N. Z.* ii. 164. Fronds $\frac{1}{4}$–$\frac{1}{2}$ in. long, $\frac{1}{20}$ in. broad, of two lateral and one antero-posterior series of cells surrounding a central cord of filiform cells; each of the marginal cells stands out like the tooth of a saw, and is terminated by an incurved bristle, which is not represented in the figure in Fl. Antarct. The fruit is immature.—Syn. Hep. 473; Fl. Antarct. 167. t. 66. f. 6.

Northern Island: probably common amongst mosses, etc., in various localities, *Colenso;* Auckland, *Lyall, Sinclair*. **Lord Auckland's** group: on the ground at the roots of tree-ferns, and amongst mosses, *J. D. H.*

28. PODOMITRIUM, Mitten.

Fronds erect from a creeping rhizome, stalked, oblong, with a stout midrib, membranous, entire. Inflorescence diœcious. Involucre from the base of the frond on the ventral surface, shortly pedicelled, with a few scales at the base. Perianth tubular, much longer than the involucre. Calyptra included, bearing a few barren pistils, subcampanulate; mouth lacerate. Capsule on a long fruitstalk, ovoid, 4-valved; elaters filiform. Antheridia crowded on short pedicelled spikes.

The only species known; it is almost impossible to distinguish its barren fronds from those of the British *Steetzia Lyallii* and *Symphyogyna subsimplex*, Mitten.

1. **P. Phyllanthus,** *Mitten in Fl. N. Z.* ii. 164. Fronds 1 in. high, oblong-lanceolate, dull green, obtuse, quite entire.—*Symphyogyna*, Fl. Antarct. 167. *Jungermannia*, Hook. Musc. Exot. t. 95. *Diplolæna cladorhizans*, Hook. f. and Tayl. in Lond. Journ. Bot. 1844, 570.

Throughout the **Northern** and **Middle** Islands: abundant on stumps of tree-ferns, rocks, etc., *Menzies*, etc. **Lord Auckland's** group, *J. D. H.* (Tasmania.)

29. STEETZIA, Lehmann.

Frond linear, more or less dichotomously branched, with a midrib. Inflorescence diœcious. Involucre at first terminal, dorsal by the subsequent elongation of the frond, seated on the midrib, cupshaped, torn. Perianth tubular; mouth toothed. Calyptra as long, torn at the apex. Capsule on a slender fruitstalk, ovoid; elaters free, with 2 fibres. Antheridia dorsal, on the midrib, covered by minute fimbriated leaves.

A considerable genus, scattered over both the temperate and tropical regions of the globe.

1. **S. Lyellii,** *Nees;—Fl. N. Z.* ii. 165. Frond 2–3 in. long, oblong or linear, crenulate or subserrate.—*Blyttia*, Syn. Hep. 475 (*Steetzia*, 785); Hook. Brit. Jung. t. 77.

Northern Island: clay banks, East and South coast, apparently common, *Colenso*. (Europe, America, Australia.)

2. **S. tenuinervis,** *Hook. f. and Tayl. Fl. N. Z.* ii. 165. Frond slender, 1–2 in. long, pea-green, tips dilated, 2-lobed; margin crenulate from the prominent cells, substance thin; midrib very slender.—*Diplolæna*, Hook. f. and Tayl. in Lond. Journ. Bot. 1844, 570.

Northern Island, *Colenso;* Auckland, *Sinclair*.—Probably a variety of *L. Lyellii*.

3. **S.? xiphioides,** *Hook. f. and Tayl. Fl. N. Z.* ii. 165. Frond bright green, 2 in. long, procumbent, $\frac{1}{12}$ in. broad, sparingly dichotomously branched; margin crenulate. Involucre immature, of about 5 lanceolate leaves.—*Diplolæna*, Hook. f. and Tayl. in Lond. Journ. Bot. 1844, 567. *Blyttia*, Syn. Hep. 476 (*Steetzia*, 785).

Northern Island: Bay of Islands, *J. D. H.;* Manawatu river, often under water, *Colenso;* Auckland, *Sinclair* —A very dubious plant, totally unlike any other *Steetzia*, and compared by Mitten with large forms of *Metzgeria furcata*.

30. SYMPHYOGYNA, Mont. and Nees.

Fronds membranous, linear, dichotomously or flabellately branched, stalked, arising from a creeping rhizome, midrib stout. Inflorescence monœcious or diœcious. Fruit from the nerves, usually at the forks. Involucre a toothed scale. Perianth 0. Calyptra smooth, exserted, coriaceous, fimbriated at the apex by abortive pistillidia. Capsule on a slender fruitstalk, 4-valved; valves often cohering by their apices; elaters with 2 spiral fibres. Antheridia in the midrib; perigonial leaves imbricate, tumid, incised, membranous.

A large tropical and south temperate genus, unknown in the north temperate hemisphere.

1. **S. flabellata,** *Montagne;—Fl. N. Z.* ii. 165. Stem 1–3 in. high. Frond orbicular or reniform, 3–5-partite in a fan-shaped manner, $\frac{1}{2}$–1 in.

broad; segments linear, 2-fid, nerved, obtuse. Fruit in the forks, involucral scales 2-lobed.—Syn. Hep. 481. *Jungermannia*, Hook. Musc. Exot. t. 13; Labill. Fl. Nov. Holl. t. 254. f. 1.

Mountainous parts of the **Northern** and throughout the **Middle** Islands: from Tehawera, *Colenso*, to Dusky Bay, *Menzies*. **Lord Auckland's** group, *Hombron, J. D. H.* (Tasmania.)

2. **S. leptopoda,** *Hook. f. and Tayl. Fl. N. Z.* ii. 165. Similar to *S. flabellata*; stipes faintly winged; frond pale green, more tender, 3-chotomous, segments broader.—Syn. Hep. 482.

Northern and **Middle** Islands: Bay of Islands, *J. D. H.*; Hokianga, *Jolliffe*; Tararua and Wairarapa, *Colenso*; Otago, *Lyall.*

3. **S. Hymenophyllum,** *Montagne*;—*Fl. N. Z.* ii. 166. Stipes 1–2 in. high. Frond flat, erect, dark green, obtruncate or obovate, twice or thrice 2-fid; segments linear, serrate, emarginate. Fruit at the union of the laciniæ; involucral scale orbicular, toothed. Calyptra cylindric.—Syn. Hep. 480. *Jungermannia*, Hook. Musc. Exot. t. 14.

Throughout the **Northern** and **Middle** Islands: common, from the Bay of Islands, *J. D. H.*, to Dusky Bay, *Menzies*. (Tasmania.)

4. **S. rhizobola,** *Nees*;—*Fl. N. Z.* ii. 166. Stipes short. Frond procumbent, lanceolate, subdivided, dark green, serrate, attenuated and rooting at the apex.—Syn. Hep. 483. *S. obovata*, Hook. f. and Tayl. in Lond. Journ. Bot. 1844, 581. *Jungermannia*, Hook. Musc. Exot. 87.

Northern Island: Bay of Islands, *J. D. H.*; Auckland, *Bolton*; East coast, Hawke's Bay, etc., *Colenso*. (Bourbon, Tasmania.)

5. **S. subsimplex,** *Mitten in Fl. N. Z.* ii. 166. Frond linear, flat, green, $1\frac{1}{2}$ in. long, $\frac{1}{6}$ broad, simple or forked, retuse, attenuate at the base, quite entire, tender; cells hexagonal. Involucral scale toothed and lacerate, sub-2-partite. Calyptra clavate, crowned with pistillidia.

Northern Island: Bay of Islands, *J. D. H.*; Auckland, *Sinclair*; East coast, on tree-ferns, *Colenso*. **Stewart's** Island, *Lyall.*

31. METZGERIA, Raddi.

Fronds more or less branched, flat, linear; midrib distinct. Inflorescence diœcious. Fruit from the lower surface of the frond, on the midrib. Involucre a ventricose 2-lipped scale. Perianth 0. Calyptra ascending, obovoid, rather fleshy. Capsule on a long fruitstalk, ovoid; elaters adhering to the tips of the valves, with one spiral fibre. Antheridia 1–3, from the midrib beneath the frond in a 1-leaved involucre. Ovoid buds occur on the narrow tips of the fronds.

A cosmopolitan and very variable plant, of which all the New Zealand forms are also European. A few other species are known in South America.

1. **M. furcata,** *Nees*;—*Fl. N. Z.* ii. 166. Fronds $\frac{1}{2}$–3 in. long, tender, linear, forked or dichotomously branched, glabrous, the margin and costa beneath setulose or naked. Calyptra ascending from the side, hairy and setose.—Syn. Hep. 502.

Abundant throughout the **Northern** and **Middle** Islands, *Colenso*, etc.; and in **Lord Auckland's** group. (Cosmopolitan.)

32. ANEURA, Dumort.

Frond pinnatifid or 2-pinnate, sinuate, rather thick, with a broad undefined nerve or 0. Inflorescence diœcious. Fruit from the margin of the frond underneath. Involucre short, cupshaped, torn. Perianth 0. Calyptra subcylindric, fleshy. Capsule ovoid or oblong, on a slender fruitstalk; elaters attached to the tips of the valves, each with a single broad fibre. Antheridia immersed in marginal receptacles.

In the New Zealand Flora, Mr. Mitten adopted the generic name of *Sarcomitrium* in preference to that of *Aneura*, because the fronds have a nerve or thickened axis, and indeed consist of little else than axis. Names are, however, as often founded on appearances as on facts, and if the appearance is obvious (as in this case) the name should be retained. A considerable genus, more common in the southern than in the northern hemisphere.

1. **A. alterniloba,** *Hook. f. and Tayl. Fl. N. Z.* ii. 167 (*Sarcomitrium*). Frond dark green, 3–4 in. long, flattened, vaguely branched; branches alternate, $\frac{1}{6}$ in. broad, obtusely lobed, crenate, tips rounded; margin minutely and remotely toothed. Calyptra setulose.—Syn. Hep. 496.

Northern Island: wet banks and watercourses, Bay of Islands, *J. D. H.*; East coast and interior, etc., *Colenso, Sinclair.*

2. **A. palmata,** *Nees;—Fl. N. Z.* ii. 167 (*Sarcomitrium*). Fronds pinnate, primary flat, procumbent; branches ascending, pinnatifidly palmate; segments linear, truncate or obtuse. Fruit lateral; involucre shallow, torn. Calyptra tubercled.—Syn. Hep. 498.

Northern Island: Beech forests on the East coast, *Colenso.* (Tasmania, Europe.)

3. **A. crassa,** *Nees;—Fl. N. Z.* ii. 167 (*Sarcomitrium*). Frond thick and leathery, almost horny, blackish-green, procumbent, pinnatifid; segments obtuse.—Syn. Hep. 500.

Northern Island: dense forests, East coast and interior, *Colenso.* (Tasmania.)

4. **A. pinnatifida,** *Nees;—Fl. N. Z.* ii. 167 (*Sarcomitrium*). Frond simple or pinnatifid, flat or channelled; branches horizontal, dilated at the apex, 2-pinnatifid or toothed, obtuse. Calyptra smooth, puberulous.—Syn. Hep. 495.

Middle Island: Southland, *Lyall.* (Europe, America, Bourbon, India, Malay Islands, Australia, Tasmania.)

5. **H. multifida,** *Dumort.;—Fl. N. Z.* ii. 167 (*Sarcomitrium*). Frond pinnately multifid or decompound; primary biconvex, rigid; branches pectinate, horizontal, linear; involucre lateral in the primary or secondary branches, turbinate, fleshy. Calyptra tubercled.—Syn. Hep. 497.

Throughout the **Northern** and **Middle** Islands: by watercourses, wet banks, etc., from Bay of Islands, *J. D. H.*, to Port Cooper, *Lyall.* **Lord Auckland's** group and **Campbell's** Island, *J. D. H.* (Europe, America, India, Australia, Tasmania, Fuegia.)

6. **A. prehensilis,** *Mitten;—Fl. N. Z.* ii. 167 (*Sarcomitrium*). Fronds loosely tufted, erect, incurved; primary branches compressed, brown, pruinose when dry, winged; lobes pinnate, secund, flat, linear, with a thick nerve. Calyptras solitary or 2, at the bases of the upper lobes, brown, scabrous, elongate-obovate, with torn involucral scales at the base.—*Metzgeria*, Hook. f. and Tayl. Fl. Antarct. 445; Syn. Hep. 505.

Northern and **Middle** Islands: Manawatu river, on dead wood, *Colenso;* Port Nicholson, *Lyall;* Otago, *Lindsay.* (Fuegia.)

7. **A. eriocaula,** *Mitten;—Fl. N. Z.* ii. 168 (*Sarcomitrium*). Frond 3–6 in. long, flexuous, pinnatifid, pubescent; primary creeping, compressed; branches alternate, ovate, 2-pinnatifid; divisions linear, obtuse, glabrous, costate, brown. Calyptra subaxillary, oblong-cylindric, fleshy, glabrous, rather rough.—*Metzgeria,* Syn. Hep. 505. *Jungermannia,* Hook. Musc. Exot. t. 72.

Middle Island: Dusky Bay, *Menzies;* Port Nicholson, *Lyall;* Otago, *Hector and Buchanan.* (Tasmania.)

8. **A. cochleata,** *Mitten in Fl. Tasman.* ii. 240 (*Sarcomitrium*). Frond short, very thick and fleshy, loosely cæspitose, procumbent, creeping, pinnatifidly irregularly lobed; lobes ovoid, cochleate, concave, very thick and fleshy, with upturned or connivent margins, 2-lobed at the apex, loosely cellular and spongy internally. Calyptra fleshy, cylindric, setulose.—*Riccia cochleata,* Hook. f. and Tayl. in Fl. Antarct. t. 66. f. 5; Syn. Hep. 612.

Lord Auckland's group: on stones, in tufts of mosses, etc., *J. D. H.* (Tasmania.)

33. PLAGIOCHASMA, Lehm. and Lindb.

Frond short, thick, rigid. Inflorescence monœcious? Fertile receptacles stalked, capitate, 2–4-lobed; lobes small, ascending. Involucres 4, large, erect, 1-capsuled, concealing the lobes, ovoid, 2-valved. Perianth 0. Calyptra fugacious. Capsule almost sessile, horizontal, globose, splitting irregularly; elaters moderately long; spores enclosed in a transparent wrinkled membrane. Antheridia immersed in sessile disks.

A genus of a dozen species or so, inhabiting warm climates.

1. **P. australe,** *Nees;—Fl. N. Z.* ii. 168. Frond 1 in. long, linear-elongate, green with purple margins; tips retuse; margins crenulate, innovations on the ventral side; scales broad, purple, acute, with hyaline tips, upper projecting beyond the margin. Fruit in the middle of the frond; peduncle short or long, with linear-lanceolate scales at the base. Female receptacle with 2–4 fruits, scaly below; subglobose, male papillose.—Syn. Hep. 515. *Fegatella australis,* Hook. f. and Tayl. in Lond. Journ. Bot. 1844, 572.

Northern Island: Bay of Islands, *J. D. H.;* Hokianga, *Jolliffe;* Tehawera and Patea, *Colenso.*

34. MARCHANTIA, Linn.

Frond broad, growing flat on the ground, branched, thick, with a broad ill-defined midrib or none, covered below with coloured imbricating scales and tubular rootlets. Inflorescence diœcious, on the surface of the frond. Receptacles peltate, stalked, rayed, fruiting on the under surface. Involucres pendulous, alternate with the rays, 2-valved, lacerate, each with 3–6 3-cleft perianths. Capsules stalked, pendulous, dehiscing at the apex by revolute valves; elaters slender; spores smooth. Antheridia immersed in the under surface of male receptacles.—Gemmæ occur abundantly in sessile cups on the surface of the frond.

A large genus, found in all parts of the world, easily recognized by forming green scale-like patches on the ground, paths, walls, and especially on the mould of green-house plants.

1. **M. tabularis,** *Nees;—Fl. N. Z.* ii. 168. Patches large, oblong, lobed; terminal lobes toothed, nerveless. Peduncles 1–4 in. long. Female receptacles with 8–9 star-like rays; involucres with 2–4 perianths, $\frac{1}{4}$ shorter than the ray, tumid, white, 4-fid; segments lacerate; central beard slender, fibrillose; fruitstalk short. Male receptacles on separate plants on shorter peduncles, orbicular, 4-lobed; lobes crenate; anthers in 8 radiating lines.—Syn. Hep. 525. *M. polymorpha*, Fl. Antarct. 168 and 446, not of Linnæus.

Throughout the **Northern** and **Middle** Islands, in wet places abundant, and in **Lord Auckland's** group and **Campbell's** Island. (S. Africa, Tasmania, Fuegia, Falkland Islands, Kerguelen's Land.)—This is the southern representative of the ubiquitous northern *M. polymorpha*, differing from that plant in the more convex fronds without a midrib, more minute cells, and more prominent pores.

2. **M. nitida,** *Lehm. and Lindenb.;—Fl. N. Z.* ii. 168. Frond 1 in. long, green, linear, dichotomous, thickened in the middle, dilated at the obcordate tips; margin undulate, crenate, under surface purple. Female receptacle excentric, convex, papillose, 8–10-rayed; peduncle paleaceous at the base, slender, flexuous, pilose; rays dilated, emarginate, crenulate, beardless, costate above; involucre 2-valved, ciliated, 2–3-fruited. Perianth plicate, purple at the mouth.—Syn. Hep. 532.

Northern and **Middle** Islands: Tuaraiawa river, *Colenso;* East Cape, *Colenso;* Canterbury, *Haast.* (India.)

3. **M. foliacea,** *Mitten in Fl. N. Z.* ii. 168. Frond 2–3 in. long, subcoriaceous, flat, smooth, glossy above; pores small, pale; below deep purple; margin entire, undulate. Female receptacle excentric, convex, umbonate, about 8-lobed; peduncle 1 in. long, with purple scales; lobes dilated, foliaceous, subcrenate; involucres pale, lacerate, ciliate. Cups with gemmæ, funnel-shaped, with toothed ciliate margins.

Northern and **Middle** Islands: Cape Turnagain and base of the Ruahine range, *Colenso;* Hokianga, *Jolliffe;* Nelson, *Mantell.* (Tasmania.)

4. **M. macropora,** *Mitten in Fl. N. Z.* ii. 169. Fronds 1–2 in. long, $\frac{1}{6}$ in. broad, dirty green, linear, dichotomous; margin flat, quite entire; pores large, with purple edges; under surface dark purple; scales few. Female receptacle subhemispherical, excentric, about 5-lobed, warted; lobes 2-partite; peduncle $\frac{1}{2}$ in. long, with a few lanceolate scales; involucres with purple lacerate margins, and clothed with lanceolate purple laciniæ at the base.

Northern Island: wet banks, Makororo river and Ahuriri, *Colenso;* Auckland, *Sinclair*, etc.

5. **M. pileata,** *Mitten in Fl. N. Z.* ii. 169. Frond 3 in. long, glaucous-green, $\frac{1}{4}$ in. broad, dichotomous, linear, flat, smooth; pores minute, margined with white; margins entire; underneath dark purple. Female receptacle subexcentric, hemispherical, tuberculate, naked below; margin crenate; peduncle $\frac{1}{2}$ in. long, smooth, purple; margin of the involucres torn.

Northern Island: wet banks, Makororo river, *Colenso.* (Tasmania.)

35. DUMORTIERA, Nees.

Habit and most of the characters of *Marchantia*, but involucres with 1 capsule, horizontal, opening by a vertical slit. Perianth 0. Capsule shortly stalked; elaters very long; spores muricate.

A small genus, scattered over various parts of the globe, growing on the ground in moist places.

1.? **D. hirsuta,** *Nees;—Fl. N. Z.* ii. 169. Frond dichotomous, bright green, tender, pellucid, with no midrib, crenulate-undulate, hairy below. Female receptacles with many fruits, convex, covered with scattered bristles; margin and involucres hirsute, bearded round the top of the peduncle.—Syn. Hep. 343. *D. dilatata*, Syn. Hep. 543. *Hygropila dilatata*, Hook. f. and Tayl. in Fl. Antarct. 168.

Northern and **Middle** Islands: abundant, from Bay of Islands, *J. D. H.*, to Akaroa, *Raoul*, in watery places, etc. **Lord Auckland's** group, *J. D. H.*—One of the largest and commonest *Hepaticæ* of New Zealand, but never found in fruit and hence doubtful as to genus and species. The specimens are much larger than any of *D. hirsuta*, from other parts of the world; the latter species is very variable and found in various parts of America and South Africa.

36. REBOULIA, Nees.

Frondose, growing on rocks or earth; habit of *Marchantia;* midrib strong, broad. Female receptacle peduncled, flat, conical, or hemispherical, 1–6-lobed, almost to the middle; lobes thick, their margins forming a 2-valved involucre, containing one capsule. Perianth 0. Calyptra ovoid, soon rupturing, leaving a cup at the base of the capsule. Capsule exserted, subglobose, membranous, lacerate or suboperculate; elaters with 2 spirals; spores tubercled. Male receptacle sessile, discoid.

A small genus of three species, scattered over the world.

1. **R. hemisphærica,** *Raddi*. Frond dichotomous, with innovations that are rounded and emarginate at the apex, green above, purple below; receptacle variable in shape; hairs at its base very slender.—Syn. Hep. 548.

Northern and **Middle** Islands: Auckland, *Knight;* Nelson, *Mantell;* Canterbury, *Travers.* (Cosmopolitan.)

37. FIMBRIARIA, Nees.

Frondose; habit of *Marchantia;* midrib keeled. Inflorescence monœcious. Female receptacle hemispherical, concave below; margins expanding and forming 4 campanulate, pendent, 1-capsuled involucres. Perianth projecting beyond the involucre, oblong, splitting into many laciniæ at the mouth. Calyptra fugacious. Capsule sessile, globose, bursting transversely; elaters short, with 1 or 2 spirals; spores muricate. Antheridia not on receptacles, but immersed in the frond.

1. **F. Drummondii,** *Tayl.;—Fl. N. Z.* ii. 169. Frond 1–1½ in. long, linear-elongate, green or purplish. Female receptacle conical, obtuse, papillose with vesicular cells, almost naked below; perianths white and purplish, broadly ovate, sub-12-fid; segments broad, flat, cohering at the apex; peduncle 2 in. long, slender, black-purple.—Syn. Hep. 566 and 791.

Northern and **Middle** Islands: heights of Cape Kidnapper and Wairarapa valley, etc., *Colenso;* Nelson, *Mantell* One of the largest species of the genus. (Swan River, Tasmania.)

2. **F. australis,** *Hook. f. and Tayl. Fl. N. Z.* ii. 170. Fronds 1 in. long, linear, dichotomous, tender, 2-lobed at the apex, below purple, with a few small lanceolate obtuse scales. Female receptacle conico-hemispherical, tubercled, sub-4-lobed, bearded below with long hairs; perianths ovate, 12–14 fid, pale; segments cohering by their tips.

Northern and **Middle** Islands: banks of Mohaka river, *Colenso;* Auckland, *Bolton;* Nelson, *Mantell.* (Tasmania.)

3. **F. tenera,** *Mitten in Fl. N. Z.* ii. 170. Frond ½ in. long, ⅙ in. broad, green, dichotomous; divisions oblong or obcordate, deeply 2-fid, very tender, veined, fertile cuneate; below green or brownish. Female receptacle small, rather convex, 3–4-lobed, naked below, rugulose; margin crenulate; perianths small, shortly conic, 8-fid; segments pale brown, ovate-lanceolate, separate at the apices.

Northern Island: clay banks at Pahawa and Patea, *Colenso.* (Tasmania.)

38. TARGIONIA, Micheli.

Fronds appressed to the ground, rooting, thick and coriaceous, linear, forked, porous on the upper surface, scaly on the lower. Involucre at the apex of the frond, 2-valved. Perianth 0. Calyptra persistent, enclosing the capsule, breaking away from over it, its bulb immersed in the frond; style deciduous. Capsule on a very short fruitstalk, membranous, irregularly torn; elaters with 2 or 3 spiral fibres. Antheridia in lateral disk-like receptacles, on proper branches. Gemmæ 0.

A small genus, scattered over the globe, growing in the ground and on tufts of mosses, etc.

1. **T. hypophylla,** *Linn.;—Fl. N. Z.* ii. 170. Frond linear-obovate or cuneate, rigid, with an undefined midrib, pores equal; scales densely imbricate, outer reaching the margin.—*T. Michelii,* Corda.; Syn. Hep. 574.

Northern Island: common on moist banks, *Colenso,* etc. (Europe, N. Africa, N. America, Tasmania.)

39. ANTHOCEROS, Micheli.

Fronds growing flat on the ground, more or less orbicular, radiating, thick, opaque, green; margins lacerate or crenate. Inflorescence monœcious. Fruit on the upper surface of the frond. Involucre tubular. Perianth 0. Calyptra conical. Capsule pedicelled, of 2 narrow linear erect lobes; columella filiform; elaters flexuous; spiral fibres imperfect or 0; spores muricate. Antheridia sessile, in cup-shaped involucres.—Gemmæ also immersed in the substance of the frond.

A very common genus in various parts of the world, forming often large green patches in moist places.

1. **A. lævis,** *Linn.;—Fl. N. Z.* ii. 171. Frond 1–2 in. long, flat, radiately branched; lobes crenate, smooth, nerveless, tender; involucre cylindric; mouth obliquely truncate, with broad scarious edges.—Syn. Hep. 586.

Dendroceros leptohymenius, Hook. f. and Tayl. in Lond. Journ. Bot. 1844, 575; Syn. Hep. 580. *Pellia carnosa*, l. c. 576; Syn. Hep. 490. *A. punctatus?*, Hook. f. and Tayl. in Fl. Antarct. i. 168.

Northern and **Middle** Islands: abundant from the Bay of Islands, *J. D. H.*, to Akaroa, *Raoul*. **Campbell's** Island, *J. D. H.* (Europe, America, Tasmania.) A very common and variable plant. Some of the numerous New Zealand plants referred to this that are not in fructification may be referable to other things.

2. **A. Jamesoni,** *Tayl.;—Fl. N. Z.* ii. 171. Frond flat, nerveless, nearly smooth, much cut into narrow laciniæ; margin lobulate; lobules undulate and crenate. Involucre obliquely truncate, smooth. Capsule elongate, valves cohering at their tips.

Northern Island: Auckland, *Bolton*. (Quito, Fuegia.)

3. **A. giganteus,** *Lehm. and Lindenb.;—Fl. N. Z.* ii. 171. Frond ½–1 in. long, linear, forked, costate, pinnatifidly laciniate; margin crisped. Involucre cylindric, obliquely truncate. Capsule 2 in. long and more, dehiscing at the apex.—Syn. Hep. 588.

Middle Island: Dusky Bay, *Menzies*; Port Cooper, *Lyall*.

4. **A. Colensoi,** *Mitten in Fl. N. Z.* ii. 171. Frond 1–2 in. broad, brownish-black, inordinately pinnatifid and pinnately lobed; lobes flabellate sinuate and crenate; lobules small, pale, pellucid, smooth above and below. Involucre rugulose, with little folds, obscurely 2-lipped. Capsule arcuate, dehiscing on one side, short, pale brown; peduncle included.

Northern Island: tops of the Ruahine mountains, *Colenso*; Auckland, *Sinclair*, *Bolton*.

40. RICCIA, Micheli.

Small, frondose, terrestrial or aquatic, often of thick consistence. Fronds simple or divided, usually orbicular or oblong, often stellate. Inflorescence monœcious or diœcious. Perianth 0. Fruit immersed in the frond. Involucre 0. Calyptra cohering with the globose capsule, and crowned with a persistent styliform apex; columella and elaters 0; spores angular; antheridia imbedded in the frond.

A considerable genus, found in clay banks, ditches, etc., in all parts of the globe.

1. **R. acuminata,** *Tayl.;—Fl. N. Z.* ii. 171. Frond 1 in. diameter, glaucous-green, solid, orbicular, stellate, naked below; segments linear, dichotomous, channelled; margins erect. Fruit tumid, crowded in the centre of the frond.—Syn. Hep. 793.

Northern Island. fern hills, Hawke's Bay, *Colenso*.

2. **R. natans,** *Linn.;—Fl. N. Z.* ii. 173. Fronds floating, pale yellow above, bordered with dull purple, obcordate, channelled, simple or proliferous from the notches, with long purple serrate fimbriæ below; substances cavernous.—Syn. Hep. 606.

Northern Island: Lake Roto-a-Kiwa, *Colenso*. (Europe, America.)

3. **R. fluitans,** *Linn.;—Fl. N. Z.* ii. 173. Fronds rather membranous, dichotomous, flat or subchannelled, green above and below, dichotomous;

branches linear-elongate, tips emarginate, 2-fid or forked; substance cavernous towards the tips.—Syn. Hep. 610.

Northern Island: deep water, head of Wairarapa valley, *Colenso*. (Europe, America, India, S. Africa.)

ORDER VI. CHARACEÆ.

Branching, submerged, slender, freshwater plants, with whorled branches, sometimes coated with carbonate of lime. Stems and branches formed of a few very long simple tubular cells placed end to end, with often smaller tubes on the surface of one central large one. Reproductive organs of two kinds; 1, solitary or clustered naked spores or *nucules*, each coated with spirally arranged cells, placed in the axils of the branches; 2, spherical *globules*, all axillary, of a bright or dull red or orange colour, consisting of 8 triangular disciform scales; in the centre of the inner surface of each scale a columnar tubular cell is fixed, which points towards the centre of the globule, and at its end bears a bundle of jointed threads, each joint (or cell) of which contains a spiral antherozoid. *Gemmæ* are often produced on the root; and stellate bodies, capable of reproducing the species, often replace the globules.

A most curious Order of plants, abounding in many sluggish freshwater streams of temperate climates, remarkable for the singular nature of its reproductive bodies, its obscure affinities, and the distinctness with which the circulation of the fluids in the stems and branches of many species may be seen with a low power of the microscope.

Each node or joint of the stem or branches consists of but one length of cells, sometimes upwards of 1 in. long, with very transparent walls (except when coated with carbonate of lime), in which the sap may be watched, forming two currents side by side, and not separated by anything, one ascending, the other descending.

The spore (or nucule) consists of a central ovoid cell, full of starch-granules, coated by 5 long cells spirally wound round it, their tips being free and having a cavity between them, down which the antherozoids are supposed to pass and impregnate the central cell. The spore germinates by the formation of a cell at its tip, from which rootlets descend and a stem ascends.

The red globules (or antheridia) have their walls divided by three circles, two vertical, and three equatorial, and consist of eight disks, each of which is in shape an equilateral spherical triangle. Each disk is formed of radiating tubular cells; a tubular cell proceeds from the centre of the disk to that of the globule, bearing at its apex a few small cells, from which the filaments, whose joints contain the spermatozoids, originate.

The various species of the genus are very fetid, giving off an odour of sulphuretted hydrogen.

I am indebted to my friend Professor A. Braun, of Berlin, for determining the New Zealand species of this Order. The specimens were in most cases insufficient for a full and satisfactory description.

1. NITELLA, Agardh.

Articulations of the stem and branches formed of 1 or more tubes, never coated with carbonate of lime. Crown of the nucule formed of about 10 cells, in 2 series, usually deciduous.

A large genus, found in all parts of the globe.

* EUNITELLA.—*Globules terminal at the forks of the branches.*

1. **N. hyalina,** *Agardh.* Monœcious; stem branched, rather stout. Whorls crowded, of many rays; longer rays 8, 3 times forked, shorter about

twice as many, intercalated in pairs, simple or forked; terminal forks rather swollen. Nucules large, subglobose, with 9 striæ.—Wallmann, Essai Fam. Charac. 14.

Northern Island: Rupahi Lake, Auckland, *Hochstetter* (*Herb. A. Braun*). (Cosmopolite.)

2 **N. Hookeri,** *Braun in Hook. Kew Journ. Bot.* i. 199. Monœcious. Whorls lax, lower remote, upper forming lax comose heads, of 6–8 rays; rays 2–3-forked to or below the middle, one of them usually further divided; ultimate articulations of about 3 joints, the first elongated, the rest shorter, forming a 2-celled mucronate apex. Nucules usually in pairs; crown short, obtuse.—*Chara australis*, Tayl. in Fl. Antarct., not of Brown.

Northern Island: still waters, Bay of Islands, *Colenso, J. D. H.*; Wellington, *Ralph.* (Tasmania, Australia, Kerguelen's Land.)

3. **N. interrupta,** *Braun, mss.* Monœcious. Sterile whorls remote, of 6 rays; rays usually 4–5 times forked to the middle, divisions 2–3-cellular; fertile whorls minute, contracted, arranged in interrupted spikes. Spores solitary on the secondary branches, brown, with 7 striæ; globules in the primary branches.—*Braun, mss.*

Northern Island: Waikate river, *Telinek.*

2. CHARA, Linn.

Articulations of the stem and branches formed of several series of cells, often coated with carbonate of lime. Crown of the nucule formed of 5 spreading cells.

Less frequent in the southern than in the northern hemisphere.

1. **C. fœtida,** *A. Braun.* Monœcious. Stems fine, striate, brittle, coated with carbonate of lime, of twice as many tubes as branches (in 2 series); upper branchlets of one tube, without external tubes. Nucules with 13 striæ, as long as 2 of the branchlets that subtend them. Stipules or short branchlets at the base of each whorl.—Wallmann, Essai Fam. Charac. 63.

Northern Island: Bay of Islands, *J. D. H.* (Europe, etc.)

2. **C. contraria,** *A. Braun.* Monœcious. Stem coated with carbonate of lime, finely striated, minutely papillose or strigillose; the primary tubes prominent, hence exposing the papillose surfaces, which in *C. fœtida*, to which this is closely allied, are hidden between the secondary tubes.—Wallmann, Essai Fam. Charac. 64.

Northern Island, *Colenso.* (Cosmopolite.)

Order VII. LICHENES.

Perennial plants, of comparatively small size, low organization, and very various forms, rarely green in colour, consisting chiefly of a horizontal ascending or erect, simple or branched, foliaceous membranous coriaceous cartilaginous or crustaceous or powdery vegetative portion, called the THALLUS, which bears organs of fructification. The THALLUS usually spreads horizontally over dead wood, bark of trees or rocks, but in many genera it is terrestrial and erect, or pendulous from rocks or branches; it may be *effuse*, that is, without determinate shape; or *effigurate*, that is, having a definite shape; or *scaly*, formed of small coriaceous scales; or *crustaceous*, of a thick crusty substance; or *powdery;* or *leprous*, when formed of minute membranous scales; or *granular;* or *foliaceous* (also called *frondose*). The attachment of the foliaceous thallus may be by the whole under-surface, or by *fibrils* (bundles of short filaments), and may be by one point or many. When the thallus is erect, it is often called a *podetium*, a name also given to the erect cylindrical portion of a horizontal thallus. Reproductive organs of four kinds. I. APOTHECIA, circular or variously-shaped shields cups slits or prominences, formed of closely-packed jointed filaments and closed tubes (*asci*), which latter contain simple or septate spores. II. SPERMAGONES, minute open cavities in the thallus, containing filaments (*sterigmata*), upon which are most minute colourless bodies (*spermatia*). III. PYCNIDIA, or superficial spermagones. IV. GONIDIA; granules often scattered like powder over the thallus, which are analogous to buds, and reproduce the species.

A large Natural Order, found in all climates and latitudes. Upwards of 1300 species are described, very many of which have most extensive ranges in distribution, from the arctic circle to the equator and in both hemispheres. Like most other Cryptogamic Orders, they chiefly affect damp temperate climates.

The internal substance of the thallus usually consists of three layers :—

1. The *cortical*, which is tough, leathery, and formed of densely-packed minute cells, with thick walls.

2. A green *gonidic* layer, formed of loose bright green or yellow globular cells, which either have a proper cellular coat, and are called *gonidia*, or have none, and are called *gonima*. These cells are almost peculiar to Lichens; they often burst through the upper layer in masses, called *soredia*, or are scattered like powder over the cortical layer, or fringe the lobes of the thallus. In the genus *Sticta*, they burst through the under surface of the thallus, and occupy small circular depressions or cups called *cyphellæ*. Many Lichens are extensively propagated by gonidia, which may be seen forming green or yellow powdery strata on bark, stones, etc.; these strata were formerly supposed to be independent Lichens, and constituted the genus *Lepraria*. Sometimes the gonidia invade the apothecium of Lichens, which gave rise to another false genus (*Variolaria*).

3. A *medullary*, spongy, filamentous or cottony layer, composed of a network of jointed delicate threads; these are sometimes developed downwards, forming rootlets or fibrils on the under surface of many horizontal thalli.

The *hypothallus* is a rudimentary horizontal stratum, from which the thallus grows, and is only distinguishable in the youngest state of the more highly organized Lichens, and lies beneath the crustaceous or granular thallus of the more lowly.

The *apothecia* are usually discoid dark-coloured bodies, occupying various portions of the thallus, rarely however (*Nephromium*) the under surface. They may be orbicular, linear (*lirellate*), sessile or stalked, superficial or sunk in the thallus, convex concave or subglobose, and of all colours. The apothecia consists of the *exciple* or *receptacle* and *thalamium*.

The *receptacle* is the enveloping part of the apothecium, and is formed of the substance of the thallus. When the apothecium is adnate to the thallus, the receptacle only borders it, and when the apothecium is peltate or stalked, the receptacle forms its under surface also; when the bordering portion is of the same colour as the thallus, it is called a *thalline* or *thallodal border;* when of a different colour, or that of the thalamium, it is called a *proper border*. The receptacle may be altogether absent, or almost enclose the thalamium, or form a globular capsule, called a *perithecium*. The *thalamium* is usually solitary in each receptacle, but sometimes there are several, which are separate or confluent. It consists of a series of vertical elongate microscopic bodies, rising from a layer of minute cells, called the *hypothecium;* and they are usually held together by a transparent gelatine. The vertical bodies are two, *paraphyses* and *thecæ* or *asci*. The paraphyses are the most numerous, they are linear, subclavate, transparent, jointed bodies, whose terminal cells are thick-walled, are often crustaceous, warted, coloured, and adhering together from the hard surface of the thalamium. The asci are shorter vertical bodies, linear clavate or ellipsoid, tapering downwards, and consist of simple, thick-walled, transparent sacs, containing one or more, usually 8, very minute spores. At maturity the ascus breaks at the top, discharges the spores, and shrivels up, when fresh asci are developed from the hypothecium, and this process goes on as long as the apothecium, which is perennial, exists.

The *spores* are usually ellipsoid, but may be of any shape, and are simple or internally 1- or many-septate, or divided both transversely and longitudinally into *sporidia*.

The *spermagones* are microscopic, simple or compound cavities in the thallus, opening by pores, whose orifice is usually dark; they are full of gelatine, and contain *sterigmata*, which are simple or branched, articulate or inarticulate filaments; short, straight, articulate sterigmata are called *arthrosterigmata*. Upon the sides or tips of the sterigmata are the *spermatia*, which are most minute, ovoid, ellipsoid, linear or acicular, straight or curved, colourless bodies. The spermatia accumulate in the cavity of the spermagone and escape by its pore; they appear to be the analogues of the antherozoids of Algæ, etc., but have no cilia, and are not developed in antheridia. They are found in maturity usually in spring or summer, long before the spores of the apothecium are mature, and the relations between these two kinds of organ is unknown.

Pycnides are larger, thicker-walled, sometimes superficial spermagones, of which the sterigmata are simple, inarticulate, tubular, tapering pedicels, called *stylospores*, that bear at their tips bodies analogous to spermatia, but larger and curved oblong and full of granular contents.

In *germination* the sporidia of a Lichen give off one or more branching filaments, whose branches interlace and form a network, the hypothallus, which again develops into the medullary stratum. A layer of colourless cells next spreads over the hypothallus, and amongst these the gonidia appear. In many Lichens, including most of the corticolous ones, there is no further development of thallus, the apothecia growing from this, and in some few the hypothallus alone is formed. In the higher forms however a cortical layer is added. The hypothallus, though often evanescent, is present as the fibrils of *Sticta*, the black fringe of *Lecidea geographica;* and in other forms or colours, in many other corticolous and rupicolous species.

The *food* of Lichens is not confined to the gaseous elements, for they take up mineral matter in abundance, especially carbonate and oxalate of lime, besides compounds of alumina, silica, iron, potash, soda, magnesia, and even of metals, as manganese and iron. Some species attain a great age, and some appear to burrow into the rock they inhabit; this burrowing has been attributed to a corroding power in the vegetable, but I should think it rather due to the action of the moisture they retain around them.

The study of Lichens is in every respect a most difficult one; the great master of the subject is Dr. Nylander, whose arrangement I have followed here, and would refer the student to his works for details I am unable here to give; these are, especially, his 'Synopsis Lichenum,' 'Énumération Générale des Lichens,' and 'Expositio Pyrenocarpeorum.'

Of British authors, Dr. Lauder Lindsay's 'Popular History of British Lichens' (12mo, with 22 plates) is a very good work, being very lucidly written and illustrated by capital figures.

The New Zealand Lichens were first hastily named and published by Dr. Taylor; since his death, my friend the Rev. C. Babington, a learned man and most sagacious Lichenist, worked them up with great care and skill for the 'New Zealand Flora;' and, more recently, Dr. Nylander has reviewed the majority of them, and verified most of Mr. Babington's determinations. A considerable number of species are introduced from a manuscript, kindly lent me by Dr. Lindsay, which contains the botanical results of his visit to the islands. Lastly, many additions are due to Dr. Knight's skilful researches amongst the corticolous genera, the results of which, by Mr. Mitten and Dr. Knight, are published in the Linnean Society's Transactions, illustrated by microscopic drawings of great beauty, made by Dr. Knight.

The following pretends to no more than a sketch of New Zealand Lichenology, and I expect that fully as many species remain to be discovered as are now known to exist in the islands. Since the following descriptions were written, Dr. Lauder Lindsay has communicated to the Botanical Society of Edinburgh, a list of Lichens found by himself at Otago, in which a good many new species are indicated but not described. With regard to the other species there mentioned, having seen no specimens, I do not venture to introduce them, except by name at the end of each genus, as many may be identical with species described under other names by Babington, myself, or others. For myself, I frankly confess that I find it impossible to determine even the foliaceous Lichens satisfactorily, except by comparison of specimens; whilst the species of the crustaceous and corticolous genera are so difficult to examine, and impossible to describe in definite language, that I doubt any two independent workers coming to a tolerably close agreement regarding their limits and nomenclature, even if they worked upon the same specimens.

The two arrangements most in vogue are those of Fée and Nylander, the latter of whom I have followed. The size of the spores affords, in many cases, excellent specific characters, but to ascertain this accurately for the New Zealand species, would take many months of microscopic study, and as the value of such characters can only be judged of after a vast number of measurements, I have not pretended to introduce them.

KEY TO NYLANDER'S ARRANGEMENT OF THE GENERA, FOLLOWED IN THIS WORK.

FAMILY I. COLLEMACEI.

Thallus black-brown or olive-green, often subgelatinous. Gonidia without a cellular membrane, usually traversing the thallus in moniliform lines. Apothecia often red, white or pale inside.

TRIBE I. **Lichinei.** *Thallus shrubby.*

1. LICHINA.

TRIBE II. **Collemei.** *Thallus usually horizontal, foliaceous, lobed.*

2. COLLEMA. Thallus without a cortical cellular layer.
3. LEPTOGIUM. Thallus with a cortical cellular layer.

FAMILY II. LICHENACEI.

Thallus variously coloured, not soft or gelatinous. Gonidia with a cellular membrane.

Series A. *EPICONIOIDEI.* Spores collected into a black powdery or crustaceous deciduous spore-mass.

TRIBE I. **Caliciei.** *Very minute. Thallus crustaceous or* 0. *Apothecia on slender, often filiform stalks.*

4. CALICIUM.

TRIBE II. **Sphærophorei.** *Thallus shrubby, with dilated or swollen tips of the branches which bear the apothecia.*

5. SPHÆROPHORON.

Series B. *CLADODIEI.* Thallus usually erect. Apothecia terminal on erect podetia, usually without a border. Spores often 8 in an ascus, oblong, rarely elongate and septate.

TRIBE III. **Bæomycei.** *Thallus horizontal, crustaceous. Apothecia pale red or brown. Spores simple or* 1–3-*septate.*

6. BÆOMYCES.

TRIBE IV. **Cladoniei.** *Thallus foliaceous or scaly, or of branched shrubby podetia. Apothecia convex, without a border. Spores simple.*

7. CLADONIA.

TRIBE V. **Stereocaulei.** *Thallus shrubby, tufted, with a solid medullary axis. Apothecia black or brown, terminal or lateral, with very rarely an obscure thalline border.*

8. STEREOCAULON. 9. ARGOPSIS.

Series C. *RAMALODEI.* Thallus shrubby or filamentous, erect or pendulous, terete compressed or angular, without any basal crust or scales, tubular or solid. Apothecia usually with a thalline border.

TRIBE VI. **Siphulei.** *Thallus of white opaque (fistular) podetia. Apothecia unknown.*

10. THAMNOLIA.

TRIBE VII. **Usneei.** *Thallus white or yellowish, branched, with a firm filiform axis. Apothecia peltate, with a thalline border which is often ciliate.*

11. USNEA.

TRIBE VIII. **Ramalinei.** *Thallus terete or compressed, erect or pendulous, with lax pith or hollow internally. Apothecia with a thalline border.*

12. ALECTORIA. 13. RAMALINA.

TRIBE IX. **Cetrariei.** *Thallus compressed, shrubby or foliaceous, lobed, shining, with a central pith. Apothecia marginal, with a thalline border. Spores small, simple.*

14. PLATYSMA.

Series D. *PHYLLODEI.* Thallus foliaceous, depressed, lobed or laciniate, with a fibrous pith. Apothecia peltate or discoid, with or without a thalline border.

TRIBE X. **Peltigerei.** *Thallus dilated, under surface naked. Apothecia usually marginal.*

15. NEPHROMA. 16. PELTIGERA.

TRIBE XI. **Parmeliei.** *Thallus dilated, rarely subterete or shrubby. Apothecia with a thalline border. Spermagonia with arthrosterigmata.*

17. STICTA. 18. RICASOLIA. 19. PARMELIA.

TRIBE XII. **Gyrophorei.** *Thallus membranous or coriaceous, usually monophyllous, attached by a point. Apothecia with a border of the substance of the disk, often complicate or gyrose.*

20. UMBILICARIA.

Series E. *PLACODIEI.* Thallus crustaceous, scaly, granular, powdery or evanescent. Apothecia with or without a thalline border, sometimes linear.

TRIBE XIII. **Lecanorei.** *Thallus various. Apothecia with a thalline border, rarely without.*

21. PSOROMA. 24. AMPHILOMA. 27. LECANORA. 30. THELOTREMA.
22. PANNARIA. 25. SQUAMARIA. 28. URCEOLARIA.
23. COCCOCARPIA. 26. PLACODIUM. 29. PERTUSARIA.

TRIBE XIV. **Lecidiei.** *Thallus various. Apothecia without a thalline border.*

31. CŒNOGONIUM. 32. BYSSOCAULON. 33. LECIDEA.

TRIBE XV. **Graphidei.** *Thallus very thin, often invisible or beneath the bark. Apothecia like cracks, flat or plicate, with or without a border.*

34. GRAPHIS. 36. PLATYGRAPHIS. 38. STIGMATIDIUM. 40. MELASPILEA.
35. OPEGRAPHA. 37. PLAGIOGRAPHIS. 39. ARTHONIA.

Series F. *PYRENODEI.* Thallus various, peltate, scaly, areolate, continuous or 0. Apothecia opening by a punctiform pore.

TRIBE XVI. **Pyrenocarpei.**

41. NORMANDINA. 42. TRYPETHELIUM. 43. ENDOCARPON. 44. VERRUCARIA.

KEY TO FÉE'S ARRANGEMENT OF THE GENERA.

I. *Thallus adherent, crustaceous, amorphous (of delicate matted filaments in* Cœnogonium *and* Byssocaulon).

a. Apothecia more or less stipitate.

FAMILY I. **Bæomyceæ.** *Apothecia subglobose, fleshy, with a solid stipes.*

6. BÆOMYCES.

FAMILY II. **Calicioideæ.** *Apothecia hollow, goblet-shaped.*

4. CALICIUM.

β. Apothecia sessile.

FAMILY III. **Graphideæ.** *Apothecia linear.*

34. GRAPHIS; 35. OPEGRAPHA; 36. PLATYGRAPHIS; 37. PLAGIOGRAPHIS; 38. STIGMATIDIUM; 39. ARTHONIA; 40. MELASPILEA.

FAMILY IV. **Verrucarieæ.** *Apothecia enclosed in tubercles of the thallus, with a small terminal opening.*

41. NORMANDINA; 42. TRYPETHELIUM; 43. ENDOCARPON; 44. VERRUCARIA (see *Thelotrema* and *Pertusaria* in Family VII.).

FAMILY V. **Leprarieæ.** *Apothecia of naked sporules.*

This family consists of rudimentary states of various Lichens.

FAMILY VI. **Variolarieæ.** *Apothecia opening into depressed hollow shields or pustules.*

Belongs to the same category as Family V.

FAMILY VII. **Lecanoreæ.** *Apothecia discoid with a more or less distinct border, sometimes thalline, and sometimes of the substance of the disk.*

a. Thallus crustaceous.

27. LECANORA; 28. URCEOLARIA; 29. PERTUSARIA; 30. THELOTREMA; 33. LECIDEA.

β. Thallus of matted microscopic filaments.

31. CŒNOGONIUM; 32. BYSSOCAULON.

II. *Thallus subfoliaceous, consisting of scales more or less combined together. Apothecia discoid, bordered, sessile.*

FAMILY VIII. **Squamaria.**

21. PSOROMA; 22. PANNARIA; 23. COCCOCARPIA; 24. AMPHILOMA; 25. SQUAMARIA; 26. PLACODIUM.

III. *Thallus shrubby or foliaceous, erect or horizontal, loosely attached by the base or by fibres from its lower surface.*

A. *Thallus horizontal, upper and under surfaces of different colour or texture, or both.*

a. Thallus not fixed by a central point, spreading diffusely.

FAMILY IX. **Parmeliaceæ.** *Thallus coriaceous or membranous. Apothecia discoid, fixed by the centre.*

17. STICTA; 18. RICASOLIA; 19. PARMELIA. (See *Platysma* in Family XIII.)

FAMILY X. **Collemateæ.** *Thallus when moist more or less gelatinous. Apothecia discoid, fixed by the centre*

2. COLLEMA; 3. LEPTOGIUM.

FAMILY XI. **Peltigereæ.** *Thallus membranous. Apothecia adnate by their whole surface to the upper or under surface of the lobes of the frond, border* 0 *or very thin.*

15. NEPHROMA; 16. PELTIGERA.

β. Thallus fixed by a central point, coriaceous.

FAMILY XII. **Umbilicarieæ.**

20. UMBILICARIA.

γ. Thallus erect, pendent or ascending, rarely foliaceous or horizontally spreading; upper and under surfaces similar.

FAMILY XIII. **Ramalineæ.** *Thallus compressed or flat, sometimes foliaceous, usually compressed and laciniate, rarely inflated or fistular. Apothecia discoid, with a thickened thalline border.*

12. ALECTORIA; 13. RAMALINA; 14. PLATYSMA.

FAMILY XIV. **Usneæ.** *Thallus terete, shrubby, erect or pendulous, with a central core or thread. Apothecia discoid, ciliated, with a very thin border or* 0.

11. USNEA.

FAMILY XV. **Cornicularieæ.** *Thallus terete or compressed, without a central thread, cottony or fistulose within. Apothecia discoid.*

Has no New Zealand representative hitherto discovered.

FAMILY XVI. **Sphærophoreæ.** *Thallus erect or suberect, shrubby, terete or compressed, solid within. Apothecia convex or subglobose, sunk in extremities of branches.*

5. SPHÆROPHORON; 8. STEREOCAULON; 9. ARGOPSIS.

1. LICHINA, ranked by Fée as an *Alga*, is referable here artificially.

FAMILY XVII. **Cladonieæ.** *Thallus of erect, fistulose podetia, often dilated into cups, with or without a horizontal scaly portion. Apothecia hemispherical, without a border.*

7. CLADONIA; 10. THAMNOLIA.

1. LICHINA, Agardh.

Thallus dark-brown, shrubby, rigid, containing beneath the cortical layer numerous naked blueish granules. Apothecia terminal, contained in globose receptacles, discoid with a thalline border; spores ellipsoid, simple.

A small genus of small Lichens, a good deal resembling *Algæ*, and found on littoral rocks.

1. **L. pygmæa,** *Ag.* Thallus about ½ in. high, tufted, branched, compressed above.—Nyland. Synops. Lich. 91. Var. *β. intermedia*, Bab. in Fl. N. Z. ii. 311. t. 128 C. Thallus ¼ in. high; branchlets slender, flattened.

Middle Island: Otago harbour. Var. *β.* on rocks, *Lyall.* (Europe.)

2. COLLEMA, Acharius.

Thallus of various forms, cartilaginous or gelatinous, dull green or brown, with no distinct cortical stratum, traversed by simple tubular threads and moniliform strings of naked granules. Apothecia rufous, with a thalline border; spores in eights, simple or variously divided.

A large genus of Lichens, found on rocks, bark, etc., in many parts of the world, somewhat resembling *Nostoc*. The New Zealand species do not appear to be taken up by *Nylander* in his 'Synopsis Lichenum.'

1. **C. flaccidum,** *Ach.;—Bab. in Fl. N. Z.* ii. 309. Thallus membranous, lobed, dirty green, blueish or greenish-brown, larger lobes expanded, flexuous. Apothecia reddish, flat, middling-sized; spores ovoid or broadly fusiform, septate.

Northern Island: common on trees, *Colenso, J. D. H.*, etc. (Europe, America, Australia, etc.)

2. **C. pulposum,** *Ach.;—Bab. in Fl. N. Z.* ii. 310. Thallus dirty green or greenish-brown, orbicular, lobed; lobes thick, crenate, subimbricate, often plaited. Apothecia middling-sized, flat, border entire; spores ovoid, usually 3-septate.—Nyland. Synops. Lich. 114.

Northern Island, *Colenso.*—Specimens imperfect, barren, hence very doubtful. I find none in the Herbarium. (Europe, America.)

3. **C. nigrescens,** *Ach.*—Var. *leucocarpum*, Bab. in Fl. N. Z. ii. 308. Thallus black-green, membranous, orbicular, depressed, lobed, radiately plaited. Apothecia lilac, glaucous, flat, crowded; border entire; spores cylindric-fusiform, usually many-septate.—*C. leucocarpum*, Tayl. *C. glaucophthalmum*, Nyland. Synops. Lich. 114, *ex ejus* Lich. Nov. Caled.

Northern Island, *Colenso.* **Middle** Island, *D'Urville;* Otago, *Lindsay.* (Found in all parts of the globe. *C. nigrescens* is cosmopolite, var. *leucocarpum* is found in Britain Mexico, and New Caledonia.)

4. **C. fasciculare,** *Ach.*—Var. *Colensoi*, Bab. in Fl. N. Z. ii. 309. Thallus suborbicular, rigid, cartilaginous, blue when dry, fixed by the centre, green when moist; lobes rounded, sinuous, crisped. Apothecia crowded on the margins of the lobes, minute, rufous, without a border.

Northern Island: on twigs, *Colenso.*—I do not find this species in the Herbarium, nor any notice of it in Nylander's Synopsis.

5. **C. plicatile,** *Ach.* Thallus greenish-brown, imbricate, lobed; lobes erect or ascending, plaited, crisped. Apothecia scattered, reddish; border entire; spores ellipsoid-fusiform, usually 3-septate.—Nyl. Synops. Lich. 109.

Auckland, *Knight* (*Mitten*). (Europe.)

6. **C. contiguum,** *Knight and Mitt. in Linn. Trans.* 23. 106. *t.* 12. *f.* 35. Thallus dull green, black when dry; lobes tufted, complicate. Apothecia small, crowded, marginal, pale brown; spores ellipsoid, multilocular.

Northern Island: Auckland, on wood, *Knight.*

3. LEPTOGIUM, Acharius.

Thallus as in *Collema*, but usually thinner, all cellular or with a cellular cortical layer. Apothecia with a thalline border, which is concolorous with, or paler than, the surface of the disc; spores in eights, ovoid or ellipsoid, variously septate or divided.

A genus with the habit, etc., of *Collema*, which it closely resembles. The species are very difficult of discrimination.

1. **L. Scoticum,** *Fries ;—Bab. in Fl. N. Z.* ii. 308. Thallus membranous, brown, laciniately lobed, plaited; lobes rounded. Apothecia small, rather concave, same colour as the thallus; spores ovoid.—Nyland. Synops. Lich. 123.

Northern Island, *Colenso*. (Europe, N. Africa.)

2. **L. tremelloides,** *Fries ;—Bab. in Fl. N. Z.* ii. 308. Thallus lead-coloured or somewhat olivaceous, glaucous, membranous, lobed; lobes sometimes imbricate or crisped. Apothecia red, usually elevate; spores ovoid or ellipsoid, 3-septate, narrowed at both ends.—Nyland. Synops. Lich. 124.

Common throughout the **Northern** and **Middle** Islands. (Cosmopolite.)

3. **L. Saturninum,** *Nyland.—Collema,* Bab. in Fl. N. Z. ii. 309. Thallus large, leaden-brown or greenish, lobed, or polyphyllous and complicate; lobes sinuate, often furfuraceous, cinereous, pubescent below. Apothecia dirty brown, flat, almost shining; border entire, cupulate; spores ellipsoid, 3-septate.—Nyland. Synops. Lich. 127.

Northern Island, *Colenso*. (Europe.)—Babington identifies this with an Antarctic, S. African, American, and Indian species, no doubt including *L. Menziesii*, Montagne, and *L. Hildenbrandii*, Nyl., under it; but Nylander separates these latter, by (confessedly) very slight characters.

4. **L. chloromelum,** *Nyland.—L. Brebissonii ?* Bab. in Fl. N. Z. ii. 307. Thallus lead-coloured and somewhat greenish, membranous, lobed, plaited longitudinally, very rugose; lobes waved. Apothecia red or red-brown, plane or concave; border thick, rugose and plaited, or granulate; spores 3–5-septate, ellipsoid or narrowed at both ends.—Nyland. Synops. Lich. 128. *C. rugatum*, Tayl.

Northern Island: Bay of Islands, on trees, *J. D. H.*, etc. (Europe, America, India.)

5. **L. bullatum,** *Nyland.* Thallus lead-coloured, opaque, closely rugulose, membranous, lobed and incised, wrinkles narrow acute undulate. Apothecia dull red, flat or tumid; border thick, rugulose, bullate, often urceolate; spores ellipsoid, 5-septate, narrowed at both ends.—Nyland. Synops. Lich. 129.

New Zealand, (*Nylander*). (Tropical Asia and America.)

4. CALICIUM, Acharius.

Thallus granular, powdery or evanescent, very rarely scaly or obsolete. Apothecia black, stipitate, rarely subsessile, in turbinate capitula, globose. Spores brown or blackish, simple or septate.

Minute, often microscopic Lichens, found on rocks and old wood.

1. **C. curtium,** *Borrer ;—Bab. in Fl. N. Z.* ii. 304. Thallus granular, cinereous or obsolete. Apothecia black; stipes often stout; capitulum cylindric-turbinate; margin white; mass of spores black; spores ellipsoid, black, 1-septate.—Nyland. Synops. Lich. 156.

Northern Island: on dead wood, *Colenso*. (Europe, N. America, N. Asia.)—A very minute plant, belonging to a genus of which the species are most difficult of discrimination.

5. SPHÆROPHORON, Persoon.

Thallus branched, tufted, shrubby, fragile, cylindric, compressed, flattened or more or less scale-like; branchlets dilated at the tips. Apothecia near the tips of the frond, terminal or on the under surface, bursting irregularly, covered with a deciduous black layer of pigment; spores black or dark violet, spherical.

A genus of 4 species, found on rocks and old wood in temperate climates.

1. **S. compressum,** *Ach.;—Bab. in Fl. N. Z.* ii. 304. *t.* 130.—*S. australe*, Laur. Thallus foliaceous, glaucous, blackish at the base, white below, much dilated above and palmately lobed; edges erose. Apothecia very large, with a broad flat thalline border, which is lacerate and scrobiculate or crested. —Nyland. Synops Lich. 170.

Northern and **Middle** Islands: abundant on the ground and stumps of trees, and **Lord Auckland's** and **Campbell's** Islands. (Australia, S. Africa, N. and S. America.)

2. **S. coralloides,** *Pers.;—Bab. in Fl. N. Z.* ii. 307. Thallus terete, pale or white, sometimes lurid, branched. Apothecia small, on globose receptacles, irregularly dehiscing above.—Nyland. Synops. Lich. 171.

Northern and **Middle** Islands, *Colenso*, etc., on rocks, etc. **Lord Auckland's** group, *J. D. H.* (Temperate and cold regions of both hemispheres.)

3. **S. tenerum,** *Laur.;—Bab. in Fl. N. Z.* ii. 304; *Fl. Antarct. t.* 197. *f.* 1. Thallus terete, slender, excessively branched, pale or white; branches dense, slender, intricate. Apothecia on the primary branches, small, the border deciduous, having a black capitulum girt by a circular rim.—Nyland. Synops. Lich. 171. *S. curtum*, Hook. f. and Tayl. in Lond. Journ. Bot. iii. 654.

Abundant throughout the islands and in **Lord Auckland's** group and **Campbell's** Island. (Chili, Antarctic Islands.)

6. BÆOMYCES, Persoon.

Thallus horizontally expanded, crustaceous, scaly powdery or granular. Apothecia sessile or stipitate, pale pink or flesh-coloured or dirty-white, without a thalline border; stipes consisting of longitudinal filaments; spores 1–3-septate. Spermagonia furnished with arthrostigmata.

A genus of some 17 species, found in various temperate climates on the ground, rotten wood, etc.

1. **B. rufus,** *DC.;—Bab. in Fl. N. Z.* ii. 298. Thallus greenish whitish or glaucous, thin, effuse, granular; granules sometimes depressed. Apothecia on long or short white stalks, flesh-coloured, convex; spores simple, ellipsoid-oblong.—Nyland. Synops. Lich. 176. *Biatora byssoides*, Fries; Bab. l. c. 299.

Northern Island: common on banks, etc., *J. D. H.*, etc. (Europe, America, Australia.)

2. **B. roseus,** *Persoon;—Bab. in Fl. N. Z.* ii. 298. Thallus white, granular, crustaceous. Apothecia pale, flesh-coloured or rosy, subglobose; stipes whitish; spores fusiform or oblong-fusiform, simple.—Nyland. Synops. Lich. 179.

Northern Island, *Colenso*. (Europe, America.)

7. CLADONIA, Hoffmann.

Thallus of horizontal scales, giving off erect tubular or laciniate podetia, or shrubby or tufted, and formed of erect podetia, which are often branched or smooth, scaly or granular and sometimes perforated, often dilated at the apex into conical cups. Apothecia terminal on the podetia or its lobes, convex, without thalline border, scarlet red-brown or whitish, rarely blackish; spores small, oblong, simple. Spermagonia with simple or sparingly branched sterigmata.

A very large genus of Lichens, most abundant on alpine and subalpine moorlands; a few species occur on old wood in damp forests.

§ 1. *Apothecia pale or brown.*

α. *Thallus foliaceous, laciniate. Podetia bearing cups.*

1. **C. pyxidata,** *Fries;—Bab. in Fl. N. Z.* ii. 297. Thallus ashy-grey, glaucous or greenish, scaly below or naked; podetia often scaly or partially powdery or furfuraceous; cups cyathiform, well developed. Apothecia pale brown or reddish; spores oblong.—Nyland. Synops. Lich. 193.

Northern and **Middle** Islands and **Lord Auckland's** group: common. (Cosmopolite.)

2. **C. fimbriata,** *Hoffm.;—Bab. in Fl. N. Z.* ii. 297. Thallus as in *B. pyxidata*, but podetia white or glaucous, altogether powdery, and the cups usually narrower.—Nyland. Synops. Lich. 194.

Northern and **Middle** Islands: common. (Cosmopolite.)

3. **C. gracilis,** *Hoffm.;—Bab. in Fl. N. Z.* ii. 297. Thallus ashy, pale or greenish, with scaly leaflets below or naked; podetia elongate, corticate, smooth, subulate or bearing cups. Apothecia sessile or stalked, brown or reddish; spores oblong.—Nyland. Synops. Lich. 196. *C. verticillata*, Flœrke; Bab. in Fl. N. Z. ii. 296. *C. sarmentosa*, Hook. f. and Tayl. in Lond. Journ. Bot. iii. 651.

Northern and **Middle** Islands and **Lord Auckland's** group: common. (Cosmopolite chiefly in cold regions.)

4. **C. decorticata,** *Fries;—Bab. in Fl. N. Z.* ii. 298. Thallus with close-set glaucous scales, which cover the podetia; lowest scales broad, crenate, white or blackish below; podetia cylindric, without cups or with narrow ones. Apothecia brown, usually numerous and confluent; spores oblong.—Nyland. Synops. Lich. 199.

Northern Island: on wood, *Colenso*. (Europe.)—I find no specimens in the Herbarium, and Nylander gives no southern habitat.

5. **C. cariosa,** *Flœrke.—C. degenerans*, Flœrke;—Bab. in Fl. N. Z. ii. 279. Thallus cinereous, glaucous, with crenate scales at the base; podetia naked below, above white granular warted and cancellate; cups unequally digitately parted; divisions fastigiate. Apothecia brown, turgid, large; spores oblong.—Nyland. Synops. Lich. 194.

Northern and **Middle** Islands: Bay of Islands, *Colenso*, to Otago, *Lindsay*. (Cosmopolite.)

β. *Cups none. Podetia imperforate, except sometimes at the axils.*

6. **C. furcata,** *Hoffm.;—Bab. in Fl. N. Z.* ii. 296. Thallus scaly or

naked at the base; podetia pale green white or brownish, smooth, branched; axils not perforate; branches subulate, fertile often fastigiate; cups 0. Apothecia brown or rufous, small; spores oblong.—Nyland. Synops. Lich. 206.

Northern and **Middle** Islands: common. **Kermadec** Islands, *Milne*. (Cosmopolite.)

7. **C. squamosa,** *Hoffm.;—Bab. in Fl. N. Z.* ii. 296. Thallus horizontal, subfoliaceous; scales crenate; podetia white, furfuraceous or scaly, branched; axils pervious, toothed and proliferous; tips forked; fertile crested, subcorymbose; cups 0. Apothecia pale or red brown.—Nyland. Synops. Lich. 209.

Northern Island, *Colenso, Jolliffe*. (Cosmopolite.)

8. **C. rangiferina,** *Hoffm.;—Bab. in. Fl. N Z* ii. 296. Thallus erect, leafless and scaleless; podetia densely tufted, ashy-white, often brown at the tips, smooth, slender, elongate, branched; axils perforate; branches short, spreading or decurved; branchlets subradiate, fertile suberect subcorymbose; cups 0. Apothecia small, brown or pale; spores oblong-fusiform.—Nyland. Synops. Lich. 211.

Abundant throughout the **Northern** and **Middle** islands, **Lord Auckland's** group, **Campbell's** and **Chatham** Islands. (Cosmopolite.)—One of the most common of Lichens in cold and temperate latitudes, often forming extended white patches, 2–4 in. high. The food of the reindeer in arctic regions.

9. **C. capitellata,** *Bab. in Fl. N. Z.* ii. 296. *t.* 130 B. Thallus leafless or with few scales; podetia white or straw-coloured, slender, smooth, straight, subulate, sparingly branched; tips pungent, brown; axils perforate, gaping; cups 0. Apothecia shortly stipitate; disk brown, reflexed, sometimes perforate in age.—*Cenomyce capillata*, Tayl. in Lond. Journ. Bot.

Northern and **Middle** Islands: not uncommon, *Colenso*, etc. (Tasmania.)—Nylander (Synops. Lich. 216) refers this very doubtfully to *C. amaurocræa*, from which, however, as he observes, it seems to differ essentially in the absence of cups and perforate axils.

β. Cups 0. Podetia naked, perforate at the sides, not at the axils.

10. **C. aggregata,** *Eschweiler;—Bab. in Fl. N. Z.* ii. 295. Thallus pale brown or dirty yellow, darker below, of erect, densely tufted, smooth, glabrous, shining, rigid, slender or stout branched podetia; sterile branches forked, with subulate tips; fertile stouter, turgid. Apothecia minute, blackish, crowded; spores oblong.—Nyland. Synops. Lich. 218. *Dufourea collodes*, Tayl. in Lond. Journ. Bot. iii. 650. *Cenomyce*, Fl. Antarct. t. 80. f. 2.

Abundant throughout the islands to **Campbell's** Island, forming large patches on rocks, trees, etc. (Mountains of West Indies, S. America, S. Africa, and S. Asia, also S. Australia.)

11. **C. retipora,** *Flœrke;—Bab. in Fl. N. Z.* ii. 295. Habit of *C. aggregata*, but the podetia forming a most beautifully reticulated open network, white or pale yellow; branchlets obtuse.—Nyland. Synops. Lich. 219.

Northern and **Middle** Islands, and as far south as **Campbell's** Island, forming large tufts on the ground. (Australia, Tasmania.)—A most beautiful Lichen.

§ 2. *Apothecia scarlet.*

12. **C. cornucopiodes,** *Fries;—Bab. in Fl. N. Z.* ii. 298. Thallus yellow, white, scaly below; scales lobed, crenate or incised; podetia smooth

or granular; cups often proliferous. Apothecia bright scarlet, often confluent.—Nyland. Synops. Lich. 220.

Northern Island: on the mountains, *Colenso*. **Middle** Island: probably common; Mount Brewster, *Haast*. (Cosmopolite in temperate and frigid regions.)

13. **C. digitata,** *Hoffm.;—Bab. in Fl. N. Z.* ii. 298. Thallus leafy below; leaflets pale green above, whitish and powdery below, lobed or crenate; podetia white, powdery, naked, cylindric, simple or branched; cups narrow or broad; margins entire toothed or digitate. Apothecia scarlet, solitary or confluent.—Nyland. Synops. Lich. 223.

Northern Island, *Cunningham*, *Colenso*, etc. (Europe, N. America, N. Asia, Tasmania.)

14. **C. macilenta,** *Hoffm.;—Bab. in Fl. N. Z.* ii. 298. Thallus with pale green crenate or incised leaflets or scales at the base; podetia cylindric, white, powdery or granular, slender, simple or divided, obtuse; cups 0 or narrow. Apothecia scarlet, confluent, tubercled.—Nyland. Synops. Lich. 223.

Northern Island, *Cunningham*, etc. (Cosmopolite.)

C. uncialis is enumerated by Montagne as a native of Lord Auckland's group, but not by Nylander, and we have seen no specimens.

8. STEREOCAULON, Schreber.

Thallus usually tufted, formed of solid podetia, with a medullary axis, shrubby, fragile, granular and usually fibrillose, sometimes also bearing cephalodia (globular bodies) on the surface. Apothecia brown, terminal or lateral, without or rarely with a thalline border; spores 8 in an ascus, cylindric-fusiform, septate. Spermagonia mixed with straight or curved subacicular spermatia.

A considerable genus, natives of cold moorlands and rocks, rarely of trunks of trees and dead wood.

1. **S. Colensoi,** *Bab. in Fl. N. Z.* ii. 295. *t.* 130 A. Podetia 1–2 in. high, stout, white, reddish below, rugulose, sparingly branched; branches often spreading, tufted and fibrillose below, naked above; cephalodia 0. Apothecia flat, black, terminal, with a thalline border.—Nyland. Synops. Lich. 232.

Northern Island: on rocks, *Colenso*.

2. **S. ramulosum,** *Ach.;—Bab. in Fl. N. Z.* ii. 295. Podetia 3–5 in. high, stout, erect, with long branches, granular and with simple or divided fibrils, also bearing spherical pedicellate pale cephalodia. Apothecia terminal, pale or brown, without a thalline border.—Nyland. Synops. Lich. 235; Fl. Antarct. t. 80. f. 1; A. Rich. Voy. Astrolabe, t. 9. f. 4.

Abundant throughout the **Northern** and **Middle** Islands, **Lord Auckland's** group, and **Campbell's** Island. (S. America, Australia, Tasmania.)

3. **S. corticulatum,** *Nyland.* Podetia small, short, tufted, branched, firm, erect, compressed, areolate on the surface, granular, powdery above; cephalodia subsessile. Apothecia brown, flat or convex, terminal, without a thalline border.—Nyland. Synops. Lich. 241.

Northern and **Middle** Islands: on calcareous rocks, *Colenso*; Otago, *Lindsay*.

S. denudatum, Flœrke, is enumerated by Babington in the 'New Zealand Flora' (vol. ii. p. 295), but barren. I find no New Zealand specimens in the Herbarium, and Nylander does not allude to it as a native of the islands. *S. corticatulum*, Nyland., is indicated in Lindsay's list of Otago Lichens.

9. ARGOPSIS, Theodore Fries.

Thallus as in *Stereocaulon*, but glabrous, or with few scattered fibrils and granules. Apothecia black, peltate, terminal, dilated, with a delicate thalline margin; spores solitary in the asci, large, murally divided.

A genus of one species, and that a very handsome plant, which Babington suspects may prove to be a form of *Stereocaulon ramulosum*.

1. **A. megalospora,** *Theodore Fries.—Stereocaulon Argus*, Fl. Antarct. 196. t. 79. f. 2. 1–4. Thallus 1–3 in. high, whitish, stout.—Nyland. Synops. Lich. 254.

Campbell's Island: on rocks, *J. D. H.*

10. THAMNOLIA, Acharius.

Thallus consisting of erect or prostrate, cylindric or subcompressed, often flexuose hollow podetia, which are simple or rarely forked, acute, imperforate, white, often rugose. Apothecia unknown. Spermagonia pale as in *Bæomyces*.

A genus of one common species, found on moors, etc., and 2 doubtful ones.

1. **T. vermicularis,** *Ach.* Thallus chalky-white, 2–4 in. long.—Nyland. Synops Lich. 265.

Middle Island, *Sinclair*, on open heathy places. (Europe, N. Asia, Himalaya, Andes, Australia.)

11. USNEA, Hoffm.

Thallus erect and shrubby or pendulous and branched, pale yellow or greenish, terete or angled; áxis solid filiform. Apothecia peltate, with a thalline border, terminal or lateral, concolorous or black. Spermagonia lateral, with simple sterigmata. Spermatia cylindric acicular.

One of the commonest genera of Lichens in all latitudes.

§ 1. *Apothecia of the same colour as the thallus.*

1. **U. barbata,** *Fries;—Bab. in Fl. N. Z.* ii. 268. Thallus white yellow pale grey or glaucous, erect or pendulous, long or short, stout or slender.—Nyland. Synops. Lich. 267.

Var. *florida.* Thallus white, stout, erect, bristly with spreading close-set fibrils.
Var. *dasypoga.* Thallus whitish, elongate, pendulous, fibrillose.
Var. *plicata.* Thallus pendulous, not fibrillose.
Var. *articulata.* Thallus white or yellowish, long, branched, fibrillose or not, jointed or constricted. Apothecia small.
Var. *ceratina.* Thallus erect or pendulous, fibrillous and rough with papillæ.
Var. *trichodea.* Thallus very slender, filiform, pendulous; branches sparsely fibrillose.

Abundant throughout the **Northern** and **Middle** Islands. **Lord Auckland's** group and **Campbell's** Island: on trees, rocks, etc. (Cosmopolite.)

2. **U. angulata,** *Ach.;—Bab. in Fl. N. Z.* ii. 269. Thallus as in *U. barbata*, var. *dasypoga*, but stems and branches angular and scabrous.—Nyland. Synops. Lich. 272.

Northern Island, *J. D. H.*, barren. (Australia, Madagascar, S. America.)

§ 2. NEUROPOGON, *Nees and Flotow.—Apothecia black.*

3. **U. melaxantha,** *Ach.;—Bab. in Fl. N. Z.* ii. 369. Thallus erect straw-coloured or orange-yellow, often banded with black, rigid, dichotomously branched; tops of branches slender, black, smooth or rough. Apothecia large, black.—*Neuropogon*, Nyland. Synops. Lich. 272.

Lofty mountains of the **Northern** and **Middle** Islands, *Colenso*, etc. (Arctic and antarctic regions, Andes and Alps of Tasmania.)

12. ALECTORIA, Ach.

Thallus terete or compressed, erect or pendulous, filiform, intricately branched; axis hollow or spongy. Apothecia usually dark-coloured, with a thalline border; paraphyses not separating; spores ellipsoid. Spermagonia with acicular spermatia.

A small temperate, alpine, and arctic genus.

1. **A. ochroleuca,** *Nyland.—Evernia*, Fries;—Bab. in Fl. N. Z. ii. 269. Thallus erect or prostrate, rigid, terete or compressed, smooth or lacimose; branches often black at the tips. Apothecia rufous or brown, nearly terminal on the subulate top of the branch, sticking out at one side of it.—Nyland. Synops. Lich. 281.

Northern and **Middle** Islands, *Colenso, Bidwill*, etc. (Australia, Bourbon, N. hemisphere generally.)

13. RAMALINA, Ach.

Thallus erect or pendulous, tufted, white or pale, compressed or leafy, laciniate, the two surfaces alike. Apothecia scattered or marginal, pale, with a thalline border; spores oblong, 1-septate, usually curved; paraphyses separate. Spermagonia with branched filaments; sterigmata with few articulations; spermatia cylindric.

A considerable genus, found in all latitudes.

1. **R. calicaris,** *Fries;—Bab. in Fl. N. Z.* ii. 270. Thallus yellow-white, rigid, compressed, linear, laciniose, unequally divided, often channelled, never rigid and cartilaginous. Apothecia subterminal, plane.—Nyland. Synops. Lich. 294.

Var. *fraxinea*. Thallus with long broad laciniæ. Apothecia shortly pedicelled, marginal or lateral.

Var. *fastigiata*. Thallus smaller; laciniæ compressed or terete, subfastigiate. Apothecia subterminal or terminal.

Var. *farinacea*. Thallus covered with soredia.

Var. *Eckloni* (var. *membranacea*, Fl. N. Z. ii. 270). Thallus glabrous, membranous. Apothecia small.

Var. *linearis* (*R. linearis*, Ach.;—Fl. N. Z. ii. 270). Thallus small, narrow, flat, glabrous.

Var. *pusilla*. Thallus short, fastigiate, subterete, inflated; branches obtuse. Apothecia terminal.—*R. inflata*, Hook. f. Fl. Antarct. 194. t. 79. f. 1.

Throughout the **Northern** and **Middle** Islands abundant. Var. *pusilla*, **Lord Auckland's** group, on maritime rocks. (Cosmopolite.)

2. **R. scopulorum,** *Ach.* Thallus pale, shining, excessively variable in stature and branching, terete or compressed, rigid and cartilaginous. Apothecia subpedicelled, subterminal or lateral.—Nyland. Synops. Lich. 292.

Northern Island, *Jolliffe*. (Cosmopolite.)

3. **R. usneoides,** *Fries;—Bab. in Fl. N. Z.* ii. 270. Thallus pale, compressed, elongate, flaccid, linear, excessively branched, and as it were longitudinally nerved, often twisted. Apothecia marginal.—Nyland. Synops. Lich. 291.

Northern Island, *J. D. H.*

14. PLATYSMA, Hoffmann.

Thallus rigid, shrubby or dilated, lobed or prostrate, erect or ascending, laciniate, rarely fistular, with a spongy axis. Apothecia marginal, with a thalline border, reddish-brown or dark; spores small, simple; paraphyses not separating. Spermagonia marginal, enclosed in spiculæ or tubercles or papillæ; spermatia various.

A considerable temperate, alpine, and Arctic genus.

1. **P. sæpincola,** *Hoffm.—Cetraria,* Ach.;—Bab. in Fl. N. Z. ii. 271. Thallus chesnut-brown or dark olive-brown, prostrate, ½–2 in. broad; laciniæ prostrate or ascending, smooth or laciniate, sinuate or crenate, paler below. Apothecia brown, rugulose beneath; border entire.

Auckland, *Sinclair*. Specimens imperfect and doubtful. (Northern hemisphere, Fuegia.)

Platysma glaucum and *Cetraria aculeata*, Fries, are enumerated by A. Richard amongs the Lichens brought by D'Urville from New Zealand, but I have seen no specimens, and Nylander does not give the islands as a habitat. They are abundant northern plants, also occurring in Fuegia.

15. NEPHROMA, Ach.

Thallus foliaceous, prostrate, dilated, upper surface shining, under soft, without veins. Apothecia adnate to the lower recurved margin of the thallus, peltate; spores 3-septate. Spermagonia marginal.

A small arctic and antarctic genus, rarer in the cold, temperate, and alpine regions.

§ 1. *Thallus, when cut across, showing a green stratum of separate granules.*

1. **N. australe,** *A. Rich.;—Bab. in Fl. N. Z.* ii. 371. Thallus whitish or straw-coloured, 2–4 in. broad, smooth, laciniate and lobed, sinuate, reddish here and there; beneath smooth, white or straw-coloured. Apothecia small, red-brown.—Nyland. Synops. Lich. 318. *N. pallens*, Nyland. Enum. Lich. 101.

Northern and **Middle** Islands: common, *D'Urville*, etc. (Tasmania.)—This is probably the *N. antarcticum*, Jacq., of Lindsay's 'Enumeration of Otago Lichens.'

2. **N. schizocarpum,** *Nyland.—N. resupinatum,* var. *pruinosa*, Mont.;—Bab. in Fl. N. Z. ii. 272. Thallus 2–3 in. broad, livid-brown, smooth,

lobed, crenate; below smooth, of the same colour, with pale margins or altogether pale. Apothecia reddish, longitudinally divided.—Nyland. Synops. Lich. 318.

Banks's Peninsula, *Hombron*, on bark.

§ 2. NEPHROMIUM, *Nylander.—Stratum of green granules in moniliform filaments.*

3. **N. lævigatum,** *Ach.—N. resupinatum*, var. *rufum*, Bab. in Fl. N. Z. ii. 272. Thallus livid, chesnut-brown, lobed or laciniate, toothed and imbricate, pale and glabrous below. Apothecia red-brown, border crenulate and fimbriate.—*Nephromium*, Nyland. Synops. Lich. 320.

Northern and **Middle** Islands, *Colenso and Lyall.* (Europe, N. America, Chili, S. Africa, Java.)

4. **N. cellulosum,** *Ach.* Thallus livid, chesnut-brown, reticulately foveolate and laciniate, 2–4 in. broad, lobed, beneath white, glabrous and bullate. Apothecia rufous.—*Nephromium*, Nyland. Synops. Lich. 321.

New Zealand, (*Nylander.*) (Tasmania, Fuegia, Juan Fernandez.)

5. **N. Lyallii,** *Bab. in Fl. N. Z.* ii. 272. *t.* 127 A. Thallus livid, brown, with a broad bluish margin, membranous, shining beneath, pale and rugulose; lobes laciniate. Apothecia pale red, with toothed fimbriate margins.—*Nephromium*, Nyland. Synops. Lich. 322.

Northern Island, *Colenso.* **Middle** Island, *Lyall.*

16. PELTIGERA, Ach.

Thallus prostrate, dilated, foliaceous, membranous, fragile, lobed, upper surface often shining, under spongy, with broad veins. Apothecia marginal on the upper surface of the lobes of the frond.

A genus of about 8 alpine arctic and temperate Lichens, unknown in tropical climates.

1. **P. rufescens,** *Hoffm.;—Bab. in Fl. N. Z.* ii. 324. Thallus small, pale reddish or ashy; margin crisped beneath, pale below, with white rootlets; veins often obscure or brown.—Nyland. Synops. Lich. 325.

Var. *spuria. P. canina*, var. *spuria*, Bab. in Fl. N. Z. ii. 324.—Thallus digitately lobed.

Common throughout the **Northern** and **Middle** islands. (Cosmopolite.)

2. **P. polydactyla,** *Hoffm.;—Bab. in Fl. N. Z.* ii. 324. Thallus pale, livid brown or glaucous, shining, smooth or impressed; beneath white, reticulated with brown nerves; fertile lobes narrow. Apothecia on ascending prolonged lobes.—Nyland. Synops. Lich. 326.

Throughout the **Northern** and **Middle** islands, and in **Campbell's** Island. (Cosmopolite.)

17. STICTA, Ach.

Thallus prostrate, spreading from a centre, foliaceous, lobed or laciniate, often very large, leathery, membranous or brittle; upper surface shining or smooth, often bearing soredia; white yellow or green internally; lower surface spongy or fibrous, rarely smooth, usually furnished with white or yellow cyphellæ (powdery or smooth, white or yellow warts or cups). Apothecia superficial, with a thalline border; paraphyses separate; spores fusiform, usually 1–3-septate. Spermagonia scattered, scarcely protruding, with arthrosterigmata.

A very large genus, temperate and tropical, not arctic nor alpine, most abundant in New Zealand on the ground and trees.

§ 1. Stictina, *Nylander.—Stratum of granules blue-green or glaucous-green; granules without a cellular membrane.*

a. Cyphellæ minute, pulverulent, white.

1. **S. argyracea,** *Delise;—Bab. in Fl. N. Z.* ii. 281. Thallus reddish or pale brown, 3–10 in. broad, smooth, laciniate; laciniæ linear, subpinnatifid, fracture white, upper surface with scattered white soredia, beneath whitish or yellowish, tomentose. Apothecia small, red-brown, marginal or submarginal; thalline border entire.—*S. aspera,* Lam. *Stictina,* Nyland. Synops. Lich. 334.

Northern and **Middle** Islands, *Colenso,* etc. (Madagascar, Java, India, S. Africa, Polynesia.)

2. **S. fragillima,** *Bab. in Fl. N. Z.* ii. 279, *excl. var. β.* Thallus yellowish or glaucous, thin, rigid, fragile, smooth or rough here and there, fracture white; laciniæ linear, pinnatifid, dichotomous; beneath ochraceous, subcostate, tomentose. Apothecia reddish or brown, scattered; thalline border toothed or crenulate.—*Stictina,* Nyland. Synops. Lich. 337.

Northern Island, *Dieffenbach,* etc. (Australia, Tasmania, Peru.)—Nylander excludes var. *glaberrima,* Bab., which has urceolate cyphellæ, and may be referable to *S. cinnamomea,* A. Rich. (see remarks at end of genus).

3. **S. Hookeri,** *Bab. in Fl. N. Z.* ii. 282. *t.* 125 B. Thallus lurid, æneous or glaucous or brownish, 3–4 in. broad, rigid, scarcely shining, scrobiculate; lobes broad, rounded, crenate, with minute furfuraceous tubercles on the costæ and margin, brownish below or blackish, with pale papilliform cyphellæ. Apothecia black, scattered; thalline border entire.—*Stictina,* Nyland. Synops. Lich. 336.

Northern Island: Bay of Islands, etc., *Colenso,* etc.

β. Cyphellæ minute, pulverulent, yellow.

4. **S. crocata,** *Ach.;—Bab. in Fl. N. Z.* ii. 275. Thallus lurid, brownish, 3–5 in. broad, opaque or shining, fracture white, broadly lobed; lobes crenate, scrobiculate, sometimes reticulately costate, with orange-yellow soredia; beneath brown or black. Apothecia scattered; thalline border crenate.—*Stictina,* Nyland. Synops. Lich. 338.

Common throughout the **Northern** and **Middle** islands. (Cosmopolite in temperate and warm damp climates.)

5. **S. carpoloma,** *Delise;—Bab. in Fl. N. Z.* ii. 276. *t.* 126. Thallus glaucous, yellowish or ashy, 3–5 in. broad, rigid, almost shining; laciniæ linear, with narrow retuse lobes, often scrobiculate; fracture white; beneath brown or pale, with small yellow cyphellæ. Apothecia black, marginal thalline border smooth.—*Stictina,* Nyland. Synops. Lich. 339.

Var. *granulata*.—*S. granulata*, Bab. l. c. 281. Thallus thicker, ashy-white; cyphelIæ sometimes white.—*Stictina*, Nyland. l. c. 340.

Northern and **Middle** Islands: not uncommon. (Bourbon, Java, Polynesia, Tasmania, Chili, Fuegia.)

γ. *Cyphellæ urceolate.*

6. **S. filicina,** *Ach.*;—*Bab. in Fl. N. Z.* ii. 276. Thallus pale yellowish, smooth, more or less stipitate, incised and lobed; lobes costate below at the base or down the middle; margin sinuous; beneath pale ochreous, tomentum thin or 0. Apothecia reddish or brown, scattered; thalline border entire or crenulate.—*Stictina*, Nyland. Synops. Lich. 349.

Var. *latifrons*. Thallus larger, brownish, stipitate.—*S. latifrons*, A. Rich, Fl. N. Z. 27. t. 8. f. 2; Bab. in Fl. N. Z. ii. 277.

Var. *Menziesii*. Thallus thick, orbicular, more stipitate, brown or blackish below.—*S. Menziesii*, Fl. Antarct. 198. *S. latifrons*, var. *Menziesii*, Fl. N. Z. ii. 277. t. 122.

Common throughout the **Northern** and **Middle** islands, in alpine and more southern localities. **Lord Auckland's** group, *J. D. H.* (S. America, Java, Tasmania.)

§ 2. STICTA, *Nylander*.—*Stratum of granules green or yellow-green; granules with a cellular membrane.*

α. *Cyphellæ urceolate, white.*

7. **S. damæcornis,** *Ach.* Thallus 4–6 in. broad, lurid or pale red-brown, more or less shining, smooth; laciniæ linear, pinnatifid, dichotomous; lobes obtuse; beneath ochraceous or blackish, tomentum often sparse or 0. Apothecia red-brown; thalline border entire or crenulate.—Nyland. Synops. Lich. 357.

Var. *sinuosa*. Thallus pale, appressed, more sinuate, pinnatifid; lobes often broader.—*S. sinuosa*, Pers.;—Bab. in Fl. N. Z. ii. 280.

Var. *macrophylla*. Thallus thicker, darker; lobes broader.

Common throughout the **Northern** and **Middle** islands. (S. America, S. Africa, Java, Polynesia, Australia.)

8. **S. variabilis,** *Ach.*;—*Bab. in Fl. N. Z.* ii. 280. Thallus 3–6 in. broad, pale glaucous or brownish, membranous, rather rigid; laciniæ sub-pinnatifid; margin sinuate-crenulate, deflected; beneath brown, with pale margins, tomentose. Apothecia red-brown, marginal or submarginal; thalline border crenulate.—Nyland. Synops. Lich. 357.

Northern and **Middle** Islands: common. **Lord Auckland's** group, (*Montagne*.) (Mauritius, Polynesia, Australia.)

9. **S. cinereo-glauca,** *Tayl.*;—*Bab. in Fl. N. Z.* ii. 283. *t.* 127. *f.* C. Thallus ashy or glaucous, about 3 in. broad, scarcely shining, lobed; lobes crowded, sinuate, crenate or repand, smooth; beneath pale, with whitish tomentum. Apothecia small, reddish; thalline border obsolete or crenulate.—Nyland. Synops. Lich. 358.

Northern Island, *Colenso*, etc.

β. *Cyphellæ punctiform, orange-yellow.*

10. **S. orygmæa,** *Ach.*;—*Bab. in Fl. N. Z.* ii. 274. Thallus 3–8 in. broad, yellowish or pale livid and glaucous, membranous or rigid, almost shining; fracture bright yellow, unequally closely scrobiculate or reticulately costate; lobes rounded, with large unequal crenatures, internally yellow,

beneath ochraceous, tomentose. Apothecia dark brown or black, glabrous below; thalline border crenulate.—Nyland. Synops. Lich. 361; Mont. in Voy. au Pôle Sud, t. 15.

Abundant throughout the **Northern** and **Middle** Islands, and in **Lord Auckland's** group and **Campbell's** Island. (Tasmania, Fuegia.)

11. **S. Urvillei,** *Delise;—Bab. in Fl. N. Z.* ii. 273. Thallus very similar to *S. orygmæa.* Apothecia with a thick rugose border.—Nyland. Synops. Lich. 360. *S. endochrysa*, Fl. Antarct. t. 195. f. 2, not of Delise.

Var. *Colensoi*, Nylander. Thallus firmer, more scrobiculate, margins granular, or lobulate and sorediiferous.

Common throughout the **Northern** and **Middle** Islands. (Tasmania, Fuegia.)

12. **S. aurata,** *Ach.;—Bab. in Fl. N. Z.* ii. 273. Thallus glaucous, testaceous or coppery-red, broad, firm, smooth, not pitted, lobed; lobes sinuate; margins often covered with golden powder; fracture golden; beneath blackish-brown, shortly tomentose. Apothecia large, blackish, marginal or submarginal; thalline border usually inflexed.—Nyland. Synops. Lich. 361. *S. angustata*, Delise, Stict. t. 3. f. 2.

Common throughout the **Northern** and **Middle** Islands, rare in fruit. (Cosmopolite.) —A beautiful Lichen.

γ. *Cyphellæ punctiform, white.*

13. **S. fossulata,** *Dufour,—S. Richardi*, Mont., and *S. foveolata*, Del.; —Bab. in Fl. N. Z. ii. 277, 278. Thallus pale, lurid or glaucous, 5–12 in. broad, scrobiculate, foveolate and transversely reticulate; laciniæ linear, subpinnatifid, intricate, retuse; beneath pale or brown, tomentum sometimes evanescent. Apothecia red-brown or black, marginal, flat; thalline border at length excluded.—Nyland. Synops. Lich. 363. *S. Billardieri*, Bab. *S. impressa*, *S. cellulifera*, and *S. linearis*, Fl. Antarct.

Var. *β. physciospora*, Nyland. Spores 2-locular, brown.

Throughout the **Northern** and **Middle** Islands, abundant, and in **Lord Auckland's** group and **Campbell's** Island.—Var. *β.* in **Lord Auckland's** group, *J. D. H.*, and Otago, *Lindsay*. (Australia, Tasmania, Chili.)

14. **S. Freycinetii,** *Delise;—Bab. in Fl. N. Z.* ii. 280. Thallus pale, usually ochraceous, 5–10 in. broad, scarcely rigid or shining, smooth or obsoletely rugose, laciniately lobed; lobes sinuate, crenate; margins often fringed with white soredia; beneath brown or ashy, tomentose. Apothecia red; thalline border inflexed when young, fimbriate or crenulate when mature. —Nyland. Synops. Lich. 365; Fl. Antarct. t. 196. *S. glaber*, Hook. f. and Tayl.

Abundant throughout the **Northern** and **Middle** Islands, and in **Lord Auckland's** group and **Campbell's** Island. (Australia, Tasmania, Chili, Fuegia.)

Besides the above, various other *Sticta* are enumerated in the 'New Zealand Flora' which are not confirmed by Nylander, as *S. Mougeotiana*, a tropical species, stated to be found by D'Urville; *S. fuliginosa*, Ach., and *S. limbata*, Ach., both Northern species, to which imperfect specimen of New Zealand plants were referred by Taylor or Babington, with more or less doubt. I do not find in Nylander's Synopsis *S. cinnamomea*, A. Rich. (Fl. N. Z. 28. t. 8. f. 3) doubtfully referred by Babington to *S. fragillima*, *β. glaberrima*.

In Lindsay's enumeration of Otago Lichens, I find the following, which I have not identified :—*S. dissimilis*, Nyland., *fuliginosa*, Dicks., *subcoriacea*, Nyland. n. sp., *dissimulata*, Nyland., and *episticta*, Nyland. n. sp.

18. RICASOLIA, De Notaris.

Thallus of *Sticta*, but usually without soredia on the upper surface, or cyphellæ on the lower (except *R. coriacea*). Apothecia with a prominent margin. Spermagonia in prominent mamillæ.

A genus of 14 tropical and temperate Lichens.

1. **R. coriacea**, *Nyland.*—*Sticta*, Tayl. ;—Bab. in Fl. N. Z. ii. 283. t. 125 A. Thallus pale or lurid yellow, 5–6 in. broad, smooth, with unequal depressions, divided and lobed; lobes lobulate, crenate and pilose; beneath white, tomentose, with white pulverulent cyphellæ. Apothecia rufous, scattered; margin fimbriate.—Nyland. Synops. Lich. 367.

Northern and **Middle** Islands : common, *J. D. H.*, etc.

2. **R. glomulifera**, *De Notaris.*—*Sticta*, Del. ;—Bab. in Fl. N. Z. ii. 284. Thallus pale, glaucous, 6–12 in. broad, opaque, thickly membranous, smooth or rugose, laciniate; lobes crowded, sinuate-lobulate; beneath pale, with brown or white rootlets. Apothecia rufous or brown; margin entire, usually inflexed.—Nyland. Synops. Lich. 369.

Northern and **Middle** Islands: on bark, *Colenso;* Otago, *Lindsay.* (Europe, America.)

3. **R. herbacea**, *De Notaris.*—*Sticta*, Del. ;—Bab. in Fl. N. Z. ii. 284. Thallus pale, lurid or glaucous, 6 in. broad and upwards, membranous, shining, smooth or rugulose; lobes crowded, crenate and undulate; beneath pale with white rootlets. Apothecia rufous, large; margin entire or obsoletely crenulate, opaque, obsoletely rugulose.—Nyland. Synop. Lich. 369.

New Zealand, *Raoul.* (Europe, America, S. Africa.)

4. **R. Montagnei**, *Nyland.*—*Sticta*, Bab. in Fl. N. Z. ii. 284. Thallus glaucous, subrufous, foliaceous, corrugated; lobes rounded, subsinuate, crenate, scaly at the margins; beneath naked, ferrugineous or blackish, with a few obsolete white cyphellæ; fracture white. Apothecia free, broad, scattered, often crowded, black; margin deeply inflexed, often lacerate, subserrate and leafy.—Nyland. Synops. Lich. 373.

Northern and **Middle** Islands : on wood, *Sinclair, Lyall.*

19. PARMELIA, Ach.

Thallus spreading from the centre, usually foliaceous, membranous or coriaceous, laciniate or lobed, rarely with terete fistulous lobes; upper surface usually shining; internal substance of matted fibrils; under surface with fibrils. Apothecia with a thalline border, scattered, often shining; paraphyses separate or not separate; spores ellipsoid. Spermagonia innate; spermatia various.

One of the largest genera of Lichens, found in all latitudes.

§ 1. PARMELIA, *Nyland.—Paraphyses not separate. Spores simple. Spermatia usually acicular.*

a. Thallus fibrillous (not densely tomentose) below, rarely glabrous.

1. **P. caperata,** *Ach.* Thallus orbicular, greenish- or yellowish-white, broad, membranous, often imbricate; lobes lobulate or granulate and crenate, rugulose or granulate here and there; beneath black and hispid. Apothecia scattered, bright chesnut; border elevated and incurved, crenulate and often powdery.—Nyland. Synops. Lich. 376.

Northern and **Middle** Islands: on trees, *Colenso,* etc. (Cosmopolite.)

2. **P. perforata,** *Ach.;—Bab. in Fl. N. Z.* ii. 285. Thallus orbicular, glaucous-green or white, membranous, naked, deeply lobed and crenulate; margin fringed with black hairs; beneath black and rough. Apothecia red-brown, concave, deep, at length perforated, margin entire.—Nyland. Synops. Lich. 377.

Northern and **Middle** Islands: on trees, abundant. **Lord Auckland's** group, *Le Guillon.* (Cosmopolite.)

3. **P. perlata,** *Ach.;—Bab. in Fl. N. Z.* ii. 284. Thallus orbicular, glaucous, with rounded sinuate crenate flattened lobes, often bordered with granules or powdery; beneath black and hairy. Apothecia elevated, olive-brown, concave; margin thin, inflexed.—Nyland. Synops. Lich. 379.

Var. *ciliata.* Thallus more naked below; margin ciliated with black hairs. Apothecia more stipitate; margin sorediate.—*P. proboscidia,* Tayl.;—Bab. in Fl. N. Z. ii. 285.

Northern and **Middle** Islands: on trees, common. (Cosmopolite.)

4. **P. tiliacea,** *Ach.;—Bab. in Fl. N. Z.* ii. 285. Thallus orbicular, membranous, grey-glaucous, smooth or rugulose, subpruinose, lobed and sinuate, crenate; beneath brown-black, shaggy. Apothecia brown; margin entire or crenate.—Nyland. Synops. Lich. 382.

Northern Island, *Colenso.* (Cosmopolite.)

5. **P. lævigata,** *Ach.* Thallus greyish-white, spreading, smooth, deeply cut into multifid lobes and segments, the ultimate broadly linear, acute, with terminal powdery warts; beneath black and shaggy. Apothecia dark chesnut, concave; margin entire, inflexed.—Nyland. Synops. Lich. 384.

Var. *sinuosa.* Thallus yellowish.

Middle Island: Otago, on gneiss and basalt (var. *sinuosa*), *Lindsay.* (Europe, etc.)

6. **P. saxatilis,** *Ach.;—Bab. in Fl. N. Z.* ii. 285. Thallus orbicular, grey, deeply lobed and sinuate; segments imbricate, retuse, rough with pits and reticulated powdery lines; beneath black and shaggy. Apothecia dark brown; margin inflexed, entire or crenate.—Nyland. Synops. Lich. 388.

Northern and **Middle** Islands, *Colenso;* trap rocks, Otago, *Lindsay.* (Cosmopolite.)

7. **P. conspersa,** *Ach.;—Bab. in Fl. N. Z.* ii. 286. Thallus whitish or pale yellow-green, orbicular, membranous, lobed and sinuate, granulate in the centre, and with scattered raised dark points; beneath brown with dark fibres. Apothecia near the centre, dark chesnut; border inflexed, entire or lobed.—Nyland. Synops. Lich. 391.

Northern and **Middle** Islands: on stones, common. (Cosmopolite.)

8. **P. olivacea,** *Ach.;—Bab. in Fl. N. Z.* ii. 286. Thallus orbicular, olive-brown, rugged in the centre and often granulated; margin lobed and crenate; lobes appressed; beneath brown and fibrous. Apothecia brown, concave, with an inflexed entire or crenated margin.—Nyland. Synops. Lich. 395.—*P. imitatrix*, Tayl.

Northern Island: on stones, *Colenso.* **Middle** Island: Otago, on basalt, *Lindsay.* (Almost cosmopolite.)

β. Thallus glabrous below, inflated at the end.

9. **P. physodes,** *Ach.;—Bab. in Fl. N. Z.* ii. 286. Thallus orbicular, stellate, glaucous-white; segments sinuate, multifid, convex, glabrous, inflated, often with powdery warts; beneath brown-black. Apothecia red-brown; margin thin, elevated.—Nyland. Synops. Lich. 401. *P. enteromorpha*, Ach.; Fl. Antarct. 532.

Common throughout the **Northern** and **Middle** Islands, and in **Lord Auckland's** group and **Campbell's** Island: on bushes and trees. (Cosmopolite.)

10. **P. pertusa,** *Schærer.—P. diatrypa*, Ach.;—Bab. in Fl. N. Z. ii. 286. Thallus substellate, greenish-grey; segments sinuate, multifid, nearly plane, smooth, with powdery warts; extremities inflated, perforated. Apothecia red-brown; margin inflexed, entire.—Nyland. Synop. Lich. 402.

Northern and **Middle** Islands: on trunks of trees and rocks, common; also in **Lord Auckland's** group, *J. D. H.* (Cosmopolite in temperate regions.)

γ. Thallus densely tomentose below.

11. **P. angustata,** *Pers.—P. moniliformis*, Bab. in Fl. N. Z. ii. 287. t. 127 B. Thallus suborbicular, yellowish and glaucous, covered with black dots, rugose and warted in age, divided almost to the centre into numerous branching narrow lobes, which are repeatedly forked, decumbent, constricted here and there; beneath tomentose with short tufts of crisped fibres, almost spongy. Apothecia large, elevated, free, chesnut-brown, villous below, with crisped hairs; margin entire when young, lobed and crisped in age.—Nyland. Synops. 403.

Northern Island: on wood, *Colenso.* (Australia, Tasmania.)

§ 2. Physcia, *Nyland.—Paraphyses separate. Spores rarely simple, usually 2-locular. Sterigmata many, articulate. Spermatia cylindric.*

a. Thallus yellow, slender, much branched, erect or ascending.

12. **P. flavicans,** *DC.—Evernia*, Fries.—Bab. in Fl. N. Z. ii. 269. Thallus erect, slender, excessively branched, tufted, tawny, compressed, angular, wavy, warted; branches divaricating, tapering. Apothecia lateral, nearly sessile, flat, orange; border pale, narrow, entire.—*Physcia*, Nyland. Synop. Lich. 406. *Borrera*, Ach.

Northern Island, *Colenso* (barren); Nelson, *Sinclair.* (Cosmopolite.)

β. Thallus prostrate.

13. **P. chrysophthalma,** *DC.;—Bab. in Fl. N. Z.* ii. 287. Thallus small, erect or ascending, tufted, branched, bright greenish or whitish-yellow, alike on both sides; branches or segments linear, multifid, fringed. Apo-

thecia abundant, terminal, orange; margin fringed.—*Physcia,* Nyland. Synops. Lich. 410. *Borrera,* Ach.

Northern Island: on branches of trees, *Colenso, Knight.* (Cosmopolite.)

β. *Thallus orange-yellow, prostrate.*

14. **P. parietina,** *Ach.;—Bab. in Fl. N. Z.* ii. 287. Thallus orbicular, small, bright yellow; lobes marginal, radiating, appressed, rounded, crenate, crisped and granulate in the centre; beneath paler and fibrillose. Apothecia deep orange, concave; margin entire.—*Physcia,* Nyland. Synops. Lich. 410.

Northern and **Middle** Islands: common on rocks, wood, bones, etc. (Cosmopolite.)

γ. *Thallus white, free, erect or decumbent, very narrow.*

15. **P. leucomela,** *Ach.;—Bab. in Fl. N. Z.* ii. 288. Thallus spreading, smooth, white on both surfaces, linear, channelled and powdery beneath; segments linear, fringed with long black slender hairs. Apothecia scattered, blue-black, with white radiating hairs on the border.—*Physcia,* Nyland. Synops. Lich. 414. *Borrera,* Ach.

Northern Island: probably common, *Colenso, Sinclair,* etc. (Cosmopolite.)

δ. *Thallus horizontally spreading, white or grey.*

16. **P. speciosa,** *Ach.;—Bab. in Fl. N. Z.* ii. 288. Thallus stellate, imbricate, greenish-white, somewhat cartilaginous, cut into numerous linear multifid segments, tips obtuse and powdery; beneath snow-white with grey fibres. Apothecia brown, with an inflexed lobed or leafy border.—*Physcia,* Nyland. Synops. Lich. 416.

On stones, wood, mosses, etc., common in the **Northern** Island: Raoul Island, *Milne.* (Cosmopolite.)

17. **P. pulverulenta,** *Ach.;—Bab. in Fl. N. Z.* ii. 287. Thallus orbicular, stellate, glaucous-green, hoary, ashy when dry, cut into numerous oblong multifid flat obtuse wrinkled lobes; beneath black, downy. Apothecia glaucous, black; border thick, inflexed, at length leafy.—*Physcia,* Nyland. Synops. Lich. 419.

Middle Island: Cook's Straits, *D'Urville.*—I have seen no New Zealand specimens. (Europe and N. America.)

18. **P. stellaris,** *Ach.;—Bab. in Fl. N. Z.* ii. 288. Thallus orbicular, stellate, pale grey; segments linear, rather convex, multifid; beneath whitish, with dark fibres. Apothecia grey-black; margin entire, elevated, inflexed.—*Physcia,* Nyland. Synops. Lich. 424.

Northern Island: common on bark of trees, *Colenso,* etc. (Cosmopolite.)

19. **P. cæsia,** *Ach.* Thallus as in *stellaris,* but the segments have grey powdery warts in the centre.—*Physcia,* Nyland. Synops. Lich. 426.

Middle Island: trap rocks, Otago, *Lindsay.* (Europe, N. America.)—I have seen no specimens from New Zealand.

20. **P. obscura,** *Schær.* Thallus as in *P. stellaris,* but livid brown or ashy-brown. Apothecia brownish-black.—*Physcia,* Nyland. Synops. Lich. 427.

New Zealand. (*Nylander.*)

21. **P. picta ?,** *Ach.;—Bab. in Fl. N. Z.* ii. 288. Thallus (almost of *Placodium*) orbicular, white, appressed; laciniæ flat, imbricate, contiguous, imbricate or subconfluent, narrow, crenate; beneath naked, opaque, black, or white at the margins only. Apothecia black, flat, small; margin slightly elevate, crenulate.—*Physcia*, Nyland. Synops. Lich. 430.

Northern Island, *Colenso*, barren. (Bourbon, Ceylon, Java, Polynesia.)—A doubtful identification.

Lindsay enumerates amongst his Otago Lichens, *P. Mougeotii*, Schærer, and *P. (Physcia) plinthiza*, Nyland. n. sp.

20. UMBILICARIA, Hoffmann.

Thallus membranous or coriaceous, of one peltate lobed lamina, upper and under surfaces opaque. Apothecia without a thalline border, often sinuous or gyrose; paraphyses separate. Spermagonia with arthrostigmata; spermatia slender, shortly cylindric, obtuse at both ends.

A considerable arctic and alpine genus, of which some species, under the name of "tripe de roche," are used for food by the Canadian hunters when famishing.

1. **U. polyphylla,** *Schrader*. Thallus naked and smooth on both surfaces, dark olive-brown, shining; beneath dark black; margin lobed, simple or many-leaved. Apothecia convex, rough and plaited.—Nyland. Lich. Scand. 119.

Middle Island: gneiss rocks, Ctago, *Lindsay*. I have seen no specimen. (Europe, Asia, America.)

21. PSOROMA, Fries.

Thallus scaly, cellular, granules large, distinct, enclosed in a cellular envelope. Apothecia usually scattered; border crenate; paraphyses separate; spores 8, large, simple. Spermagonia with arthrostigmata.

A considerable temperate genus.

1. **P. subpruinosum,** *Nyland*. Thallus broad, leafy, lobed, appressed; border free, sinuate, subcrested with granular scales, above reddish-green, woolly on the margins, subsilky beneath. Apothecia with a chesnut-brown disk and subfoliaceous crenulate margin, granular and corrugated beneath.—*P. rubiginosa*, var. *araneosa*, Bab. in Fl. N. Z. ii. 289.

Northern and **Middle** Islands: on mosses, wood, and rocks, *Colenso, J. D. H.*; Otago, *Lindsay*.

2. **P. hypnorum,** *Fries*. Thallus spreading, of small greenish- or yellow-brown scales, which are rounded, crenulate, and often granular on the margin. Apothecia red or brown, flat, with a thin inflexed lobed or crenate border.—Nyland. Lich. Scand. 121.

Var. *coralloideum*, Nyland. Enum. 109.

Northern Island, *Colenso*. **Campbell's** Island, *J. D. H.*, var. *corallinum*. (*L. hypnorum* is cosmopolite, the var. *coralloideum* is only known from New Zealand.)

3. **P. sphinctrinum,** Nyland.—*Parmelia rubiginosa*, *β*, Bab. in Fl. N. Z. ii. 289. Thallus stellately laciniate, dull red-brown; laciniæ sublinear,

inciso-multifid, flattish, scaly, granular and crenulate at the edges; beneath brown, adhering by the whole surface by means of short matted fibrils. Apothecia crowded; thalline border crenulately striate. *Parmelia sphinctrina*, Mont. Voy. au Pôle Sud, t. 15. f. 3.

Common throughout the **Northern** and **Middle** Islands, on rocks, wood, and the ground, and in **Lord Auckland's** group. (Chili, Australia.)—I can hardly distinguish this from *Pannaria rubiginosa.*

4. **P. xanthomelanum,** *Nyland. in Herb. Hook., and in Dist. Psorom. and Parmel.*, without description.

In various parts of New Zealand, *Lindsay and Colenso.*

P. euphyllum, Nyland. Dist. Psorom. and Parm. I find no published description of this, which is stated to be remarkable for its laciniate thallus.

22. PANNARIA, Delise.

Thallus laciniate and radiate, or granular and scaly, cellular; granules without a cellular coat. Apothecia with or without a border; spores 8, almost always simple. Spermagonia with arthrostigmata.

A genus of temperate and cold regions.

1. **P. Gayana,** *Nyland.—Parmelia*, Mont.;—Bab. in Fl. N. Z. ii. 288. Thallus foliaceous, membranous, leaden-ashy, laciniate at the circumference; laciniæ broad, rounded, subentire, concentrically furrowed; beneath blueish-white, tomentose. Apothecia scattered and crowded, reddish, with a single or double border, one thalline, the other of the substance of the disk.—*Parmelia*, Montagne in Ann. Sc. Nat. ser. 3. ii. 58.

Northern Island, *Colenso.* (Chili.)

2. **P. rubiginosa,** *Delise;—Bab. in Fl. N. Z.* ii. 289. Thallus orbicular, livid-red, cut towards the circumference into broad notched imbricate lobes; beneath blue-back, spongy with fibres. Apothecia rusty-red, nearly flat, with a thick inflexed crenated border.—Nyland. Lich. Scand. 122. *Parmelia Femsjonensis*, Fries?; Bab. l. c. 291.

Var. *conopsea.* Thallus thinner, with scattered blue gonidia (*P. conopsea*, Bab. in Fl. N. Z. ii. 290).

Northern Island: probably common, on moss, bark, etc. **Lord Auckland's** group: common. Apothecia blacker. (*P. Mariana*, Fries?, Bab.) (Cosmopolite.)

3. **P. nigrocincta,** *Nyland.—Parmelia*, Mont.;—Bab. in Fl. N. Z. ii. 290. Thallus of close-set membranous reddish scales radiating from a centre, rounded, incised; beneath bluish-black. Apothecia red-brown, flat; border pale, entire.—*Parmelia*, Mont. in Ann. Sc. Nat. ser. 2. iii. 91.

Northern Island, *Colenso.* (Tropical America, Africa, Java, Juan Fernandez.)

4. **P. pholidota,** *Nyland.—Parmelia*, Mont.;—Bab. in Fl. N. Z. ii. 290. t. 128 A. Thallus suborbicular, 1–2 in. broad, of a dull leaden or blueish-green, imbricate, lobulate and crenulate scales, appressed; beneath dull. Apothecia orange-red, scattered, with a thick inflexed crenulate border.—*Parmelia*, Mont. in Ann. Sc. Nat. ser. 2. iii. 91.

Northern Island: on bark, *Colenso.* (Juan Fernandez, Chili, tropical S. America.)—Babington finds this scarcely distinguishable from the *P. Saubinetii* of France, and the following.

5. **P. microphylla,** *Massolonghi*. Thallus livid, cervine or ashy, rather crustaceous, of closely imbricated crenate scales; beneath black. Apothecia brown or reddish, often convex, pale or whitish within.—Nyland. Lich. Scand. 124.

Northern Island, *Colenso*. (Cosmopolite.)

6. **P. triptophylla,** *Nyland.—Parmelia*, Fries;—Bab. in Fl. N. Z. ii. 290. Thallus ashy-brown or lead-coloured, forming a thin spreading coralloid, scaly crust; scales dissected; beneath bluish-black. Apothecia small, flat, brown or red-brown, with a paler border.—Nyland. Lich. Scand. 125.

Common in the **Northern** Island: on bark, etc., *Colenso*, etc. (Cosmopolite.)

7. **P. muscorum,** *Nyland.—Parmelia*, Ach.;—Bab. in Fl. N. Z. ii. 291. Thallus of flat pale brown or flesh-coloured imbricating scales, with blueish dilated mealy lobed and crenated edges; beneath whitish. Apothecia brown or orange-brown, flat or concave; border slight.—Nyland. Lich. Scand. 127.

Northern Island: Bay of Islands, *J. D. H.* (Cosmopolite.)

Dr. Lindsay enumerates as natives of Otago, *P. immixta*, Nyland. n. sp.; *P. leucosticta*, Tuck.; *P. gymnocheila*, Nyland. n. sp.; and *P. nigra*, Huds.

23. COCOCARPIA, Persoon.

Thallus almost leafy. Apothecia adnate, without a thalline border, red or brown.

1. **C. molybdæa,** *Persoon*, var. *plumbea, Nyland.—Parmelia*, Bab. in Fl. N. Z. ii. 289. Thallus orbicular, livid, lead-colour, broadly lobed and notched, thick; centre often granulated; margin imbricate; beneath with blue-black spongy fibres. Apothecia central, small, flat, rusty-red; border very obscure.—Nyland. Lich. Scand. 128.

Northern Island: Bay of Islands, *J. D. H.*, barren and doubtful. (Europe.)

24. AMPHILOMA, Fries.

Thallus soft, membranous; surface powdery. Granules without a cellular coat. Apothecia unknown.

1. **A. lanuginosum,** *Fries.—Parmelia*, Bab. in Fl. N. Z. ii. 290. Thallus orbicular, yellow-white, pulverulent; lobes rounded, plane, imbricate, slightly crenate; beneath grey-black, downy. Apothecia reddish; border powdery.—Nyland. Lich. Scand. 129.

Northern Island, *Colenso*, barren and doubtful. (Europe.)

25. SQUAMARIA, De Candolle.

Thallus radiating, scaly or cartilaginous. Apothecia with a thalline border paraphyses slender, separate; spores simple, ellipsoid. Spermagonia immersed, orifice of the same colour as the thallus; spermatia very long, arcuate or nearly straight, cylindric-acicular.

1. **S. gelida,** *Delise.—Parmelia*, Ach.;—Bab. in Fl. N. Z. ii. 291.

Thallus orbicular, dirty white, adnate, radiated, lobed and laciniate, crenulate, smooth with large brown central radiated fleshy warts. Apothecia concave, rose-coloured, with a thick elevated entire border.—Nyland. Lich. Scand. 134. *L. macrophthalma* and *L. marmorea*, Tayl.

Northern and **Middle** Islands: on rocks, etc., *Colenso*, etc.; Otago, *Lindsay*. (Europe, America.)

Dr. Lindsay enumerates *S. galactina*, Ach., var. *dispersa* amongst his Otago Lichens, also a *Placopsis perrugosa*, Nyland. n. sp.

26. PLACODIUM, De Candolle.

Thallus radiating, rarely laciniate. Apothecia with a thalline border, rarely without one; paraphyses separate; spores 8, ellipsoid, 2-locular, rarely simple. Spermagonia immersed, orifice coloured; spermatia shortly cylindric, slender, attached to arthrosterigmas.

A considerable genus in temperate and cold climates. The New Zealand species are all orange-yellow.

1. **P. fulgens,** *DC.—Parmelia*, Ach.;—Bab. in Fl. N. Z. ii. 291. Thallus orbicular, lemon-coloured, white when dry, adnate, lobed, waved; lobes somewhat imbricate, crenate and lobulate. Apothecia deep orange, at length convex, the border becoming obliterated.—Nyland. Lich. Scand. 137.

Northern Island, *Colenso*. (Europe.)

2. **P. elegans,** *DC.—Parmelia*, Ach.;—Bab. in Fl. N. Z. ii. 291. Thallus orbicular, hard, deep orange, adnate, plaited or rugged; lobes linear, compound, convex, wavy. Apothecia central, concave, same colour as the crust; border somewhat inflexed, entire.—Nyland. Lich. Scand. 136. ? *Parmelia aurea*, A. Rich.; Fl. N. Z. 23. t. 8. f. 1.

Northern Island: on pebbles, etc., *Colenso*. (Temperate arctic and antarctic regions.) —I find no allusion to *P. aurea*, A. Rich., in Nylander's works, and do not know the species; from the description it chiefly differs from *P. elegans* in the yellow colour being broken up into patches.

3. **P. murorum,** *DC.* Thallus orbicular, bright yellow, adnate, cracked, plaited and lobed; segments linear. Apothecia central, crowded, sessile, flattish, orange-yellow; border slightly waved.—Nyland. Lich. Scand. 136.

Var. *miniatum*. Thallus a brighter and deeper orange-red.

Middle Island: Otago, *Lindsay*, on trap rocks, both states. (Cosmopolitan.)

27. LECANORA, Nylander.

Thallus crustaceous, granular or smooth or leprous, rarely radiating. Apothecia with a thalline border; paraphyses separate; spores 8 or more, rarely septate. Spermagonia with arthrosterigmata, or with very long curved acicular spermatia, or with straight simple sterigmata.

* *Apothecia orange or yellow, reddish in* L. ferruginea.

1. **L. cerina,** *Ach.—Parmelia rupestris*, DC.;—Bab. in Fl. N. Z. ii. 293. Thallus greyish-white, somewhat granulated, unequal, thin. Apothecia scattered, elevated, flat, at length convex, yellow, warted; border inflexed, somewhat pruinose.—Nyland. Lich. Scand. 144.

Northern Island, *Colenso*: Bay of Islands, *J. D. H.* (Europe, Africa, America.)

2. **L. chyrosticta,** *Tayl.—Parmelia*, Bab. in Fl. N. Z. ii. 293. Thallus thin, white, studded with large granular buds, which are sometimes tipped with orange. Apothecia crowded, concave; disk yellow, pruinose; thalline border crenulate.—Nyland. Enum. Lich. 115.

Northern Island: on bark, Bay of Islands, *J. D. H.*, *Colenso*.

3. **L. aurantiaca,** *Ach.;—Bab. in Fl. N. Z.* ii. 292. Thallus granulated, pale lemon-coloured, often rugose. Apothecia sessile, rather convex, orange-yellow, with a yellow waved border.—Nyland. Lich. Scand. 142.

Var. *erythrella*. Thallus rugose and rimose. Apothecia without a border.
Var. *erythrella*. **Northern** Island: on stones, *Colenso*. (Cosmopolite.)

4. **L. ferruginea,** *Nyland*. Thallus spreading, thin, rugged, granular, greyish-white, sometimes evanescent. Apothecia rusty-orange or red-brown, at length convex, with a waved border of the same colour.—Nyland. Lich. Scand. 143.

Middle Island: on trap rocks, Otago, *Lindsay*.

5. **L. vitellina,** *Ach*. Thallus leprous, granulated, indeterminate, bright yellow-green, crenulate. Apothecia clustered, sessile, flat, tawny-yellow, at length convex and brownish; border elevated, crenate.—Nyland. Lich. Scand. 141.

Northern Island: on rails and other wood, Auckland, *Knight*. (Cosmopolite.)

** *Apothecia black brown or pale.*

6. **L. sophodes,** *Ach*. Thallus orbicular, granulated, dull greenish ash-coloured; apothecia clustered, black-brown when moist, with an elevated entire border.—Nyland. Lich. Scand. 148.

Var. *exigua*, Nyl.;—Bab. in Fl. N. Z. ii. 292. Thallus whiter. Apothecia with a thin, sometimes crenulate border.—*L. exigua*, Hook.
Northern Island: Bay of Islands, on wood, *J. D. H.* (Cosmopolite.)

7. **L. cinerea,** *Sommerf*. Thallus grey or ashy white, areolate, rugged and cracked, with a broad greenish undulate border. Apothecia immersed, black, flat, with an entire border.—Nyland. Lich. Scand. 153. *Urceolaria*, Ach.

Middle Island: trap rocks, Otago, *Lindsay*.

8. **L. parella,** *Ach.—Parmelia pallescens*, Fries;—Bab. in Fl. N. Z. ii. 292. Thallus dirty white, determinate, plaited and warted. Apothecia scattered, thick; disk concave, of the same colour as the thallus, flat, often angular; border tumid, thick.—Nyland. Lich. Scand. 156.

Common throughout the **Northern** and **Middle** Islands, on rocks, wood, the ground, etc., **Lord Auckland's** group, *J. D. H.* (Cosmopolite.)

9. **L. tartarea,** *Ach*. Thallus crustaceous, thick, granular, tartareous, greyish-white. Apothecia scattered; disk convex, at length plain or tumid, yellow-brown or flesh-coloured; border thick, inflexed, at length wavy; spores ellipsoid or subgranular.—Nyland. Lich. Scand. 157. *Porina granulata*, Tayl. in Fl. Antarct.

Probably common throughout the islands, **Lord Auckland's** group, *J. D. H.* (Cosmopolite.) Scarcely to be distinguished from *L. parella*, but by the thicker thallus and larger browner apothecia (*Nyl.*).

10. **L. parellina,** *Nyland.* Undescribed.

Northern Island, *Colenso.* (Java, Chili.)

11. **L. verrucosa,** *Lam.*—*Parmelia,* Fries;—Bab. in Fl. N. Z. ii. 293. Thallus filmy, elliptical, whitish, surrounded by an obsolete border of the same colour. Apothecia blackish, naked or pruinose, rather concave, with a thick entire border.—Nyland. Lich. Scand. 156. *Urceolaria,* Acharius.

Northern Island, *Colenso.* Specimen imperfect. (Europe, N. America.)

12. **L. glaucoma,** *Ach.* Thallus tartareous, uneven, hard, greyish-white, cracked and areolate. Apothecia crowded, depressed, livid-black or brown, glaucous, often deformed, flat or convex, with a thin entire border.—Nyland. Lich. Scand. 159.

Otago, on trap rocks, *Lindsay.* (Cosmopolite.)

13. **L. subfusca,** *Ach.*—*Parmelia,* Bab. in Fl. N. Z. ii. 292. Thallus continuous, thin, smoothish, brownish-white or ashy. Apothecia pale or dark brown or black, sessile, slightly convex; border tumid, entire; spores ellipsoid.—Nyland. Lich. Scand. 160.

Northern Island: probably common, on wood, stones, etc., *Colenso, Knight.* (Cosmopolite.)

14. **L. varia,** *Ach.*—*Parmelia,* Fries;—Bab. in Fl. N. Z. ii. 292. Thallus scattered, thin, granulate, pale yellow-green. Apothecia crowded, flattened, buff or brown; border waved, inflexed, entire; spores ellipsoid, simple.—Nyland. Lich. Scand. 163.

Northern Island: on stones, *Colenso.* **Lord Auckland's** group: on bark, *J. D. H.* (Europe, Africa, America.)

15. **L. argopholis,** *Nyland.* Thallus pale straw-coloured, yellowish or white, firm, warted and granular, granules crenulate. Apothecia black-brown, with an entire or crenate border.—Nyland. Lich. Scand. 166.

Var. *thiodes.*—*Parmelia frustulosa,* Fr., var. *thiodes,* Bab. in Fl. N. Z. ii. 292. Thallus more yellow, with the granular more crowded.

Northern Island: var. *thiodes,* on stones, *Colenso.* (Europe, America.)

16. **L. atra,** *Ach.*—*Parmelia,* Bab. in Fl. N. Z. ii. 292. Thallus rugged, subdeterminate, often cracked here and there, white, granulate. Apothecia nearly flat, very black; border white, elevated, at length flexuous and notched; spores ellipsoid.—Nyland. Lich. Scand. 170.

Northern and **Middle** Islands: probably common on bark and stones, *Colenso, Lindsay,* etc. (Cosmopolite.)

17. **L. punicea,** *Ach.*—*Parmelia,* Bab. in Fl. N. Z. ii. 292. Thallus thin, white, rather glaucous, bordered with black, smooth, not shining. Apothecia rounded, appressed, angular when crowded; disk flattish; margin thin, white, finally crenulate, bright red, convex when old.—Mont. Crypt. Cuba. 208; Eschweiler in Mont. Fl. Bras. i. 191.

Northern and **Middle** Islands: on wood, *Colenso;* Otago, *Lindsay.* (Tropical America and Africa, Cape of Good Hope, Java.)

Dr. Lindsay enumerates the following as natives of Otago:—*L. pyracea,* Ach.; *L. um-*

brina, Ehr.; *L. simplex*, Dev., and three new species of Nylander's, viz. *L. homologa*, *peloleuca*, and *thiomela*.

28. URCEOLARIA, Acharius.

Thallus crustaceous. Apothecia urceolate; spores 8, brown, septate. Spermogonia with branched sterigmata; spermatia cylindric.

1. **U. scruposa,** *Ach.—Parmelia*, Fries;—Bab. in Fl. N. Z. ii. 293. Thallus greyish-white, rugose, granulated, continuous or areolate. Apothecia black, concave, often glaucous; border thick, incurved, rugose and crenulate; spores 5-septate.—Nyland. Lich. Scand. 176.

Northern Island, *Colenso*. (Cosmopolite.)

Nylander enumerates *N. stictica*, Krb., as a native of Otago.

29. PERTUSARIA, De Candolle.

Thallus crustaceous, continuous, irregularly warted or nearly smooth. Apothecia with or without a thalline border, immersed in warts of the thallus; spores 1–8, ellipsoid, large, with a thickish margin, simple. Spermatia acicular.

A genus of temperate and cold country Lichens, rare in the tropics.

1. **P. subverrucosa,** *Nyland.—P. communis*, DC.;—Bab. in Fl. N. Z. ii. 306. Undescribed.

Northern Island: on bark and stones, *Colenso*, *Raoul*.

2. **P. subglobulifera,** *Nyland.* Undescribed.

Northern Island, *Colenso*.

3. **P. cucurbitula,** *Mont.—Porina*, Bab. in Fl. N. Z. ii. 306. Thallus radiating, orbicular, cream-coloured, warty, beneath filmy-white, silvery. Apothecia numerous, small, sessile, prominent, subconfluent, of the same colour as the thallus, at first closed, then expanding and margining the apiculate nucleus; nucleus wavy within, at length falling away; asci large, clavate; spores 1-celled.—Mont. in Ann. Sc. Nat. ser. 3. 18, 312.

Northern Island: on bark, *Colenso*. (Chili.)

Dr. Lindsay adds from Otago, *P. velata*, Turn.; *perrimosa*, Nyland. n. sp., and *perfida*, Nyland. n. sp.

30. THELOTREMA, Ach.

Thallus thin, crustaceous, continuous. Apothecia on tubercles or warts, with a double margin, the external thalline, the internal of the substance of the apothecium; spores large, fusiform, multilocular.

A large, chiefly tropical genus.

1. **T. lepadinum,** *Ach.;—Bab. in Fl. N. Z.* ii. 294. Crust small, cream-coloured; warts of the apothecia smooth, conical, truncate. Apothecia 1-2 in a wart, pale brown, with a thin inflexed edge.—Nyland. Lich. Scand. 185.

On bark of trees throughout the islands, and in **Lord Auckland's** group. (Cosmopolitan.)

Dr. Lindsay enumerates *T. monosporum*, Nyland., as a native of Otago.

31. CŒNOGONIUM, Ehrenb.

Thallus effuse, of subcontinuous, pellucid, obscurely-articulate filaments, forming a cottony, loosely-interwoven, greenish web. Apothecia orbicular, substipitate, without a border, or margined in the young state only, the disk bearing asci.

A small genus common in warm climates, on bark of trees, remarkable for the Conferva-like structure of the thallus.

1. **C. Linkii,** *Ehrenb.;—Bab. in Fl. N. Z.* ii. 310. Thallus suborbicular, glaucous green. Apothecia saffron-yellow; spores elliptic-ovate, in one series within the filiform subclavate asci.—Fée, Meth. Lich. 63. t. 2.

Northern and **Middle** Islands: common on trees, etc., *Menzies, D'Urville,* etc.

(Cosmopolite in tropical and subtropical countries.)

C. inflexum, Nyland., is enumerated amongst Lindsay's Otago Lichens.

32. BYSSOCAULON, Mont.

Thallus filiform, spreading, formed of excessively-branched fibres, which are not articulate, and are crisped when dry.

A very anomalous genus, recently regarded by its author, Montagne, as a form of *Pannaria.* (See Mont. Sylloge, Plant. Crypt. 293.)

1. **B. filamentosum,** *Nyland.;—Parmelia gossypina,* var. *filamentosa,* Bab. in Fl. N. Z. ii. 288. Thallus forming a white cottony stratum.

Northern Island, *Colenso,* on moss, barren. (Juan Fernandez.) (I find no specimens in the Herbarium.)

33. LECIDEA, Acharius.

Thallus crustaceous, scaly, granular, powdery or 0. Apothecia with a margin of their own substance or 0. Spermagonia with straight (rarely curved) or acicular or shortly cylindric spermatia.

A very large genus, found in all parts of the world,

§ 1. *Apothecia concave, bright-coloured* (Gyalecta, Ach.).

1. **L. cupularis,** *Ach.—Gyalecta,* Bab. in Fl. N. Z. ii. 294. Thallus pale, thin, scattered, ashy or whitish. Apothecia subglobose, salmon-coloured, at length urceolate, with a pale, flesh-coloured, thick, elevated, inflexed, often crenate and pulverulent border; spores ellipsoid, 3-septate.—Nyland. Lich. Scand. 190.

Northern Island: on stones, probably common, and as far south as **Lord Auckland's** group. (Europe, America, Tasmania.)

2. **L. carneola,** *Ach.* Thallus mealy, thin, white, or obsolete. Apothecia small, dusky or brownish, suburceolate, with a thick elevated even smooth paler border; spores acicular, many-septate.—Nyland. Lich. Scand. 191.

Northern Island: on bark, *Colenso.* (Europe, America.)

§ 2. *Apothecia flat or convex, rarely black* (BIATORA, Fries).

a. Crust scaly, spores simple.

3. **L. parvifolia,** *Pers.—Biatora,* Mont.;—Bab. in Fl. N. Z. ii. 299. Thallus glaucous, scaly; scales lobulate or incised. Apothecia pale red or testaceous, at length convex; border indistinct; spores oblong, simple.—Nyland. Enum. Lich. 120.

Northern Island, *Colenso.* (Cosmopolite in the tropics.)

β. Crust granular, powdery or evanescent. Spores simple, oblong or ellipsoid.

4. **L. cinnabarina,** *Sommerfeldt.—Biatora,* Fries;—Bab. in Fl. N. Z. ii. 300. t. 129 C. Thallus thin, white, effuse, of minute unequal granules. Apothecia scarlet, with an obtuse margin or 0; spores small, simple, fusiform.—Nyland. Lich. Scand. 194.

Northern Island: on branches of shrubs, *Colenso.* (Europe, America, Australia, Tasmania.)

5. **L. intermixta,** *Nyland.* Thallus thin, white, effuse or evanescent. Apothecia brownish, flattish; border obtuse or 0; spores ellipsoid, 1-septate.—Nyland. Lich. Scand. 194.

Kermadec Island, *Milne,* on bark. (Cosmopolite.)

6. **L. flavo-pallescens,** *Nyland.* Undescribed.

Northern Island, *Colenso.*

7. **L. pyrophthalma,** *Nyland.—Parmelia,* Bab. in Fl. N. Z. ii. 293. t. 129 A. Thallus very thin, membranous, olive-green, effuse. Apothecia at first globose, yellow, at length plane; disk bright orange; border quite entire, paler; asci filiform; spores cymbiform, 1-septate.—Nyland. Enum. 121. *Biatora,* Mont. Sylloge, 339.

Northern Island, *Colenso,* on bark. (Chili.)

8. **L. vernalis,** *Ach.—Biatora,* Fries;—Bab. in Fl. N. Z. ii. 300. Thallus thin, powdery, ashy or greenish-white or obsolete. Apothecia clustered, rusty flesh-coloured, convex, white internally, at length often globose; spores oblong.—Nyland. Lich. Scand. 200. *Biatora anomala,* Fries?; Bab. in Fl. N. Z. l. c.

Northern Island: on moss, etc., *Colenso, Sinclair.* (Europe, etc.)

9. **L. decolorans,** *Flœrke.* Thallus leprous, grey, with white granulations, effuse, thin. Apothecia pale brown, blackish or brick-red, opaque, flat, with a pale appressed border, white within; spores ellipsoid, oblong.—Nyland. Lich. Scand. 197.

Northern Island, *Colenso.* (Europe, N. America.)

10. **L. coarctata,** *Nyland.—Parmelia,* Ach.;—Bab. in Fl. N. Z. ii. 291. Thallus spreading, thin, cracked, greyish, unequal. Apothecia with the disk somewhat immersed, at length elevated, flat, black; border elevated, inflexed, coarctate, irregular, pulverulent.—Nyland. Lich. Scand. 196.

Northern Island, *Colenso.* (Cosmopolite.)

β. *Thallus grey or white, granular, powdery or evanescent. Spores acicular or oblong, 1-many-septate.*

11. **L. rosella,** *Ach.—Biatora*, Fries;—Bab. in Fl. N. Z. ii. 300. Thallus ashy, thin, unequal or granular. Apothecia rose-red or flesh-coloured, rather powdery; border obtuse, white within; spores acicular, many-septate.—Nyland. Lich. Scand. 209.

Northern Island, *Colenso.* (Europe.)

12. **L. ceroplasta,**—*Biatora*, Bab. in Fl. N. Z. ii. 300. Thallus white, waxy, thickish, granular, subpellucid; granules minute. Apothecia large, appressed, flattish, at length convex, red, not pruinose, internally flesh-coloured; border thin, flesh-coloured; spores not found.

Northern Island, *Colenso.*—Compared with *L. rosella* by Babington, but of a peculiarly waxy hue. Omitted in Nylander's 'Enumeratio Licheuum.'

13. **L. marginiflexa,** *Tayl.—Biatora*, Bab. in Fl. N. Z. ii. 299. t. 129 B. Thallus white, glaucous, thin, minutely cracked, leprous, bordered with black. Apothecia rather large, crowded, purplish when moist; disk pruinose, convex; border thin, flexuous; spores oblong, 1-septate.

Northern Island: common on mosses, bark, etc. (Australia, India.)—Perhaps not distinct from *L. tuberculosa*, Fée, and if so perhaps no *Biatora*. The apothecia are very concave and dark, as in *Gyalecta*, and the disk dark, almost black. Nylander places it in his section *Biatora.*

14. **L. tuberculosa,** *Fée, Essai sur Cryptolg. des Écorces Officin.* 107. *t.* 27. *f.* 1. Thallus crustaceous, tuberculate, pale sulphur-coloured, indefinite. Apothecia scattered, black, flattish; margin slender, entire, evanescent, horny internally; spores many-septate.—Nyland. Enum. 123.

Northern Island, *Colenso*, on bark. **Lord Auckland's** group, *J. D. H.* (Tropical America, Madagascar, Australia.)

15. **L. Domingensis,** *Ach.?—Parmelia gyrosa?*, Mont.;—Bab. in Fl. N. Z. ii. 293. Thallus orbicular, glaucous, imbricate; laciniæ flat, palmately multifid at the tips; margins repand, toothed, powdery, beneath white, fibrillose. Apothecia scattered, brownish, pruinose; border inflexed, at length crenulate; spores many-septate.—Nyland. Enum. 123. *Parmelia*, Mont. Crypt. Cuba, 212.

Northern Island: on bark, mosses, etc., *Colenso.*—Specimens imperfect. (W. Indies.)

§ 3. *Apothecia black.*

α. *Thallus scaly.*

16. **L. decipiens,** *Ach.—Biatora*, Fries;—Bab. in Fl. N. Z. ii. 299. Thallus of roundish subimbricate scales or concave lobes, flesh-coloured or red, at length brown; margins white. Apothecia marginal, convex or subglobose, brown-black inside; border obsolete; spores ovoid or elliptic.—Nyland. Lich. Scand. 214.

Northern Island: on rocks, *Colenso.* (Cosmopolite.)

17. **L. mamillaris,** *Dufour;*—Bab. in Fl. N. Z. ii. 300. Thallus white, turgidly areolate, somewhat lobed and plaited, cancellated or cracked. Apo-

thecia on the margins of the areolæ; spores oblong.—Nyland. Prod. Lich. Gall. and Alg. 120.

Northern Island: on rocks and earth, *Colenso*. (South Europe.)

18. **L. vesicularis,** *Ach.*—Bab. in Fl. N. Z. ii. 301. Thallus of irregular imbricated tumid powdery greyish warts or scales. Apothecia irregular, black, flattish, marginal on the scales, at length hemispherical; spores narrow, fusiform or subacicular.—Nyland. Lich. Scand. 214.

Northern Island, *Colenso*. (Cosmopolite.)

19. **L. Colensoi,** *Bab.*—*Biatora*, Bab. in Fl. N. Z. ii. 298. Thallus scaly; scales subimbricated, ascending, lobed, erose, glaucous, ferruginous below, with a subferruginous medullary stratum. Apothecia ferruginous, black, with paler margins, at first concave, at length confluent, deformed, ferruginous internally, glaucous and pruinose.

Northern Island: on earth, *Colenso*.

β. *Thallus granular, powdery or evanescent.*

20. **L. parasema,** *Ach.*—Bab. in Fl. N. Z. ii. 301. Thallus crustaceous, thin, greyish, uninterrupted, somewhat granulated, edged with black. Apothecia sessile, flat, opaque, black, at length convex; border smooth, black; spores ellipsoid, simple.—Nyland. Lich. Scand. 217? *L. albido-plumbea*, Tayl. Fl. N. Z. l. c. 302. *L. geomæa*, Tayl. Fl. Antarct. (*L. papillata*, Fries, ibid. in Suppl. p. 547.)

Northern Island: on smooth bark of trees, common, *Colenso*, *Knight*, etc. **Lord Auckland's** group: on bark and the ground, *J. D. H.* (Cosmopolite.)

21. **L. atro-alba,** *Flotow;—Bab. in Fl. N. Z.* ii. 301. Thallus crustaceous, grey or white, spreading, cracked and granular, somewhat mealy. Apothecia sunk to a level with the crust, small, crowded, black, pruinose, ashy within; spores ellipsoid, 1-septate.—Nyland. Lich. Scand. 233.

Northern Island: on stones, etc., common, *Colenso*, etc. (Cosmopolite.)

22. **L. contigua,** *Fries;—Bab. in Fl. N. Z.* ii. 301. Thallus crustaceous, ashy or white, thin or rather thick, continuous, cracked or areolate. Apothecia sessile, black within and without, often pruinose; spores ellipsoid, simple.—Nyland. Lich. Scand. 224.

Var. *platycarpa*. Thallus obsolete. Apothecia larger, pruinose; disk white within, on a black substratum.—*L. platycarpa*, Ach. Bab. l. c. *L. petræa*, Tayl.

Northern and **Middle** Islands: on calcareous and other rocks, *Colenso*, *J. D. H.*; Otago, *Lindsay*. (Cosmopolite.)

23. **L. lapicida,** *Fries;—Bab. in Fl. N. Z.* ii. 301. Thallus crustaceous, glaucous white, tessellated and granulated. Apothecia sessile, plane or convex, angular, black, with a narrow elevated border, black or dark grey within.—Nyland. Lich. Scand. 225.

Northern Island: on rock, *Colenso*. (Europe.)

24. **L. abietina,** *Ach.?—Bab. in Fl. N. Z.* ii. 302. Thallus crustaceous, spreading, smooth, erose, pale glaucous or leprous, effuse or evanescent. Apothecia sessile, flattish, black, pruinose, with a black border; spores fusiform, 3-septate.—Nyland. Lich. Scand. 241.

Northern Island: on sandstone rocks, *Colenso*. (Europe.)—A doubtful identification.

25. **L. disciformis,** *Fries.* Thallus crustaceous, thin, white, continuous, determinate, unequal, often evanescent. Apothecia flat or subconvex, black, bordered; spores brown, ellipsoid or oblong, 1-septate,—Nyland. Lich. Scand. 236.

Var. *albula*, Nylander.

Northern Island: on quartz rocks, *Colenso.*

γ. *Thallus yellow, crustaceous.*

26. **L. geographica,** *Schærer.* Thallus crustaceous, bright yellow, smooth, cracked, tessellated; areolæ edged with black. Apothecia flat, black, often confluent; spores brown or black, 3- or more-septate.—Nyland. Lich. Scand. 248.

Middle Island: Otago, on rocks, *Lindsay.* (Cosmopolite in cold regions.)

27. **L. citrinella,** *Ach.*—*L. flavo-virescens*, Fries, Bab. in Fl. N. Z. ii. 301. Thallus crustaceous, thin, bright yellow, continuous. Apothecia appressed, black outside and within; spores acicular.—Nyland. Lich. Scand. 248.

Northern Island, *Colenso*, on rocks. (Europe.)

28. **L. pachycarpa,** *Dufour.*—*Biatora*, Fries, Bab. in Fl. N. Z. ii. 229. Thallus leprous, mealy, soft, uneven, greyish-green. Apothecia scattered, sessile, brown, with a pale brown, even, smooth border; spores many-septate.—Nyland. Prod. Lich. Gall. and Alger. 118. *L. incana*, Ach.; Fl. Antarct.

Northern Island: on rocks, trees, and banks, *Colenso.* **Lord Auckland's** group, on the ground, *J. D. H.* (Europe.)

29. **L. tuberculosa,** *Fée, Essai sur Cryptol. des Écorces Officin.* 107. *t.* 27. *f.* 1. Thallus crustaceous, tuberculate, pale sulphur-coloured, indefinite. Apothecia scattered, black, flattish; margin slender, entire, evanescent, horny internally; spores many-septate.—Nyland. Enum. 123.

Northern Island, *Colenso*, on bark. **Lord Auckland's** group, *J. D. H.* (Tropical America, Madagascar, Australia.)

δ. *Parasitic species, consisting of Apothecia growing on other Lichens.*

30. **L. oxyspora,** *Nyland.* Apothecia black or brown, minute, flat or slightly convex, brownish within; spores fusiform, ellipsoid.—Nyland. Lich. Scand. 247. *Abrothallus*, Tulasne in Mem. Lich. 116. t. 16. f. 27.

Middle Island: Otago, parasitic on *Parmelia conspersa*, Lindsay. (Europe, Asia.)

The following are enumerated in Dr. Lindsay's Otago catalogue:—*L. furfuracea*, Pers.; *subsimilis*, Nyl.; *sabuletorum*, Flk.; *trachona*, Flot.; *allotropa*, Nyl., n. sp.; *melanotropa*, Nyl., n. sp.; *millegrana*, Tayl.; *pulverea*, Bow.; *leucothalamia*, *Otagensis*, *amphitropa*, and *flavido-atra*, all n. sp. of Nylander; *arceutina*, Ach.; *fusco-atra*, Ach.; *petræa*, Flot.; *stellulata*, Tayl.; *myriocarpa*, DC.; *lenticularis*, Ach.; *grossa*, Pers.; and *Curreyi*, Lindsay (a species of *Abrothallus*).

34. GRAPHIS, Ach.

Thallus very thin, continuous, often scarcely visible above or below the bark on which the plant grows. Apothecia linear, black, simple or forked, with a raised thallodal border; paraphyses slender, distinct; spores oblong, many-celled.

A large tropical and subtropical, rarely temperate genus of Lichens, growing on the bark of trees, comparatively rare in temperate climates.

1. **G. scripta,** *Ach.;—Bab. in Fl. N. Z.* ii. 302. Thallus thin, membranous, smooth, somewhat shining, grey-white, indistinctly bordered with black. Apothecia partly immersed, black, naked or pruinose, flexuose, simple or branched; branches subparallel; spores oblong, 7–9-celled.—Nyland. Lich. Scand. 251.

Northern Island: on bark, *Colenso, Knight.* (Cosmopolite.)

2. **G. anguina,** *Mont. Sylloge*, 352 (Ustalia). Thallus between membranous and crustaceous, white, minutely granular. Apothecia immersed, minute, crowded, linear, flexuose, simple and forked; border prominent; disk brown, concave, convex when moist; spores oblong-cylindric, multicellular.

Northern Island: Auckland, *Knight*, on bark. (Cosmopolite.)

3. **G. elegans,** *Ach.* Thallus orbicular, granulated, white. Apothecia immersed, scattered, divaricating, usually simple, with a grooved as well as thallodal border; spores cylindric-fusiform, about 10-septate.—Nyland. Prodr. Lich. Gall. et Alger. 151.

Northern Island: Auckland, *Knight*, on bark. (Almost cosmopolitan.)

4. **G. scalpturata,** *Ach.* Thallus yellow, membranous. Apothecia short or long, simple forked or branched, flexuose; margined by a thickening of the thallus; disk black, deciduous; spores large, brown, oblong, 10-septate and usually divided.—Mont. in Ann. Sc. Nat. ser. 2. xviii. 274.

Northern Island: Auckland, *Knight*, on bark. (Tropical America.)

5. **G. confinis,** *Knight and Mitten in Linn. Trans.* xxiii. 102. *t.* 11. *f.* 20. Thallus cream-coloured or grey. Apothecia immersed, elongate, wavy, sometime forked or branched, ends acute, closed by the inflexed border; perithecium entire below or dimidiate; spores 6–8-septate, interspaces transversely oval, 0·00034 in. broad, 0·0011 in. long.

Northern Island: Auckland, on bark of trees, *Knight.*—Near *G. scripta,* but thallus paler and border of perithecium narrower and thicker.

6. **G. (Fissurina) insidiosa,** *Knight and Mitten, l. c. t.* 12. *f.* 21. Thallus thick, uneven, warty, dull green, brownish when dry. Apothecia crowded, deeply immersed in warts of the thallus, simple or branched, closed; lips paler; spores in single series, ovoid or obovate, yellow, 0·0003 in. broad, 0·0007 long.

Northern Island: Auckland, on bark of trees, *Knight.*—Apothecia invisible to the naked eye.

7. **G. (Fissurina) inquinata,** *Knight and Mitten, l. c. t.* 12. *f.* 22. Thallus cartilaginous, polished, ashy-grey or brown, uneven. Apothecia deeply immersed, narrow, elongate, wandering, variously branched and wavy, almost covered by the darkened border of the thallus; spores oblong or obovate, yellow, 3–4-septate, 0·00040 in. broad, 0·00084 long.

Northern Island: Auckland, on bark, *Knight.*—Browner than *G. insidiosa,* thallus less warty, apothecia faintly visible by a dark line. It may be a form of *G. nitida,* Mont.

35. OPEGRAPHA, Ach.

Thallus excessively thin or almost evanescent. Apothecia superficial or with the base innate, black, linear-lanceolate rounded or linear, elongate, flexuose or branched, with a proper border; epithecium plane or like a narrow slit; spores colourless or brownish, fusiform, few-septate. Spermatia slender, cylindric straight or curved.

A very large tropical or subtropical genus, rarer in temperate regions.

1. **O. varia,** *Persoon.* Thallus thin, powdery, white or brown, dispersed. Apothecia sessile, prominent, scattered, roundish oval or oblong, wavy; disk plane, at length convex, hemispherical, somewhat tuberculose; border subevanescent; spores fusiform, 3–9-septate.—Nyland. Lich. Scand. 252.

Var. Thallus olive-green; apothecia superficial, slender; spores 5–9-septate, 0·0003 in. broad, 0·00115 long.—Knight and Mitt. in Trans. Linn. Soc. 23. 101. t. 11. f. 15.

Var. Thallus olive-brown, tending to grey; apothecia more or less immersed when moist; margin separating in the middle; spores $\frac{1}{4}$–5-septate, 0·00024 in. broad, 0·0009 long.—Knight and Mitt. l. c. f. 16.

Var. Thallus dusky-grey; apothecia immersed when moist, oblong, open; spores 3–4-septate, 0·00025 in. broad, 0·0007 broad.—Knight and Mitt. l. c. f. 17.

Northern Island: Auckland, on bark, *Knight.* (Cosmopolite.)

2. **O. atra,** *Ach.* Thallus very thin, membranous, smooth, whitish. Apothecia sessile, various in shape, the smaller globose or oblong, the larger long, narrow, terete, subrugulose, coal-black, simple or divided; spores oblong-ovoid, 3-septate.—Nyland. Lich. Scand. 254.

Northern Island: Auckland, on bark, *Knight.* **Lord Auckland's** group, *J. D. H.* (Cosmopolite.)

3. **O. herpetica,** *Ach.* Thallus continuous, limited, very thin, reddish ash-coloured or brownish, slightly rugged. Apothecia immersed, oblong linear rounded or elliptic, somewhat curved, the disk broader than the border; spores fusiform, 3-septate.—Nyland. Lich. Scand. 255.

Var. *rufescens.* Apothecia flexuose, sometimes divided.

Northern Island: Auckland, on bark, *Knight.*

4. **O. cinerea,** *Knight and Mitt. in Linn. Trans.* xxii. 101. *t.* 12. *f.* 18. Thallus ashy-grey, pulverulent, sometimes obsolete. Apothecia superficial, simple, oblong, straight or curved, closed; spores brown, 3-septate or 1-septate and contracted in the middle, frequently murally divided, 0·00025 in. broad, 0·00075 long.

Northern Island: Auckland, on bark, *Knight.*—The short thick closed apothecia at once distinguish this from *O. varia.*

5. **O. prominula,** *Knight and Mitt. l. c.* 102. *t.* 11. *f.* 19. Thallus white, uneven or minutely warted, moderately thick, containing extensive patches of green round gonidia. Apothecia superficial, crowded, subparallel or disposed in all directions, straight or curved, closed; spores ellipsoid or oblong, 7-septate, 0·00015 in. broad, 0·0008 in. long.

Northern Island: Auckland, on bark, *Knight*.—Resembles *O. varia*, but the thallus is even and whiter, the apothecia usually entirely closed by the more prominent border.

Dr. Lindsay, in his 'Catalogue of Otago Lichens,' adds three new species of Nylander,—*O. agelæoides*, *spodopolia*, and *subeffigurans*.

36. PLATYGRAPHIS, Nyland.

Thallus thin or obsolete. Apothecia black, flat, simple, no proper border, but sometimes a thin thallodal border. Spores colourless, fusiform, septate. Spermatia shortly cylindric, straight or slightly curved.

A considerable tropical and temperate genus of corticolous Lichens.

1. **P. microsticta,** *Knight and Mitt. in Linn. Trans.* xxiii. 103. *t.* 12. *f.* 23. Thallus dusky-grey. Apothecia round or oblong, flexuose, simple or branched, broader at one end; ends rounded. Apothecia dark brown; hypothecium black; spores fastigiate, fusiform, 3-septate, 0·00010 in. broad, 0·00140 long.

Northern Island: Auckland, *Knight*, on bark.—Thallus covered everywhere at equal distances by the apothecia, which resemble specks to the naked eye.

2. **P. inconspicua,** *Knight and Mitt. l. c. t.* 24. Thallus brown, with a lilac tinge when dry. Apothecia immersed, very minute, roundish or oblong, one end acuminate, blackish brown; spores fastigiate, ? not septate, fusiform, 0·00120 broad, 0·00175 in. long.

Northern Island: Auckland, on bark, *Knight*. Remarkable for its lilac colour. Apothecia invisible till magnified, then like black dots.

3. **P. tumidula,** *Knight and Mitt. l. c. f.* 25. Thallus dark yellow-brown. Apothecia flat, surrounded by the somewhat swollen thallus, elongate or short-curved, defined by a narrow dark line; hypothecium brownish-black; spores fastigiate, without septa, fusiform, 0·0001 in. broad, 0·0014 long.

Northern Island: Auckland, *Knight*, on bark.—Thallus colour of *Lecidia parasema*. Apothecia conspicuous, variable in form and size.

4. **P. occulta,** *Knight and Mitt. l. c.* 104. *t.* 12. *f.* 26. Thallus reddish-brown, obscure. Apothecia irregularly branched, subradiate, angular; edges torn; thallodal border raised, powdery, white; epithecium blackish-brown, flat; hypothecium black; spores fastigiate, fusiform, 4–?-septate, 0·00005 in. broad, 0·00160 long.

Northern Island: Auckland, on bark, *Knight*.—Apothecia invisible till magnified.

Dr. Lindsay enumerates *P. longifera*, Nyland. n. sp.

37. PLAGIOGRAPHIS, Knight and Mitten.

Thallus very thin, obscure. Apothecia elongate, like cracks, surrounded by a perithecium, which is covered by the thallus, connivent above, divaricated and oblique below; hypothecium free. Asci pyriform; spores 2-locular.

A genus only known from New Zealand.

1. **P. devia,** *Knight and Mitt. in Linn. Trans.* xxiii. 104. *t.* 12. *f.* 27. Thallus smooth, yellow or grey. Apothecia blackish-brown, elongate,

flexuose or branched, contracted here and there; thallodal margin very thin, at length erect; epithecium nearly covered by the perithecial walls, dark-coloured; spores contracted in the middle, 1-septate, upper cell larger, 0·00030 in. broad, 0·00065 long.

Northern Island: Auckland, on smooth bark, *Knight*.—Apothecia externally like black pecks. At first sight this resembles an *Arthonia*.

2. **P. rubrica,** *Knight and Mitt. l. c. f.* 28. Thallus corneous, polished, red or reddish-brown. Apothecia round oblong or elongate and nerved, without thallodal border, blackish-brown, slightly prominent, open; perithecium extending far below the thallus; spores 1-septate, 0·00020 in. broad, 0·00053 long.

Northern Island: Auckland, on smooth bark, *Knight*.—Similar to *P. devia*, but differently coloured, and with open apothecia.

38. STIGMATIDIUM, Meyer.

Thallus thick, crustaceous or cartilaginous. Warts containing the apothecia crowded, punctiform, solitary or in series, immersed, indehiscent, without a border.

A small genus of tropical and temperate corticolous Lichens, found in various parts of the world.

1. **S. crassum,** *Duby*. Thallus cartilaginous, undulated, olive-brown, smooth, black-edged. Warts of the apothecia large, irregular, many-celled, with numerous, black, slightly depressed, often confluent and then linear curved points.—Nyland. Prodr. Lich. Gall. and Alger. 163.

Northern Island: on bark, *Colenso*. (West Europe, N. Africa.)

39. ARTHONIA, Ach.

Thallus thin, superficial or beneath the bark, or evanescent or none. Apothecia flat, without a border, separate paraphyses none; asci pyriform, easily separable from the apothecium; their walls thickened at the tip; spores 4–8, ovoid, septate. Spermatia cylindric, straight or curved.

A large genus of Lichens, growing on wood and bark in all latitudes.

1. **A. lurida,** *Ach.—Myriangium inconspicuum*, Bab. in Fl. N. Z. ii. 310. t. 128 B. Thallus somewhat tartareous, thin, cracked, uneven, white. Apothecia crowded, immersed, thin, flat, confluent, brownish lead-coloured, pruinose, internally of the same colour; spores ovoid, 1-septate.—Nyland. Lich. Scand. 268.

Northern Island: on dead leaves, *Colenso*. (Europe.)

2. **A. pruinosa,** *Ach.—Lecanactis impolita*, Fries;—Bab. in Fl. N. Z. ii. 303. Thallus white, thin, cracked, somewhat uneven. Apothecia immersed, flat, confluent, brownish lead-coloured, pruinose, same colour within; spores ovoid, 3-septate.—Nyland. Lich. Scand. 258.

Northern Island: Auckland, on bark, *Knight*. (Europe, N. America.)

3. **A. astroidea,** *Ach.* Thallus beneath the bark, indicated by a white patch outside, more or less determinate. Apothecia appressed, lobed, divided, substellate, ashy-grey within; spores ovoid, 3-septate.—Nyland. Lich. Scand. 259.

Northern Island: Auckland, on bark, *Knight.* (Cosmopolite.)

4. **A. minutula,** *Nyland.* Thallus indicated by a white spot on the back. Apothecia black, slender, flexuose, irregular; spores ovoid, 1-septate. —Nyland. Prodr. Lich. Gall. et Alger. 169.

Northern Island: on bark, *Colenso.* (Europe.)

5. **A. lobulata,** *Knight and Mitt. in Linn. Trans.* xxiii. 104. *t.* 12. *f.* 29. Thallus ochreous or grey. Apothecia angular, lobed or subradiate, plane; margin defined by a narrow line, deep brown, almost black; spores 1-septate, oblong, light yellow, 0·0002 in. broad, 0·0006 long.

Northern Island: Auckland, on bark, *Knight.*—Differs from the other New Zealand species in the clavate asci, with the spores in parallel lines.

6. **A. indistincta,** *Knight and Mitt. l. c.* 105. *t.* 12. *f.* 30. Thallus grey, thin. Apothecia irregularly elongate-oblong, simple or branched; border ragged, obscure, grey black, convex; spores 3-septate, oblong, constricted in the middle, terminal cells much smallest.

Northern Island: Auckland, *Knight,* on bark.—Apothecia mere specks, when magnified the borders seem to fade into the thallus.

7. **A. albida,** *Knight and Mitt. l. c. t.* 12. *f.* 31. Thallus white or dull white; edges filamentous. Apothecia prominent, smooth, lobed or angular; thallodal border thin, erect, lilac-brown, powdery; hypothecium pale, with a black spot sometimes at the sides; spores 3-septate, obovate, yellow; upper cell largest.

Northern Island: Auckland, on bark, *Knight.*—Resembles *A. cinnabarina,* but apothecia smaller.

8. **A. ramulosa,** *Knight and Mitt. l. c. t.* 12. *f.* 32. Thallus dull white, thin, brown. Apothecia superficial, black, dark brown at the base and sides, much branched, bordered by a narrow undulating line; spores 3-septate, oblong or obovate, brown; upper cell largest, 0·00020 in. broad, 0·00045 in. long.

Northern Island: Auckland, on bark, *Knight.*—Apothecia more branched and prominent than in *A. astroidea.*

9. **A. ampliata,** *Knight and Mitt. l. c.* 106. *t.* 12. *f.* 33. Thallus grey, 1 in. and more diameter. Apothecia copious, scattered or parallel, roundish or oblong, partly veiled by the ruptured thallus, blackish-brown; spores brown, 5–6-septate; upper cell largest, others often longitudinally divided, 0·0005 in. long, 0·0013 long.

Northern Island: Auckland, on bark, *Knight.*

10. **A. nigro-cincta,** *Knight and Mitt. l. c. t.* 12. *f.* 34. Thallus moderately thick, whitish-brown, with a lilac tinge, edged with dark. Apothecia thinly scattered, superficial, simple, irregular, roundish, oblong, bent or

curved, one end larger, blackish-brown, internally brown; spores obovate, 4-septate, light yellow, 0·00030 in. broad, 0·00084 long.

Northern Island: Auckland, on bark, *Knight.*—The apothecia have no dark marginal line as in most of its allies.

11. **A. polymorpha,** *Ach.;—Bab. in Fl. N. Z.* ii. 302. Thallus sub-membranous, ashy-white, variegated with olive-green. Apothecia large, sub-orbicular, pruinose, glaucous-blue, when moist, tumid, tubercular and black; spores murally divided.—Eschweiler in Mart. Ic. Sel. Crypt. Bras. 14. t. 9. f. 3 (not in Nyland. Enum.).

Northern Island: on bark, *Colenso.* (Brazil.)

The following *Arthoniæ* are enumerated in Dr. Lindsay's 'Catalogue of Otago Lichens:' *A. excidens,* Nyl.; *conspicua,* Nyl.; and *platygraphella,* Nyl. n. sp.

40. MELASPILEA, Nyland.

Thallus excessively thin or obsolete. Apothecia as in *Arthonia,* black, superficial; paraphyses distinct; spores 8, usually colourless, 1-septate. Spermatia straight.

A small cosmopolitan genus.

1. **M. deformis,** *Nyland.* Thallus white, very thin or obsolete. Apothecia smallish, rounded or deformed, often confluent into a rugulose mass; border obtuse or 0; spores ovoid.—Nyland. Lich. Scand. 263. *Opegrapha gregaria,* Ach.

Northern Island: Auckland, on bark, *Knight.* (Europe.)

41. NORMANDINA, Del.

Thallus of thin orbicular scales. Apothecia black, closed, with a punctiform aperture, immersed.

A genus of 2 species. (European and American.)

1. **N. Jungermanniæ,** *Delise.—Coccocarpia? pulchella,* Bab. in Fl. N. Z. ii. 273. Thallus glaucous, tips of the scales reflexed. Apothecia in tubercles of the thallus; perithecium black; epithecium black, prominent; spores oblong-cylindric, usually at most 7-septate.—Nyland. Prodr. Lich. Gall. and Alg. 174.

Northern Island: Bay of Islands, *J. D. H.* (Europe.)

42. TRYPETHELIUM, Ach.

Thallus imperfectly developed or beneath the bark. Apothecia compound, many together being immersed in a depressed or elevated impérfect thallus, black, thin; spores usually 8, rarely 2 or 4; paraphyses slender, mostly branching.

A genus of corticolous, chiefly tropical Lichens.

1. **T. madreporiforme,** *Eschweiler;—Bab. in Fl. N. Z.* ii. 305. Thallus

faintly indicated, yellowish-green, somewhat shining and wavy. Apothecia several together, in convex brown tubercles, with a slightly depressed pore; spores oblong.—Nyland. Exp. Synops. Pyren. 78.

Var. *β. obscurius*, Bab. l. c. Thallus white, subpapillose, evanescent; thallus brownish within.

Northern Island: var. *β*, on old bark, *Colenso*. (Var. *α*, tropical America and Africa.)

43. ENDOCARPON, Hedwig.

Thallus peltate or scaly or areolate. Apothecia pale, rarely brown, closed, with a punctiform orifice; paraphyses 0; spores 8, simple, colourless, elliptic-oblong. Spermagonia with arthrosterigmata.

A considerable genus, found on stones and rocks in all latitudes.

1. **E. hepaticum,** *Ach.—E. pusillum*, Hedw.;—Bab. in Fl. N. Z. ii. 306. Thallus of red-brown, subcartilaginous, adnate, simple, grey-brown scales; margin blackish, fibrillose. Apothecia with brownish-black protruded tips.—Nyland. Lich. Scand. 265.

Northern Island: on the ground, *Colenso*. (Cosmopolite.)

2. **E. fluviatile,** *De Cand.;—Bab. in Fl. N. Z.* ii. 306. Thallus peltate, lurid-brown, broadly expanded, polyphyllous, blackish below.—Nyland. Lich. Scand. 265.

Northern Island, *Colenso*. (Europe.)

44. VERRUCARIA, Persoon.

Thallus scaly, areolate, continuous, powdery, or beneath the bark or obsolete. Apothecia black or pale, closed, with a punctiform orifice; spores various. Spermagonia with simple sterigmata.

A very large, chiefly temperate genus of minute Lichens, found on rocks, stones, earth, bark, etc.

Series 1. *Species found on stone.*

1. **V. rupestris,** *Schrad.—V. immersa*, Pers., and *V. muralis*?, Ach.;—Bab. in Fl. N. Z. ii. 307. Crust indeterminate, thin or evanescent, whitish, smooth. Apothecia small, black, globose, sunk in a hollow of the crust and of the stone; upper part pale, sunken part black; spores ellipsoid.—Nyland. Lich. Scand. 275.

Northern Island: on calcareous and sandstone rocks. Identification doubtful. (Cosmopolite.)

2. **V. umbrina,** *Wahl.;—Bab. in Fl. N. Z.* ii. 307. Thallus brown or blackish, thin, opaque, smooth or unequally granular, cracked or areolate, often widely extended. Apothecia immersed in mamillary tubercles of the thallus, pale, of the same colour as the thallus; spores murally divided.—Nyland. Lich. Scand. 269.

Northern Island: on pebbles, *Colenso*; in a bad state. (Cosmopolite.)

3. **V. maura,** *Wahl.;—Bab. in Fl. N. Z.* ii. 307. Thallus thin, continu-

ous, imperfectly defined, smooth, coal-black, with innumerable minute cracks. Apothecia black, immersed in slightly swollen papillæ of the thallus, dot-like; spores ellipsoid, simple.—Nyland. Lich. Scand. 275.

Northern and **Middle** Islands: on stones on the beach, *Colenso;* Otago, on trap rock, *Lindsay*. (Cosmopolite ?)

Series 2. *Species growing on bark.*

a. Spores usually 1-septate.

4. **V. gemmata,** *Ach.—V. alba ?*, Schrad.;—Bab. in Fl N. Z. ii. 307. Thallus almost filmy, whitish, indeterminate, continuous or somewhat cracked, nearly smooth. Apothecia large, prominent, hemispherical or deformed, naked or invested with a thin film; asci cylindric; spores ellipsoid.—Nyland. Exp. Synops. Pyren. 53.

Northern Island: Bay of Islands, on bark, *J. D. H.* (Europe, N. America.)

5. **V. epidermidis,** *Ach.* Thallus very thin, spreading, cream-coloured. Apothecia black, very minute, roundish, convex, circumference depressed, with a hemispherical central mamilla; spores oblong or ovoid-oblong.—Nyland. Exp. Synops. Pyren. 58. *V. punctiformis*, DC.; Fl. Antarct.

Var. *gemellipara*, Knight in Linn. Trans. 23. 99. t. 11. f. 3. Thallus yellowish, with a black border. Apothecia crowded, when moist hidden by the epidermis; spores obovate; upper cell largest.

Var. *pseudo-punctiformis*, Knight, l. c. t. 11. f. 4. Thallus fulvous. Apothecia crowded, more or less superficial, but prominent when dry, dimidiate. Spores obovate fusiform, sometimes 5-septate.

Northern and **Middle** Islands: Auckland, on bark, *Knight*. (Cosmopolite.)

6. **V. minutella,** *Knight, l. c. t.* 11. *f.* 1. Thallus thin ashy, smooth, effuse. Apothecia, when moist immersed, dry, innate, very minute, crowded, entire, globose; asci elliptic-oblong; spores yellow, clavate, 0·00014 in. broad, 0·0006 in. long.

Northern Island: Auckland, *Knight*.

7. **V. binucleolata,** *Knight, l. c. t.* 11. *f.* 2. Thallus brown, effuse. Apothecia small, superficial, dimidiate, spreading at the base; asci elongate, cylindric; spores brown, 2-nucleolate, obovate, 0·00034 in. broad, 0·00065 in. long.

Northern Island: Auckland, on bark, *Knight*.

8. **V. magnospora,** *Knight, l. c. t.* 11. *f.* 5. Thallus ashy-black, contiguous, effuse, with a black border. Apothecia naked, large, with a white border, dimidiate, spreading at the base; asci clavate; spores very large, at length pale brown, constricted in the middle, with a double hyaline border, 0·00086 in. broad, 0·018–0·0224 in. long.

Northern Island: Auckland, on bark, *Knight*.

β. Spores usually 3-septate.

9. **V. nitida,** *Schrad.;—Bab. in Fl. N. Z.* ii. 307. Thallus determinate, somewhat tartareous, continuous, smooth, waxy-brown, marked with minute pale dots and swellings. Apothecia rather large, hemispherical, black,

immersed, at length partially exposed; spores dark brown, fusiform-ellipsoid, with 4 minute cells.—Nyland. Exp. Syn. Pyren. 45.

Var. *pseudo-nitidella*, Knight, l. c. t. 11. f. 6. Thallus ferruginous, effuse. Apothecia small, naked.

Northern and **Middle** Islands: Bay of Islands, on bark, *J. D. H.* (Cosmopolite.) Var. *pseudo-nitidella*, Auckland, on bark, *Knight*; Otago, Lindsay.

10. **V. glabrata,** *Ach.* Thallus effuse, white or evanescent. Apothecia dimidiate, somewhat prominent, in other respects as in *V. nitida*.—Nyland. Exp. Syn. Pyren. 47.

Var. *α*. Knight l. c. t. 11. f. 7. Thallus white or waxy, effuse or limited. Apothecia large, naked or veiled by the epidermis.

Var. *β*. *cinereo-alba*, Knight and Mitt. l. c. Thallus membranous, ashy-white, contiguous. Apothecia small, hemispheric at first, covered by the thallus; spores 2-seriate.

Var. *γ*. *deprimens*, Knight, l. c. t. 11. f. 8. Thallus ashy, effuse. Apothecia entire or dimidiate; spores sometimes 1–2-septate; upper cell much the largest.

Var. *δ*. *homalisma*, Knight, l. c. t. 11. f. 9. Thallus yellow-brown, effuse, smooth. Apothecia very large, quite flat, entire, widely spreading.

Northern and **Middle** Islands: probably common, from Auckland, on bark, *Knight*, to Otago, *Lindsay*. (Europe, etc.)

γ. Spores many-celled or muricate.

11. **V. moniliformis,** *Knight, l. c. t.* 11. *f.* 10. Thallus olive-green, effuse, unequal. Apothecia entire, sinuate or superficial, spreading at the base; nucleus subglobose, yellow; asci cylindric; spores 7-septate, brown, ellipsoid; nucleoli often moniliform.

Northern Island: Auckland, on bark, *Knight.*

12. **V. deliquescens,** *Knight, l. c. t.* 11. *f.* 11. Thallus effuse, yellow, glabrous. Apothecia more or less hid by the epidermis, entire, thick, spreading at the base; spores somewhat curved, 1-septate or 10-nucleolate, brown; nuclei lenticular-elliptic, often deliquescent.

Northern Island: Auckland, on bark, *Knight.*

13. **V. pyrenastroides,** *Knight, l. c. t.* 11. *f.* 12. Thallus yellow or grey, effuse. Apothecia entire, confluent, more or less hidden by the epidermis, at length exposed; ostioles distinct, surrounded by a pale areola; spores elliptic-oblong, large, brown, many celled, 8–10-annulate; rings formed of quadrate cells.

Northern Island: Auckland, on bark, *Knight.*

14. **V. cellulosa,** *Knight, l. c. t.* 11. *f.* 13. Thallus effuse, olive-green. Apothecia excessively minute, entire, cellular, superficial; nucleus grey-black; asci subrotund; spores large, elliptic-oblong, 2-seriate, brown, muricate.

Northern Island: Auckland, on bark, *Knight.*

15. **V. Haultaini,** *Knight, l. c. t.* 11. *f.* 14. Thallus olive-green or ochraceous, unequal, effuse. Apothecia excessively minute, immersed, when moist dimidiate; nucleus globose; asci clavate; spores large, subclavate, muricate, 0·00025 in. broad, 0·0009 long.

Northern Island: Auckland, on bark, *Knight.*

Dr. Lindsay, in his 'Catalogue of Otago Lichens,' enumerates *V. pallida*, Ach.

ORDER VIII. FUNGI.

Cellular plants, composed of short or long cells or filaments, variously combined, usually growing on decayed or living substances, or in earth impregnated with organic matter. Fructification of spores, that are either naked or enclosed in membranous tubes (*asci*), then called *sporidia*, or terminating filaments or spicular cells (*sporophores* or *basidia*). Spores germinating by elongation of the outer covering or protrusion of the lining membrane. Fungi absorb oxygen and exhale carbonic acid, like animals, and abound like them also in nitrogen.

A vast, complicated, and varied Order, approaching Lichens through *Peziza*, *Sphæria*, and other *Ascomycetes*, and differing from *Algæ* chiefly in growing upon land, in deriving their nourishment through rooting filaments, and in their chemical constituents.

The species hitherto collected in New Zealand give a most imperfect idea of the extent, variety, and structural characters of this vast Order, which are of themselves the study of a lifetime. New Zealand no doubt contains several thousand species.

A multitude of terms have been applied to designate the various modifications of the organs of growth, vegetation, and reproduction of Fungi. I have reduced them to a minimum here, and doubt not they might be still further reduced with advantage. The most useful may be thus defined:—Fungi have no root proper, but usually have a *mycelium*, which consists of branching or simple white or dark filaments, that ramify through the ground, deriving nourishment from it (as the spawn of mushrooms), or in dead wood, hastening its decay, or in living animal and vegetable tissues; these mycelia usually produce chemical change (fermentation, etc.) where they are developed. The "Dry-rot," the "Vinegar-plant," the "Yeast-plant," "Amadou," and the white spreading fibrous growths found in cellars and pits, are all mycelia of various Fungi. When the mycelium casts off fine matted filaments, it is called "*byssoid*."

The surface on which the mycelium or Fungus grows, is called the *matrix*.

The unusual terms applied to the forms of Fungi are:—*determinate*, when the form is defined; *effuse*, when spread over the matrix; *emergent* or *erumpent*, when developed below the cuticle or bark of a plant and bursting through it; *immersed*, when sunk in the matrix; *resupinate*, when, as it were, turned upside down, the hymenium being upwards,—a term applied only to those tribes in which it is normally downwards.

When the substance of the Fungus forms a sort of amorphous mass, in which closed sacs containing the spores are sunk, it is called a *stroma*.

In the more perfect determinate-formed Fungi, the fructification is spread over the surface and exposed; the fructifying surface is then called *hymenium*, and the layer beneath it the *hymenophorum*; sometimes this *hymenium* is spread over both surfaces of thin membranous plates, or the surfaces of pores, as in mushrooms, Polypori, etc., when the intermediate substance of such plates, etc., is called a *trama*.

The *hymenium* may consist of spicular cells (*basidia* or *sporophores*) tipped by spores; or of *asci*, containing *sporidia*, exactly as in Lichens; or of a nucleus or *capillitium* of mixed filaments and spores.

When the hymenium or spores are enclosed in proper membranes, these are called *perithecia* or *peridia*; the former generally applying to separate sacs imbedded in a common stroma, the latter to a membrane covering the whole Fungus.

When the fruit-bearing portion of the Fungus takes a definite shape, different from and usually larger than the stem, it is called a *pileus*, as the cap of a mushroom.

For all we know of the fungology of New Zealand, we are indebted to the Rev. M. J. Berkeley, the most eminent English author, who worked up this Order for the Floras of the Antarctic Expedition. From that of New Zealand, the following descriptions are chiefly compiled, and I am further indebted to that gentleman's good offices for revising this compilation from his own labours.

The principal books that should be procured to pursue this difficult study, are Berkeley's

'Introduction to Cryptogamic Botany' and 'Outlines of British Fungi,' and Fries's 'Epicrisis.' A good compound microscope is essential, as also some skill as a dissector and draughtsman, to make any progress.

SYNOPSIS OR KEY TO THE SUBORDERS, TRIBES, AND GENERA.

SUBORDER I. HYMENOMYCETES.—Hymenium exposed, consisting of closely packed cells, of which the fertile (sporophores) bear naked, usually 4-nate spores on distinct spicules.

TRIBE I. **Agaricini.** *Hymenium on the under surface of a pileus, spread over the surface of radiating flat gills. (Mushrooms, toadstools, etc.)*

The most highly organized Fungi, of very various form and consistence, usually of considerable size and definite shape; but presenting every variation from the symmetrical mushroom and woody *Polyporus*, to the incoherent shapeless gelatinous mass of *Tremelia.*

α. Pileus fleshy; gills soft, brittle.

1. AGARICUS. Gills membranous, persistent.
2. COPRINUS. Gills membranous, deliquescent.

β. Pileus fleshy; gills coriaceous.

3. HYGROPHORUS. Hymenium waxy.
4. MARASMIUS. Hymenium dry, extending over gills and their interstices.
5. LENTINUS. Gills toothed or lacerate.
6. SCLEROMA. Gills entire; trama indistinct.
7. PANUS. Gills unequal, entire; trama distinct, fibrous.

γ. Pileus hard or tough.

8. SCHIZOPHYLLUM. Gills splitting longitudinally.
9. LENZITES. Gills anastomosing behind.

TRIBE II. **Polyporei.** *Hymenium on the under surface of a pileus, lining the cavity of tubes or pores, which are sometimes broken up into teeth or concentric plates.*

10. POLYPORUS. Pores minute at first. Contiguous walls separable.
11. DÆDALEA. Pores labyrinthiform.
12. FAVOLUS. Pores large at first. Contiguous walls inseparable.

TRIBE III. **Hydnei.** *Hymenium on the under surface of the pileus, spread over the surface of spines or papillæ, not lining pores or tubes.*

13. HYDNUM. Spines more or less conical, short or elongated.
14. IRPEX. Spines flat, more or less divided, connected at the base.

TRIBE IV. **Auricularini.** *Hymenium on the under surface of the pileus, which is, however, often resupinate, at first even or rarely veined and commonly remaining so, or forming obscure folds or granulations, not prickly or tubular.*

15. THELEPHORA. Pileus coriaceous, of matted fibres, without cuticle.
16. STEREUM. Pileus coriaceous, with an adherent cuticle.
17. CORTICIUM. Pileus fleshy, swollen when moist.
18. CYPHELLA. Pileus submembranous, cup-shaped.
19. GUEPINIA. Pileus gelatinous, swelling when moist.

TRIBE V. **Clavariei.** *Erect Fungi, usually terete or clavate. Hymenium vertical, scarcely distinct from the hymenophorum, extending over the apex of the plant, even, or at length wrinkled.*

20. CLAVARIA. Substance fleshy.
21. PISTILLARIA. Substance waxy, finally hoary; cellular internally.

TRIBE VI. **Tremellini.** *Gelatinous Fungi, with sometimes a more solid nucleus, variously shaped, often lobed, convolute or discoid; sporophores large, simple or divided. Spicules elongated into threads, which are not compacted into a true hymenium.*

22. HIRNEOLA. Substance tough, externally hispid.

SUBORDER II. GASTEROMYCETES.—Hymenium concealed within the substance of the plant, exposed only by the rupture or decay of its walls (*peridium*), consisting of closely-packed cells, of which the fertile bear naked spores on distinct spicules.

A large suborder, of high development, remarkable for the drying up of the hymenial tissues of many, whence their cavities contain a dusty mass of spores. Spiral threads occur, mixed with the spores, of some.

TRIBE VII. **Phalloidei.** *Young pileus enclosed in a gelatinous stratum, and this in a globular volva, which bursts irregularly and exposes it. Hymenium deliquescent. (Fetid Fungi.)*

23. ASEROE. Pileus cylindric, branching at the summit into forked horizontal rays, sheathed with the volva at its base.
24. ILEODICTYON. Pileus forming a globular reticulated branched network, the branches tubular.

TRIBE VIII. **Trichogastres.** *Usually globose dry Fungi. Hymenium enclosed in a single or double coat (peridium), at length drying up into a dusty mass of microscopic threads and spores (capillitium). (Chiefly terrestrial, puff-balls, etc.)*

25. SECOTIUM. Pileus subglobular, stalked; stalk sheathed with the volva at the base.
26. PAUROCOTYLIS. Globose. Peridium of closely interwoven flocci.
27. GEASTER. Peridium splitting into reflexed segments.
28. TRICHOSCYTALE. Cylindric. Outer peridium hard, surrounding the base of the capillitium with a volva.
29. BOVISTA. Globose. Peridium like parchment, cracking off in large flakes.
30. LYCOPERDON. Globose or pyriform. Peridium papery, persistent, warted or tubercled towards the apex.
31. SCLERODERMA. Subglobose. Peridium corky, bursting by an indefinite aperture.

TRIBE IX. **Myxogastres.** *Usually minute Fungi, pulpy when young, of various forms. Peridium usually globose, single or double, containing a dusty mass of flocci, mixed with spores.*

32. ÆTHALIUM. Indeterminate, adnate, globose or oblong, pulpy masses. Outer peridium floccose.
33. DIDERMA. Minute. Peridium globose, double, outer polished.
34. DIDYMIUM. Minute. Peridium subglobose, single, outer furfuraceous.
35. STEMONITIS. Minute. Peridium globose or cylindric, single, evanescent, traversed by the stipes.

TRIBE X. **Nidulariacei.** *Peridium of various forms, bursting at the apex horizontally, containing separate sporangia, in which the spores are formed.*

36. CYATHUS. Peridium of 3 membranes.
37. CRUCIBULUM. Peridium of a uniform spongy consistence.

SUBORDER III. CONIOMYCETES.—Very minute Fungi, including rusts, bunt, smut, various mildews, etc. Hymenium 0. Spores abundant, conspicuous, often large, surrounded by a perithecium or naked, terminating inconspicuous threads. Threads often arising from a creeping mycelium. Peridium (*perithecium*) when present very delicate and evanescent.

This suborder is distinguished by the relative predominance of the reproductive bodies, which are very fugacious and soil the hand. It contains many obscure and most curious plants, whose development and organization have taxed the skill and patience of the most eminent naturalists. Some species present different forms, arising from different forms of spore; some are supposed to originate infectious diseases in the animal kingdom; others attack and destroy living plants.

TRIBE XI. **Sphæronemei.** *Perithecium more or less distinct.*

38. LEPTOSTROMA. Perithecium thin, falling off by a transverse rupture.

39. PHOMA. Perithecium punctiform or pustular, with a minute orifice. Spores minute, simple.

40. HENDERSONIA. Perithecium subglobose. Spores 2-multiseptate.

41. ARCHERSONIA. Perithecium fleshy, lobed, multicellular. Spores simple or septate.

42. PHLYCTÆNA. Perithecium spurious; spores very slender, on short sporophores.

43. PILIDIUM. Perithecium flat, shield-like, smooth, variously ruptured. Spores linear, sessile.

44. ASTEROMA. Perithecia arising from creeping filaments, flat, with no determinate orifice.

TRIBE XII. **Torulacei.** *Perithecium altogether wanting. Spores compound, moniliform or arising from repeated division, rarely reduced to a single cell.*

45. GYMNOSPORIUM. Spores superficial, conglobate.

TRIBE XIII. **Pucciniæi.** *Parasitic on living plants. Peridium* 0. *Spores producing secondary spores in germination, usually oblong and septate.*

46. PUCCINIA. Spores naked, 1-septate, on a distinct peduncle.

47. UREDO. Spores sessile, each contained in a cavity of a multicellular stroma.

48. UROMYCES. Spores stalked, each contained in a cavity of a multicellular stroma.

49. USTILAGO. Spores deep-seated, simple, on delicate threads, or breaking up into a powdery mass.

TRIBE XIV. **Æcidiacei.** *Peridium distinctly cellular. Mycelium traversing the tissues of living plants.*

50. ÆCIDIUM. Peridium bursting open. Spores concatenate.

SUBORDER IV. HYPHOMYCETES.—Filamentous or floccose Fungi. Filaments naked, simple or branched, free or united below so as to form a distinct stem with free branches. Spores terminating the filaments.

Most of the species of this vast tribe come under the common term of *moulds*, whose ravages no animal or vegetable matter escapes. Coincident with their attack a chemical change takes place in the substance attacked, known as fermentation. They are developed with extreme rapidity, and are often poisonous. The various mildews of the Vine, Hop, and silkworm belong here, as do the potato disease, and *oïdium* of the grape.

TRIBE XV. **Stilbacei.** *Stem or stroma compound. Spores collected into a globose head, usually terminating a distinct stalk, subgelatinous, diffluent.*

51. STILBUM. Stem firm, long. Head globose, deciduous. Spores small, involved in gluten.

52. EPICOCCUM. Head subglobose, studded with large spores.

TRIBE XVI. **Dematiei.** *Filaments free, more or less corticated and carbonized. Spores often compound and cellular.*

53. ŒDEMIUM. Filaments free, dark, flexuose, with reticulate large spores on their sides.

54. MACROSPORIUM. Filaments slender, evanescent. Spores erect, stipitate, multiseptate.

55. Cladosporium. Filaments flexible, branched, jointed. Spores short and septate.

Tribe XVII. **Mucedines.** *Filaments not coated with a membrane, distinct, white or coloured. Spores mostly simple.*

The moulds which belong to this tribe have never been collected in New Zealand.

Tribe XVIII. **Sepedoniei.** *Spores collected on a floccose mycelium, very abundant. Fertile filaments scarcely distinct from the mycelium.*

56. Sepedonium. Spores large, simple, globose or appendiculate.

Tribe XIX. **Trichodermacei.** *Spores enclosed within or covered by a floccose mycelium, which at length bursts.*

57. Pilacre. Stipes solid. Heads of flexuose, branched threads. Spores forming a dusty mass.

Suborder V. Ascomycetes.—Fruit consisting of asci, containing sporidia, and springing from a naked or enclosed nucleus or hymenium, which is often spread over a receptacle.

Morells and Truffles belong to this tribe, and are doubtless to be found in New Zealand, as is *Cyttaria*, a curious genus found on Beech-trees in Tasmania and Fuegia, but not hitherto in these islands.

Tribe XX. **Elvellacei.** *Substance soft, fleshy or waxy. Hymenium more or less exposed.*

58. Leotia. Receptacle pileate, stipitate, covered everywhere with the smooth rather viscid hymenium.

59. Geoglossum. Receptacle clavate. Hymenium surrounding the club.

60. Peziza. Receptacle cup-shaped, its disk naked. Asci fixed.

61. Patellaria. Receptacle patellæform, margined. Hymenium dusty from the spores breaking up.

62. Cenangium. Receptacle coriaceous, at first closed, then open and margined, covered with a thick cuticle.

Tribe XXI. **Phacidiacei.** *Perithecium coriaceous or carbonaceous; outer coat or perithecium bursting and exposing the disk, which is surrounded by an obtuse or inflected margin.*

63. Hysterium. Perithecium labiate; border entire; orifice narrow. Asci elongate.

64. Ailographium. Perithecium branched; orifice narrow. Asci subglobose.

65. Asterina. Perithecium suborbicular, in a byssoid mycelium, splitting irregularly. Asci short.

66. Excipula. Perithecium carbonaceous, spherical; mouth orbicular; nucleus gelatinous.

Tribe XXII. **Sphæriacei.** *Perithecium carbonaceous or membranous, pierced at the apex. Hymenium diffluent. Asci usually springing from the walls.*

67. Cordiceps. Stroma vertical, erect, coriaceous or fleshy. Receptacle distinct. Sporidia submoniliform.

68. Hypocrea. Stroma horizontal. Perithecia tender, hyaline or coloured.

69. Xylaria. Stroma vertical, fleshy or corky, with a black or rufous bark. Receptacle stipitate.

70. Hypoxylon. Stroma brittle or corky, flat or convex, at first clothed with a floccose veil, then with a black crust. Perithecia vertical or divergent.

71. Diatrype. Stroma confluent with the matrix. Perithecia sunk, elongated into a distinct neck.

72. Dothidea. Stroma with globose cavities containing nuclei, which have a decided neck and papillæform aperture. Perithecium 0.

73. Nectria. Stroma 0, or bearing the naked coloured perithecia on its surface.

74. Sphæria. Stroma 0. Perithecia black, pierced at the apex, superficial or erumpent.

75. Capnodium. Mycelium of black, branched, jointed filaments. Perithecia elongate, of confluent filaments, with often free tips.

76. Micropeltis. Perithecium free, carbonaceous, flattened, dimidiate, scutate. Asci clavate. Sporidia septate.

76. Pemphidium. Perithecia spurious, convex, black, formed of the epidermis. Nucleus gelatinous. Asci perfect. Sporidia elongate.

Tribe XXIII. **Perisporiacei.** *Perithecia free, subglobose, always closed, except when decaying, membranous or carbonaceous. Nucleus never diffluent. Asci springing from the base.*

77. Erysiphe. Mycelium cobwebby. Perithecia soft; appendages floccose, simple or branched.

78. Chætomium. Perithecium thin, brittle. Asci linear, with dark mostly lemon-shaped sporidia.

79. Meliola. Perithecia carbonaceous, from a strigose mycelium. Asci broad.

Suborder VI. Physomycetes.—Filaments free or slightly matted, bearing vesicles which contain indefinite sporidia.

To this tribe belong the true moulds (*Mucor*, etc.) of which species must occur abundantly in New Zealand, though they have never been collected. *Antennaria* is an anomalous member of it.

Tribe XXIV. **Antennariei.** *Filaments black, matted, often moniliform.*

80. Antennaria. Filaments articulated. Walls of sporangia mostly cellular. Spores chained together, immersed in gelatinous pulp.

1. AGARICUS, L.

Fleshy putrescent Fungi. Hymenium inferior, spread over the surface of membranous persistent gills, formed of closely-packed cells. Gills radiating from the centre of a pileus, with shorter ones between, formed of 2 separate membranes, their edges acute. Substance of the gills internally filamentous, continuous with that of the pileus.

An immense ubiquitous genus, containing upwards of 1000 species, including the Mushroom, Toadstool, etc. The species are very variable and difficult to determine from dried specimens; they should be drawn, coloured, and described when gathered, dried by slicing, and the colour of the spores noted.

KEY TO THE SUBGENERA.

Series I. **Leucospori.** *Spores white.*

Subgenus I. Amanita. Pileus centrical. Young plant enclosed in a membranous free volva, through which the pileus pushes. Fully-formed plant with a veil extending from the circumference of the pileus to near the top of the stem. Gills not decurrent on the stem.

Subgenus II. Lepiota. Pileus centrical. Veil enclosing the young plant, connate with the cuticle of the pileus; when burst, leaving a ring on the middle of the stem. Gills not decurrent on the stem.

Subgenus III. Tricholoma. Pileus centrical. Veil obsolete or floccose, and adherent to the circumference of the pileus. Stem fleshy. Gills emarginate behind, not decurrent on the stem.

Subgenus IV. CLITOCYBE. Pileus centrical. Veil 0. Gills more or less decurrent on the stem. Stem elastic, with a fibrous outer coat.

Subgenus V. OMPHALIA. Pileus centrical. Veil 0. Gills adnate and decurrent. Stem cartilaginous.

Subgenus VI. PLEUROTUS. Pileus excentric, unequal or lateral. Stem 0 or solid and firm. Gills acute behind, unequal.

Series II. **Dermini.** *Spores rusty-coloured, tawny or brownish (not purple-brown nor pink).*

Subgenus VII PHOLIOTA. Veil dry, forming a ring round the more or less scaly stem. Gills unequal, juiceless.

Subgenus VIII. HEBELOMA. Veil, when present, floccose, not interwoven. Stem fleshy. Gills sinuated.

Subgenus IX. FLAMMULA. Veil fugacious. Stem fleshy, firm, fibrillose. Gills adnate or decurrent.

Subgenus 10. GALERA. Veil floccose, fugacious. Stem slender, cartilaginous externally. Pileus campanulate; margin straight. Gills adnate.

Series III. **Pratella.** *Spores purple-brown or pink-brown.*

Subgenus 11. PSALLIOTA. Stem with a ring formed by the veil.

Subgenus 12. HYPHOLOMA. Veil forming a fugacious web, that adheres to the circumference of the pileus.

1. **A. (Amanita) phalloides,** *Fries;—Berk. in Fl. N. Z.* ii. 173. Pileus at first campanulate, viscid when moist, often green; margin even, regular. Volva free above, bulbous. Gills rounded, ventricose.—Berk. Outlines, 89. t. 3. f. 1.

Northern Island: in woods, *Colenso.* (Europe.)

2. **A. (Lepiota) clypeolarius,** *Bull.;—Berk. in Fl. N. Z.* ii. 173. Sweet-scented. Pileus fleshy, umbonate, white pink red brown or yellow; surface at first even, then scaly. Stem fistulose, the ring floccose and scaly. Gills free, approximate.—Berk. Outlines, 94.

Northern Island: on the ground, *Colenso.* (Europe.)

3. **A. (Lepiota) exstructus,** *Berk. in Fl. N. Z.* ii. 173. Pileus campanulate, fleshy, 1 in. high and 2 across or more; surface broken up into warts, each of which is clothed at the top with an angular portion of smooth cuticle. Stem nearly equal, 3 in. high, not penetrating the substance of the pileus; ring superior. Gills narrow, very distant.

Northern Island: Bay of Islands, on the ground, *J. D. H.*—A beautiful species.

4. **A. (Tricholoma) brevipes,** *Bulliard;—Berk. in Fl. N. Z.* ii. 173. Pileus fleshy, at first rigid, then soft smooth; umbo evanescent. Stem solid, firm, rigid, very short, rather thickened below, brown. Gills crowded, ventricose, brownish, then dirty white.

Northern Island: on the ground, *Colenso.* (Europe.)

5. **A. (Tricholoma) cartilagineus,** *Bulliard.* Cartilaginous, elastic, brittle. Pileus fleshy, convex, gibbous, undulated, smooth; cuticle cracked, finely dotted with black. Stem stout, equal, stuffed, striate, rather mealy. Gills adnate, slightly emarginate, crowded, pallid.—Berk. Outlines, 101.

Middle Island: Beech forests, amongst moss, *Haast.* (Europe.)

6. **A. (Tricholoma) carneus,** *Bulliard.* Pileus rufous-pink, fleshy, obtuse, even, nearly smooth, becoming pale. Stem short, axis stuffed, rigid, reddish, thickened upwards, pruinose. Gills very wide behind, rounded, crowded, white.—Berk. Outlines, 103.

Middle Island, *Sinclair and Haast.* (Europe.)

7. **A. (Clitocybe) infundibuliformis,** *Schæffer;—Fl. N. Z.* ii. 173. Pileus usually pale reddish, fleshy, convex, then broadly funnel-shaped, umbonate, clothed with minute down, flaccid. Stem soft, elastic, axis stuffed. Gills decurrent, white, moderately distant.—Berk. Outlines, 110.

Northern Island: on the ground, *Colenso.* (Europe.)

8. **A. (Omphalia) pyxidatus,** *Bulliard.* Pileus red-grey, at first umbilicate, then funnel-shaped, watery when fresh; margin striate. Stem even, at length fistulose. Gills decurrent, rather distant, narrow, reddish-grey.—Berk. Outlines, 131.

β. *hepaticus*, liver-coloured.—*Berk. in Fl. Antarct.* 169.

Lord Auckland's group, *J. D. H.*—Specimens very imperfect and identification doubtful.

9. **A. (Omphalia) Colensoi,** *Berk. in Fl. N. Z.* ii. 173. Small. Pileus $\frac{1}{2}$ in. across, smooth, minutely striate when dry, deeply umbilicate; margin involute. Stem $\frac{1}{2}$–$\frac{3}{4}$ in. high, slender, narrowed downwards, furfuraceous, at length smooth. Gills moderately broad, decurrent.

Northern Island: on sand, logs, etc., *Colenso.*

10. **A. (Omphalia) umbelliferus,** *Linn.;—Berk. in Fl. N. Z.* ii. 173. Pileus variable in colour, membranous, plano-convex, obconic, brittle, radiato-striate, pale when dry, even, slightly silky; margin first inflexed, crenate. Stem equal, downy below. Gills very distant, thick, decurrent, very broad behind.—Berk. Outlines, 132.

Northern Island: on the ground, *J. D. H., Colenso.* (Europe and other countries.)

11. **A. (Pleurotus) novæ-Zelandiæ,** *Berk. in Fl. N. Z.* ii. 174. Pileus gelatinous and thin, watery when fresh, $2\frac{1}{2}$ in. broad, white, flabellate, reniform, fixed by the elongated vertex, which forms a little round disk, smooth in front, minutely scabrous behind. Stem obsolete. Gills broad, distant, thin, interstices veiny.

Northern Island: on dead wood, *Colenso.*

12. **A. (Pleurotus) cocciformis,** *Berk. in Fl. N. Z.* ii. 174. Minute. Pileus $\frac{1}{12}$–$\frac{1}{6}$ in, broad, at first cup-shaped, then reflexed, thickly clothed with fawn-coloured hairs. Stem 0. Gills narrow, setulose, pale tan-coloured.

Northern Island: on dead wood, *J. D. H.*

13. **A. (Pholiota) erebius,** *Fries;—Berk. in Fl. N. Z.* ii. 174. Gregarious, fragile. Pileus thin and fleshy, rugulose, lurid, striate at the margin. Stem smooth, hollow, pale; veil subcampanulate. Gills rather distant, adnate.

Northern Island: on the ground, Ahuriri, *Colenso.* (Europe.)

14. **A. (Pholiota) adiposus,** *Fries;—Berk. in Fl. N. Z.* ii. 174.

Gregarious, yellow. Pileus compact, plano-convex, obtuse, glutinous, rough, with evanescent concentric darker scales, as is the stem, which is somewhat bulbous below; axis stuffed. Gills adnate, broad, yellow, then ferruginous. —Berk. Outlines, 151. t. 8. f. 2.

Northern Island: Cape Turnagain, on trees, *Colenso*. (Europe.)

15. **A. (Hebeloma) strophosus,** *Fries;—Berk. in Fl. N. Z.* ii. 174. Pileus fleshy, convex, rather flattened; disk darker, subumbonate, silky towards the even margin. Stem hollow, short, equal silky, with a floccose ring at the apex. Gills adnexed, then free, crowded, ventricose, pallid, then of a watery cinnamon.

Northern Island: grassy spots, Wairarapa, *Colenso*. (Europe.)

16. **A. (Flammula) sapineus,** *Fries;—Berk. in Fl. N. Z.* ii. 174. Pileus fleshy, yellowish, plano-convex, even. Stem subsolid, striate, pale. Gills dirty white or yellowish or tawny-cinnamon.

Northern Island: on dead Coniferæ, *Colenso*. (Europe.)

17. **A. (Psalliota) campestris,** *Linn.;—Berk. in Fl. N. Z.* ii. 174. Pileus fleshy, plano-convex, dry, floccose or scaly. Stem even, white; axis stuffed; ring about the middle, torn. Gills free, approximate, ventricose, subdeliquescent, flesh-coloured, then brown.—Berk. Outlines, 165. t. 10. f. 2.

Northern Island: on the ground, *Colenso*. (The common "mushroom," found in all parts of the world.)

18. **A. (Psalliota) arvensis,** *Schæffer;—Berk. in Fl. N. Z.* ii. 174. Pileus fleshy, obtusely conico-campanulate, then expanded, at first floccose, then smooth, even or cracked. Stem hollow, with floccose pith; ring broad, pendulous, double, outer split into rays. Gills free, wider in front, dirty white. Stem brown, tinged with pink.—Berk. Outlines, 166. t. 10. f. 4.

Northern Island: on the ground, *Colenso*. (Common.) Grows in rings (fairy rings). Edible.

19. **A. (Psalliota) campigenus,** *Berk. in Fl. N. Z.* ii. 174. Small. Pileus 1 in. high, campanulate, obtuse, silky, with a few scattered scales, rather fleshy, reddish when dry. Stem $1\frac{1}{2}$ in. high, $\frac{1}{6}$ thick, thickened at the base, slightly furfuraceous; ring broad, near the top. Gills rather narrow, attenuated behind, adnexed or adnate. Spores pale red-brown. obliquely obovate.

Northern Island: on the ground amongst grass, *Colenso*.

20. **A. (Psalliota) semiglobatus,** *Batsch*. Pileus slightly fleshy, hemispherical, even, glutinous and yellowish. Stem slender, smooth, straight, fistulose, glutinous, yellowish. Gills broad, adnate, plane, clouded with black, —Berk. Outlines, 169.

Middle Island: on dung, *Sinclair and Haast*. (Europe.)

21. **A. (Hypholoma) fascicularis,** *Hudson; Berk. in Fl. N. Z.* ii. 175. Bitter. Pileus fleshy, yellow within, thin, subumbonate, smooth. Stem hollow, thin, flexuose, fibrillose, yellow. Gills adnate, very crowded,

linear, subdeliquescent, sulphur-coloured, then greenish.—Berk. Outlines, 169. t. ii. f. 1.

Northern Island: on the ground, *Colenso.* (Europe, Tasmania.)

22. **A. (Hypholoma) stuppeus,** *Berk. in Fl. N. Z.* ii. 175. Pileus 2 in. broad and more, fleshy, convex, expanded, clothed with towy fascicled fibrous scales especially towards the margin. Stem 1½ in. high, ⅓ in. thick, fibrillose, thickened at the base, and attached to abundant mycelium. Gills crowded, moderately broad, adnate, umber.

Northern Island: on the ground, *Colenso.*

23. **A. (Hypholoma) appendiculatus,** *Bulliard;—Berk. in Fl. N. Z.* ii: 175. Pileus between fleshy and membranous, ovate, expanded, smooth, watery when fresh, wrinkled and sparkling with atoms when dry; margin fringed by the remains of the veil. Stem fistulose, equal, smooth, white, pruinose above. Gills somewhat adnate, crowded, dirty white, then rosy-brown.—Berk. Outlines, 170. t. ii. f. 3, 4.

Northern Island: on decayed stumps, *Colenso.* (Europe.)

2. COPRINUS, Persoon.

Habits and characters of *Agaricus,* but the gills deliquesce, and are membranous. Spores black.

Fugacious toadstools, growing on rotten wood and manured soil, etc.

1. **C. Colensoi,** *Berk.;—Fl. N. Z.* ii. 175. Small, subfascicled; stem slender, ¼–¾ in., tomentose. Pileus snow-white, cylindric, obtuse, at length campanulate, furfuraceous. Gills linear. Spores small, oblong, $\frac{1}{3500}$ in. long.

Northern Island, *Colenso,* on dung, very near the European *C. niveus.*

2. **C. fimetarius,** *Fries.* Stem solid, scaly, thickened at the base. Pileus submembranous, clavate-conic, soon torn and revolute, at first rough with floccose scales, then naked, longitudinally cracked, even at the apex. Gills free, black, lanceolate, linear and flexuose.—Berk. Outlines, 179.

Middle Island: Canterbury province, on dungheaps, *Sinclair and Haast.* (Europe, and almost ubiquitous.) Very variable.

3. HYGROPHORUS, Fries.

Habit, etc., of *Agaricus.* Pileus fleshy; hymenophorum continuous with the stem, descending without change into the sharp-edged gills; hymenium waxy.

A considerable European etc. genus.

1. **H. cyaneus,** *Berk. mss.* Azure blue. Stem 4–5 in. high, stout, nearly ½ in. diam., irregularly hollow, thickened at the base. Pileus 1½ in. high, 2 in. diam., conic acuminate. Gills broad, very obtuse in front, narrowed posteriorly.

Middle Island: beech forests, amongst moss, Nelson Province, *Haast* (drawing only). A very fine species.

4. MARASMIUS, Fries.

Habit, etc., of *Agaricus*. Pileus fleshy or membranous; hymenium dry, extending over both gills and their interstices. Gills thick, tough, coriaceous, edge acute.

A large European and exotic genus.

1. **M. caperatus,** *Berk. in Hook. Kew Journ. Bot.* iii. 44;—*Fl. N. Z.* ii. 175. Snow-white, delicate, inodorous. Stem short, furfuraceous. Pileus 1–1½ in. broad, membranous, plicate and corrugate, smooth. Gills distant, broad, adnate.

Northern Island: on dead wood, Wairarapa, *Colenso*. (Himalaya mountains.)

5. LENTINUS, Fries.

Habit, etc., of *Agaricus*. Pileus fleshy or leathery, or hard. Gills tough, with toothed or lacerated acute edges; hymenium of the same consistence as the stem.

A large genus, especially in warm countries.

1. **L. novæ-Zelandiæ,** *Berk.;—Fl. N. Z.* ii. 176. Subimbricated, small. Stem obsolete. Pileus thin, brown, 1 in. broad and upwards, flabelliform suborbicular or reniform, clothed behind with velvety olive down. Gills narrow, thin, decurrent behind, of the same colour as the pileus; edges thin, torn.

Northern Island: on dead wood, *Colenso*. M. Haast sends a sketch of a very different and much larger species, but indeterminable.

6. SCLEROMA, Fries.

Characters of *Lentinus*, but the edges of the gills are quite entire. Trama indistinct.

Tropical and subtropical Fungi, growing on dry soil in patches.

1. **S. pygmæum,** *Berk. in Fl. N. Z.* ii. 176. Small, dirty white. Stem 1–2 in. high, very slender, powdery, rooting, reddish. Pileus 1 in. diam., umbilicate, thin, glabrous, striate. Gills rather distant, decurrent far down the stem, interstices smooth.

Northern Island: Tehawera forests, on decayed logs, *Colenso*. The smallest species of the genus.

7. PANUS, Fries.

Pileus tough but fleshy, drying up. Gills tough, unequal, with quite entire acute edges. Trama distinct, fibrous.

A considerable tropical and temperate genus, distinguished from *Scleroma* by the fibrous trama.

1. **P. stypticus,** *Fries;—Berk. in Fl. N. Z.* ii. 176. Stem short, thick, lateral, dilated upwards. Pileus reniform, coriaceous; cinnamon-coloured, vinous when young; surface breaking up into mealy scales. Gills

determinate, thin, crowded, connected by veins; cinnamon-coloured.—Berk. Outlines, 217.

Northern Island: common on dead wood, *Colenso.* (Europe, etc.)

2. **P. maculatus,** *Berk. in Fl. N. Z.* ii. 176. Closely imbricate. Stems connate, scarcely visible. Pileus broad, reniform, convex, fleshy, at first tomentose, then breaking up into reflexed scales, at length smooth; margin slightly involute. Gills rather distant, broad, decurrent, crisped when dry; margin entire; spores white, oblong, $\frac{1}{3000}$ in. long.

Northern Island: on dead trunks, *Colenso.*

8. SCHIZOPHYLLUM, Fries.

Pileus hard, dry. Gills coriaceous, branched in a flabellate manner, longitudinally splitting along the trama, the divisions becoming revolute.

Chiefly a tropical genus, abounding on dead trunks.

1. **S. commune,** *Fries;—Berk. in Fl. N. Z.* ii. 177. Pileus adnate behind, simple and lobed. Gills grey, then brownish-purple, the divided surface villous.—Berk. Outlines, 228.

Abundant on dead wood. (Ubiquitous.)

9. LENZITES, Fries.

Pileus corky or coriaceous. Gills firm, often anastomosing, and forming spurious pores; edge entire.

A common tropical and temperate genus.

1. **L. repanda,** *Fries.* Pileus plane, corky, very broad, zoned, smooth, pallid, whitish; margin curved. Gills anastomosing, crowded, somewhat toothed.

Northern Island: on dead wood, *J. D. H.*

10. POLYPORUS, Fries.

Fleshy, coriaceous, or woody Fungi, often projecting horizontally from the trunks of decaying trees, consisting of a stalked or sessile pileus, which is usually marked with concentric rings on the upper surface and studded with pores on the lower, sometimes the pileus is adnate to the wood by one surface. Hymenium concrete with the substance of the pileus, consisting of subrotund pores with thin simple dissepiments. Hymenophorum descending into the trama of the pores, which are not easily if at all separable, and changed with them into a distinct substance.

A very large genus found in all parts of the world. The species are extremely variable in form and size, and are difficult of definition.

KEY TO THE SECTIONS.

A. Favolus. Pores large, 4–6-angled, like honeycomb.
B. Microporus. Pores minute, rounded.

1. *Mesopus.* Stem distinct, simple, lateral or nearly central. Substance corky or fleshy.
2. *Pleuropus.* Stem lateral, simple, substance hard.
3. *Merisma.* Stem etc. more or less divided.
4. *Apus.* Pileus sessile, lateral.
5. *Resupinatus.* Pileus resupinate, effuse; pores on various surfaces.

1. **P. (Favolus) arcularius,** *Fries;—Berk. in Fl. N. Z.* ii. 177. Pileus yellowish, corky; margin deflexed, hispid. Stem central, smooth. Pores rhomboid, white.

Northern Island: on dead wood, *Sinclair, Colenso.* (Europe.)

2. **P. (Mesopus) oblectans,** *Berk.;—Fl. N. Z.* ii. 177. Pileus thin, coriaceous, depressed, 1½ in. diam., incised, repand, zoned especially towards the centre, strigose-striated, shining, cinnamon-coloured. Stem central, velvety, red-brown. Pores small, toothed, cinnamon-coloured.

Northern Island: on the ground, *Colenso.* (Swan River, Tasmania.)

3. **P. (Mesopus) lucidus,** *Fries;—Berk. in Fl. N. Z.* ii. 177. Pileus corky, flabelliform, grooved and rugose, yellow, then blood-red and chesnut, varnished. Stem lateral. Pores minute, white, long, determinate.—Berk. Outlines, 240. t. 16. f. 2.

Northern Island: on dead wood, *Colenso.* (Ubiquitous.)

4. **P. (Pleuropus) phlebophorus,** *Berk. in Fl. N. Z.* ii. 177. *t.* 105. *f.* 3. Small, white. Pileus flabellate, contracted at the top into a short stem, glabrous, veined, undulate; cuticle gelatinous. Pores minute, subirregular, with thin toothed dissepiments.

Northern Island: on dead wood, Tehawera, *Colenso.*

5. **P. (Pleuropus) xerophyllus,** *Berk. in Fl. N. Z.* ii. 178. *t.* 105. *f.* 2. Pileus flabellate, 1 in. across, suborbicular, brown, with a strong rufous tinge, marked with raised radiating lines and small wrinkles, minutely scabrous; margin thin, crenate. Stem ⅓ in. long, black, rugose, minutely velvety. Pores white, just visible; dissepiments variable in thickness; edges rather obtuse.

Northern Island: on dead wood, *Colenso.*

6. **P. (Merisma) Colensoi,** *Berk. in Fl. N. Z.* ii. 178. Much branched, 1 foot across. Stem distinct, slender, elongated, repeatedly dichotomous. Pilei very numerous, flabellately expanded, depressed above, brownish, nearly smooth, with a few raised lines. Pores pale, decurrent, often much elongated; dissepiments thin; edges very acute, often toothed, sublamellæform.

Northern Island: forests of Tehawera, *Colenso.*

7. **P. (Apus) adustus,** *Fries;—Berk. in Fl. N. Z.* ii. 178. Pileus fleshy, tough, firm, thin, villous, ashy; margin straight, at length black, effuso-reflexed behind. Pores short, minute, round, obtuse, dirty white and powdery, then cinereous-brown.—Berk. Outlines, 243.

Northern and **Middle** Islands: on dead wood, *Colenso, Lindsay.* (Europe, etc.)

8. **P. (Apus) dichrous,** *Fries;—Berk. in Fl. N. Z.* ii. 178. Imbricate, white, very effuse. Pileus fleshy, tenacious, soft, reflexed, silky, obsoletely zoned. Pores short, equal, rounded, obtuse, cinnamon-brown.

Northern Island: on dead wood, *J. D. H.*, *Colenso.*

9. **P. (Apus) igniarius,** *Fries;—Berk. in Fl. N. Z.* ii. 179. Perennial. Mycelium and spores white. Pileus even, often resupinate, flocculent, substance hard and thick, opaque, ferruginous, changing to brownish-black; margin rounded, surface uneven, flesh zoned. Pores very minute, first whitish, then cinnamon-coloured, convex.—Berk. Outlines, 246.

Northern Island: trunks of trees, Bay of Islands, *J. D. H.* (Europe, etc.)

10. **P. (Apus) salicinus,** *Fries;—Berk. in Fl. N. Z.* ii. Large, perennial, often resupinate, very effuse. Pileus woody, very hard, undulate, smooth; margin short, obtuse, spreading, cinnamon, then brown; crust black. Pores very small, round, ferruginous and cinnamon.—Berk. Outlines, 246.

Middle Island: Dusky Bay, *Menzies.* (Europe and various countries.)

11. **P. (Apus) iridioides,** *Berk.;—Fl. N. Z.* ii. 179. Large. Pileus hard, between woody and corky, reniform, convex above, flat below, cinnamon-coloured, obscurely zoned, ferruginous-grey, rhubarb-coloured within, minutely rugulose; margin velvety; behind very scabrous, with elongate bristly nodules. Pores very minute, punctiform.

Northern Island: Bay of Islands, on wood, *J. D. H.* (South Africa Australia.)

12. **P. (Apus) scruposus,** *Fries;—Berk. in Fl. N. Z.* ii. 178. Pileus corky, subtriquetrous, rough with little points, zoneless, reddish-grey, at length sulcate near the margin; substance rhubarb-coloured. Pores minute round, equal, ferruginous.

Northern Island: on dead wood, *Colenso.* (Europe.)

13. **P. (Apus) plebeius,** *Berk. in Fl. N. Z.* ii. 179. Pale, imbricate. Pileus corky, 2 in. across, dimidiate, pulvinate, zoneless externally and internally, usually even, minutely pubescent, rigid, under surface concave; margin obtuse. Pores minute, punctiform; edges obtuse.

Northern Island: on wood, *Colenso.* (A var. with acute margins is found in the Himalayas.)

14. **P. (Apus) australis,** *Fries;—Berk. in Fl. N. Z.* ii. 179. Pileus very hard, woody, plano-convex, tuberculate and undulate, glabrous, brownish-black, under surface concave, with a broad poreless band. Pores minute, rounded, equal, obtuse, whitish.

Northern Island: on dead wood, *J. D. H.* (Pacific Islands.)

15. **P. (Apus) hemitrephius,** *Berk. in Fl. N. Z.* ii. 179. Pileus firm, corky, zoned, ungulate, with 2 or 3 deep, concentric furrows, minutely tomentose, within zoned, wood-coloured, tawny beneath the hardened cuticle; under surface concave, white. Pores punctiform, substratified.

Northern Island: on trunks of trees, *Colenso.*

16. **P. (Apus) borealis,** *Fries.* Imbricate. Pileus white, at length yellowish, between fibrous and corky, thick, convex above, flat below, velvety. Pores slender, rounded, white, forming long tubes; edges lacerate.

Middle Island: Otago, *Lindsay.* (Europe.)

17. **P. (Apus) hirsutus,** *Fries;—Berk. in Fl. N. Z.* ii. 179. White. Pileus coriaceous, corky, often imbricate, reniform, almost plane on both surfaces, strigose, zoned. Pores equal, white internally, obtuse, brown or ashy externally.

Northern Island: on dead wood, *J. D. H., Colenso.* (Europe.)

18. **P. (Apus) velutinus,** *Fries;—Berk. in Fl. N. Z.* ii. 179. White, at length yellowish. Pileus corky-coriaceous, flat on both surfaces, velvety, slightly zoned; margin acute, attenuated. Pores minute, round, white, equal. —Berk. Outlines, 248.

Northern Island: on dead wood, *Colenso.* (Europe, America.)

19. **P. (Apus) versicolor,** *Fries;—Berk. in Fl. N. Z.* 179. White yellowish or ashy-blue, tufted, subimbricate. Pileus thin, coriaceous, rigid, flattened, depressed behind, velvety, shining in parts, variegated with different-coloured zones. Pores minute, round; dissepiments acute, torn, white, at length pallid —Berk. Outlines, 248.

Northern Island: on rotten logs, *Colenso.* (Ubiquitous.)

20. **P. (Apus) tabacinus,** *Montagne;—Berk. in Fl. N. Z.* ii. 178. Imbricate, rusty-brown. Pileus thin, coriaceous, rigid, effuse, reflexed, shell-like, tomentose, zoned; margin acute, pale. Pores middling-sized; dissepiments toothed and lacerate.

Northern Island: on dead wood, *J. D. H., Colenso.* (Chili, Juan Fernandez.)

21. **P. (Apus) cinnabarinus,** *Fries;—Berk. in Fl N. Z.* ii. 179. Cinnamon-red. Pileus subcoriaceous or corky, thickish, slightly convex, obsoletely zoned, rugulose, when young pubescent. Pores rounded, conspicuous.

Northern Island: on dead wood, *J. D. H., Colenso.* (Europe, N. America.)

22. **P. (Apus) sanguineus,** *Fries;—Berk. in Fl. N. Z.* ii. 178. Shining, vermilion-red. Pileus subreniform, coriaceous, thin, obsoletely zoned, flat, glabrous. Pores very minute, vanishing towards the margin.

Northern and **Middle** Islands, *J. D. H., Lindsay.* (America, Tasmania, S. Africa.)

23. **P. (Resupinatus) catervatus,** *Berk. in Fl. N. Z.* ii. 180. *t.* 105. *f.* 1. Forming dull white patches of small laterally confluent stipitate pilei, which form together a flat thin stratum. Pileus rather membranous, silky; margin often lobed. Pores small, irregular; dissepiments thin, toothed.

Northern Island: on *Podocarpus spicata*, Bay of Islands, *Colenso.* A very curious species.

23. **P. (Resupinatus) leucoplacus,** *Berk. in Fl. N. Z.* ii. 180. White, effused. Pileus nearly smooth, thin, rigid; margin free. Pores

small, punctiform; dissepiments thick, with flattened obtuse pulverulent edges.

Northern Island: on sticks, *Colenso.*

24. **P. (Resupinatus) diffusus,** *Berk. in Fl. N. Z.* ii. 180. Bright red, effuse, at length tearing away from the matrix, and leaving part of its substance behind. Pores small; edges thin, membranous, slightly toothed.

Northern Island: on charred wood, *Colenso.* Probably an imperfect form.

25. **P. (Resupinatus) vaporarius,** *Fries;—Berk. in Fl. N. Z.* ii. 180. Firm, persistent, with difficulty torn away, thin, dry, whitish; mycelium creeping in the tissue of the wood. Pores large, angular, white, crowded into a close firm persistent stratum; edges torn.—Berk. Outlines, 252.

Northern Island: on dead wood, *Colenso.* (Europe, Tasmania.)

11. DÆDALEA, Persoon.

Habits of *Polyporus.* Hymenophorum descending unchanged into the trama. Pores, when fully formed, torn, toothed or labyrinthiform.

A large genus, both tropical and temperate.

1. **D. confragosa,** *Pers.;—Fl. N. Z.* ii. 180. Stem 0. Pileus between corky and coriaceous, somewhat zoned, scabrous, brownish-red, subferruginous within. Hymenium porous, at length reddish-brown, labyrinthiform and torn.—Berk. Outlines, 254.

Northern Island: Bay of Islands, on dead wood, *J. D. H.* (Europe, etc.)

2. **D. pendula,** *Berk. in Fl. N. Z.* ii. 181. *t.* 105. *f.* 4. Stem 0. Pileus imbricate, coriaceous, 1½ in. long, 1 in. broad, pendulous, pocket-shaped, attached by the back and vertex, pale reddish-grey, sparingly zoned, clothed with short strigose matted brown hairs; margin tomentose. Hymenium tinged with lilac and red-grey. Pores sparing, irregular, with finely setulose tooth-like septa.

Northern Island: on dead wood, *Colenso.* Berkeley remarks that the hymenium is more that of a *Radulum.*

12. FAVOLUS, Fries.

Habit of *Polyporus.* Pileus fleshy, flexible. Hymenium reticulated, cellular, alveolate. Pores radiating, elongated, formed by densely anastomosing laminæ with double walls.

A small tropical genus, allied closely to *Polyporus,* but the pores are large from the first, not minute and afterwards dilating.

1. **F. intestinalis,** *Berk. in Hook. Kew Journ. Bot.* iii. 167; *Fl. N. Z.* ii. 181. White, turning olive- or dirty yellow in drying. Pileus thin, soft, subreniform, lobed, narrowed into an obsolete stem, almost transparent when dry. Pores hexagonal, ¼ in. diam. Spores white, broadly elliptic, with a small nucleus.

Northern Island: on dead wood amongst moss, *Colenso*. A singular plant, resembling a piece of tripe when fresh, probably esculent. (Himalaya mountains.)

13. HYDNUM, Linn.

Habit of *Polyporus*. Pileus fleshy, coriaceous, or hard. Hymenium of the same substance as the pileus, spread over the surface of awl-shaped or compressed spines that are free at the base.

A large genus, especially in hot climates. Some species are esculent. and excellent when cooked.

1. **H. clathroides,** *Pallas;—Fl. N. Z.* ii. 181. Much branched from the base, ashy-grey, soft; branches anastomosing, above muricate and papillose, beneath covered with long crowded filiform spines.

Northern Island: Manawata (on *Knightia*), *Colenso*. Specimens imperfect, and identification hence doubtful; nat. name, "Pekepeke Rione" (*Col.*)..

2. **S. scopinellum,** *Berk. in Fl. N. Z.* ii. 181. Widely effuse, white; spines tomentose at the base, pencilled at the tips.

Northern Island: on dead wood, *Colenso*.

14. IRPEX, Fries.

Habit of *Polyporus*, and intermediate in character between it and *Hydnum*. Hymenium at first toothed; teeth firm, subcoriaceous, acute, disposed in rows, or like network, connected together at the base by porose or plaited folds.

A small temperate genus.

1. **I. brevis,** *Berk. in Fl. N. Z.* ii. 181. Loosely distantly imbricate; stem 0; mycelium forming patches on bark. Pileus dimidiate, pendulous, ½ in. long, 1 in. broad, at first white, then red-brown, somewhat zoned, slightly rugose, fibrillous; teeth elongate, compressed, often again toothed.

Northern Island: dead bark, Bay of Islands, *J. D. H.*

15. THELEPHORA, Fries.

Pileus without a cuticle, consisting of matted fibres. Hymenium tough, at length rigid, finally collapsing and flocculent, costate, striate, or papillose.

A very large genus, especially in temperate climates.

1. **T. vaga,** *Berk. in Fl. N. Z.* ii. 182. Resupinate; mycelium forming stringy reticulated patches on wood. Hymenium pale clay-coloured, pulverulent but not setulose.

Northern Island: on dead wood, *Sinclair*.

2. **T. pedicellata,** *Schwein.;—Fl. N. Z.* ii. 182. Widely effused, of a byssoid texture, tawny-cinnamon; margin waved white, rooting beneath, blackish from scattered papillæ.

Northern Island: Bay of Islands, *Colenso, J. D. H.*—A curious plant which, as Mr. Berkeley observes, may easily be taken for a Lichen.

16. STEREUM, Fries.

Pileus coriaceous, sometimes velvety, with an adherent cuticle. Hymenium coriaceous, rather thick, concrete with the middle stratum of the pileus, even, veinless, unchangeable.

A large chiefly tropical genus, of which some species are also found in temperate climates.

1. **S. Sowerbei,** *Berk. in Fl. N. Z.* ii. 182. Corky, pale, dull, not shining, gregarious. Stem distinct, rather villous. Pileus funnel-shaped, lacerate, fimbriate, scaly, beneath smooth.—*Helvella pannosa*, Sowerby. *Thelephora*, Fries.

Northern Island: on the ground, *Colenso.* (Europe, Tasmania.)

2. **S. lobatum,** *Kunze;—Berk. in Fl. N. Z.* ii. 183. Pileus coriaceous, rigid, undulated, villous, light red, marked with red-brown smooth zones. Hymenium even, smooth, somewhat cinnamon-coloured.

Northern and **Middle** Islands: on dead wood, *J. D. H.*, etc. (Tasmania.)

3. **S. cinereo-badium,** *Fries;—Berk. in Fl. N. Z.* ii. 183. Pileus dimidiate, sessile, robust, coriaceous, tomentose, margined, chesnut-brown; zones smooth, black. Hymenium smooth, glaucous, flesh-coloured.

Northern Island: on dead wood, *Colenso.* (S. America.)

4. **S. vellereum,** *Berk. in Fl. N. Z.* ii. 183. Adnate to dead twigs, dirty white, border free, broad, lobed, ochreous, zoned above, clothed with coarse towy fibres; edge ciliated. Hymenium ochraceous, smooth.

Northern and **Middle** Islands: on twigs, *Colenso; Lyall.*

5. **S. hirsutum,** *Fries;—Berk. in Fl. N. Z.* ii. 183. Pileus coriaceous, effused, reflexed, strigose-hirsute, somewhat zoned, turning pallid; margin obtuse, yellow. Hymenium even, smooth, naked, juiceless, bright tawny yellow, unchanged when bruised.—Berk. Outlines, 270. t. 17. f. 7.

Northern and **Middle** Islands: on dead wood, *J. D. H., Lindsay.* (Europe.)

6. **S. phæum,** *Berk. in Fl. N. Z.* ii. 183. Pileus zoned, 1–2 in. broad, sessile, dimidiate, rather coriaceous, thin, dark bay, clothed with short matted hairs. Hymenium zoned, bay, with a ferruginous bloom, setulose.

Northern Island: on dead wood, *J. D. H.*—A very beautiful species.

7. **S. rugosum,** *Fries;—Berk. in Fl. N. Z.* ii. 183. Pileus corky, rigid, effuse and shortly reflexed, obtusely margined, velvety, at length smooth, bright brown. Hymenium dull, pruinose, blood-stained when wounded.—Berk. Outlines, 271.

Northern Island: on dead wood, *Colenso.* (Europe.)

8. **S. papyrinum,** *Mont.;—Berk. in Fl. N. Z.* ii. 183. Pileus very thin, between coriaceous and papery, very broad, effuse, reflexed, strigose-hirsute, ashy, concentrically furrowed; margin acute, fulvous. Hymenium umber-purple, with a velvety pubescence.

Northern Island: on dead wood, *Colenso.* (Cuba.)

9. **S. latissimum,** *Berk. in Fl. N. Z.* ii. 183. Forming broad very thin chalk-white patches, minutely subtomentose; margin abrupt.

Northern Island: on bark of trees, *Sinclair.*

17. CORTICIUM, Fries.

Coriaceous or fleshy moist Fungi, spreading over dead wood in patches; margin free, often fringed with filaments (byssoid). Hymenium soft, swollen, undulate and papillose when moist, collapsing when dry, then even, often rimose, never flocculent and deliquescing.

A large genus, chiefly confined to temperate regions of the globe, difficult of determination because of their texture and often imperfect development.

1. **C. læve,** *Fries;—Berk. in Fl. N. Z.* ii. 184. Membranous, smooth, pale, effuse, villous beneath; circumference byssoid, not radiating. Hymenium even, smooth, pinkish and livid.—Berk. Outlines, 273.

Northern Island: on twigs and dead bark, *Colenso.* (Europe, Tasmania.)

2. **C. viride,** *Berk. in Fl. N. Z.* ii. 184. Olive-green, crustaceous, effuse, cracked; margin very thin, membranous, scarcely byssoid, livid. Hymenium cracked. Spores large, elliptic or subglobose.

Northern Island: on bark and dead wood, *Colenso.*

3. **C. tenerum,** *Berk. in Fl. N. Z.* ii. 184. Effuse, resupinate, at first even, partly brown with a vinous tint, then cracked into areolæ with often defined borders; margin very narrow, not distinctly byssoid.

Northern Island: on bark of *Knightia, Colenso.*

4. **C. polygonium,** *Fries;—Berk. in Fl. N. Z.* ii. 184. Patches adnate, determinate, hard, cartilaginous, flesh-coloured or rich red-brown, covered with large polygonal crowded tubercles; margin concolorous. Hymenium continuous, red, coated with meal.—Berk. Outlines, 276.

Northern Island: on dead wood, *Colenso.* (Europe, N. America.)

5. **C. rhabarbarinum,** *Berk. in Fl. N. Z.* ii. 184. Effuse, resupinate, forming a tawny uniform stratum; margin paler, delicately and shortly byssoid. Hymenium even, not cracked, distinctly setulose.

Northern Island: on bark, *Colenso.*

6. **C. ochroleucum,** *Fries.* Coriaceo-membranaceous, flaccid, effuso-reflexed or resupinate, silky; zones and thin margin at length smooth. Hymenium smooth, yellowish, at length cracked.

Middle Island: Otago, *Lindsay.* (Europe, Tasmania.)

18. CYPHELLA, Fries.

Pileus submembranous, cup-shaped, elongated at the point of attachment behind, often pendulous. Hymenium distinctly inferior, completely confluent with the pileus.

A small genus, resembling *Peziza,* but distinguished by having sporophores and not asci.

1. **C. densa,** *Berk. in Fl. N. Z.* ii. 184. Patches of numerous fascicled pendulous elongated fawn-coloured bodies, attached by a narrow apex. Pileus pruinose, pubescent above, the pubescence consisting of short obtuse flocci. Hymenium of the same colour, even.

Northern Island: on live bark of *Corynocarpus*, Cape Kidnapper, *Colenso.*

19. GUEPINIA, Fries.

Pileus between cartilaginous and gelatinous, folded and twisted, swelling when moist. Hymenium distinct, inferior or at first superior, unchanged, persistent.

Handsome tropical and temperate Fungi, intermediate between *Tremellini* and *Auricularini.*

1. **G. spathularia,** *Fries;—Fl. N. Z.* ii. 185. Very delicate, gelatinous, cæspitose, suberect. Pileus dimidiate-spathulate, and stem pubescent and glaucous. Hymenium orange-yellow, plaited.

Northern Island: on logs of wood, *Colenso.* (N. America.)

2. **G. pezizæformis,** *Berk. in Fl. N. Z.* ii. 185. Minute, bright orange-red. Stem short. Pileus lateral, externally and stem minutely velvety. Hymenium obliquely cup-shaped, slightly lobed, sparingly wrinkled and pitted within. Spores oblong, sometimes curved.

Northern Island: Bay of Islands, on dead wood, *J. D. H.*—Habit of a *Peziza.* (Australia, Tasmania.)

20. CLAVARIA, Linn.

Herby, terete or clavate, simple or branched erect Fungi, without a distinct stem. Hymenium dry.

A vast temperate and tropical genus of Fungi, readily recognized by their forming straight, fleshy, variously coloured, erect, finger-shaped bodies.

1. **C. lutea,** *Vittadini;—Berk. in Fl. N. Z.* ii. 185. Yellow. Stem thick, repeatedly divided above dichotomously, ultimate division acute, subfastigiate.

Northern Island?, on the ground, *Bidwill.* (Europe.)

2. **C. Pusio,** *Berk. in Fl. N. Z.* ii. 185. Stem 1½ in. high, slender, thickened above, divided into a few acute cylindrical ascending branches about as long as the stem, and making an acute angle with each other; branches rarely subdivided (rufous when dry).

Northern Island: on the ground, *Colenso.*

3. **C. flagelliformis,** *Berk. in Fl. N. Z.* ii. 186. About 2 in. high, divided from the base; branches forming a fastigiate flagelliform mass, elongate, acute, cylindric, forked; tips undivided.

Northern Island: Bay of Islands, on the ground, *J. D. H.*

4. **C. arborescens,** *Berk. in Fl. N. Z.* ii. 186. About 2 in. high, scattered, bright amethyst when fresh. Stem slender, 1 in. high, simple,

slightly thickened upwards, spreading into a few forked branches that bear fastigiate branchlets at their tips.

Northern Island: Bay of Islands, on the ground, *Colenso.*

5. **C. flaccida,** *Fries;—Bab. in Fl. N. Z.* 186. Slender, flaccid, ochraceous, much branched; stem smooth; branches crowded, unequal, converging, acute.—Berk. Outlines, 230.

Northern Island: on dead leaves and twigs, *Colenso.* (Europe.)

6. **C. crispula,** *Fries;—Berk. in Fl. N. Z.* ii. 186. Much branched, tan-coloured, then ochraceous; stem slender, villous, rooting; branches flexuous, multifid; branchlets concolorous, divaricate.—Berk. Outlines, 281.

Northern Island: on rotten wood, *Colenso.* (Europe.)

7. **C. Colensoi,** *Berk. in Fl. N. Z.* ii. 186. Small, 1 in. high, attached by towy fibres; stem mostly compressed, branched from the base; branches repeatedly forked to the apex, very acute, subfastigiate, delicate.

Northern Island: on dead wood, *Colenso.*

8. **C. inæqualis,** *Fries;—Berk. in Fl. N. Z.* ii. 186. Yellow, brittle, gregarious, subfasciculate; stem simple, bearing numerous simple or forked clubs.—Berk. Outlines, 282.

Northern Island: on the ground, Bay of Islands, *J. D. H.* (Europe, Tasmania.)

21. PISTILLARIA, Fries.

Club-shaped, erect Fungi, pink, waxy, becoming horny; cellular internally. Hymenium even.

A considerable genus, in all climates.

1. **P. ovata,** *Fries;—Fl. N. Z.* ii. 186. Small, white, obovate, hollow; stem very short, pellucid, sometimes lobed.

Northern Island: on dead stems of herbaceous plants, *Colenso.* (Europe.)

22. HIRNEOLA, Fries.

Gelatinous, cup-shaped or ear-shaped Fungi, horny when dry, velvety externally. Hymenium often more or less wrinkled, even, without papillæ.

A small genus, found in various climates.

1. **H. auricula-Judæ,** *Berk.* Thin, concave, flexuous, at length black, veined and plaited externally and internally, tomentose beneath.—Berk. Outlines, 289. t. 18. f. 7.

Northern Island: Auckland, *Sinclair.* (Cosmopolite.)

2. **H. polytricha,** *Mont.* Subhemispherical, cup-shaped, expanded, lobed, densely villous externally with grey hairs, disk purplish-brown.—Mont. Sylloge, 181.

Northern Island: Waikehi, in damp places, *Milne.* (East Indies, Java.)

3. **H. hispidula,** *Berk. in Ann. Nat. Hist.* iii. 396.—*Exidia,* Fl. N. Z.

ii. 187. Globose-campanulate, oblique, brown-black, externally fawn-coloured, clothed with short bristly down.

Northern and **Middle** Islands: common on trunks of trees, especially overhanging water. (S. America, Mauritius.)

23. ASEROE, Labill.

Volva globose, gelatinous within. Pileus stalked, divided at the summit into long radiating simple entire or forked horizontal arms. Hymenium at the base of the arms.

A curious genus, the arms of whose pileus somewhat resemble a star-fish. Found in New Zealand, Ceylon, and Australia.

1. **A. rubra,** *Labill. Fl. N. Holl.* ii.;—*Berk. in Fl. N. Z.* ii. 187. 2 to 4 in. high; stem as thick as the thumb, even. Rays of the pileus about 8, bright red, long, subulate, 1–2 in. long, split to the base, continuous with the stem, not divided from them by a deep groove.—Berk. in Hook. Lond. Journ. Bot. iii. 192. t. 5 A.

Northern and **Middle** Islands: probably common, Auckland and Nelson, *Sinclair, Haast.* (Australia.)

2. **A. Hookeri,** *Berk. in Fl. N. Z.* ii. 187. t. 105. f. 13. Much smaller than *A. rubra*, about 1 in. high, dark green or red. Stipes ½ in. thick, corrugated, Rays of the pileus about 8, split to beyond the middle, divisions filiform, subulate, separated from the stipes by a deep groove.—*A. viridis*, Berk. in Hook. Lond. Journ. Bot. iii. 192. t. 5 B.

Northern Island: clay banks, Bay of Islands, *J. D. H., Colenso.*

24. ILEODICTYON, Tulasne.

Volva globose, gelatinous internally. Pileus forming a globose network of anastomosing soft corrugated branches, which are hollow internally. Hymenium adhering to the inner walls of the network.

A genus of 2 or 3 species, found in Australia, New Zealand, and Chili. The volva is esculent, and was much used by the natives.

1. **I. cibarium,** *Tulasne;—Berk. in Fl. N. Z.* ii. 188. Branches of the network stout, corrugated.

Northern and **Middle** Islands: on the ground, probably common, *Raoul*, etc.

2. **I. gracile,** *Berk.* Branches of the network slender, smooth and even.

Middle Island: Otago, *Lindsay.* (Australia, Tasmania.)—Much smaller than *S. cibarium.*

25. SECOTIUM, Kunze.

Volva clothing the whole pileus, at length bursting away from it. Pileus stipitate, hollow, bearing a subglobular or conical hymenium on its apex, which is internally full of labyrinthiform cells.

A very small genus, found in various countries, one is European. An Australian one is edible.

1. **S. erythrocephalum,** *Tulasne ;—Berk. in Fl. N. Z.* ii. 187. Stem glabrous, white or reddish. Pileus globose, umbonate, plicate below, scarlet. —Tulasne in Ann. Sc. Nat. ser. 3. ii. 115, and iv. t. 9. f. 5–17.

Northern and **Middle** Islands : common in grassy places, *Raoul*, etc.

2. **S. lilacensis,** *Berk. mss.* Azure-blue. Stem short, pale, fibrous, pruinose. Pileus subglobose, lilac-coloured, spotted, $\frac{1}{2}$ in. high.

Middle Island : drift-wood, Buller river, *Haast.* (Drawing only.)

26. PAUROCOTYLIS, Berk.

Globose. Peridium simple, thin, hard, rigid, consisting of compactly interwoven flocci ; nucleus formed of a floccose mass of loosely interwoven membranous bodies, enclosing flexuose sinuses. Spores large, pedicelled, seated on the surface of the sinuses.

A very singular monotypic Fungus.

1. **P. pila,** *Berk. in Fl. N. Z.* ii. 188. *t.* 105. *f.* 9. Crimson or globular, sessile, somewhat sinuated, contracted and waved when dry. Spores pale, tan-colour in mass, $\frac{1}{1250}$ in. diam., endochrome with a globose nucleus.

Northern Island : on the ground, Tehawera, *Colenso.* (Only one specimen seen.)

27. GEASTER, Micheli.

Globose. Peridium double, persistent, the outer splitting from the base to the apex into longitudinal segments, which are spread out like a star, or are turned back, supporting the nucleus, which bursts by a terminal pore.

A common genus of Puff-balls.

1. **G. fimbriatus,** *Fries ;—Berk. in Fl. N. Z.* ii. 188. Outer peridium multifid, expanded, flaccid, inner sessile ; mouth of nucleus indeterminate, fimbriate and pilose.—Berk. Outlines, 300. t. 20. f. 4.

Northern Island : on the ground, *Colenso.*

28. TRICHOSCYTALE, Corda.

An erect cylindric Fungus. Outer peridium between woody and corky, rugose, at first globose and enclosing the inner, which is floccose, and bursts through it ; inner tubular, enveloping a cylindrical mass of vertical flocculent filaments mixed with spores.

A most remarkable monotypic Fungus.

1. **T. paradoxa,** *Corda ;—Berk. in Fl. N. Z.* ii. 189. Sulphur-coloured, gregarious, about 1 in. high.—*Trichocoma paradoxum*, Junghuhn Præmissa, p. 9.

Northern Island : Bay of Islands, on dead wood, *J. D. H.* (Himalaya, Java, Carolina.)

29. BOVISTA, Dillenius.

Globose puff-balls, often of great bulk. Peridium double, persistent, like paper or pasteboard, forming a distinct bark, which shells off from the nucleus (capillitium). Nucleus composed of a mass of brown filaments attached to the peridium and pedicelled spores.

Universally distributed; the young of various species are eatable sliced and cooked.

1. **B. brunnea,** *Berk. in Fl. N. Z.* ii. 189. Globose, 1 in. diameter, attached by a central point, opening by a narrow aperture; outer peridium papery, subsinuate, evanescent; inner brownish-umber; capillitium and spores olive-brown.

Northern Island: Manawata river, *Colenso,* amongst moss.

30. LYCOPERDON, Tournefort.

Globose or obovoid or pyriform puff-balls. Peridium membranous, evanescent or flaccid above, forming a persistent bark which breaks up into scales or warts. Nucleus (capillitium) soft, dense, compact below, adnate to the peridium and base of the Fungus internally.

A very large and widely diffused genus, the young of which are excellent eating when sliced and cooked.

1. **L. giganteum,** *Batsch.* Usually very large. Peridium obtuse and brittle above, cracking, areolate, evanescent, opening widely; bark floccose, rather distinct; nucleus disappearing with the olive spores.—Berk. Outlines, 302.

Middle Island: Otago, *Lindsay.* (Europe.)

2. **L. Fontanesei,** *Durieu and Léveillé;—Berk. in Fl. N. Z.* ii. 189. Very large, turbinate. Peridium areolate, with depressed stellate warts on the areolæ, rupturing. Nucleus with forked flocci and minute sessile spores.

Northern Island: Waihake, on the ground, *Colenso.* (Algeria.)—Called by the natives "Pukuvau."

3. **L. cælatum,** *Fries;—Berk. in Fl. N. Z.* ii. 190. Large, obovoid. Peridium flaccid above, collapsing, obtuse, bursting at the apex, at length open and cup-shaped. Barren stratum cellular; nucleus distinct all round. Spores dingy yellow.—Berk. Outlines, 302. t. 20. f. 7.

Northern and **Middle** Islands: on the ground, *Colenso, Lindsay.* (Europe, etc.) —Forming fairy rings.

4. **L. pusillum,** *Fries;—Berk. in Fl. N. Z.* ii. 190. Small. Peridium entirely flaccid, persistent, obtuse, bursting by a narrow aperture; bark even, then cracked, with appressed scales; barren stratum obsolete, continuous with the nucleus. Spores olive.—Berk. Outlines, 302.

Northern Island: on the ground, *Colenso.* (Europe, Java, etc.)

5. **L. novæ-Zelandiæ,** *Léveillé;—Berk. in Fl. N. Z.* ii. 190. Globose, sessile, papery, thin, fragile; bark white, shining, sprinkled with gra-

nular warts, plaited and laciniose at the base; mouth wide. Nucleus and spores violet.

Middle Island: Akaroa, *Raoul.*—Mr. Berkeley, who has not seen this, observes that, according to the description, it much resembles *Bovista fragilis*, Vitt.

6. **L. reticulatum,** *Berk. in Fl. N. Z.* ii. 190. Globose, 1 in. or more diameter, narrowed below into a short stem. Peridium finely reticulated, at length smooth and shining; orifice small, irregular, terminal. Nucleus olive, at length bleached and french-grey; barren stratum confluent with the nucleus.

Northern Island: on the ground, *Colenso.*

7. **L. microspermum,** *Berk. in Fl. N. Z.* ii. 190. Subglobose, obtuse, $\frac{1}{3}$–1 in. diameter, reddish-brown. Peridium all flaccid, persistent, obscurely cracked, rough with minute scales, at length pale, smooth below; mouth small, circular. Nucleus uniform, green; barren stratum very small. Spores globose, minute.

Northern Island: on the ground, *Colenso.* (Himalaya, Java, etc.)

8. **L. gemmatum,** *Fries;—Berk. in Fl. N. Z.* ii. 190. Subglobose, sessile or narrowed at the base. Peridium membranous, persistent; bark farinaceous, adnate, covered with spinulose warts; mouth umbonate. Nucleus with a sort of central columella. Spores yellow or greenish.—Berk. Outlines, 302.

Northern Island: on the ground, Tehawera, *Colenso.* (Europe, Tasmania.)

9. **L. pyriforme,** *Schæffer;—Berk. in Fl. N. Z.* ii. 190. Pyriform. Peridium membranous, persistent, umbonate; bark not separable, covered with minute fugacious scales. Nucleus greenish-yellow, with a conical columella.—Berk. Outlines, 303.

Northern Island: on decayed wood, *Colenso.* (Europe, Java, etc.)

31. **SCLERODERMA,** Persoon.

More or less globose, sessile or substipitate Fungi. Peridium firm, hard, with an innate bark, bursting irregularly. Nucleus (capillitium) with flocci adhering on all sides to the peridium and forming distinct veins in the central mass of spores. Spores large, granulated.

A widely diffused genus. In a young state some species are subterranean, and mistaken for Truffles.

1. **S. Geaster,** *Fries;—Fl. N. Z.* ii. 190. Subglobose, unequal, as large as the fist, pale. Peridium bursting at top in a stellate manner. Nucleus dirty-blue. Sporidia brown.

Northern Island: Bay of Islands, *J. D. H.* (Europe, N. America, Australia.)

2. **S. vulgare,** *Fries;—Fl. N. Z.* ii. 190. Irregular, nearly sessile. Peridium corky, bursting by an indefinite aperture. Nucleus bluish-black.—Berk. Outlines, 303. t. 15. f. 4.

Northern Island: on the ground, *Colenso.* (Europe.)

32. **ÆTHALIUM,** Link.

Subglobose or oblong, sessile, amorphous, simple or lobed, pulpy Fungi. Peridium indeterminate, outer layer forming a floccose, evanescent bark, fragile, internally divided into cells by membranous strata of confluent interwoven flocci.

A curious plant; common on rotten wood, tan, etc., often inside houses, and very destructive of hothouse timber.

1. **Æ. septicum,** *Fries;—Fl. N. Z.* ii. 191. Forming simple or lobed patches, one or more inches long; colour variable, yellow, violet, etc.

Northern Island: on dead wood, moss, etc., common, *Colenso*, etc. (Ubiquitous.)

33. **DIDERMA,** Persoon.

Usually minute, globose, sessile or stalked Fungi, found on bark, wood, mosses, etc. Peridium double; outer membranous or crustaceous, glabrous, polished, fragile, dehiscent; inner membranaceous, evanescent. Nucleus (capillitium) of flocci, adnate at the base, or fixed to a central columella, traversing the mass of spores in all directions.

A large and common genus, in all parts of the world.

1. **D. Hookeri,** *Berk. in Fl. N. Z.* ii. 191. *t.* 105. *f.* 12. About $\frac{1}{12}$ in. high, white; stem slender, as long as the globose peridium, dilated below, slightly veined. Peridium evanescent above; outer membranous, subplicate; inner colourless. Nucleus adhering to the columella and peridium, branched, enclosing triangular lacunæ. Spores black in mass.

Northern Island: on fronds of *Hymenophyllum*, *Colenso*.

34. **DIDYMIUM,** Schrader.

Habit and characters of *Diderma*, but the two peridia are adnate and the outer furfuraceous or mealy.

A very common genus.

1. **D. australe,** *Berk. in Fl. N. Z.* ii. 191. Peridium subglobose, on a short nearly black, even stipes, which rises from a thin pale mycelium. Flocci crisped. Spores globose, black in mass.

Northern Island: on naked soil, *Colenso*.

2. **D. cinereum,** *Fries;—Fl. N. Z.* ii. 192. Solitary or confluent, adnate or shortly stipitate. Peridium subglobose, dirty white, scurfy; flocci reticulated, white, flat, irregularly reticulate. Spores mixed with minute, white, angular bodies, black.

Northern Island: on the naked soil, *Colenso*. (Europe, America, etc.)

35. **STEMONITIS,** Gleditsch.

Characters of *Diderma*, but peridium globose or cylindric, very delicate, simple, evanescent, with the stipes passing through it. Nucleus (capillitium) reticulate, springing from the stem.

A considerable tropical and temperate genus.

1. **S. ferruginea,** *Ehrb.;—Berk. in Fl. N. Z.* ii. 192. Fascicled. Hypothallus persistent. Peridia fugacious; capillitium cylindric. Sporidia rusty-purple, large.

Northern Island: on dead wood, *Colenso.* (Europe.)

2. **S. typhoides,** *DC.;—Fl. N. Z.* ii. 192. Gregarious, minute. Hypothallus evanescent. Stipes black. Peridium cylindric. Spores brown, small.

Northern Island: Hawke's Bay, on a decayed *Polyporus, Colenso.* (Europe, etc.)

36. CYATHUS, Persoon.

Small Fungi, like diminutive birds'-nests. Peridium of three closely connected membranes, at length bursting at the apex, and the orifice closed by a white membrane. Sporangia plane, umbilicate, attached to the peridium by elastic cords.

A curious genus, from the resemblance of the species to birds'-nests with eggs; found in all parts of the world.

1. **C. novæ-Zelandiæ,** *Berk. in Fl. N. Z.* ii. 192. Narrow, elongate, obconical, about ½ in. high. Peridium tomentose externally, grooved and striate internally; margin indistinct, tomentose, not produced into a corona. Sporangia thick, black, glabrous, smooth.

Middle Island: Banks's Peninsula, on dead wood, *Raoul.*

2. **C. Colensoi,** *Berk. in Fl. N. Z.* ii. 192. Densely crowded, cup-shaped, ¼ in. high, thin, flexible, dirty umber-coloured, pubescent externally, within even, brownish. Sporangia irregular, brown, thick-walled. Spores very minute, ovoid.

Northern Island: on the ground, *Colenso.*

37. CRUCIBULUM, Tulasne.

Habit of *Cyathus.* Peridium at first globose, at length cupped and obconic, of a uniform, spongy, fibrous, felted consistence. Orifice closed by a flat furfuraceous cover of the same colour. Sporangia plane, attached by a long cord to a nipple-like tubercle.

A genus of two species, one widely distributed.

1. **C. vulgare,** *Tulasne;—Berk. in Fl. N. Z.* ii. 193. Peridium dirty yellow, finally white. Spores minute, ovate.—Berk. Outlines. 312. t. 2. f. 1.

Northern and **Middle** Islands: common on twigs, etc. (Universally distributed.)

2. **C. emodense,** *Berk. in Hook. Kew Journ. Bot.* vi. 204?—*Fl. N. Z.* ii. 193 (*Crucibulum*). White, bell-shaped, narrowed below, subsessile, broadly open above, not striate, externally fasciculately tomentose; margin stellate, ciliate. Sporangia short, umber.

Middle Island: Nelson, on wood, *Munro.*—A very fine species, doubtfully identified with the Himalayan one by Mr. Berkeley.

38. LEPTOSTROMA, Fries.

Minute, obscure Fungi, consisting of depressed, irregular or orbicular masses, covered with a thin perithecium that falls off by a transverse rupture, leaving a thin lower half. Spores simple, minute, borne on sporophores.

Little-known Fungi, of which some species are supposed to be forms of other polymorphous genera.

1. **L. litigiosum,** *Desmazières;—Fl. N. Z.* ii. 193. Perithecia subrotund, very minute, punctiform, scattered or conglomerate, brown-black, at length entirely splitting off.

Northern Island: on dead Fern stems. A doubtful identification. Mr. Berkeley adds that there are two forms, one smooth, the other strongly granulate. (Europe.)

39. PHOMA, Fries.

Minute Fungi, forming pustules on wood, leaves, stems, etc. Perithecium punctiform or subglobose, often spurious, discharging the minute simple spores by a small orifice at the apex.

A genus of numerous species.

1. **P. fallax,** *Berk. in Fl. N. Z.* ii. 193. Perithecia forming scattered or crowded, minute, discoid spots, when dry black, when moist brown, thin and transparent above. Spores oblong, subcymbiform; hyaline $\frac{1}{1750}$ in. long.

Northern Island: on berries of *Rhipogonum, J. D. H.*

2. **P. acmella,** *Berk. in Fl. N. Z.* ii. 193. *t.* 106. *f.* 10. Perithecia forming brown specks on both sides of the leaf, depressed, brown, darkest in the centre. Spores about $\frac{1}{3250}$ in. long, oblong, attenuated, but obtuse at each end.

Northern Island: on leaves of *Corynocarpus, J. D. H.*, with *Sphæria acetabulum*, of which it is perhaps a form.

40. HENDERSONIA, Berk.

Microscopic Fungi. Perithecia subglobose, with often a terminal pore. Spores 2-multiseptate.

A considerable British and exotic genus.

1. **H. hyalospora,** *Berk. in Fl. N. Z.* ii. 194. *t.* 106. *f.* 8. Perithecia forming minute, black, shining dots upon bark, sometimes slightly papillate, generally without a pore. Spores $\frac{1}{1250}$ in. long, hyaline, linear, obtuse at each end, 3-septate.

Northern Island: on bark of *Olea, Colenso.*

2. **H. microsticta,** *Berk. in Fl. Antarct.* 170. *t.* 68. *f.* 1. Perithecia minute, black, globose, covered with the epidermis. Spores pellucid, irregular, generally lanceolate, acute at both ends, 3-septate.

Lord Auckland's group and **Campbell's** Island: on scapes of *Anthericum, J. D. H.*

41. ASCHERSONIA, Mont.

Small, yellow or pallid, fleshy, superficial Fungi, growing on living vegetable substances. Cells tubular, interlaced, arranged concentrically. Spores upon filiform sporophores, simple or septate, escaping by broad open pores.

A small tropical and subtropical genus.

1. **A. duplex,** *Berk. in Fl. N. Z.* ii. 194. Minute, $\frac{1}{20}$–$\frac{1}{12}$ in. broad, yellow, convex, lobed like a raspberry, rarely simple, convex, and containing only one cell; cells variable in size, one or more in each lobe. Spores oblong, hyaline, $\frac{1}{3500}$ in. long.

Northern Island: on leaves of *Astelia, Colenso.*

42. PHLYCTÆNA, Desmazières.

Minute, epiphytic Fungi. Perithecium spurious, formed of the blackened cuticle of the plant, convex, opening by a pore, containing a gelatinous nucleus. Spores curved, elongate, spindle-shaped, upon very short sporophores.

A temperate and tropical genus.

1. **P. dissepta,** *Berk. in Fl. N. Z.* ii. 194. *t.* 106. *f.* 14. Spots scattered, variable in form, in patches $\frac{1}{12}$–1 in. long, surrounded by a black, flexuous, irregular line, black or colourless. Spores very slender, elongate, strongly curved at the apex, $\frac{1}{600}$ in. long, on slender sporophores.

Northern Island: on dead stems of *Rhipogonum, Colenso.* Possibly a form of *Pemphidium opacum.*

43. PILIDIUM, Kunze.

Minute epiphyllous Fungi. Perithecium flat, shield-like, smooth, shining, tough, variously ruptured. Spores simple, linear, curved, without sporophores, that is with the sporophores sunk in the perithecium.

A small temperate and tropical genus.

1. **P. coriariæ,** *Berk. in Fl. N. Z.* ii. 195. Perithecia black, rather rigid, seldom collapsed, at length perforate in the centre. Spores linear, slightly curved, obscurely narrowed at either end, $\frac{1}{1500}$ in. long.

Northern Island: on dead leaves of *Coriaria sarmentosa, Colenso.*

44. ASTEROMA, DC.

Minute epiphyllous Fungi. Perithecia attached to crisped branching threads, flat, with no determinate orifice. Spores simple or 1-septate.

A considerable genus of temperate Fungi.

1. **A. dilatatum,** *Berk. in Fl. Antarct.* 173. *t.* 68. *f.* 7. Forming superficial, suborbicular, olive-black patches, $\frac{1}{8}$ in. broad, with lobed edges; lobes dilated, consisting of serpentine filaments, so crowded together as to form a membrane. Spores unknown.

Northern Island, *Colenso*, and **Lord Auckland's** group, *J. D. H.*, in both places on leaves of *Panax simplex*. This is perhaps only a rudimentary state of a *Collema*.

45. GYMNOSPORIUM, Corda.

Pulverulent Fungi, forming strata or vegetable substances. Spores superficial, conglobate, simple, arising directly from the matrix, smooth.

Usually found on grass-stems.

1. **G. culmigenum,** *Berk. in Fl. N. Z.* ii. 197. Spots $\frac{1}{12}$ in. and upwards long, half as broad, deep black. Spores elliptic or subglobose, with a thin epispore.

Northern Island: on dead grass, *Colenso*.

46. PUCCINIA, Persoon.

Minute mildews. Spores crowded in patches, breaking through the epidermis of the plant they infest, and growing from a mycelium, 1-septate.

The mildews are the most common and destructive of Fungi, and are found in all parts of the world.

1. **P. compacta,** *Berk. in Fl. N. Z.* ii. 195. Patches orbicular, pale, of various sizes. Sori solitary and crowded, depressed; mass of spores compact, spongy. Spores elongate, subapiculate, $\frac{1}{300}$–$\frac{1}{400}$ in. long, often oblique; nucleus distinct.

Southern Island: on leaves of *Myosotis capitata, Lyall.*

2. **P. graminis,** *Persoon;—Berk. in Fl. N. Z.* ii. 195. Spots pale, diffuse; sori linear, confluent. Sporidia at length black.

Probably abundant on corn and all grasses: on *Triticum*, *Colenso*.

47. UREDO, Léveillé.

Minute epiphyllous Fungi. Receptacle formed of several superpose irregular cells, each containing a single simple sessile or very shortly stalked spore.

A large genus of Fungi, inhabiting all parts of the world, and very destructive of vegetable tissues.

1. **U. antarctica,** *Berk. in Fl. Antarct.* 170. *t.* 68. *f.* 2. Spots small, opposite on both sides the leaf, rounded, purplish, pale beneath. Sori bullate. Spores large, smooth, broadly obovate, brown, with a central oily globule, very shortly stalked.

Campbell's Island: on leaves of *Luzula crinita, J. D. H.*

48. UROMYCES, Léveillé.

Habit and characters of *Uredo*, but receptacle formed of a single layer of cells; spores stalked.

1. **U. scariosa,** *Berk. in Fl. N. Z.* ii. 195. Spots scattered, distinct,

on the under surface of the leaf; stroma prominent. Sori surrounded with the persistent cuticle. Spores obovate, echinulate.

Northern Island: on *Geranium* leaves, *Colenso.*

2. **U. citriformis,** *Bab. in Fl. N. Z.* ii. 195. Spots obliterated. Sori scattered, surrounded with the persistent cuticle. Spores large, linear-shaped, brown, shortly stalked.

Northern Island: on *Microtis?* or *Thelymitra?* leaves, *Colenso.*

49. USTILAGO, Link.

Very minute Fungi, deep-seated in the tissues of the plants they infest, and occupying the whole of the part. Spores simple, arising from delicate threads or produced in the form of closely-packed cells that break up into a powdery mass.

This genus forms dusty masses on the seed, stems, and other parts of plants, and is found in all parts of the world. The *U. segetum* is the smut of wheat, oats, and barley.

1. **U. Candollei,** var. *a. Tulasne;—Berk. in Fl. N. Z.* ii. 196. Forming elongated masses on *Polygonum* peduncles. Spores elliptic-ovate, smooth, violet-coloured.

Northern Island: on the inflorescence of *Polygonum prostratum, Colenso.*

2. **U. endotricha,** *Berk. in Fl. N. Z.* ii. 196. *t.* 106. *f.* 4. Forming oblong black bodies on *Gahnia* peduncles. Spores subglobose, mixed with crisp fibres, minutely granulate.

Northern Island: on the inflorescence of *Gahnia, Sinclair.*

3. **U. bullata,** *Berk. in Fl. N. Z.* ii. 197. *t.* 106. *f.* 12. Forming bullate black long spots on the inflorescence of grasses. Spores irregularly subglobose, mixed with threads, obscurely rough, nucleus pale.

Northern and **Middle** Islands: on *Triticum scabrum, Raoul, Colenso.*

50. ÆCIDIUM, Persoon.

Forming dark spots or pustules on leaves and other parts of living plants. Peridium membranous, with a lacerated or toothed reflected orifice. Spores concatenate, collected into sori.

A vast genus, found in all parts of the globe, often very destructive to vegetable tissues.

1. **Æ. Ranunculacearum,** *DC.;—Berk. in Fl. N. Z.* ii. 196. Spots obliterated. Peridia densely crowded into irregular heaps, orange-yellow.

Northern Island: on leaves of *Ranunculus rivularis, Colenso.* (Europe, Tasmania, etc.)

2. **Æ. monocystis,** *Berk. in Fl. N. Z.* ii. 196. *t.* 105. *f.* 15. Peridia large, solitary, persistent on the upper surface of the leaves towards their tips, surrounded with a tough border. Spores pale orange-yellow.

Northern Island: on the tips of the leaves of *Helophyllum Colensoi, Colenso.*

51. STILBUM, Tode.

Minute Fungi, consisting of a globose deciduous head on an erect stalk. Stalk solid. Heads deciduous, gelatinous. Spores minute.

Chiefly a temperate genus, with a few tropical species.

1. **S. lateritium,** *Berk.;—Fl. N. Z.* ii. 197. Gregarious, erumpent, pale brick-red. Stems $\frac{1}{12}$ in. high, thickest at the base, often confluent and flattened, pruinose from the presence of curved obtuse flocci. Capitula ovate or hemispherical, minutely setulose. Sporidia oblong.

Middle Island: on bark, *Lyall, Bidwill.* (S. America.)

52. EPICOCCUM, Link.

Minute Fungi, forming raised specks on leaves, etc. Spores large, sub-globose, simple or compound, springing from the surface of a cushion-like receptacle.

All the known species are natives of temperate climates.

1. **E. pallescens,** *Berk. in Fl. N. Z.* ii. 198. *t.* 105. *f.* 14. Specks fawn-coloured, at first covered by epidermis. Receptacula of several layers, the lowest with fimbriated edges. Spores shortly stalked, cellular, enclosing many sporidia.

Northern Island: on dead leaves of *Earina, Colenso.*

53. ŒDEMIUM, Link.

Small Fungi, forming black felted masses of floccose filaments. Filaments rigid, opaque, protruding large globose usually reticulated spores at their sides.

A small genus of temperate climates, probably allied to *Antennaria.*

1. **Œ. robustum,** *Berk. in Fl. N. Z.* ii. 198. Forming a stratum of rigid erect bristly filaments, $\frac{1}{666}$ in. long, jointed below, above bearing a few divided curved branches.

Northern Island: on bark, *Colenso.*

54. MACROSPORIUM, Fries.

Plants forming a thin floccose stratum, covered with erect, clavate, multi-septate spores.

A little-studied genus of microscopic Fungi, often parasitic on other Fungi.

1. **M. obtusum,** *Berk. in Fl. N. Z.* ii. 198. Forming an olivaceous coat on the mouths of a *Hypoxylon.* Filaments obsolete. Spores $\frac{1}{750}$ in. long, clavate, obtuse, 4–5-partite, each division with a globose nucleus.

Northern Island: on the mouths of a *Hypoxylon, Colenso.*

55. CLADOSPORIUM,

Fungi consisting of a minute filamentous mycelium, the filaments divided into short branches, which bear short 1-septate deciduous spores.

The only well-known species of this genus is the commonest of parasitic Fungi, infesting all sorts of decaying substances, and sometimes smothering living trees.

1. **C. herbarum,** *Link;—Fl. Antarct.* 170. Mycelium dense, soft, green, then olive-black; filaments pellucid, collapsing. Spores olive.

Lord Auckland's group and **Campbell's** Island: on *Carex* leaves, *J. D. H.* (Ubiquitous.)

56. SEPEDONIUM, Link.

Minute Fungi, consisting of a copious filamentous mycelium, bearing at the tips of its divisions myriads of large globose spores.

Abundant Fungi, preying on the large fleshy species, through whose soft tissues their mycelium spreads. Some species certainly rudimentary parasitic *Sphæriæ.*

1. **S. chrysospermum,** *Fries;—Berk. in Fl. N. Z.* ii. 199. Filaments white. Spores golden-yellow, globose, echinulate when dry.

Northern Island: on *Boleti*, Bay of Islands, *J. D. H.* (Probably ubiquitous.)

57. PILACRE, Fries.

Very minute Fungi, consisting of globose stipitate heads. Stem solid, cylindric, simple or branched. Head globose, of branched radiating flexuous threads. Spores at the tips of the branches, forming a dusty mass.

A small genus of temperate Fungi.

1. **P. divisa,** *Berk. in Fl. N. Z.* ii. 197. Stems $\frac{1}{12}$–$\frac{1}{6}$ in. high, brown, subfascicled, compressed below, forked above, when young pale and tomentose below. Heads globose, clay-coloured. Spores subglobose or elliptic, $\frac{1}{5000}$ in. long, nucleus distinct.

Northern Island: on bark, *Colenso.*

58. LEOTIA, Hill.

Stalked fleshy pileate Fungi. Receptacle capitate, covered everywhere with the asci; margin revolute.

Natives of temperate regions, growing on moss and on the ground amongst grass, etc.

1. **L. lubrica,** *Persoon.* Subgelatinous, 1–2 in. high. Pileus tumid, repand, olive-green. Stem at length hollow.

Var. *β*, *Berk.* Stipes green, thickened downwards.

Var. *β*. **Middle** Island: Canterbury, *Haast*, on the ground. Identified by Mr. Berkeley from a drawing. (The original plant is European.)

59. GEOGLOSSUM, Persoon.

Fleshy erect Fungi. Receptacle stalked, clavate, covered with the hymenium. Asci elongate.

A large genus in temperate climates, growing in shade, on the ground, moss, etc.

1. **G. hirsutum,** *Persoon;—Berk. in Fl. N. Z.* ii. 199. Erect, hairy, black, 1–2 in. high.

Northern Island, *Colenso.* (Europe.)

60. **PEZIZA,** Dillenius.

Cup-shaped Fungi. Receptacle fleshy or subfleshy, at first closed, then open. Hymenium persistent. Asci distinct, fixed, mixed with paraphyses, ejecting the spores elastically.

A vast genus, found in all parts of the world, on the ground, rotten wood, etc. etc. Many species resemble the apothecia of *Lichens.*

§ 1. ALEURIA.—*Fleshy, or between fleshy and membranous, externally pruinose or furfuraceous and floccose.*

1. **P. miltina,** *Berk. in Fl. N. Z.* ii. 199. Cup $\frac{1}{3}$ in. across, depressed, expanded, crimson, irregular, fixed to the soil by its whole surface; margin free, paler beneath. Asci linear. Sporidia globose, nucleus single.

Northern Island: Hawke's Bay, amongst moss, *Colenso.*

2. **P. endocarpoides,** *Berk. in Fl. N. Z.* ii. 199. *t.* 105. *f.* 8. Cup $\frac{1}{3}$ in. diameter, fixed by a floccose mass; margin free, concave, at length expanded and convex, obscurely floccose externally. Asci cylindrical. Sporidia 8, globose, each with a large nucleus.

Northern Island: on the ground, *Colenso.*

3. **P. rhytidia,** *Berk. in Fl. N. Z.* ii. 200. *t.* 105. *f.* 6. Cup 1 in. and more diameter, fuliginous, hemispherical, nearly sessile, deeply incised; margin narrow, inflected; outer coat wrinkled, of netted fibres, inner more compact, giving rise to the long asci and slender paraphyses. Sporidia elliptic-oblong, subcymbiform.

Northern Island: on the ground, *J. D. H., Colenso.*

4. **P. campylospora,** *Berk. in Fl. N. Z.* ii. 200. Cup 1 in. diameter, oblique, lobed, wrinkled externally, clothed with myceloid filaments; stem short, wrinkled. Hymenium fuliginous, vinous-red, white internally and composed of intricate threads. Asci linear. Sporidia strongly curved.

Northern Island: on dead wood, *Colenso.*

§ LACHNEA.—*Waxy, externally pilose or villous.*

5. **P. stercorea,** *Fries;—Berk. in Fl. N. Z.* ii. 200. Gregarious. Cup sessile, concave, $\frac{1}{6}$–14 in. diameter, yellow; margin ciliated with straight bristles.—Fl. Antarct. 451. t. 163. f. 4.

Northern Island: on horsedung, *Colenso.* (Europe, America, Tasmania, etc.)

6. **P. Kerguelensis,** *Berk.;—Fl. N. Z.* ii. 201. Cup $\frac{1}{2}$–$\frac{3}{4}$ in. broad, scarlet, flat, adnate; margin free, ciliated with spreading bristles. Asci linear, obtuse. Sporidia broadly elliptic; nucleus solitary, globose.—Fl. Antarct. 451. t. 164. f. 3.

Northern Island: on the ground, *J. D. H.* (Kerguelen's Land.)

7. **P. calycina,** *Fries;—Berk. in Fl. N. Z.* ii. 200. Cups $\frac{1}{12}$–$\frac{1}{4}$ in. diameter, stipitate, gregarious, erumpent, funnel-shaped, externally white and villous; disk flat, pale orange-colour.

Northern Island: on dead twigs, *Sinclair.* (Europe.)

8. **P. Colensoi,** *Berk. in Fl. N. Z.* ii. 200. *t.* 105. *f.* 5. Cup $\frac{1}{8}$ in. diameter, pale tan-colour, funnel-shaped, with a short thick stipes, fixed by an orbicular disk, base plicate, minutely downy above and on the margin, not ciliate. Hymenium even. Asci nearly equal. Sporidia oblong, elliptic, or subfusiform; nucleus contracted into 2–4 masses, but not septate.

Northern Island: on dead sticks, *Colenso.*

9. **P. chrysotricha,** *Berk. in Fl. N. Z.* ii. 201. *t.* 105. *f.* 7. Cups $\frac{1}{12}$ in. diameter, golden-yellow, sessile, at first globose then subhemispherical, edge sometimes membranous, clothed with short matted hairs, somewhat pulverulent. Hymenium closed when dry, seated on a stratum of large unequal cells. Asci cylindric, rather large. Sporidia cymbiform, narrowed at either end.

Northern Island: on dead twigs, *Colenso.*

§ 3. PHIALEA.—*Waxy or membranous, quite glabrous. Cup at first closed.*

10. **P. montiæcola,** *Berk. in Fl. N. Z.* ii. 201. Cups minute, black, soon expanding; margin narrow, sometimes flexuous. Asci large, clavate, shorter than the long paraphyses. Sporidia subcymbiform; nucleus 2-partite.

Northern Island: on decaying *Montia, Colenso.*

61. PATELLARIA, Fries.

Flat, discoid, small, tough Fungi. Receptacle saucer- or cup-shaped, open, never closed. Hymenium rather powdery. Asci connate, without paraphyses.

A small genus, closely allied to *Peziza.* The species are chiefly temperate.

1. **P. nigro-cinnabarina,** *Schweinitz;—Berk. in Fl. N. Z.* ii. 201. Hymenium vermilion-coloured.

Northern Island: on dead branches, *Colenso.* (N. America.)

2. **P. atrata,** *Fries;—Berk. in Fl. N. Z.* ii. 201. Cups flattened, black.

Northern Island: on dead bark, *Colenso.* (Europe.)

62. CENANGIUM, Fries.

Form of *Peziza,* but firm, tough and coriaceous. Receptacle or cup closed, opening late. Hymenium smooth, persistent.

A small European genus.

1. **C. Colensoi,** *Berk. in Fl. N. Z.* ii. 201. Scattered, oblong, at first covered with the cuticle, then expanding, and orbicular, brown. Hymenium pale, pinkish. Asci oblong. Sporidia elliptic, hyaline, $\frac{1}{4000}$ in. long.

Northern Island: on dead leaves, *Colenso.*

63. HYSTERIUM, Tode.

Minute, tumid, elliptic or elongate Fungi. Perithecia bursting by a simple longitudinal fissure, hence 2-lipped; borders entire. Asci elongate.

A large genus in temperate regions, growing on leaves, bark, etc.

1. **H. breve,** *Berk. in Fl. Antarct.* i. 174. *t.* 68. *f.* 8. Perithecia short, black, elliptic, prominent, orifice very narrow. Asci linear, mixed with paraphyses. Sporidia filiform.

Campbell's Island: on dead leaves of *Uncinia, J. D. H.*

64. AILOGRAPHUM, Libert.

Habit of *Hysterium*, but perithecia often branched and asci subglobose.

A small temperate genus, usually epiphyllous.

1. **A. Bromi,** *Berk. in Fl. Antarct.* i. 174. *t.* 68. *f.* 9. Perithecia simple, black, shining, mixed with filaments. Asci very short, obovate. Sporidia 8, elliptic-oblong, constricted in the middle, 1-septate.

Lord Auckland's group: on leaves of *Danthonia, J. D. H.*

65. ASTERINA, Léveillé.

Minute, flattened, epiphyllous Fungi. Perithecia fragile, formed of a fimbriated mycelium. Asci perfect. Sporidia 4–8.

A small tropical, N. America, and N. Zealand genus; only one species is known in Europe.

1. **A. torulosa,** *Berk. in Fl. N. Z.* ii. 208. Mycelium very sparse. Perithecia in scattered groups. Asci elongate. Sporidia filiform, 3–5-septate, contracted at the septa, greenish.

Northern Island: on leaves of *Piper, Colenso.*

2. **A. sublibera,** *Berk. in Fl. N. Z.* ii. 208. *t.* 106. *f.* 1. Mycelium sparse, in little patches. Asci elongate. Sporidia 4, oblong, 1-septate, hyaline.

Northern Island: on leaves of *Metrosideros diffusa, Colenso.*

3. **A. fragilissima,** *Berk. in Fl. N. Z.* ii. 208. Mycelium forming spots scarcely $\frac{1}{12}$ in. broad, sparse. Perithecia crowded, very brittle. Asci globose. Sporidia obovate-oblong, 1-septate.

Northern Island: on *Veronica* leaves, *Colenso.*

66. EXCIPULA, Fries.

Minute epiphyllous Fungi, forming spots or patches. Perithecium carbonaceous, spherical, free, at first closed, then opening; mouth orbicular, entire; nucleus naked, gelatinous, turgid. Asci perfect.

Mr. Berkeley limits this genus as defined by Fries, in the 'Summa Vegetabilium Scandinaviæ.' The species are natives of temperate regions.

1. **E. nigro-rufa,** *Berk. in Fl. N. Z.* ii. 202. *t.* 106. *f.* 11. Subglobose, black without, red within; mouth inflexed. Asci broad, clavate,

obtuse. Sporidia hyaline, obovate, oblong, obtuse, with about 4 transverse septa.

Northern Island: on the under surface of the leaves of *Pittosporum crassifolium*, *Colenso.*

2. **E. gregaria,** *Berk. in Fl. N. Z.* ii. 202. Minute, crowded on a brownish spot. Cups black without, pale within. Asci broad, clavate. Sporidia obovate-oblong, obscurely 2-partite internally.

Northern Island: upper surface of *Gnaphalium* leaves, *Colenso.*

67. CORDICEPS, Fries.

Erect, columnar, rather rigid, fleshy or corky Fungi, growing often on animal matter, as dead caterpillars. Perithecia crowded round the axis or its branches, which form the fruit-bearing head or heads, hyaline or coloured. Sporidia submoniliform, repeatedly divided.

A very curious genus, found in both tropical and temperate climates. The ergots of Rye and other grasses are an imperfect state of species of *Cordiceps.*

1. **C. Robertsii,** *Berk. in Fl. N. Z.* ii. 202. Slender, 2–6 in. high. Stem rigid, cylindric. Receptacle very little broader than the stipes, 1–2 in. long, cylindric, often flexuous.—*Sphæria Robertsii,* Hook. Ic. Pl. t. 11. *S. Hugelii,* Corda.

Northern and **Middle** Islands: on the caterpillar of *Hepialus virescens, Roberts,* etc. Caterpillar Fungus.)

2. **C. Sinclairii,** *Berk. in Fl. N. Z.* ii. 338. Yellowish, ¾–1 in. high. Stem cylindric, simple or forked, divided into many simple or lobed heads, clothed with oblong conidia.

Northern Island: on the larva of an orthopterous insect, *Archdeacon Williams.*—The perithecia are absent, and hence the species is not well established. Mr. Travers sends a similar plant growing from a coleopterous larva, likewise destitute of perithecia.

68. HYPOCREA, Fries.

Fleshy or gelatinous horizontal Fungi, growing on wood, etc. Perithecia tender, hyaline or coloured. Sporidia indefinite.

A tropical and temperate genus.

1. **H. gelatinosa,** *Fries;—Berk. in Fl. N. Z.* ii. 202. Convex, equal, opaque, variable in colour without, dirty white within. Perithecia prominent, darker than the stroma.

Northern Island: on wood, *Colenso.* (Europe.)

69. XYLARIA, Fries.

Club-shaped, somewhat corky, often friable Fungi, usually distinctly stipitate, covered with a rufous or black bark. Perithecia crowded on the surface of the receptacle. Sporidia 8.

A very large, especially tropical genus, growing on vegetable substances, or dung.

1. **X. Hypoxylon,** *Fries;—Berk. in Fl. N. Z.* ii. 203. Corky, 2–4 in. high, simple or branched, compressed, at first pulverulent with white meal, then naked. Stem villous.

Northern Island: on dead wood, *Colenso.* (Europe, Tasmania.)

2. **X. multiplex,** *Kunze;—Berk. in Fl. N. Z.* ii. 203. Corky, tufted, brown-black, divided above into even somewhat cylindrical compressed branches; stem elongate, rough with leprous down.

Northern Island: on dead wood, *Colenso.* (Juan Fernandez.)

3. **X. anisopleuron,** *Mont.;—Berk. in Fl. N. Z.* ii. 205. Solitary, woody. Receptacle oblique, obconic, leprous, opaque, pale within. Perithecia in the upper subhemispheric part, globose, papillate in the centre.

Northern Island: on dead wood, *Colenso*, specimen young. (Guyana.)

4. **X. castorea,** *Berk. in Fl. N. Z.* ii. 204. *t.* 105. *f.* 10. Stem $\frac{1}{4}$ in. high, longitudinally wrinkled, downy, then glabrous. Receptacle ovate, compressed, obtuse, 1 in. long, $\frac{1}{2}$–$\frac{2}{3}$ broad, dotted with the mouths of the perithecia. Asci slender. Sporidia subelliptic.

Northern Island: on dead wood, *Colenso.*

5. **X. tuberiformis,** *Berk. in Fl. N. Z.* ii. 204. *t.* 105. *f.* 11. Subglobose, sessile or shortly stipitate. Receptacle $\frac{1}{4}$–$\frac{1}{3}$ in. diameter, minutely cracked, white and corky within. Perithecia elliptic; mouth large. Sporidia large, cymbiform.

Northern Island: on dead wood, *Colenso.*

70. HYPOXYLON, Bulliard.

Characters of *Xylaria*, but horizontal. Stroma not confluent with the substance of the plant on which it grows.

A tropical and temperate genus.

1. **H. concentricum,** *Fries;—Berk. in Fl. N. Z.* ii. 204. Large, brownish subglobose, at length black, concentrically zoned within.

Northern and **Middle** Islands: common on dead wood, from the Bay of Islands, *J. D. H.*, to Otago, *Lindsay.* (Cosmopolite.)

2. **H. annulatum,** *Montagne;—Berk. in Fl. N. Z.* ii. 204. Mycelium thick, abundant, of branched threads, the ultimate branchlets zigzag, with often 2-fid tips. Stroma convex or effuse and confluent, blackish outside and in. Perithecia globose with a depressed ring round the mouth, at last nearly free.

Northern Island: on dead bark, *Colenso.* (N. and S. America.)

71. DIATRYPE, Fries.

Characters of *Hypoxylon*, but the stroma is confluent with the wood of the plant on which the Fungus grows. Perithecia with a more or less obvious neck.

Both tropical and temperate Fungi.

1. **D. glomeraria,** *Berk. in Fl. N. Z.* ii. 205. *t.* 106. *f.* 13. Pustules thickly scattered, small, breaking through the cuticle, black, opaque, $\frac{1}{24}$ in. long. Perithecia crowded, subglobose; neck short; mouth obscure. Asci with 8 linear curved sporidia.

Northern Island: on branches of *Rhipogonum, Colenso.*

2. **D. lata,** *Fries;—Berk. in Fl. N. Z.* ii. 205. Effuse, unequal. Stroma thin, black. Perithecia sunk in the wood; mouth conical.

Northern Island: on sticks, *Colenso.* (Europe, Tasmania.)

72. DOTHIDEA, Fries.

Characters of *Sphæria*, but perithecium 0. Stroma with numerous cells, opening by a simple pore.

A very large tropical and temperate genus of small Fungi, found on leaves, twigs, etc.

1. **D. Ribesia,** *Fries;—Berk. in Fl. N. Z.* ii. 207. Subelliptic, breaking through the cuticle, depressed, black inside and out; cells small, white, almost superficial.

Northern Island: on gooseberry twigs, *Colenso.* (Europe.)

2. **D. filicina,** *Mont.;—Berk. in Fl. N. Z.* ii. 207. Hypophyllous; stroma thin, areolate, black; cells shallow; ostiola minute. Sporidia uniseptate.

Northern Island: on under surface of fern fronds. (Otaheite.)

3. **D. Colensoi,** *Berk. in Fl. N. Z.* ii. 207. Orbicular, $\frac{1}{8}$ in. broad, black, granulate, fertile on either side; cells minute, white within.

Northern Island: on dead leaves, *Colenso.*

4. **D. hemisphærica,** *Berk. in Fl. Antarct.* 172. *t.* 67. *f.* 2. Solitary, black, carbonaceous, breaking through the under surface of the leaf; floccose at the base; cells elliptic, obtuse. Asci short. Sporidia 8, oblong, 1-septate, contracted in the middle.

Lord Auckland's group: on leaves of *Veronica odora, J. D. H.*

5. **D. spilomea,** *Berk. in Fl. Antarct.* 173. *t.* 67. *f.* 2. Gregarious on the under surface of the leaf, thin, depressed, orbicular, often confluent, minutely granular, shining; cells globose, orifice punctiform. Asci clavate. Sporidia oblong, 1-septate, constricted in the middle.

Lord Auckland's group and **Campbell's** Island: on leaves of *Veronica elliptica, J. D. H.*

73. NECTRIA, Fries.

Small Fungi, growing on wood, bark, etc. Perithecia free or seated on a mycelium, thin, rarely thick, vertical, brightly coloured. Sporidia 8, translucent.

A large temperate genus with few tropical species, often brightly coloured.

1. **N. polythalama,** *Berk. in Fl. N. Z.* ii. 203. *t.* 106.*f.* 15. Tufted, bright scarlet. Perithecia ovate, umbilicate at the apex or depressed. Sporidia many-septate, oblong.

Northern Island: on dead bark, *J. D. H.*

2. **N. illudens,** *Berk. in Fl. N. Z.* ii. 203. Scattered or crowded, but not tufted, globose, bright ochreous or cinnabar-red, dimpled at the apex, warted, the warts answering to large cells. Sporidia broad, elliptic, 1-septate, as broad as long.

Northern Island: on dead bark, *J. D. H.*

74. SPHÆRIA, Haller.

Stroma 0, or spurious and formed of mycelium. Perithecia various, vertical, firm, black or dark, often with a bark. Asci perfect. Sporidia 8.

An immense genus, found in all parts of the world and on various substances, some even under salt-water.

§ 1. CÆSPITOSÆ.—*Perithecia cæspitose, superficial, free, seated on a mycelioid stroma that bursts through the cuticle.*

1. **S. fragilis,** *Berk. in Fl. N. Z.* ii. 205. *t.* 106. *f.* 7. Forming scattered clusters of minute, brownish, opaque, obtuse, brittle perithecia, breaking away when still full of spores; mouth obsolete. Sporidia shortly fusiform, hyaline, often broader at one end, 1-septate.

Northern Island: on the under surface of leaves of *Eurybia furfuracea, Colenso.*

2. **S. pullularis,** *Berk. in Fl. N. Z.* ii. 205. *t.* 106. *f.* 6. Forming specks of a few crowded subglobose opaque dark brown perithecia; mouth 0, or a papilliform prominence. Asci subcylindric. Sporidia of 2 opposed cones.

Northern Island: on leaves of *Leucopogon Fraseri, Colenso.*

3. **S. rasa,** *Berk. in Fl. N.* Z. ii. 205. Forming neat round clusters. Perithecia oblong, crowded below, above free, convex, not collapsing, opaque as if minutely downy, but rough from the prominence of the septa of the cells; orifice minute. Asci cylindric, slightly attenuate downwards. Sporidia very numerous, curved.

Northern Island: on *Weinmannia* leaves, *Colenso.*

4. **S. Saubinetii,** *Mont. and Durieu;—Berk. in Fl. N. Z.* ii. 206. Erumpent. Perithecia solitary or slightly aggregate, brown, globose, at length collapsed and umbilicate. Asci diffluent. Sporidia fusiform, lunulate, 3–5-septate.—*Gibbera Saubinetii,* Mont. Syllog. 252.

Northern Island: on Monocotyledonous leaves, *Colenso.* (Algeria.)

5. **S. pulicaris,** *Fries ;—Berk. in Fl. N. Z.* ii. 206. Minute, tufted, forming a blackish stroma with the bark. Perithecia stipitate, superficial, crowded, opaque, flat, finally sub-cupshaped.

Northern Island: on dead sticks, *Colenso.* (Europe.)

§ 2. SERIATÆ.—*Perithecia disposed in parallel lines.*

6. **S. nebulosa?,** *Pers.;—Berk. in Fl. Antarct.* 171. Forming grey spots on a subcuticular black crust. Perithecia gregarious, very minute, free, disposed in irregular series; mouth rather prominent, subacute.

Lord Auckland's group and **Campbell's** Island: on *Anthericum, J. D. H.* Specimen imperfect and identification doubtful.

§ 3. OBTECTÆ.—*Perithecia immersed, often erumpent mouth dilated, produced into a neck, which is immersed in the matrix.*

7. **S. livida,** *Fries;—Berk. in Fl. N. Z.* ii. 206. Forming elongated subdeterminate white or grey spots on wood. Perithecia scattered, subglobose, thin, black, sunk in a prominent elliptic grey elevation of the wood; mouth short, imbedded.

Northern Island: on dead bleached twigs, *Colenso.* (Europe.)

§ 4. CAULICOLÆ.—*Perithecia at first covered with the epidermis, through which they break.*

8. **S. herbarum,** *Persoon;—Berk. in Fl. N. Z.* ii. 206. Subgregarious, minute, black. Perithecia partially covered, depressed, globose, smooth; mouth rather prominent, punctiform.—Fl. Antarct. 170. t. 68. f. 3.

Throughout the **Northern** and **Middle** Islands, and in **Lord Auckland's** group and **Campbell's** Island: on stems of herbaceous plants, abundant. (Ubiquitous.)

9. **S. coffeata,** *Berk. in Fl. N. Z.* ii. 206. *t.* 106. *f.* 3. Perithecia scattered, subglobose, indicated by black specks on the cuticle. Asci cylindric. Sporidia filiform.

Northern Island: on sheaths of grasses, *Colenso.*

10. **S. Lindsayi,** *Currey.* Perithecia very small, round, rupturing the epidermis by a circular rimose or radiate fissure. Sporidia 8, 2-seriate, colourless, irregularly cymbiform, 0·0014 to 0·002 in. long.

Middle Island: dead leaves of *Phormium*, Otago, *Lindsay.*—Plant when dry somewhat resembling *S. nebulosa, Fries.*

11. **S. nigrella?,** *Fries;—Berk. in Fl. Antarct.* 171. Perithecia subglobose, smooth, black, superficially innate, on an elongate determinate spot, white within.

Lord Auckland's group and **Campbell's** Island: on stems of *Anthericum, J. D. H.*—Very imperfect, and doubtfully referred to the European plant.

§ 5. FOLIICOLÆ.—*Perithecia concrete with the matrix, covered, without any surrounding discoloured spot.*

12. **S. phæosticta,** *Berk. in Fl. Antarct.* 171. *t.* 68. *f.* 4. Gregarious, appearing in punctiform brown spots, under which is a perithecium. Perithecia globose, black, with a punctiform mouth. Asci at first short, then linear. Sporidia at first in two series, elliptic, afterwards brown, cymbiform, and in one series; nucleus large, globose.

Lord Auckland's group and **Campbell's** Island: on leaves of *Hierochloe Brunonis, J. D. H.*

13. **S. depressa,** *Berk. in Fl. Antarct.* 172. *t.* 68. *f.* 5. Gregarious,

minute, black, appearing as black specks on the cuticle. Perithecia covered by the cuticle, subglobose, smooth, not produced into a neck, depressed when dry. Asci linear, mixed with slender paraphyses. Sporidia 2-seriate, lanceolate; nucleus 2-partite but not septate.

Campbell's Island: on leaves of *Luzula crinita, J. D. H.*

14. **S. Acetabulum,** *Berk. in Fl. N. Z.* ii. 206. *t.* 106. *f.* 2. Minute, black, covered by the cuticle. Perithecia slightly irregular, depressed, convex below, concave above, adnate with the cuticle in the centre, without a trace of orifice. Asci oblong, narrowed and truncate above. Sporidia oblong, immature.

Northern Island: on *Corynocarpus, Colenso.*

15. **S. cryptospila,** *Berk. mss.* Microscopic, punctiform, black, shining. Asci short. Sporidia elongate, linear, curved, 3–4-septate.

Northern Island: on leaves of *Hypericum, Sinclair.*

75. CAPNODIUM, Montagne.

Mycelium of black jointed moniliform filaments, mixed with other simple ones, enclosing elongate sometimes branched perithecia. Perithecia containing sometimes asci with sporidia, at others only spores.

A remarkable genus, perhaps a more perfect form of *Antennaria.*

1. **C. fibrosum,** *Berk. in Fl. N. Z.* ii. 209. Forming a rigid bristly stratum, $\frac{1}{8}$ in. thick. Perithecia elongate, crowded, fasciculately branched.

Northern Island: on bark, *Colenso.*

76. PEMPHIDIUM, Montagne.

Perithecium spurious, convex, shield-like, black, formed of the diseased epidermis, closed or terminating in a papilla or 2-lipped pore; nucleus gelatinous. Asci perfect. Sporidia elongate.

A peculiar genus, of one species, the following, and a native of Guyana.

1. **P. opacum,** *Berk. in Fl. N. Z.* ii. 207. *t.* 106. *f.* 9. Patches $\frac{1}{12}$–$\frac{1}{6}$ in. broad, round, black, opaque, beneath which is an ovate transparent mass of asci. Asci cylindric or swollen in the middle. Spores fusiform, narrowed to a thread at either end, not septate, but the endochrome is divided irregularly into 2 or 3 masses.

Northern Island: on stems of *Rhipogonum, Colenso.*

77. ERYSIPHE, Hedwig.

Perithecia free, globose, without an opening, giving off at the base flexuous, equal, simple or sparingly branched filaments. Asci saccate.

An immense genus of minute Fungi, natives of temperate regions, where they are the pest of agriculturists, etc.

1. **E. densa,** *Berk. in Fl. N. Z.* ii. 208. *t.* 106. *f.* 16. Mycelium dense, cobwebby, persistent. Perithecia scattered, with flexuose somewhat forked filamentose appendages.

Northern Island: on leaves of *Aristotelia, Colenso.*

78. CHÆTOMIUM, Kunze.

Perithecium brittle, membranous, without an opening, clothed with opaque hairs. Asci gelatinous, evanescent. Sporidia brown.

Minute parasitic Fungi.

1. **C. amphitrichum,** *Corda;—Berk. in Fl. N. Z.* ii. 209. Perithecia half-immersed; threads smooth, slightly branched; apices curved. Sporidia subglobose.

Northern Island: on damp paper, *Colenso.* (Europe.)

2. **C. elatum,** *Kunze;—Berk. in Fl. N. Z.* ii. 209. Perithecia free; threads branched, minutely scabrous. Sporidia lemon-shaped.

Northern Island: on damp straw, *Colenso.* (Ubiquitous.)

79. MELIOLA, Fries.

Perithecia carbonaceous, fragile, without a pore, developed from a strigose mycelium. Asci broad. Sporidia few, large.

Minute parasitic Fungi, natives of warm climates.

1. **M. amphitricha,** *Fries;—Berk. in Fl. N. Z.* ii. 209. Superficially innate, black, forming spots $\frac{1}{4}$–$\frac{1}{2}$ in. diameter. Perithecia crowded, surrounded with erect simple bristles.

Northern Island: on living leaves, *Colenso.* (Subtropics of Old and New World.)

80. ANTENNARIA, Link.

Black Fungi, consisting of black jointed moniliform filaments, bearing here and there spore-cases full of granules.

Found in all climates, infesting living plants, often smothering them in a black mass of filaments.

1. **A. Robinsonii,** *Mont.;—Berk. in Fl. N. Z.* ii. 209. Forming a dense black coat on leaves, consisting of a close gelatinous web, traversed by moniliform threads, from which arise branched filaments, bearing lateral or subterminal spore-cases.—Fl. Antarct. 175. t. 67.

Northern and **Middle** Islands: on leaves and twigs, *Colenso, Lyall,* etc. **Lord Auckland's** group and **Campbell's** Island, *J. D. H.*

The genus *Sclerotium,* which consists of imperfect states of various Fungi, is now suppressed. A Lord Auckland's group species was published in Fl. Antarct. i. 175, as *S. durum,* Pers. It was found on the capsules of *Gentiana,* but of what Fungus it is a state is unknown.

ORDER IX. ALGÆ.

Plants consisting either of simple cells or of threads of cells, or of cellular fronds, almost invariably aquatic and chiefly marine; usually fixed by a root (which does not nourish the plant); imbibing nutriment through their whole surface; supposed to be always 2-sexual, though this is not yet demonstrated in many genera; very many are propagated also by division of the plant, or by buds from it, or by the division of individual cells. Reproductive organs of various kinds, on one or on different plants, naked on the surface of the frond, or buried in its substance, or contained in proper cavities or sacs (conceptacles), which again may either be superficial or immersed, or contained in special branches of the frond. *Female fructification*, consisting of simple or compound spores, which originate as cells that are fertilized by antherozoids, or by the contact of other cells that transmit the fertilizing matter; these spores may be naked or clothed in a gelatinous envelope, or contained in tubes (*asci*). *Male fructification* essentially and usually of microscopic antherozoids, contained in sacs or cells, and consisting of a nucleus with vibrating threads of infinite delicacy. In the unicellular species and those formed of threads of cells, the sexes may be represented by a single cell that divides in the middle, or by 2 cells, from the union of which the spore is developed.

The vast Order of *Algæ* is now often regarded as consisting of several Orders, which I have here distinguished as Suborders, there being so many genera whose position is doubtful amongst these Suborders, and all being so obviously modifications of one common plan, that I think *Algæ* may still be regarded with propriety as a Natural Order, equivalent to such others as *Gramineæ*, *Lichens*, *Fungi*, *Compositæ*, etc. Though so different in habit, locality, and general appearance from Lichens, the absolute difference between these Orders is reduced to a very little in *Lichina* and others. With Fungi, also, their affinity is very close; and, finally, some of the unicellular and filamentous genera present appearances at certain stages of their development that render it impossible to say whether they should be referred to the animal or vegetable kingdom.

Marine *Algæ* occur at all depths in the ocean between high tide or its spray, and upwards of 100 fathoms, but they abound most between tidal limits or just beyond them. As a rule, the bright green kinds occur nearest to high-water mark and extend to fresh water; the dark olive-green abound between tidal marks; and the bright red affect deeper water; but to this there are many exceptions. The great seaweed that forms olive-green floating patches girdling the New Zealand harbours (*Macrocystis pyrifera*) I have seen off the Crozet Island, where the soundings gave 40 fathoms, and which there is reason to suppose may hence have been 700 feet long (Fl. Antarct. 464).

Iodine, kelp, and soda are the products of this Order, which also contains several plants that are eatable in a raw or cooked state, as *Luver* and *Dulse*, and a copious jelly in the case of the "Carrageen moss" of Ireland, the produce of species of *Chondrus*, *Gracilaria*, and *Gigartina*. Others are used as manure and as fodder for cattle; a valuable cement is made by the Chinese from *Gracilaria tenax*; and knife-handles of the stems of various *Laminarieæ* by whalers and others.

New Zealand is very rich in *Algæ*; its deep waters and quiet sounds especially abound in species of this interesting and beautiful Order, which have been most admirably collected by my old friend Dr. Lyall. Amongst the microscopic unicellular and filamentous tribes, there must still be many hundreds, perhaps thousands, of species to be discovered in the sea and in freshwater ponds, rivers, and the islands. These are subjects of special research, far beyond the scope of a student of general botany or of a handbook like the present. To study them advantageously would require a powerful microscope and a very good library, infinite patience, and considerable manual skill. To those who would undertake these branches I would recommend Hassall's 'British Freshwater Algæ,' Ralfs's 'British Desmidieæ,' and Kützing's 'Tabulæ Phycologicæ.'

Of more general works on *Algæ* there are Harvey's 'Manual of British Algæ' and 'Phycologia Britannica,' but especially his 'Nereis Australis' and 'Phycologia Australis,' both admirable works full of beautiful plates of southern *Algæ*. In my Antarctic, New Zealand, and Tasmanian Floras also, many New Zealand species are figured. The best general work, though still incomplete, is J. Agardh's 'Systema Algarum,' an excellent book in all respects.

To my late friend Dr. Harvey I am indebted for the determination and description of the New Zealand *Algæ* in my Flora of these islands. His widely and deeply deplored death during the present year has deprived me and this work of the benefits of his revisal of the following compilation.

KEY TO THE SUBORDERS, TRIBES, AND GENERA.

SUBORDER I. MELANOSPERMEÆ OR FUCOIDEÆ.

(*Olive-green or brown* Algæ.)

Olive-brown, more rarely olive-green or black, marine *Algæ*, usually large, coarse, much-branched or horizontally expanded. *Frond* continuous or articulate, simple lobed or branched, often bearing distinct leaves, branches, bladders, and receptacles of fructification. *Fructification* on the surface of the frond, or in cavities in its substance; 1, obovoid or ellipsoid, simple or 4-parted *spores*, enclosed in a hyaline membrane, naked, or surrounded by jointed filaments (*peranemata*); 2, oblong *antheridia*, terminating jointed filaments, and containing moving spores (*antherozoids*); 3, buds or leaflets (*propagula*), developed on the surface or edges of the frond, capable of becoming new plants.

Series A. Frond thick, not articulate, its outer stratum formed of minute densely-packed cells.

TRIBE I. **Fuceæ.** *Spores and antheridia in spherical cavities (conceptacles), either sunk promiscuously in the surface of the frond, or confined to proper receptacles.*

§ 1. *Conceptacles in distinct receptacles, not promiscuously sunk in the frond or in its joints.*

* *Frond with distinct stem, branches, leaves, and bladders (except* Turbinaria).

a. Receptacles very different from the leaves.

† *Frond branching from both sides of the stem.*

Receptacles axillary. Bladders distinct from the leaves . . . . 1. SARGASSUM.
Bladders undistinguishable from the leaves, which are turbinate and peltate . . . . 2. TURBINARIA.
Receptacles marginal, racemose, cylindric, warted . . . . . . 3. CARPOPHYLLUM.

†† *Branches from one edge only of a flattened frond.*

Receptacles marginal . . . . 4. MARGINARIA.

β. Receptacles obviously formed of leaves.

Leaves distinct from the branches, marginal, distichous . . . . 5. PHYLLOSPORA.
Leaves distinct from the branches, spirally disposed round the stem 6. SCABERIA.
Leaves passing into branches . . . . 7. CYSTOPHORA.

** *Frond with distinct stem and leaves, but no bladders* . . . 8. LANDSBURGIA.

*** *Frond flabellately dichotomously branched, without leaves or bladders* . . . . 9. FUCODIUM.

§ 2. *Conceptacles scattered over the frond, or sunk in its joints.*

a. Frond with distinct stem and branches.

Frond moniliform, internodes inflated . . . . 10. HORMOSIRA.
Frond cylindric, proliferously branched . . . . 11. SPLACHNIDIUM.
Frond filiform, irregularly branched . . . . 12. NOTHEIA.

β. Frond without stem and branches, flabellately expanded and palmately lobed . . . . 13. D'URVILLÆA.

TRIBE II. **Sporochnoideæ.** *Spores superficial, collected on proper branches of the frond (receptacles).*

Frond filiform, cylindric. Receptacles terminated by a pencil of filaments . . . . 14. SPOROCHNUS.
Frond filiform, compressed or flat. Receptacles conical or mitriform . . . . 15. CARPOMITRA.
Frond compressed or flat, linear; branches pinnate, opposite . . 16. DESMARESTIA.

TRIBE III. **Laminarieæ.** *Spores superficial, forming cloudy sori of indefinite forms on the frond.*

Frond terete, slender, undivided, of immense length, bearing at the upper part a series of lanceolate leaves with a bladder at their base . . . . 17. MACROCYSTIS.
Frond with an erect, solid, dichotomous trunk, bearing branches and lanceolate leaves without bladders . . . . 18. LESSONIA.
Frond broad, pinnatifid, its stipes solid or hollow . . . . . . 19. ECKLONIA.

TRIBE IV. **Dictyoteæ.** *Spores superficial, forming dot-like or small sori of definite forms on the frond.*

Frond flat, dichotomous, flabellate, concentrically striated . . . 20. ZONARIA.
Frond flat, dichotomous, linear . . . . 21. DICTYOTA.
Frond cylindric, membranous, branched . . . . 22. DICTYOSIPHON.
Frond saccate, membranous, unbranched . . . . 23. ASPEROCOCCUS.

Series B. Frond not articulate, its outer stratum formed of radiating articulated filaments.

TRIBE V. **Chordarieæ.**

* *Frond tubular or saccate, unbranched, filled with fluid.*

Frond cylindric, septate internally . . . . 24. CHORDA.
Frond saccate . . . . 25. ADENOCYSTIS.

** *Frond solid, cylindric, branched; axis of interlaced filaments; periphery of radiating filaments immersed in gelatine.*

Axis of longitudinal filaments, radiating outwards to form the periphery . . . . 26. SCYTOTHAMNUS.
Axis cartilaginous, dense; filaments of periphery unbranched . . 27. CHORDARIA.
Axis gelatinous, loose; filaments of periphery branched . . . . 28. MESOGLOIA.

*** *Frond solid, hemispherical or tuberous* . . . . 29. LEATHESIA.

Series C. Frond filiform, branched, of superimposed single cells, or of longitudinal series of cells of equal length, hence articulate.

TRIBE VI. **Ectocarpeæ.**

Articulations of 3 or more parallel cells . . . . 30. SPHACELARIA.
Articulations of solitary cells . . . . 31. ECTOCARPUS.

SUBORDER II. RHODOSPERMEÆ OR FLORIDEÆ.

(*Red or purple* Algæ.)

TRIBE I. **Rhodomeleæ.**

Rosy-red or purple, rarely brown-red or greenish-red marine *Algæ*, flat or filiform. *Fructification* diœcious, organs of three kinds:—1. *Spores*, contained in external or immersed conceptacles, or densely crowded and dispersed in masses (*nuclei*) through the frond. 2. *Tetraspores*, red or purple, external or immersed in the frond, rarely contained in proper receptacles, each enveloped in a pellucid covering, and at maturity separating into 4 sporules, which are arranged crosswise or transversel . 3. *Antheridia*, filled with hyaline or yellow corpuscles (*antherozoids*).

Series A. Spores arranged on tufts of threads (articulated filaments), which spring together from a basal parietal or central receptacle (placenta) within the conceptacle. Single spores formed in all the cells of the thread, or in the terminal cell only.

I. Tufts of spore-threads contained in an external ovate or spherical conceptacle.

a. Placenta on the base of a hollow conceptacle. Spores pyriform or obconic, formed in the terminal cell of the spore-thread.

TRIBE I. **Rhodomeleæ.** *Frond articulate or furnished with an articulated axis of many tubes; surface areolated. Spore-threads simple. Tetraspores in series in the branchlets, or in pod-like bodies* (Stichidia).

* *Frond flat, proliferous* (*see* Amansia *in* **).

Frond membranous, costate, obliquely striate . . . . 32. LENORMANDIA.

** *Frond flat, lobed or pinnatifid.*

Frond costate, transversely striate. Stichidia on the costa . . . 33. EPINEURON.
Frond costa, transversely zoned. Stichidia marginal . . . . . 34. AMANSIA.

*** *Frond and branches filiform or somewhat compressed.*

† *Frond inarticulate externally; axis articulate.*

Frond laxly reticulate, closely transversely striate above; cortical cells polygonal . . . . 35. ALSIDIUM.
Frond opaque, often compressed; branchlets filiform, transversely striate; cortical cells minute, polygonal . . . . 36. RYTIPHLŒA.
Frond opaque, not transversely striate. Stichidia formed of swollen branchlets . . . . 37. RHODOMELA.

†† *Frond articulate* (*inarticulate in some* Dasyæ).

a. Frond an articulate tube, surrounded by tessellated cells. Tetraspores 2-seriate in terminal stichidia . . . . 38. BOSTRYCHIA.

Frond opaque or reticulate; joints longitudinally striate, internodes hyaline. Tetraspores 1-seriate . . . . 39. POLYSIPHONIA.

β. Frond with articulate unicellular branches. Tetraspore 2–∞-*seriate in stichidia* . . . . 40. DASYA.

†† *Frond minute, filiform, articulate, bearing uniform angular toothed leaflets* . . . . 41. POLYZONIA.

TRIBE II. **Laurencieæ.** *Frond inarticulate, solid, tubular or septate; surface-cells minute. Spore-threads simple or branched. Tetraspores irregularly scattered through the branchlets.*

* *Frond hollow, full of fluid; tube septate; septa close together* 42. CHAMPIA.

** *Frond solid, cellular, flat or compressed; branchlets obtuse or constricted at the base.*

Frond terete or compressed . . . . 43. LAURENCIA.
Frond filiform, articulate . . . . 44. CHONDRIA.
Frond flat, membranous . . . . 45. CLADHYMENIA.

*** *Frond solid, cellular, cylindric or compressed; branchlets distichous, subulate or setaceous or serrate.*

Frond decompound; branches setaceous, subarticulate. Conceptacles scattered, on long stalks . . . . 46. ASPARAGOPSIS.
Frond decompound; branches subulate. Conceptacles sessile on the tips of the branches . . . . 47. DELISEA.
Frond decompound, serrate. Conceptacles terminal on the tips of branches. . . . . 48. PTILONIA.

b. Placenta on the base of a hollow conceptacle. Spores roundish or elliptic, in moniliform threads, all the cells of the spore-thread being changed into spores.

TRIBE III. **Corallineæ.** *Frond calcareous, its cells secreting carbonate of lime. Spore-threads separating into 4 spores.*

* **Corallines.** *Frond branched, filiform, articulate.*

Conceptacles conical, on the disk of the articulations . . . . . 49. AMPHIROA.
Conceptacles urceolate, terminal, smooth . . . . 50. CORALLINA.
Conceptacles turbinate or urceolate, 2–4-horned . . . . 51. JANIA.

** **Nullipores.** *Frond crustaceous or foliaceous, inarticulate.*

Frond rigid, crust-like, or branching with the branches stony . . 52. MELOBESIA.

TRIBE IV. **Sphærococcoideæ.** *Frond cartilaginous or membranous. Spore-threads separating into many spores.*

* **Delesserieæ.** *Tetraspores divided into threes, disposed in definite sori.*

† *Frond foliaceous, symmetrical, costate.*

Costa running through the frond . . . . 53. DELESSERIA.
Costa slender, vanishing below and above . . . . 54. HEMINEURA.

†† *Frond foliaceous; costa* 0 *or faint.*

55. NITOPHYLLUM.

** **Sphærococceæ.** *Tetraspores immersed in the cortical cells, divided cruciately or transversely.*

† *Tetraspores transversely divided.*

§ *Frond with a central articulate tube* . . . . 56. PHACELOCARPUS.

§§ *Frond without a central tube.*

Frond of 2 strata of cellular tissue . . . . 57. CALLIBLEPHARIS.
Frond of 3 strata of cellular tissue . . . . 58. SARCODIA.

†† *Tetraspores cruciately divided* (*unknown in* Melanthalia).

Frond ecostate, of two strata of tissue . . . . 59. GRACILARIA.
Frond dichotomously flabellate, of three strata of tissue . . . . 60. MELANTHALIA.

c. Placenta axial or parietal, or suspended by filaments in the cavity of the external or half-immersed conceptacles.

TRIBE V. **Gelidieæ.** *Frond inarticulate, cartilaginous or horny, opaque, the axis (at least) formed of elongate confervoid filaments (axis articulate in* Caulacanthus *when young).*

* *Conceptacles 2-celled* . . . . 61. GELIDIUM.

** *Conceptacles 1-celled, traversed by filaments, hemispherical, or, if immersed, in swollen branchlets (unknown in* Caulacanthus).

Frond subtubular; axis articulate when young . . . . 62. CAULACANTHUS.
Frond cellular, solid. Spore-mass suspended . . . . . . 63. HYPNEA.
Frond cellular, solid. Spore-mass lateral . . . . 64. PTEROCLADIA.

*** *Conceptacles immersed, scattered over the frond. Spore-mass parietal* . . . . 65. APOPHLŒA.

II. Tufts of spore-threads not lodged in a hollow conceptacle.

* *Tufts of spore-threads in wart-like excrescences or tubercles.*

TRIBE VI. **Squamarieæ.** *Frond cartilaginous or membranous, horizontally expanded and rooting from the under surface. Spores in moniliform strings within the wart.*

Frond red-brown, flat, fan-shaped . . . . 66. PEYSSONNELIA.

** *Tufts of spore-threads immersed in the frond.*

TRIBE VII. **Helminthocladieæ.** *Frond cylindric, gelatinous or submembranous, almost wholly composed of filaments set in loose gelatine. Tufts of spores spherical, of branching moniliform spore-threads springing from a central point.*

Outer stratum of moniliform filaments lying in gelatine . . . . 67. NEMALION.
Outer stratum membranous, thin, of angular cells . . . . 68. SCINAIA.

*** *Tufts of spore-threads naked, surrounded with an involucre of branchlets of the frond.*

TRIBE VIII. **Wrangelieæ.** *Frond filiform, of one articulate tube, which is naked or coated with small cells. Spores pyriform, in the terminal cells of branching spore-threads.*

Frond filiform, clothed at the nodes with flaccid capillary filaments 69. WRANGELIA.

Series B. Spores in simple or compound globose masses, enclosed in a hyaline mucous or membranous coat. Spore-masses naked or immersed in the frond, or contained in conceptacles.

I. Frond inarticulate, flat or cylindric.

TRIBE IX. **Rhodymenieæ.** *Spore-masses in globose conceptacles (sunk in the frond in* Dasyphlœa). *Spores first developed in the cells of moniliform branching filaments issuing from a centre, at length aggregated without order.*

* *Frond broad, leafy, dichotomous.*

Conceptacles hemispheric, scattered. Tetraspores crucially divided 70. RHODYMENIA.
Conceptacles spherical, submarginal. Tetraspores transversely divided . . . . 71. RHODOPHYLLIS.
Conceptacles sunk in the frond . . . . 72. DASYPHLŒA.

** *Frond slender, pectinate-pinnatifid* . . . . 73. PLOCAMIUM.

TRIBE X. **Cryptonemieæ.** *Spore-masses in conceptacles or sunk in the frond. Spores developed within solitary or clustered detached cells, at length aggregated without order.*

Subtribe GIGARTINEÆ. *Spore-masses compound.*

§ 1. **Tylocarpeæ.** *Frond rigid, compact; inner stratum of roundish polygonal cells outer of densely packed vertical filaments.*

Frond flat, leaf-like. Conceptacles linear, rib-like . . . . . 74. STENOGRAMMA.
Frond linear, subterete, subcartilaginous . . . . 75. GYMNOGONGRUS.

§ 2. **Kallymenieæ.** *Frond membranous or coriaceous; inner stratum of longitudinal filaments or of polygonal cells surrounded by these; outer of roundish or polygonal cells, smaller outwards; the cortical of minute vertical filaments.*

Frond flat, dichotomous; inner stratum of large roundish cells in a network of filaments . . . . 76. CALLOPHYLLIS.
Frond flat, subsessile, expanded; inner stratum of longitudinal interlaced fibres . . . . 77. KALLYMENIA.

§ 3. **Eugigartineæ.** *Frond cartilaginous, wholly composed of slender anastomosing longitudinal filaments set in firm gelatine. Tetraspores in sori.*

* *Spore-mass surrounded by a coat of densely interwoven filaments.*

Frond terete or flat. Spore-mass in an external conceptacle . . . 78. GIGARTINA.
Frond flat, simple or vaguely cleft. Spore-mass imbedded in the frond . . . . 79. IRIDÆA.

** *Spore-mass immersed in the frond, with no definite coat.*

Frond flabelliform, dichotomous . . . . 80. CHONDRUS.

Subtribe CRYPTONEMEÆ. *Spore-masses, simple, rounded.*

§ 4. **Gastrocarpeæ.** *Inner stratum of anastomosing filaments or polygonal cells; outer of rows of smaller cells, the outermost smallest and coalescing.*

a. Conceptacles external, with a defined pericarp.

Frond flat, forked, ribbed below; inner stratum cellular. Conceptacles on proper leaflets, springing from the lamina . . . . 81. EPYMENIA.
Frond linear, tubular, constricted at intervals, thus divided into water-sacs, traversed by longitudinal filaments . . . . 82. CHYLOCLADIA.
Frond compressed, tubular, not constricted, full of water and traversed by longitudinal filaments . . . . 83. CHRYSIMENIA.

β. Spore-mass immersed in the frond.

Frond flat or compressed, dichotomous, pinnatifid . . . . . . 84. HALYMENIA.

§ 5. **Nemastomeæ.** *Frond composed of two strata of filaments (with rarely an intermediate cellular one); inner stratum longitudinal, interlaced; outer vertical, dichotomously fastigiate. Spore-masses in the N. Z. genera dispersed through the frond (not in separate branches).*

a. Frond compact, solid, compressed or flat (tubular and soft in some Nemastomæ).

Frond with a stratum of roundish cells between the inner and outer filamentous ones . . . . 85. PRIONITIS.
Frond of two filamentous strata.
Frond flat, pinnated. Favellidia with a pore . . . . 86. GRATELOUPIA.
Frond compressed, dichotomous or pinnate. Favellidia immersed . . . . 87. NEMASTOMA.

β. Frond cylindric, tubular, continuous or constricted at intervals.

Spore-masses in ovoid or subglobose branchlets . . . . 88. CATENELLA.
Spore-masses scattered, in the inner surface of the frond . . . . 89. DUMONTIA.

II. Frond filiform, articulate, of one tube. Articulations naked or coated with small cells.

TRIBE XI. **Spyridieæ.** *Spore-masses in an external, closed, cellular conceptacle, consisting of many lesser masses, each formed by the evolution of paniculately branched spore-threads. Spores at length aggregated without order.*

Frond filiform, pinnately decompound . . . . 90. SPYRIDIA.

TRIBE XII. **Ceramieæ.** *Spore-masses single, naked or involucrate (by small branchlets). Spores aggregated without order within a hyaline membranous cell, developed externally.*

§ 2. *Tetraspores formed of the cortical cells, more or less sunk in the frond.*

Frond coated with roundish, irregularly placed cells . . . . . . 91. CERAMIUM.
Frond coated with rectangular cells, in longitudinal series . . . 92. CENTROCERAS.

§ 2. *Tetraspores formed of whole branchlets, or external cells of these, hence external, stalked or sessile.*

* *Spore-masses involucrate.*

Frond linear, compressed, opaque, pellucid and articulate at the tips, pinnate and pectinate . . . . 93. PTILOTA.
Frond filiform, pellucid, trichotomous . . . . 94. GRIFFITHSIA.
Frond filiform, rigid, pellucid, pinnately decompound and plumose 95. BALLIA.

** *Spore-masses not involucrate.*

Frond articulate, pellucid, or main stems etc. coated with cells . . 96. CALLITHAMNION.

SUBORDER III. CHLOROSPERMEÆ.

(*Green* Algæ.)

Plants almost always grass-green, rarely olive-green or purple, still more rarely red. Propagation by division of the cells, or by the transformation of the cell-contents into spores (called *zoospores*), or more rarely by spores developed in proper spore-cases. Antheridia containing antherozoids have been found in some.—Marine or freshwater plants, or inhabiting damp places.

TRIBE I. **Siphonieæ.** *Frond green, rooting or fixed by its base, simple or compound, of a single filiform branching cell, or of many such uniting to form a spongy frond.—Marine or freshwater.*

Rhizome prostrate, rooting; branches erect, membranous, unicellular; cells filled with a network of branching fibrils . . . 97. CAULERPA.
Frond a matted mass of tubules, full of granular green matter, without fibrils . . . . 98. CODIUM.
Frond of tufted filaments, matted below, irregularly branched above 99. VAUCHERIA.
Frond of free, tufted or solitary filaments, which are pinnately branched . . . . 100. BRYOPSIS.

TRIBE II. **Dasycladeæ.** *Frond green, rooting, of simple or branched, inarticulate, axial threads, with whorled articulate branchlets. Spores spherical, in proper fruit-cells.—Marine, sometimes crusted with lime.—No species has hitherto been sent from New Zealand.*

TRIBE III. **Valonieæ.** *Frond green, rooting, of large bladdery cells, full of coloured fluid, often united into filaments or a network.—Marine.—Not yet sent from New Zealand.*

TRIBE IV. **Ulveæ.** *Frond green or purple, fixed by its base, membranous, flat or tubular, formed of minute quadrate cells. Fructification, zoospores formed in the cells.*

Frond leaf-like, purple, membranous . . . . 101. PORPHYRA.

Frond leaf-like, green, membranous . . . . 102. ULVA.
Frond tubular, membranous, green . . . . 103. ENTEROMORPHA.
Frond filiform or expanded, purplish . . . . 104. BANGIA.

TRIBE V. **Batrachospermeæ.** *Frond blackish or olive-green or purple, filiform, inarticulate, branching, composed of small cells; branchlets* 0 *or moniliform and whorled. Fructification moniliform strings of naked spores, forming external tufts or concealed in a tubular frond.—Freshwater.*

Frond nodose, with whorled moniliform branchlets . . . . 105. BATRACHOSPERMUM.

TRIBE VI. **Conferveæ.** *Frond green, of articulated filaments, consisting of cylindric cells, that are usually longer than broad, and are full of coloured matter. Antherozoids minute, swarming in the cells.—Marine and freshwater.—There must be an immense number of New Zealand species to be discovered.*

Filaments tufted, branched . . . . 106. CLADOPHORA.
Filaments unbranched . . . . 107. CONFERVA.
Filaments joined by transverse tubes . . . . 108. TYNDARIDEA.

TRIBE VII. **Zygnemaceæ.** *Frond green, of simple articulated floating threads of excessive tenuity, composed of cylindric cells; cell contents of definite form. Spores large, green orange or vermilion-coloured, solitary, formed of the union of the contents of two cells, or the division of the contents of one.—Marine and freshwater Algæ.—Very numerous, not hitherto collected in New Zealand.*

TRIBE VIII. **Hydrodictyeæ.** *Frond green, of cylindrical cells, uniting and forming a saccate network, with polygonal meshes; each side of the mesh formed of a single cell. Cell contents resolving into antherozoids, which arrange themselves and form a new network within the parent cell. Network hence viviparous.—This beautiful freshwater group has hitherto not been detected in New Zealand.*

TRIBE IX. **Oscillatorieæ.** *Fronds green, attached or floating, rarely olive-brown blue or purple, forming a stratum composed of simple or sparingly branched very minute threads; each thread consists of a membranous pellucid sheath, enclosing an annulate axis of short cells.—Marine or freshwater minute Algæ.—Very numerous in species, but hitherto hardly at all collected in New Zealand.*

Filaments attached, short, tufted, erect, branched . . . . 109. CALOTHRIX.
Filaments rigid, usually floating, lying in a mucous matrix . . . 110. OSCILLATORIA.
Filaments free, branched . . . . 111. TOLYPOTHRIX.

TRIBE X. **Nostochineæ.** *Fronds olive or bright green, forming gelatinous masses, traversed by moniliform filaments of globose or oval cells, with here and there a larger cell than the others, of whose function nothing is known.—Freshwater pools and on damp ground, often appearing suddenly after rain.—The species are very numerous, and have never been collected or studied in New Zealand.*

Filaments unbranched . . . . 112. NOSTOC.

TRIBE XI. **Desmidieæ.** *Microscopic unicellular green Algæ. Cell-wall membranous. Propagation by the cell contents becoming divided in the middle, and a new half-cell being formed at the medial line.—Very minute marine or freshwater Algæ, either parasitic or forming floating masses, usually occurring amongst other* Confervæ.*—The tribe is a vast one.*

TRIBE XII. **Diatomeæ.** *Microscopic unicellular, yellow-brown Algæ; cell orbicular triangular or of very various symmetric forms; cell-wall siliceous. Propagation as in* Desmidieæ.*—Microscopic Algæ, marine or freshwater; the siliceous coat of the cell is often exquisitely sculptured, and is a beautiful microscopic object.—The genera and species are extremely numerous, and subject of special study. Dr. Lauder Lindsay has published a catalogue of New Zealand species (see Journ. Linn. Soc.* ix. 129).

Frustules few, convex, connate into a stalked flag-shaped frond . 113. Achnanthes.
Frustules in longitudinal series . . . . 114. Schizonema.

Tribe XIII. **Palmelleæ.** *Frond green red yellow or orange, composed of separate globose or ellipsoid cells, free or in a gelatinous matrix. Propagation by division of the cell contents.—Some of the most simple forms of vegetables belong to this tribe, including the Red Snow of the Arctic regions, which has not hitherto been found in the southern hemisphere. The species have never been collected in New Zealand.*

SUBORDER I. MELANOSPERMEÆ.

1. SARGASSUM, Agardh.

Roots scutate. Frond olive-brown, pinnately decompound, with distinct stem, midribbed branches, leaves, bladders, and receptacles. Bladders stalked, supra-axillary, simple, usually pointed or terminated by a leaf. Receptacles pod-like, axillary, solitary or fascicled, tubercled or moniliform; tubercles porose, answering to the immersed conceptacles, which contain *tetraspores* and tufted *antheridia*.

A large tropical genus, rarer in temperate and colder seas. The Gulf-weed (*S. bacciferum*) belongs to it.

§ 1. Pterophycus, *J. Ag.—Stems flattened, ribbed. Leaves parallel to the stem.*

1. **S. longifolium,** *Agardh;—Fl. N. Z.* ii. 212. Frond 4 feet and upwards long. Stem flat, spirally twisted at the base. Leaves marginal, 5–6 in. long, linear-lanceolate, serrate, costate, lower dichotomously pinnatifid. Bladders spherical or ellipsoid, terminated by a leaf. Receptacles panicled, ovoid, unarmed.—J. Ag. Sp. Alg. i. 283. *Anthophycus*, Kuetzing. *Fucus*, Turn. Hist. Fuc. t. 104.

Shores of New Zealand, *Banks and Solander, D'Urville.* (South Africa, Indian Ocean.)

§ II. Arthrophycus, *J. Ag.—Stems flattened, angular or 2–3-edged, branches bent down at their insertion. Leaves horizontal, owing to a twist at their base.*

2. **S. plumosum,** *A. Rich.;—Fl. N. Z.* ii. 212. Frond 1–2 feet long. Stem flat, distichously pinnate, hardly costate. Leaves marginal, of two forms (on the same or different plants); the broader falcate pinnatifid; segments costate, linear-falcate, $\frac{1}{12}$ in. broad; narrower leaves dichotomously decompound; segments capillary incurved. Bladders size of a pea, stalked, younger mucronate. Receptacles panicled, cylindric, torulose, $\frac{1}{10}$–$\frac{1}{8}$ in. long. —J. Ag. Sp. Alg. i. 286. *S. pennigerum* and *S. capillifolium*, A. Rich. l. c. t. 5 and 6. *S. flexuosum*, Kuetzing?

Common on all the coasts.—Kuetzing has published a *S. flexuosum*, Hook. f., of which I know nothing. Harvey suspects it to be a state of this.

3. **S. Raoulii,** *Hook. f. and Harv. Fl. N. Z.* ii. 212. Stem 2–4 ft. long, slender, smooth, compressed, zigzag, excessively branched. Leaves filiform, flat, dichotomously multifid, ribless. Bladder spherical or ellipsoid, obtuse, broad, solitary at the base of the leaves. Receptacles racemose, smooth, cylindric.—J. Ag. Sp. Alg. i. 288; Harv. Phyc. Austr. t. 110.

Banks's Peninsula, *Raoul.* (Tasmania.)

4. **S. adenophyllum,** *Harv. in Fl. N. Z.* ii. 212. Frond 1–2 feet long. Stem 1–2 in., dividing into many slender very long compressed smooth flexuous branches, $\frac{1}{10}$ in. diameter. Leaves alternate, distichous, long-petioled, dichotomously multifid, lower broader membranous ribbed, upper filiform, with several large prominent glands. Bladders on slender stalks, spherical, shortly mucronate. Receptacles unknown.

Shores of New Zealand, *Lyall.* (Australia.)

5. **S. scabridum,** *Hook. f. and Harv. Fl. N. Z.* ii. 211. Stems angled, muricate. Leaves oblong-lanceolate, acuminate, toothed, upper very narrow. Bladders scattered, stalked, globose, smooth, not mucronate. Receptacles racemed, stalked, lanceolate, smooth; racemes axillary, shorter than the leaves.—J. Ag. Sp. Alg. i. 347.

Bay of Islands, *J. D. H.;* Houraki Gulf, *Lyall.*—Specimens imperfect.

6. **S. Sinclairii,** *Hook. f. and Harv. Fl. N. Z.* ii. 211. Stems semi-terete, angled below, compressed and filiform above. Leaves lanceolate, narrowed at the base, rib vanishing; lower larger, 3–4 in. long, incised and toothed; upper toothed or subentire. Bladders few, shortly stalked, bearing a leaf. Receptacles very short, axillary, sparingly divided, subtended by a minute leaf; lobes smooth, turbinate, abruptly 3–4-horned.—J. Ag. Sp. Alg. i. 300.

Common along the shores from the Bay of Islands to Port Cooper, *Sinclair,* etc.

§ III. EUSARGASSUM, *J. Ag.—Stem flat or terete. Branches ascending (not bent down at their insertion). Leaves horizontal.*

7. **S. bacciferum,** *Agardh;—Fl. N. Z.* ii. 211. Stem much branched, filiform, smooth. Leaves linear-lanceolate, acutely often doubly-serrate, costate, slightly glandular or eglandular. Bladders spherical on a stalk of their own diameter, mucronate. Receptacles $\frac{1}{4}$ in. long, axillary, in forked racemes, cylindric, warted, unarmed.—J. Ag. Sp. Alg. i. 344. *Fucus natans,* Linn.; Turn. Hist. t. 47.

Shores of New Zealand, *D'Urville, Lesson, Sinclair* (all tropical and subtropical seas). —This is the Sargasso-weed (or Gulf-weed) of the Atlantic, which forms free floating patches in the ocean, and increases by accidental division of the frond.

Various other species of *Sargassum* are enumerated as natives of New Zealand, but none with any probability of being really so; such are *S. vulgare*, Ag., a plant confined to the Atlantic; *S. granuliferum*, Ag., a native of the Indian Ocean; *S. droserifolium*, Bory, an imperfectly-described New Ireland species; *S. crassifolium*, J. Ag., found in the ocean between New Zealand and New Ireland; and *S. duplicatum*, a most imperfectly-described plant found in the ocean between New Zealand and Tahiti. On the other hand, no doubt many Australian species are washed on the New Zealand coasts, and that the number of indigenous and transported species to be found will prove to be large.

2. TURBINARIA, Lamouroux.

Root branching. Frond olive-brown, alternately decompound, with distinct stem, leaves or bladders, and receptacles. Leaves confluent with the bladders, which are stalked, turbinate, crowned with a peltate lamina. Receptacles axillary, pod-like, dichotomously branched, tubercled, the tubercles answering to conceptacles containing obovoid spores.

A genus of one or two species, natives of the Southern Ocean.

1. **T. ornata,** *J. Agardh;—Fl. N. Z.* ii. 212. Frond 6–8 in. long, between fleshy and coriaceous. Stem simple. Leaves on bladders, peltate on a triquetrous stalk, rather concave, ½ in. long, crowned with a double series of stout teeth, one series marginal, the other within it.—J. Ag. Sp. Alg. i. 266. *Fucus turbinatus* var. *ornatus*, Turner, Hist. Fuc. t. 24. f. *e–h*.

New Holland, *D'Urville.* (Australia, Chili, Pacific Islands.)

3. CARPOPHYLLUM, Greville.

Frond olive-brown, alternately pinnately decompound, with an obscure midrib, with distinct marginal stem, leaves, bladder, and receptacles. Leaves undivided, vertical. Bladder apiculate. Receptacles in alternate fascicles, simple or forked, cylindric, warted, the warts answering to spherical conceptacles containing obovoid spores.

A small genus, natives of the south temperate ocean.

1. **C. Phyllanthus,** *Hook. f. and Harv. Fl. N. Z.* ii. 212. Stem flat, many feet long, ⅙ in. broad, uniformly distichously pinnate. Leaves membranous, lower 3 in. long, linear, entire or sinuate-serrate, upper 1–1½ in. long, serrate. Bladders solitary, ovoid, mucronate, as large as a hazel-nut. Receptacles on the marginal teeth, $\frac{1}{12}$ in. long.—J. Ag. Sp. Alg. i. 263. *C. flexuosum*, Grev. *Fucus*, Turn. Hist. t. 206.

Common on all the shores, *Banks and Solander*, etc.

2. **C. Maschalocarpus,** *Hook. f. and Harv. Fl. N. Z.* ii. 212. More robust and coriaceous than *C. Phyllanthus.* Leaves ecostate, acuminate, quite entire. Bladders pyriform, mucronate. Receptacles very minute. —J. Ag. Sp. Alg. i. 264. *Phycobotrys*, Kuetz. *Sargassum Phyllanthum*, A. Rich. Fl. Nov. Zel. t. 7 et 7 bis. *Fucus*, Turn. Hist. f. 205.

Common on all the shores, *Banks and Solander*, etc.

3. **C. macrophyllum,** *Montagne in Voy. au Pôle Sud*, 76. Stem 14 in. long, flat, thin, winged. Frond broadly lanceolate from a very narrow base, serrate and pinnately laciniate. Bladders spherical, very large, leaf-bearing and axillary, as are the minute dichotomously paniculate receptacles.

Shores of **Lord Auckland's** Island, *D'Urville.*—I have seen no specimens.

4. MARGINARIA, A. Richard.

Frond olive-green, unilaterally flabellately pinnate. Leaves, bladders, and receptacles distinct. Leaves subconfluent with the stem, dichotomously semi-flabellate, vertical. Bladders in series on the inner upper margins of the leaves. Receptacles in series with the bladders, unilateral, subsimple, terete or compressed, containing spherical conceptacles with obovoid spores.

A small genus, confined to New Zealand and its islands.

1. **M. Boryana,** *A. Rich.;—Fl. N. Z.* ii. 213. Frond many feet long, naked below; pinnæ linear, very long, $\frac{1}{6}$–$\frac{1}{4}$ in. broad, ribless, with hooked serratures. Bladders elliptic-obovoid, as large as a hazel-nut, subapiculate. Receptacles cylindric, 1 in. long, acuminate, simple or sparingly spinous.—Mont. in Voy. au Pôle Sud, t. 2 and 3. f. 2; J. Ag. Sp. Alg. i. 250.

Common on the shores, *D'Urville*, etc.

2. **M. Urvilleana,** *A. Rich.;—Fl. N. Z.* ii. 213. A smaller plant than *M. Boryana*, but hardly distinct specifically, the pinnæ are 1 foot long, gradually dilated, simple or flabellately branched on one side. Bladders smaller, subspherical, not apiculate.—Mont. in Voy. au Pôle Sud, t. 3. f. 1. *M. Urvilliana* et *Gigas*, A. Rich. Fl. N. Z. t. 3 and 4. *Sargassum Lessonianum* and *Urvilleanum*, A. Rich., Sert. Astrolab. 138.

Shores of New Zealand, *Lesson*, etc. **Lord Auckland's** group, *D'Urville.*

5. PHYLLOSPORA, Agardh.

Root fibrous. Frond olive-brown, pinnately decompound, with flat ribless stem, branches, leaves, bladders, and receptacles which are formed of leaves. Bladders marginal, stalked, simple, terminated by a leaf. Fruit diœcious. Conceptacles immersed in small marginal leaves, containing obovoid spores, or antheridia.

A gigantic weed of the Southern Ocean.

1. **P. comosa,** *Agardh;—Fl. N. Z.* ii. 214. Frond solitary, 10–30 ft. long. Stem and branches $\frac{1}{4}$–$\frac{1}{2}$ in. broad, compressed or flat, 2-edged. Leaves variable, 4–8 in. long, lanceolate or linear, remotely serrate, intermixed with spinous processes.—Harvey, Phyc. Austr. iii. t. 153. *Macrocystis*, Agardh, Sp. Alg. i. 253. *Fucus*, Labill. Fl. Nov. Holl. t. 258; Turn. Hist. t. 142.

Common on the shores, *Banks and Solander*, etc. (Australia.)

6. SCABERIA, Greville.

Root discoid. Frond olive-brown, dendroid, irregularly branched, having distinct stem, leaves, bladders, and receptacles which are often undistinguishable from the leaves. Leaves spirally disposed, peltate, warted, fleshy. Bladders formed of an inflated leaf. Conceptacles immersed in the transformed leaves, containing obovoid spores and tufted antheridia.

The following is the only species.

1. **S. Agardhii,** *Grev.* Frond several feet long, irregularly or alternately branched. Stem below filiform, flexuous, above covered with crowded leaves and bladders. Leaves $\frac{1}{6}$–$\frac{1}{3}$ in. diameter, fleshy, with short excentric petioles, vertically compressed, warted above, smooth below. Bladders large, globose, warted.—J. Ag. Sp. Alg. i. 352; Harv. Phyc. Austr. iii. t. 164. *Castraltia Salicorrioides*, A. Rich. Fl. N. Zeal. ii. 143.

Below low-water mark, common, *Lesson*, etc. Omitted in the 'Flora Novæ-Zelandiæ.' (Australia, Tasmania.)

7. CYSTOPHORA, J. Agardh.

Root scutate. Frond olive-brown or -green, dendroid, with a distinct stem, branches, and branchlets, the latter more or less leaf-like, or passing into leaves. Bladders (if present) simple. Receptacles pod-like, moniliform or torulose, forming part of the branchlets. Conceptaclès containing pyriform spores, and branched filaments bearing antheridia.

A large tropical and southern genus.

§ 1. *Branches from the margin of the stem, bent suddenly down at their insertion.*

1. **C. monilifera,** *J. Agardh.* Stem flat, several feet long, decompound, pinnate, $\frac{1}{8}$–$\frac{1}{4}$ in. broad; pinnæ inserted on the flat side of the stem, bent down at their insertion with very short truncate alternate branches at the base; pinnules dichotomously pinnate, the ultimate transformed into moniliform filiform apiculate receptacles, $\frac{1}{2}$–1 in. long. Bladders stalked, globose, size of a pea, placed towards the bases of the primary pinnæ.—J. Ag. Sp. Alg. i. 241; Harv. Phyc. Austr. v. 245. *Blossevillea*, Decaisne. *Cystoseira retroflexa*, A. Richard. *Fucus*, Turn. Hist. Fuc. t. 155, not of Labillard.

Shores of New Zealand, *D'Urville*. (Australia, Tasmania.)

2. **C. retroflexa,** *J. Agardh, Sp. Alg.* i. 242;—*Fl. N. Z.* ii. 214. Very similar to the preceding, but more robust. Bladders obovoid. Receptacles ensiform, subtorulose—*Fucus*, Labill. Fl. Nov. Holl. t. 260. *Blossevillea caudata*, nob. in Lond. Journ. Bot. vi. t. 414. *B. retroflexa*, Kuetzing.

Abundant on the shores as far south as **Lord Auckland's** group. (Australia, Tasmania.)

3. **C. retorta,** *Agardh*;—*Fl. N. Z.* ii. 214. Lower part of frond as in *C. retroflexa*, upper pinnæ with a nearly terete rachis, subdichotomously branched; axils rounded. Receptacles 2–3 in. long, terete, usually incurved, obscurely torulose.—J. Ag. Sp. Alg. i. 243. *Blossevillea*, Mont.

Shores of New Zealand, *Raoul*. **Lord Auckland's** group, *Hombron*. (Australia, *Ag.*)

4. **C. torulosa,** *Agardh*;—*Fl. N. Z.* ii. 214. Stem compressed, 1 foot long, nearly simple, alternately pinnate; pinnæ bent down, pinnules subfascicled, passing into clavate torulose obtuse receptacles. Bladders ellipticspherical.—J. Ag. Sp. Alg. 243; Harv. Phyc. Austr. t. 123. *Blossevillea*, Decaisne. *Fucus*, Turn. Hist. t. 157.

Shores of New Zealand, *D'Urville*. Banks's Peninsula, *Lyall*. (Australia, Tasmania.)

§ 2. *Branches distichously pinnate, not bent down.*

5. **C. Platylobium,** *J. Agardh, Sp. Alg.* i. 245.—*C. Lyallii*, Harv. in Fl. N. Z. ii. 214. t. 108. Stem 2–3 feet long, very stout, compressed, grooved, flexuous, 2–3-pinnate; pinnæ from the margins of the frond, naked below, with alternate tubercles; pinnules flat, 2–3 in. long, alternately toothed, barren with obtuse simple teeth, fertile with the teeth prolonged into stalked receptacles which are 1 in. long, ensiform, compressed, elliptic-lanceolate, long acuminate, torulose, with 2 rows of submarginal cavities. Bladders nearly globose, stalked, as large as hazel-nuts.

Foveaux Straits, *Lyall*. Otago, *Lindsay*. (Tasmania.)

8. LANDSBURGIA, Harvey.

Root scutate. Frond olive-brown, with a distinct stem and leaves, but no bladders; stem filiform, alternately branched. Leaves linear, flat, with a stout midrib, pinnatifid to the middle. Receptacles formed of small oblong toothed leaves near the tips of the branches, containing densely crowded conceptacles on their thickened surfaces; conceptacles with stalked spores and antheridia, mixed with simple filaments.

The following is the only species.

1. **L. quercifolia,** *Fl. N. Z.* ii. 213. *t.* 107. Root solid, 1–2 in. diameter. Stem 3–4 ft. long, ¼ in. diameter at the base; branches filiform, naked and tubercled below. Leaves distichous, alternate, 2–4 in. long, petioled, ¼–½ in. broad, pinnatifid to the middle; lobes sometimes toothed, between membranous and coriaceous, midrib produced beyond the middle. Fruiting leaves ¼–½ in. long, serrate.—*Phyllospora quercifolia*, nob. in Lond. Journ. Bot. 4. 525 (*not Fucus quercifolius*, Turner.)

Bay of Islands and probably elsewhere, abundant, *D'Urville*, etc.

9. FUCODIUM, J. Agardh.

Frond olive-brown or green, often black when dry, dichotomously or subpinnately branched, cylindric compressed or flat, ribless, leafless, with distinct receptacles.. Bladders 0 in the N. Z. species (in the middle of the frond in others). Receptacles terminal or lateral. Conceptacles containing ellipsoid spores surrounded by simple-jointed filaments or fascicled or racemose antheridia, or both.

A small genus, native of both temperate zones. The N. Z. species belong to a peculiar southern section without bladders.

1. **F. gladiatus,** *J. Agardh, Sp. Alg.* i. 202. Frond 3–4 in. long, as thick as a swan's quill, terete below, compressed above, dichotomously decompound; segments spreading, truncate or 2-lobed. Receptacles 6 in. long and more, $\frac{1}{12}$–$\frac{1}{6}$ in. broad, dichotomously branched; ultimate segments 4 in. long, ensiform.—*Fucus*, Labill. Fl. N. Holl. t. 256; Turn. Hist. t. 240. *Hermanthalia*, Kuetzing. *Xiphophora Billardieri*, Mont. in Voy. au Pôle Sud, t. 7. f. 1; Fl. N. Z. ii. 215; Fl. Antarct. t. 69. f. 3.

Abundant on rocky shores as far south as **Lord Auckland's** group. (Tasmania, Australia.)

2. **F. chondrophyllus,** *J. Agardh, Sp. Alg.* i. 203.—*Xiphophora*, Mont.; Fl. N. Z. ii. 215. Frond 12 in. long; segments linear, erecto-patent, lower ¼–⅓ in. broad, upper $\frac{1}{12}$ in. Receptacles terminal, their ultimate segments very short, ⅙–⅛ in. long.—*Fucus*, Brown in Turn. Hist. t. 222.

New Zealand, *D'Urville*; Banks's Peninsula, *Lyall*; Otago, *Lindsay*. (Australia, Tasmania.)

10. HORMOSIRA, Endlicher.

Root discoid. Frond olive-brown, without distinct organs, dichotomously branched, moniliform; internodes inflated, fertile. Fruit diœcious. Conceptacles sunk in the periphery of the internodes, containing subsessile narrow-pyriform spores and unbranched peranemata.

A small genus, confined to Australia and New Zealand.

1. **H. Billardieri,** *Mont.;—Fl. N. Z.* ii. 215. Frond 6–18 in. long, very variable in size and robustness; internodes obconical, wingless.—J. Ag. Sp. Alg. i. 199. *Moniliformia*, Bory. *Monilia*, A. Rich.

Var. *α. Banksii;* Harv. Ner. Aust. iii. 135. f. 2. Frond divaricate, with spreading branches and wide angles; internodes cuneate, depressed at the summit, the terminal ovoid.

Var. *β. Labillardieri*, Harv. l. c. p. 3. Frond elongate, 2–3-chotomous, with narrow angles; internodes ovoid, rounded at both ends, terminal, sometimes cylindric.—*Hormosira Billardieri*, Mont.; Fl. N. Z. ii. 215.

Var. *γ. Sieberi*, Harvey, l. c. p. 4. Frond short, fastigiate, dichotomous; internodes obconic, truncate at the top, base truncate or tapering, terminal ovoid.—*Hormosira Sieberi*, Decaisne;—Fl. N. Z. l. c. *H. obconica*, Kuetzing.

Common along all the coasts on tidal rocks, between half and high-water marks, *Banks and Solander*, etc. (Australia, Tasmania.)

11. SPLACHNIDIUM, Greville.

Root a disk. Frond olive-green, cylindric, proliferously branched; branches saccate, full of mucilage and branched filaments; walls thin, membranous. Fruit diœcious. Conceptacles scattered over the whole frond, attached to the inner surface of its walls; spores linear-oblong, subsessile; peranemata simple.

The following is the only species.

1. **S. rugosum,** *Grev.;—Fl. N. Z.* ii. 215. Fronds 4–8 in. high; main axis stout, cylindric or clubshaped, ¾ in. diameter; branches sac-like, truncate, 1–2 in. long, surface covered with mamillæ, each furnished with a pore that opens into the spore-cavity beneath.—J. Ag. Sp. Alg. i. 186; Harv. Ner. Austr. i. t. 14.

Common on rocks near low-water mark from Auckland to Banks's Peninsula, *Lesson*, etc. (Australia, S. Africa.)

12. NOTHEIA, Bailly and Harvey.

Frond olive-green, parasitic, filiform, irregularly branched, proliferous, solid, with distinct stem and branches, but no bladders or leaves. Conceptacles scattered over the whole frond under the surface, containing linear-obovate spores and simple peranemata.

A most curious plant, abounding on the *Hormosira*, and to all outward appearances a part of that plant.

1. **N. anomala,** *Bailly and Harvey;—Fl. N. Z.* ii. 216. *t.* 109 A. Fronds solitary, 3–8 in. long, growing from the conceptacles of *Hormosira Banksii*, excessively branched, bushy, cylindric; branchlets narrow, spindle-

shaped, axis of solid interwoven filaments; periphery of radiating coloured filaments.

Probably common, first found by Wilkes's Expedition; Parimahu, *Colenso;* Bank's Peninsula, *Lyall.* (Tasmania, Australia.)

13. D'URVILLÆA, Bory.

Root scutate. Frond stalked, dark olive-brown or black, flat, expanded, very thick and coriaceous or honeycombed transversely internally, palmate or pinnate, without distinct organs. Fruit diœcious. Conceptacles scattered over the whole frond in the cortical stratum, containing either obovoid subsessile spores or branched filaments bearing obovoid antheridia.

A genus of a few southern huge black *Algæ*, one often forming a load for a man, which are thrown up after every storm on the rocky coasts of the Southern Ocean.

1. **D. utilis,** *Bory in Voy. Coquil. t.* i. *and* ii. *f.* 1;—*Fl. N. Z.* ii. 216. Frond dark brown or black, often 30 ft. long, forming an immense flabellate palmately lobed laciniated lamina, contracted at the cuneiform base into a short stipes as thick as the wrist; segments or thongs often 1 in. thick, honeycombed internally.—J. Ag. Sp. Alg. i. 188. *Fucus Antarcticus,* Chamisso.

Common on the coasts, especially southwards as far as **Campbell's** Island, and floating in the ocean as far as lat. 62° S. (Chili, Fuegia, Falkland's Islands, Kerguelen's Land.)

14. SPOROCHNUS, Agardh.

Frond dull olive-brown or yellow-green, filiform, pinnately decompound, solid, cellular; axis more dense. Receptacles on long slender stalks, narrow cylindric, crowned with a soft long pencil of filaments, surface covered with branching sporiferous filaments; spores obovoid, attached to the sides of the filaments.

A genus found in both temperate zones.

1. **S. stylosus,** *Harv. in Fl. N. Z.* ii. 216. *t.* 109 B. Frond pale olive-green, 6–8 in. long. Stem filiform, laterally branched; branches scattered or fascicled, simple, elongate. Receptacles subsessile, cylindric or elliptic-oblong, crowned by a long style-like process ½ to ⅔ its own length.

Foveaux Straits and Otago Harbour, *Lyall.*

15. CARPOMITRA, Kuetzing.

Frond pale olive-green, linear, filiform, compressed or flat and midribbed, irregularly branched, cellular; axis dense. Receptacles ovoid conical or mitriform, terminating the branches, composed of branching filaments whorled round a vertical axis and developing elliptic-oblong spores.

A European as well as southern genus.

1. **C. Cabreræ,** *Kuetz.;—Fl. N. Z.* ii. 217 Root of matted fibres. Frond 6–8 in. long, much branched from the very base; branches 2-edged,

terete below, compressed or flat above, much pinnately divided. Receptacles short, conic, sessile on truncate branchlets.—J. Ag. Sp. Alg. i. 177; Harv. Phyc. Brit. t. 14. *Fucus*, Turner, Hist. t. 140.

Probably common, Cook's Straits, *Lyall;* Hawke's Bay, *Colenso.* (Tasmania, Europe.)

2. **C. Halyseris,** *Hook. f. and Harv. Fl. N. Z.* ii. 216. *t.* 110 A. Frond much broader than in *C. Cabreræ*, $\frac{1}{4}$–$\frac{1}{3}$ in. broad, linear, flat, with a distinct midrib.—J. Ag. Sp. Alg. i. 179.

Bay of Islands, *Cunningham*, etc. Probably only a broad form of *H. Cabreræ.*

16. DESMARESTIA, Lamouroux.

Root a disk. Frond dull olive-green, linear or filiform, flat or compressed, distichously branched; structure cellular, surrounding in one species a single slender-jointed tube, which traverses the axis; branches when young producing marginal tufts of very slender flaccid branching filaments. Fructification unknown.

A genus of several species, inhabiting both temperate zones.

1. **D. ligulata,** *Lamouroux;—Fl. N. Z.* ii. 217. Frond very variable, a foot or more long, flat, membranous, costate, pinnate, with a central articulate tube; pinnæ oblong or linear-lanceolate, decompound, ultimate serrate or ciliate.—J. Ag. Sp. Alg. i. 169; Harv. Phyc. Brit. t. 115. *Fucus*, Turn. Hist. t. 98.

East coast, *Colenso;* Akaroa, *Lyall.* (Australia, Chili, Fuegia, S. Africa, N. Atlantic.)

2. **D. viridis,** *Lamouroux;—Fl. Antarct.* i. 178. Frond a foot long and upwards, slightly compressed, pinnately decompound, solid, without a central articulate tube, ecostate, ultimate branches capillary, passing into pencils of jointed filaments.—J. Ag. Sp. Alg. i.; Harv. Phyc. Brit. *Fucus*, Turn. Hist. t. 97. *Dichloria*, Greville.

Lord Auckland's group, *J. D. H.* (Kerguelen's Land, N. and S. Pacific, N. Atlantic.)

17. MACROCYSTIS, Agardh.

Root branching, giving off immensely long slender simple stems, which bear leaves at the surface of the water. Leaves formed by the continued splitting of a primary terminal leaf, developed in secund order along the lengthening floating stem, each lanceolate, serrate, ribless, undulate, with a pyriform oblong or subcylindric bladder at its base. Spores superficial on submerged radical leaves, forming clouded sori, ellipsoid, with a hyaline coat, surrounded by densely packed inarticulate clavate peranemata.

A most wonderful well-known southern *Alga*, forming a breakwater of matted fronds in deep water. Only one species is known, the following, which includes all the species of J. Agardh, Sp. Alg. i. 155–158, except *M. obtusa* (which is *Phyllospora Menziesii*).

1. **M. pyrifera,** *Agardh;—Fl. N. Z.* ii. 217. Stems 50 to perhaps 700 ft. long or upwards. Fronds extremely variable in length and breadth, 2–4 ft. long, 2–6 in. broad, ciliate-serrate.—*Fucus*, Turn. Hist. t. 110.

Var. *Dubenii*, Harv. Phyc. Austr. iv. t. 202. Bladders subcylindric, 6–8 in. long.—*M. Dubenii*, Areschoug.

Abundant on all the rocky coasts. (Throughout the Southern Ocean and along the west coasts of N. and S. America, also found in floating masses to lat. 65° S.)

18. **LESSONIA,** Bory.

Root scutate or branched. Frond with a more or less distinct stem, or sometimes a tree-like trunk, dichotomously branched and bearing leaves, but no bladders. Leaves geminate, formed from the fission of one, between coriaceous and membranous, ovate ensiform or lanceolate, ribless, entire or somewhat toothed. Sori superficial in the middle of the leaf, of narrow ellipsoid spores mixed with clavate inarticulate filaments.

A genus of often huge erect subdendroid *Algæ*, abounding in the Southern Ocean.

1. **L. fuscescens,** *Bory in Voy. Coquil. t. 2. f. 2. et t. 3;—Fl. N. Z.* ii. 217. Gregarious, forming submarine miniature forests; trunks sometimes 10 ft. long, cylindric, as thick as the thigh, bearing towards the top short branches with pendulous foliage. Leaves 2–3 ft. long, 1–2 in. broad, linear-lanceolate, toothed, older sinuate.—J. Ag. Sp. Alg. i. 151; Fl. Antarct. t. 167–8 A and 171 D.

East coast, *Colenso;* Cook's Straits, *Lyall.*—Probably common in all rocky bays. (Chili, Fuegia, Falkland Islands, Kerguelen's Land.)

19. **ECKLONIA,** Hornemann.

Root scutate or dividing into short fibres. Frond olive-green, pinnatifid, ecostate; segments produced from the magnified teeth of a simple lamina, which is contracted to a solid or inflated stem at the base. Sori superficial on the lower part of the pinnæ, of narrow ellipsoid spores, mixed with clavate inarticulate filaments.

Often gigantic weeds, natives of the Southern Ocean, of which one species inhabits also the Canary Islands, and the other, with a long trumpet-like stem, inhabits the Cape of Good Hope.

1. **E. radiata,** *J. Agardh;—Fl. N. Z.* ii. 217. Frond 1–2 ft. long; stem solid or sparingly inflated.

Var. α. Frond palmately pinnatilobate, smooth; lobes subcuneate, spinulose, toothed.—J. Ag. Sp. Alg. i. 146. *Fucus*, Turn. Hist. t. 134. *Capea radiata*, Endl.

Var. β. *exasperata. E. exasperata*, J. Ag. l. c. 146; Fl. N. Z.—Surface of the frond spinulose.—*Capea biruncinata*, Montagne in Phytogr. Canariens. Crypt. t. 7; Fl. Antarct. 160; *Laminaria*, Bory in Voy. Coquil. t. 10.

Var. γ. *Richardiana. E. Richardiana*, J. Agardh, l. c. 147; Fl. N. Z. l. c.—Frond elongate, pinnately decompound, smooth; lobes minutely toothed.—*Capea*, Kuetzing.

Var. δ. *flabelliformis. E. flabelliformis*, J. Ag. l. c. 147; Fl. N. Z. l. c.—Stem somewhat inflated, 2 feet long; frond pinnately decompound, surface rugose, plaited.—*Capea*, nob. in Lond. Journ. Bot. 4, 528; *Laminaria*, A. Richard, Fl. Nov. Zel. t. 1, 2.

Common along the shores of the **Northern** and **Middle** Islands, *Banks and Solander*. etc. (Australia, Tasmania, Chili, Canary Islands.)

20. ZONARIA, Agardh.

Root woolly. Frond dull yellow or olive-green, flat, ribless, coriaceo-membranous, flabelliform, entire or vertically multifid, concentrically striate, surface cells in radiating lines. Spores pyriform, superficial, scattered or collected into sori, mixed with simple-jointed filaments.

A genus of both tropical and temperate zones.

1. **Z. Sinclairii,** *Hook. f. and Harv. Fl. N. Z.* ii. 218. Fronds tufted, erect, 2–4 in. high, much branched; stem and branches filiform, clothed with rusty tomentum, ending in flat cuneate laminæ 1–1½ in. long, with pale edges, obsolete zones, and fine radiating lines; sides of the laminæ entire or cut.—J. Ag. Sp. Alg. i. 111; Harv. Phyc. Austr. i. t. 49. *Stypopodium*, Kuetzing.

New Zealand, *Sinclair.* (N. S. Wales.)

2. **Z. interrupta,** *Agardh;—Fl. N. Z.* ii. 218. Frond tufted, erect, 4–6 in. high, fastigiately branched, flabellate; stem terete or winged, woolly; segments long, narrow, obtuse, irregularly toothed or pinnatifid; tips cuneate, radiatingly striate.—J. Ag. Sp. Alg. i. 111; Harv. Phyc. Austr. iv. 190. *Z. flava*, nob. in Lond. Journ. Bot. 4. 529 (excl. syn.). *Fucus*, Turn. Hist. t. 245. *Phycopteris*, Kuetzing.

Common on the shores of the **Northern** and **Middle** Islands, *Colenso*, etc. (Australia, Tasmania, S. Africa, Madagascar.)

3. **Z. velutina,** *Harv. in Fl. N. Z.* ii. 218. Fronds densely tufted, 1–2 in. high, sparingly branched; stem densely woolly, widening into a broad flabellate entire or divided lamina, upper surface quite smooth, under covered except round the margin by a dense velvety black-brown mat of fibres. Fructification unknown.

East coast, *Colenso;* Milford Haven and Port Cooper, *Lyall.*

21. DICTYOTA, Lamouroux.

Root covered with woolly fibres. Frond linear, pale olive-green or brown, flat, ribless, membranous, cellular, dichotomous or pinnatifid, surface cells parallel. Spores superficial, collected in minute sori, or scattered singly over both surfaces of the frond.

Natives of both temperate zones.

1. **D. Kunthii,** *Agardh;—Fl. N. Z.* ii. 219. Frond 1 foot high, distantly dichotomously decompound, narrowed at the nearly glabrous base, sinus acute; segments linear-elongate, quite entire, $\frac{1}{4}$–$\frac{1}{3}$ in. broad, rounded at the apex. Spores scattered in small superficial spathulate leaflets.—J. Ag. Sp. Alg. i. 94. *Zonaria*, Agardh, Icon. Ined. t. 15.

Abundant on the coasts, *J. D. H.*, etc. (Chili.)

2. **D. dichotoma,** *Lam.;—Fl. N. Z.* ii. 219. Frond 6–10 in. long,

membranous, dichotomously decompound, narrowed at the somewhat woolly base; segments linear, quite entire, sinus rather acute, segments quite entire; apex obtuse or 2-fid. Sori or spores scattered over the whole surface of the frond except the margin.—J. Ag. Sp. Alg. i. 92. *Zonaria*, Harv. Phyc. Brit. t. 103. *Dichophyllum*, Kuetzing.

Hawke's Bay, *Colenso*; Queen Charlotte's Sound, *Lyall*. (Australia, Tasmania, S. Africa, Red Sea, Gulf of Mexico, North Atlantic.)

22. DICTYOSIPHON, Greville.

Frond olive-green or brown, tubular, branched, of 2–3 series of cells; inner cells elongate, cylindric; superficial minute, coloured. Spores scattered over the whole surface of the frond.

A genus of only two species, the following and a European one.

1. **D.? fasciculatus,** *Hook. f. and Harv. Fl. Antarct.* i. 178. *t.* 69. *f.* 1. Frond 4–8 in. long, hardly coriaceous, lanceolate in outline; stem filiform, clothed with numerous short quadrifarious cylindric divided fascicled attenuate or secund branches, 1–2 in. long; branchlets setaceous, erect, all acute, narrowed at both ends. Spores scattered over the whole frond, semi-immersed.—J. Ag. Sp. Alg. i.

Lord Auckland's group, *J. D. H.* (Kerguelen's Land, Falkland Island.)

23. ASPEROCOCCUS, Lamouroux.

Root scutate. Frond pale olive-green, tubular, cylindric or saccate, rarely compressed, continuous, membranous, its walls thin. Spores scattered over the whole frond, in minute sori, roundish, mixed with clavate filaments.

A northern as well as southern temperate genus.—*Encœlium*, Agardh.

1. **A. sinuosus,** *Bory;—Fl. N. Z.* ii. 219. Frond forming hemispherical lobed expanded sinuous masses, from the size of a hazel-nut to that of the human head, finally cavernous and often laciniose.—J. Ag. Sp. Alg. i. 75.

East coast, Capes Kidnapper and Turnagain, *Colenso*; Otago, *Lyall*. (Australia, Tasmania, Falkland Islands, N. and S. Atlantic, Indian Ocean, and Red Sea.)

2. **A. echinatus,** *Greville;—Fl. Antarct.* i. 180. Fronds tufted, 2 in. and upwards long, very variable in diameter, cylindric, clavate, much attenuated downwards. Sori oblong.—J. Ag. Sp. Alg. i. 76.

Lord Auckland's group, *J. D. H.* (Australia, Europe.)

24. CHORDA, Stackhouse.

Root scutate. Frond olive or yellow-brown or greenish, very long, simple, cylindric, tubular, its walls rather thick and fleshy, divided internally transversely by membranous septa. Sori covering the whole surface, consisting of stalked obconic spores and ellipsoid autheridia.

A northern genus, rare in the Southern Ocean.

1. **C. lomentaria,** *Lyngbye;—Fl. N. Z.* ii. 218. Frond 1 foot and more long, membranous, almost cylindric, often as thick as the little finger, constricted here and there.—*Scytosiphon,* J. Ag. Sp. Alg. i. 126.

East coast, *Colenso;* Waitemata harbour, *Lyall;* **Lord Auckland's** group, *J. D. H.* (Tasmania, Falkland Islands, N. Atlantic and Pacific Oceans.)

25. ADENOCASTIS, Hook. f. and Harv.

Root a small disk or shield. Frond a dull green or olive-brown, membranous pyriform sac, on a slender short stalk, hollow or full of water, coated with a thin layer of vertical clavate articulate filaments. Spores pedicelled, pyriform, attached to the base of the filaments and scattered over the whole frond.

The following is the only species.

1. **A. Lessonii,** *Hook. f. and Harv. Fl. N. Z.* ii. 218. Frond 1–3 in. high, coriaceous or of a firm membranous texture; colour dark brown when dry and not adhering to paper.—J. Ag. Sp. Alg. i. 124; Fl. Antarct. t. 69. f. 2; Harv. Ner. Austr. i. t. 48. *Asperococcus,* Bory in Voy. Coquil. t. 11. f. 2.

Common on tidal rocks from the Bay of Islands to **Lord Auckland's** group and **Campbell's** Island, *J. D. H.* (Fuegia, Kerguelen's Land, Tasmania.)

26. SCYTOTHAMNUS, Hook. f. and Harv.

Frond blackish-olive, cylindric, branched, fibrous, nearly solid; axis of densely interwoven filaments, which radiate outwards and form a dense cortical substance of minute coloured cells immersed in gelatine. Spores oblong, immersed amongst the cortical cells.

The only species is the following.

1. **S. australis,** *Hook. f. and Harv. Fl. N. Z.* ii. 219. Frond 2–12 in. long, very variable, simple or excessively branched. Stem simple, at the base as thick as a crowquill, terete; branches more slender, acute, firm, rather rigid, ultimate acuminate.—J. Ag. Sp. Alg. i. 64.

Common on tidal rocks, *J. D. H.*, etc.

27. CHORDARIA, Agardh.

Root discoid. Frond olive-green, filiform, much branched, cartilaginous; axis of densely-packed longitudinal interlaced cylindrical filaments; periphery of simple clavate horizontal whorled filaments, and long byssoid gelatinous fibres. Spores obovoid, scattered amongst the filaments of the periphery.

Medium-sized *Algæ,* natives of both temperate zones.

1. **C. sordida,** *Bory;—Fl. N. Z.* ii. 219. Frond 1–2 ft. long, flaccid; stem short, soon divided, or long and laterally branched; branches long, simple or with horizontal branchlets; young frond densely clothed with

slender articulate filaments, which are thrown off as the peripheric cells are developed. Spores small, ellipsoid.—*Myriocladia Capensis*, J. Ag. Sp. Alg. i. 54.

On rocks, probably common. Cook's Straits, *Lyall;* East coast, *Colenso;* Otago, *Lindsay.* (E. and W., tropical, and S. Africa, S. America.)

2. **C. flagelliformis,** *Agardh;—Fl. Antarct.* i. 180. Frond 1–2 ft. long, as thick as a crowquill, filiform, equal throughout, excessively branched; branches dense, elongate, simple or forked, scarcely attenuate, often a foot long; filaments of the periphery clavate.—J. Ag. Sp. Alg. i. 66; Harv. Phyc. Brit. t. iii.; Turn. Hist. Fuc. t. 85.

Campbell's Island, *J. D. H.* (Chili, S. Africa, N. Atlantic and N. Pacific Oceans.)

28. MESOGLOIA, Agardh.

Frond olive-green or brown, filiform, much branched, gelatinous; axis of longitudinal subsimple interlaced fibres, immersed in gelatine; periphery of radiating dichotomous coloured filaments. Spores ovate or elliptic, olivaceous, attached to the branches of the peripheric filaments.

Medium-sized flaccid *Algæ*, natives of both temperate zones.

1. **M. intestinalis,** *Harvey in Fl. N. Z.* ii. 220. Frond 1–2 ft. long, or more, soft, clothed with villous filaments, simple or sparingly divided, $\frac{1}{12}$ in. at the base, gradually widening to $\frac{1}{2}$ in., then elongate and cylindric; branches few, irregularly inserted, obtuse; axial filaments loose, branched, peripheric fascicled, cylindric, not moniliform, widening upwards, outer joints as long as broad. Spores oblong, 2 or more in each fascicle.

Cook's Straits, Auckland, and Otago, *Lyall, Lindsay.*

29. LEATHESIA, Gray.

Frond olive-green, solid, forming globose or lobed masses, between fleshy and cartilaginous in texture, formed of jointed colourless dichotomous filaments radiating from the points of attachment, the tips closely packed and coloured forming the surface. Spores pyriform or ovoid, attached to the coloured tips of the filaments.

A small genus of annual *Algæ*, natives of both temperate zones.—*Corynophlæa*, Kuetzing.

1. **L. Berkeleyi,** *Harvey, Phyc. Brit. t.* 176;—*Fl. N. Z.* ii. 220. Frond dark brown, gregarious, convex, depressed, solid, 1–2 in. diameter, soft and fleshy; filaments very densely packed. Spores pyriform.—J. Ag. Sp. Alg. i. 51.

Cape Kidnapper, *Colenso.* (Europe.)

30. SPHACELARIA, Lyngbye.

Frond olive-green, rigid, filiform, excessively and often fastigiately branched, jointed throughout; branches distichous, pinnate, rarely dichotomous; joints of numerous longitudinal cells; tips of the branches distended, membranous, containing a dark granular mass. Spores ovoid or globose, borne on the branchlets.

Small perennial *Algæ*, natives of all seas.—*Stypocaulon* and *Halopteris*, Kuetzing.

§ 1. *Frond pinnate, woolly at the base.*

1. **S. paniculata,** *Suhr;—Fl. N. Z.* ii. 221. Frond variable in size, aspect, and ramification, 3–6 in. long, pinnately-decompound, woolly, caulescent, pinnæ fascicled; fascicles more or less corymbose. Spores in terminal spikelets, or clustered; dense tufts of grumous cells sometimes occur in the axils of the uppermost branchlets.—J. Ag. Sp. Alg. i. 46. *S. hordeacea,* Harv. in Hook. Ic. Pl. t. 614. *S. virgata,* nob. in Lond. Journ. Bot. iv. 530.

Abundant on all the coasts. (Australia, Tasmania, S. Africa.)

2. **S. funicularis,** *Mont. in Voy. au Pôle Sud, t.* 14. *f.* 1;—*Fl. N. Z.* ii. 221. Shorter and stouter than *S. paniculata.*—J. Ag. Sp. Alg. i. 38; Fl. Antarct. i. 180.

Akaroa, *Hombron, Lyall;* East coast, *Colenso.* **Lord Auckland's** group, *J. D. H.*

§ 2. *Frond dichotomous, not woolly.*

3. **S. botryoclada,** *Hook. f. and Harv. Fl. N. Z.* ii. 221. *t.* 110 B. Frond slender, several inches long, filiform, sparingly branched, inarticulate, bearing dense spherical tufts of fertile filaments or branchlets throughout its whole length and on all sides; branchlets of the tufts incurved, dichotomous, bearing lateral sessile spores.

East coast and Cook's Straits, *Lyall.*—Harvey suspects that this curious plant, of which we have imperfect specimens only, may prove a state of *S. paniculata.*

§ 3. *Frond parasitic, very minute.*

4. **S. pulvinata,** *Hook. f. and Harv. Fl. N. Z.* ii. 221. *t.* 110 C. Fronds forming dense hemispherical or globular brown or olive-green patches $\frac{1}{12}$–$\frac{1}{8}$ in. diameter; filaments rigid, arcuate, simple or sparingly branched, brown at the tips; branchlets erect, often secund; articulations of about 3 cells. Spores secund, rarely opposite, pedicelled, ellipsoid-oblong.

Parasitic on *Carpophyllum Maschalocarpus, Colenso.*

31. ECTOCARPUS, Lyngbye.

Frond flaccid, olive or brown, capillary, jointed; joints of a single cell, not striated. Spores spherical or elliptic, external or imbedded. Lanceolate linear or conical bodies (silicules) also occur, and granular masses formed in connective cells of the branches.

A very large genus in the northern hemisphere, of small annual *Algæ*.

1. **E. granulosus,** *Agardh;—Fl. N. Z.* ii. 222. Fronds olive-green or yellowish, 2–8 in. high, rigid, decompound; branches and branchlets opposite, ultimate usually secund pectinate; articulations $1\frac{1}{2}$ as long as broad. Spores obovoid, secund on the upper branches.—J. Ag. Sp. Alg. i. 21; Harv. Phyc. Brit. t. 200. *E. ochraceus,* Kuetzing.

East coast, *Colenso,* parasitic. (N. and S. Atlantic Oceans.)

2. **E. siliculosus,** *Lyngbye;—Fl. N. Z.* ii. 222. Fronds $\frac{1}{12}$ in. long, olive-yellow, soft, decompound; branches and branchlets alternate, ultimate elongate, distant, often secund. Silicules stalked, subulate, or conical and very acute, subsecund.—J. Ag. Sp. Alg. i. 22; Harv. Phyc. Brit. t. 162.

Common on all the coasts, *J. D. H.*, etc. (Tasmania, N. and S. Atlantic Oceans.)

3. **E. confervoides,** *Harv. in Fl. N. Z.* ii. 222. Fronds tufted, 2–4 in. high, rather coarse, dense, not adhering to paper; stems rigid, distantly branched; branchlets few, simple, scattered, erect or patent, axils all acute. Silicules elongate, terminal, or imbedded in the branches.

Cook's Straits and Otago, *Lyall.*

4. **E.? pusillus,** *Griffiths;—Fl. N. Z.* ii. 222. Fronds parasitic, tufted, pale brown, capillary, matted; stems slender, long, simple or slightly branched; branches distant, naked, horizontal; articulations twice as long as broad in European specimens, much shorter in N. Z. ones. Spores (not seen in N. Z. plants) abundant, sessile, oblong.—Harv. Phyc. Brit. t. 153.

Parasitic on Corallines. Hawke's Bay, *Colenso.*—Doubtful, no fruit having been found. (Britain.)

SUBORDER II. RHODOSPERMEÆ.

32. LENORMANDIA, Sonder.

Frond dull red, leaf-like, proliferous. Leaf-like branches flat, membranous, simple; midrib evident, surface striated obliquely; substance honeycombed with rhomboidal cavities; surface-cells minute. Conceptacles scattered, stalked, enclosing pyriform spores. Stichidia lanceolate, containing 3-partite tetraspores.

A small Australian and New Zealand genus.

1. **L. Chauvinii,** *Harv. in Fl. N. Z.* ii. 222. Leaves $1\frac{1}{2}$–3 in. long; $\frac{1}{12}$–$\frac{1}{2}$ in. wide, dark purple when fresh, thickish, rigid, and semi-opaque when dry, shortly stalked, broad, linear-oblong, obtuse, quite entire; nerve slender, evanescent, proliferous. Stichidia thickly fringing the margin or on the nerve, lanceolate, simple or multifid.

East coast, *Colenso;* Otago, *Lyall.*—Harvey says of this, that he first received it from New Holland, but I do not find it in his enumeration appended to the 'Phycologia Australis.'

33. EPINEURON, Harvey.

Frond dark red, flat, linear, toothed or ciliated, membranous or horny; midrib evidently distichously branched, transversely striate; interior of large angular colourless cells; cortical of many series of minute coloured cells. Stichidia lanceolate, incurved or circinate, placed on the costa, containing 2 rows of tetraspores.

A small genus, of medium-sized southern *Algæ.*

1. **E. Colensoi,** *Hook. f. and Harv. Fl. N. Z.* ii. 223. Frond red-brown, very narrow, $\frac{1}{12}$ in. broad, rigid, alternately branched above; branches

3–4 in. long, simple or 1–2-pinnate, linear, deeply toothed, transversely striate; tips involute when dry.—Harv. Ner. Austr. xxvi. t. 10. *Vidalia*, J. Ag. Sp. Alg. ii. 1127.

East coast, *Colenso*; Bay of Islands, *Lyall*.

2. **E.? lineatum,** *Hook. f. and Harv. Fl. N. Z.* ii. 223. Frond pink, membranous, forked, 4 in. and upwards long; stem cylindric at the base, winged above and passing into the flat frond; branches ½–1 in. long, linear, ultimate fringed with spreading teeth.—Harv. Ner. Austr. 27. *Fucus*, Turn. Hist. Fuc. t. 201. *? Amansia multifida*, J. Ag. Sp. Alg. ii. 1112.

New Zealand, *Banks and Solander*.—Described by Harvey from the figure in Turner's work; the plant has been found by no recent collector. J. Agardh refers it doubtfully to a tropical Atlantic plant.

34. AMANSIA, Lamouroux.

Frond dull red, flat, membranous, pinnatifid, transversely zoned and striate; midrib evident; structure of a single layer of oblong hexagonal cells of equal length, giving the striated appearance. Conceptacles ovate or globose, enclosing a tuft of pyriform spores. Stichidia simple or branched, marginal or superficial, containing a double row of tetraspores.

A small tropical and southern genus.

1. **A.? marchantioides,** *Harv. in Fl. N. Z.* ii. 223. Frond in flat flabellate patches, 1 in. and upwards across, rooting from the under-surface, lobed or pinnatifid; lobes overlapping, ascending, crenate and toothed, variable in size, membranous, nerveless, transversely zoned, and with radiating striæ; areolate with oblong cells.

Hawke's Bay and Cape Kidnapper, on tidal rocks, *Colenso*.—An immature plant of very doubtful genus.

35. ALSIDIUM, Agardh.

Frond red, filiform or compressed, cartilaginous, opaque, pinnately or irregularly decompound; axis articulated, of numerous tubular cells, coated with cortical layers of numerous small coloured irregular cells; branches alternate, subulate, acute, transversely striate. Conceptacles containing a tuft of pyriform spores, enclosed in a membrane. Stichidia axillary or terminal, containing 1 or 2 rows of 3-parted tetraspores.

1. **A. triangulare,** *J. Ag.;—Fl. N. Z.* ii. 223. A native of the West Indies; has, probably erroneously, been said to have been brought from New Zealand by Banks.

36. RYTIPHLŒA, Agardh.

Frond dull red, usually wrinkled when dry, filiform or compressed, pinnate, reticulate; axis articulate, of a circle of large oblong cells surrounding a central cell; walls of several series of minute angular cells, the outer coloured. Fructification diœcious. Conceptacles ovate, containing a tuft of pyriform spores. Stichidia lanceolate, containing 3-partite tetraspores (tetraspores sometimes scattered in the ultimate branchlets).

A genus of both temperate zones.—Scarcely different from *Rhodomela*, or from the opaque series of *Polysiphonia*, Harv. Ner. Austr. 31.

1. **R. delicatula,** *Hook. f. and Harv. Fl. N. Z.* ii. 224. *t.* 112 D. Frond tufted, setaceous, 2 in. high, distichously pinnate above, pellucidly striate; branches corymbose and fastigiate, again pinnate or 2-pinnate, margined with subulate spreading branchlets, which are crowded towards the tips; axis of 4 large tubes surrounding a central one, all coated with smaller cells. Conceptacles ovoid, sessile on the larger pinnæ.

Cook's Straits and Akaroa, *Lyall.*

R. pinastroides, J. Ag.—A European plant, has been erroneously stated to be a native of New Zealand. See Fl. N. Z. ii. 222.

37. RHODOMELA, Agardh.

Frond dull red, terete, dendroid, inarticulate, solid, coated with minute polygonal coloured cells; axis articulate, of many large colourless tubular cells. Conceptacles enclosing a tuft of pyriform spores. Stichidia formed of swollen branchlets, and enclosing one or several rows of tetraspores.

Natives of both temperate zones.

1. **R. Gaimardi,** *Ag.;—Fl. N. Z.* ii. 225. Frond cylindric, setaceous, dark coloured, 4–6 in. high, simple below, above divided into 3–4 principal branches, which are subdichotomous or irregular, repeatedly 2-fariously branched; secondary and tertiary long, subsimple, loosely set with short branchlets, which are $\frac{1}{6}$–$\frac{1}{4}$ in. long, slender, and often secund.—Fl. Antarct. 481. t. 184; Harv. Ner. Austr. 35. *R. Hookeriana,* J. Ag. Sp. Alg. ii. 880.

Hawke's Bay, *Colenso,* to Akaroa, *Lyall,* probably common. (Fuegia and the Falkland Islands.)

2. **R. cæspitosa,** *Harv. in Fl. N. Z.* ii. 225. Frond 2 in. high and upwards, densely tufted; stem naked below, setaceous, very slender, forked; branches alternately decompound above, subfastigiate, arcuate: branchlets scattered, secund, or alternate; ultimate linear, subulate, arcuate, acute, coated with small cells, opaque under the microscope. Axial tubes 4, their joints visible externally. Stichidia densely crowded on the sides of the upper branches, linear-subulate, simple or branched, with one series of tetraspores.

Rocks near low-water mark, Parimahu, etc., *Colenso.*—" Colour bright emerald-green," *Col.*

3. **R. concinna,** *Hook. f. and Harv. Fl. N. Z.* ii. 225. *t.* 111. Root a woolly disk. Frond 8–12 in. high, opaque, as thick as a sparrow's quill, ovate in outline, repeatedly distichously pinnate; main lower branches sub-horizontal, upper spreading, pinnately decompound; penultimate pinnæ $\frac{1}{2}$–$1\frac{1}{2}$ in. long, close-set with uniform alternate dichotomously multifid pinnules, $\frac{1}{12}$–$\frac{1}{6}$ in. long, which are inserted distichously, but themselves branch in all directions, the branchlets spinous. Stichidia sessile, oblong, acute. Tetraspores in several series.

Foveaux Straits and Chalky Bay, *Lyall.*

4. **R. glomerulata,** *Mont. in Voy. au Pôle Sud,* 141. Frond red, cartilaginous, rigid, 2 ft. long, terete, filiform, inarticulate, flexuose, excessively branched from the base, longitudinally striate when dry, dark brown; branches corymbose, lateral branchlets simple or forked, bearing tufts of sessile oblong or ovate-lanceolate stichidia in the fork; tubes about 7. Conceptacles minute, ovoid.—Harv. Ner. Austr. 36; J. Ag. Sp. Alg. 889. *R.? Gaimardi,* Mont. in Voy au Pôle Sud, 140, not of Agardh. *R. ? botryocarpa,* J. Ag. l. c. 882. *Polysiphonia botryocarpa,* Fl. Antarct. t. 70; Harv. Ner. Austr. 57; Fl. N. Z. ii. 230.

Lord Auckland's group, *D'Urville, J. D. H.*; Akaroa, *Raoul*; Otago and Foveaux Straits, *Lyall, Lindsay.*

38. BOSTRYCHIA, Montagne.

Frond dull red-purple, filiform, inarticulate or obscurely articulate with short joints, irregularly pinnately branched; axis an articulate tube, surrounded by one or more series of large coloured surface cells; branchlets hooked. Conceptacles terminal, ovoid, enclosing a tuft of pyriform spores. Stichidia terminal, spindle-shaped, containing a double row of tetraspores.

A beautiful genus of many small tropical and temperate *Algæ.*

1. **B. mixta,** *Hook. f. and Harv. Fl. N. Z.* ii. 225. Frond ½ in. long, dull dark purple, rather rigid, pinnate; pinnæ patent, simple or forked, with alternate subulate divaricate pinnules; tips straight, of the younger incurved; surface-cells hexagonal; axile in one row. Stichidia curved.—Harv. Phyc. Austr. iii. 176 A; Ner. Austr. 70.

On rocks near high-water mark. Bay of Islands, *J. D. H.* Otago, *Lyall.* (Tasmania, S. Africa.)

2. **B. Harveyi,** *Montagne;—Fl. N. Z.* ii. 225. Fronds 1–3 in. long, deep purple, cartilaginous, capillary, 2–3-pinnate, angularly flexuous; pinnæ distichous, alternate, patent; pinnules 2–3-fid or multifid, terminal involute; surface-cells minute, quadrate; axile in several rows.

On rocks near high-water mark. Paterson's Harbour, *Lyall.* (Australia, Chili.)

3. **B. distans,** *Harv. in Fl. N. Z.* ii. 226. Very near *B. Harveyi,* but laxer and more straggling; branchlets simple or nearly so, more distant; tips less incurved. Perhaps only a variety.

Freshwater streams, Kowhaia, *Colenso.* Wellington and Banks's Peninsula, *Lyall.* (Tasmania.)

4. **B. Arbuscula,** *Harv. in Fl. N. Z.* ii. 226. Frond 1 in. high, robust, much thicker than a hog's bristle, erect, densely tufted, compressed. Stem naked below or with broken branchlets only, above densely clothed with short, erect, closely 2-pinnate branches; pinnæ erect, subulate, acute or mucronate; tips strict; axils acute. Structure solid, of 8 primary tubes surrounded with densely-packed minute cells in several series. Fructification unknown.

Otago, *Lyall, Lindsay.*

39. POLYSIPHONIA, Greville.

Frond, rose-red or purplish, filiform or capillary, articulate; joints striate, of few or many long tubular cells surrounding one or a few smaller central ones, the whole sometimes coated with a series of minute cortical cells. Conceptacles ovate or urceolate, enclosing a tuft of pyriform spores. Tetraspores sunk in swollen branchlets.

An immense genus, found in all oceans, but chiefly the temperate; the species are most difficult of discrimination and of definition: very many more are to be found in New Zealand than are here recorded. The most important characters for sectional purposes, are whether the primary stem creeps or is erect; whether the articulations are naked or coated with cortical cells; and the number of parallel tubes in each joint surrounding the central one. Secondary characters, all very variable, are afforded by size of frond, habit, ramification, colour, and comparative rigidity or flaccidity of stem and branches. The fructification is very uniform. Some species are with difficulty distinguished from *Dasya* and *Rhodomela*. I have adopted Agardh's classification as far as I can. Harvey divides the genus as follows:—

HARVEY'S CLASSIFICATION.

Subgenus I. **Oligosiphonia.** *Primary tubes* 4, *rarely* 5.

§ 1. ELONGATÆ. Stem opaque, inarticulate. Branches articulate.—10, *dumosa;* 19, *Lyallii.*

§ 2. DICHOTOMEÆ. Frond all pellucid, articulate, subdichotomous, decompound.—11, *strictissima;* 13, *abscissa;* 9, *rudis;* 7, *implexa;* 8, *macra;* 17, *brachygona;* 14, *variabilis;* 15, *nana;* 16, *rhododactyla;* 12, *amphibia;* 1, *Colensoi.*

Subgenus II. **Polysiphonia.** *Primary tubes numerous.*

§ 5. BYSSOIDEÆ. Frond alternately branched, with 1-tubed branchlets.—23, *australis.*

§ 6. PUNICEÆ. Frond purplish, vaguely branched; lateral branchlets dichotomous, many-tubed. Tubes 6–8.—18, *P. Brodiæi?;* 6, *P. neglecta;* (*P. punicea* see *Dasya Berkeleyi*).

§ 7. PENNATÆ. Frond rose-colour or purple, often distichously pinnate; branchlets simple, subulate, alternate or 4-farious. Tubes 12–16.—2, *dendritica;* 3, *ceratoclada;* 4, *Sulivanæ;* 5, *pennata.*

§ 8. CANCELLATÆ. Frond brown, black when dry, shrubby, furrowed, vaguely branched; branches decompound. Tubes 7, rarely 8 or 9.—25, *cancellata;* 24, *decipiens;* 26, *aterrima;* 27, *ramulosa.*

§ 9. ATRORUBESCENTES. Fronds red-brown or blackish, darker when dry, cylindric, vaguely or pinnately branched; branchlets decompound. Tubes 10–16 or more.—20, *corymbifera;* 21, *comoides;* 22, *isogona;* 28, *nigrescens.*

§ 11. BOTRYOCARPÆ. Frond more or less inarticulate, tall, red, brown when dry, cells anastomosing. (*P. botryocarpa*, see *Rhodomela glomerulata.*)

J. AGARDH'S CLASSIFICATION.

§ I. **Ptilosiphonia.**—Primary stem erect or creeping, distichously or 2-fariously pinnate. Articulations naked or coated with small cells. Tubes 4–16.

a. DENDRITICÆ. Small, parasitic; primary stem creeping; pinnæ appressed or ascending; articulations naked. Tubes 4–8 or more.

1. *P. Colensoi;* 2, *dendritica.*

β. Pectinatæ. Small, parasitic; primary stem creeping; pinnæ vertical, 2-farious; articulations naked. Tubes 8–16.

3. *P. ceratoclada;* 4, *P. Sulivanæ.*

γ. Pennatæ. Stem erect or creeping, growing on rocks, distichously or pinnately decompound; articulations naked (or coated with small cells). Tubes 5–16.

5. *P. pennata.*

§ II. **Herposiphonia.**—Usually small. Primary stem creeping, secondary erect, branched all round. Articulations naked. Tubes usually many.

α. Obscuræ. Fronds small, forming spreading tufts. Primary stem creeping, branches erect.

6. *P. neglecta.*

β. Intricatæ. Fronds less tufted, creeping amongst other *Algæ* or floating, intricate; threads often rooting.

7. *P. implexa;* 8, *macra;* 8, *rudis;* 10, *dumosa.*

§ III. **Oligosiphonia.**—Fronds large, branched on all sides; articulations naked or coated with small cells. Tubes 4.

α. Urceolatæ. Tufts stemless, erect; branches and branchlets soft, flaccid, twiggy above; articulations naked (or the lower only coated with cells).

11. *P. strictissima;* 12, *amphibia;* 13, *abscissa;* 14, *variabilis;* 15, *nana;* 16, *rhododactyla.*

β. Violaceæ. Caulescent. Stem firmer, clothed with flaccid fascicled or penicillate branches; lower or all the articulations coated with small cells.

17. *P. brachygona;* 18, *Brodiæi?*

γ. Hystrices. Fronds subpinnately decompound; branches rather rigid, ultimate subulate or spinescent; articulations coated with small cells.

19. *P. Lyallii.*

§ IV. **Polysiphonia.**—Fronds erect, branching all round; articulations with 5 or more tubes, rarely coated with small cells (never in the N. Z. species).

α. Atro-rubescentes. Fronds red-brown or purplish; branches dichotomous, attenuated upwards. Tubes 9–14.

20. *P. corymbifera;* 21, *comoides;* 22, *isogona.*

β. Byssoideæ. Fronds pinnately decompound; branches various, branchlets soft, of a single tube; articulations. Tubes 7.

23. *P. australis.*

γ. Cancellatæ. Frond pinnately decompound; branches all similar, or the shorter soft or spinescent; articulations very soft. Tubes 7–10.

24. *P. decipiens;* 25, *cancellata;* 26, *aterrima;* 27, *ramulosa.*

δ. Nigrescentes. Frond subpinnately compound; branches all similar, soft, twiggy. Tubes 12–20.

28. *P. nigrescens.*

1. **P. Colensoi,** *Hook. f. and Harv. Fl. N. Z.* ii. 229. *t.* cxii. C. Frond erect, dark coloured when dry. Stem 1 2 in. long, stout; articulations everywhere pellucid, 4-angled, branched from every side; branches alternately decompound, giving off simple and pinnate branchlets from each node; articulations of 4 tubes, which are square in outline, their walls thick. Conceptacles on the larger branches, ovate-globose, pedicelled.—J. Ag. Sp. Alg. 915.

Parasitical on *Fucoideæ*, *Colenso*.—Habit and ramification of *P. ceratoclada*.

2. **P. dendritica,** *Hook. f. and Harv. Fl. N. Z.* ii. 232. Frond 1½–2 in long, ovate. Stem prostrate, attached by its whole length, compressed, un divided, 3-pinnately branched; branches alternately simple and 2-pinnate long, spreading; pinnules likewise alternately simple and 2-pinnate; articulations very short, of many tubes, hence striated. Conceptacles terminal globose-urceolate, oblique, with a prominent mouth.—J. Ag. Sp. Alg. 916 Harv. Ner. Austr. 47. *Hutchinsia*, Ag.

Parasitic on various *Fuci*, abundant. (Australia, S. America.)

3 **P. ceratoclada,** *Mont.;—Fl. N. Z.* ii. 232. Frond 1–4 in. high rigid, red-brown, erect from creeping filaments, compressed, undivided below, then giving off alternate lateral branches. Stem and branches furnished throughout with frequent subulate patent or recurved acute branchlets; articulations with 12–16 tubes. Conceptacles solitary, large, sessile or shortly stalked.—Mont. Voy. au Pôle Sud, i. 130. t. 5. f. 2; Fl. Antarct. t. 76. f. 2; Harv. Ner. Austr. 48; J. Ag. Sp. Alg. 923.

Parasitical on other *Algæ*, from the east coast, *Colenso*, to **Lord Auckland's** group, *D'Urville*, etc.

4. **P. Sulivanæ,** *Hook. f. and Harv. Fl. N. Z.* ii. 232. Frond slender, 1–2 in. high, red-brown, much flabellately branched, flaccid. Stem a creeping thread; branches fastigiate, alternately decompound, with 3 subulate simple branchlets from as many successive nodes, and then a pinnated branchlet from the following node and so on; branchlets patent, curved, alternate or secund, terminal almost circinate; articulations with 12 tubes, of the branches twice as long as broad, of the branchlets shorter.—Fl. Antarct. ii. t. 182. f. 4; Harv. Ner. Austr. 48; J. Ag. Sp. Alg. 923.

Stewart's Island, *Lyall*. (Falkland Islands.)

5. **P. pennata,** *Agardh;—Fl. N. Z.* ii. 231. Frond 1–2 in. high, dark red, blackish when dry. Stem branched above the middle, setaceous, compressed; branches erecto-patent, simple or alternately decompound, all pinnated with distichous subulate erecto-patent branchlets; articulations as broad as long, with 8–10 tubes.—J. Ag. Sp. Alg. 928.

Auckland, *Lyall*; Cape Kidnapper, *Colenso*. (Australia, Atlantic and Mediterranean Seas.)

6. **P. neglecta,** *Harv. in Trans. Irish Academy*, xxii. 541. Fronds forming dense tufts 1 in. diameter. Stems decumbent, rooting below; tips ascending, sparingly branched; branches ascending, then erect, sparingly divided, attenuated at the apex; articulations twice as long as broad, upper as long as broad. Tubes about 8–9.—J. Ag. Sp. Alg. 942.

New Zealand, *Sinclair*. (Australia.)

7. **P. implexa,** *Hook. f. and Harv. Fl. N. Z.* ii. 229. Frond small, tufted, 1 in. high, purple. Stems creeping, intricate; branches alternate, spreading, divided at the apex; branchlets subulate, spreading, subsimple; lower articulations once and a half as long as broad, upper as long as broad. Tubes 4.—J. Ag. Sp. Alg. 946.

Akaroa, *Raoul*; Cape Kidnapper and Parimahu, *Colenso*.

8. **P. macra,** *Harv. in Fl. N. Z.* ii. 229. A small and imperfectly-known plant, forming matted patches on rocks, similar in many respects to *P. implexa*, but the branches are more simple; the articulations are similar and shorter.

Akaroa, *Raoul*; Hawke's Bay, on tidal rocks, *Colenso*.

9. **P. rudis,** *Hook. f. and Harv. Fl. N. Z.* ii. 229. Fronds small, 1–1½ in. long, densely tufted, dark-brown, rigid, obovate in outline. Stems erect from matted creeping filaments, slender, irregularly branched; branches erect, subfastigiate, alternate; lower long, simple; upper pinnated above; pinnules erect, subulate, elongate; articulations pellucid at the nodes, of the branches 2–3 times longer than broad, of the branchlets 1½ times. Tubes 3–4. Tetraspores in one row in the swollen tips of the branches.—Harv. Ner. Austr. 44; J. Ag. Sp. Alg. 946.

Cook's Straits, *Lyall*; Akaroa, *Raoul*; **Lord Auckland's** group, *J. D. H.*

10. **P. dumosa,** *Hook. f. and Harv. Fl. Antarct.* 182. *t.* 75. *f.* 1. Fronds tufted, dark red-brown. Stems erect, rigid, 1–3 in. long, flabellately-branched; branches fastigiate, simple or forked, clothed with very spreading squarrose multifid branchlets $\frac{1}{12}$–$\frac{1}{6}$ in. long; pinnules divaricate, subulate; articulations about as long as broad; cells about 5, lower coated with a band of smaller cells.—Harv. Ner. Austr. 42; J. Ag. Sp. Alg. 950.

Parasitic on larger *Algæ*, **Campbell's** Island, *J. D. H.*—J. Agardh suspects that this, which Harvey places near *P. Lyallii*, is better placed here.

11. **P. strictissima,** *Hook. f. and Harv. Fl. N. Z.* ii. 227. Fronds densely tufted, dark red, flaccid. Stems 4–6 in. long, setaceous and rigid below, capillary and excessively branched above; branches fastigiate, lower spreading with obtuse axils, upper erect or appressed, with very acute axils; lower articulations about as long as broad, upper 5 times as long. Tubes 4.—Harv. Ner. Austr. 42; J. Ag. Sp. Alg. 962.

New Zealand, *Raoul*; Port Underwood, *Lyall*.

12. **P. amphibia,** *Harv. in Fl. N. Z.* ii. 229. Fronds in dense intricate red-brown tufts, 2–3 in. high. Stems capillary, rigid, much-branched, attenuated upwards, decompoundly-branched; branches and branchlets irregularly alternate; axils spreading; branchlets often multifid at the tips; articulation about twice as long as broad, upper much longer. Tubes 4. Tetraspores small.

Massacre Bay, in brackish water when covered at high tide by the sea, *Lyall*.—Allied to the European *P. urceolata*, which also inhabits brackish water.

13. **P. abscissa,** *Hook. f. and Harv. Fl. N. Z.* ii. 227. Fronds densely tufted, 3–6 in. high, bright-red. Stems capillary, flaccid, alternately decompound; smaller branches naked below, fastigiately-branched and truncate at the top; tips naked or fibrilliferous; articulations of the branches much longer than broad, of the branchlets twice as long as broad. Tubes 4. Conceptacles ovoid-globose, stalked, often secund; tetraspores small.—*P.*

abscissa and *microcarpa*, Fl. Antarct. 479, 480. t. 182. f. 3; Harv. Ner. Austr. 42, 43; J. Ag. Sp. Alg. 974.

Cook's Straits to Stewart's Island, *Lyall;* Akaroa, *Raoul.* (Tasmania, Fuegia.)

14. **P. variabilis,** *Harv. in Fl. N. Z.* ii. 228. Fronds red-brown or purple, loosely tufted, flaccid, 2–4 in. high, setaceous, irregularly dichotomous below, variously branched above; branches naked or with multifid branchlets, or excessively divided; branchlets erect; articulations very variable in length, all short, the lower shorter than broad, and middle about twice as long, or all very much longer than broad; tubes 4, broad, spirally twisted. Conceptacles broadly ovoid, subsessile.—J. Ag. Sp. Alg. 975.

Common on the coasts, especially of the **Middle** Island, on shells, other seaweeds, etc, *Lyall*, etc.

15. **P. nana,** *Harv. in Fl. N. Z.* ii. 228. Very similar to the preceding, but much smaller, not 1 in. high, covered with ovoid conceptacles.

Auckland, on Corallines, *Lyall.*

16. **P. rhododactyla,** *Harv. in Fl. N. Z.* ii. 228. Fronds tall, stout, 8 in. long; articulate throughout, setaceous below, flaccid above, dichotomously branched; branches bright rose-red, forked, with spreading axils; branchlets dicho omously multifid, acute or acuminate, ultimate subsecund or fascicled, with narrow axils; lower articulations very short, coated with small cells; middle ones as long as broad; upper of the larger branchlets shorter than broad, of the upper extremely short; tubes 4, broad.

D'Urville Island, in five fathoms water, *Lyall.*

17. **P. brachygona,** *Harv. in Fl. N. Z.* ii. 228. Fronds tufted, rigid, purple, 3–4 in. high, setaceous, divided from the base into alternate decompound branches, which are again divided and taper to slender points, secondary and ultimate branches capillary, irregular, alternate, secund or subdichotomous; axils patent; lower articulations very short, coated with cells; middle ones 1½–3 times as long as broad; upper very short; tubes 4.

Stewart's Island, *Lyall.*—Allied to the *P. Griffithsiana* of Europe.

18. **P. Brodiæi ?,** *Greville;—Fl. N. Z.* ii. 230. Fronds tall, tufted, 10–18 in. high, lanceolate. Stem often as thick as a sparrow-quill below, simple, forked or alternately branched, or clothed with branchlets, 2 in. long, which are fascicled, multifid, pellucid; articulations 1½ times as long as broad, all coated with small cells; tubes 6–8; conceptacles urceolate-ovoid, sessile on the branches.—Harv. Phyc. Brit. t. 195; J. Ag. Sp. Alg. 993.

Apparently common from the East coast to Stewart's Island, *Lyall.* (N. Atlantic Ocean.)

19. **P. Lyallii,** *Hook. f. and Harv. Fl. N. Z.* ii. 230. Frond erect, 4–5 in. long, brownish-red, darker when dry, elongate, cartilaginous, setaceous, inarticulate, simple or divided below, branched throughout its length; branches simple, 4-farious, densely clothed with multifid, short, imbricate, dichotomous spreading branchlets $\frac{1}{12}$–$\frac{1}{6}$ in. long, ultimate subulate; articulations about as long as broad, visible on the branchlets only, coated with

small cells; tubes 4.—Fl. Antarct. t. 74. f. 1; Harv. Ner. Austr. 42. *P. Mallardiæ*, Harvey in part (N. Z. specimens only).

Hawke's Bay and elsewhere, on the East coast, *Colenso;* Otago and Stewart's Island, *Lyall, Lindsay.* **Lord Auckland's** group, *J. D. H.*

20. **P. corymbifera,** *Agardh;—Fl. N. Z.* ii. 231. Fronds loosely or densely tufted, red-brown, 6–8 in. long, setaceous, attenuated, many-times dichotomous, striate; upper branches alternately divided, bearing at intervals short lateral alternate branchlets, $\frac{1}{4}$–$\frac{1}{2}$ in. long, which are obovate or corymbose in outline; branchlets dichotomous; articulations hyaline, of the main stem 2–3 times as long as broad, of the branches shorter, of the ultimate shorter than broad; tubes 10–14. Tetraspores in distorted branchlets.—Harv. Ner. Austr. 54; J. Ag. Sp. Alg. 1039.

Maketu, *Chapman.* (South Africa.)

21. **P. comoides,** *Harv. in Fl. N. Z.* ii. 231. Fronds densely tufted, red-brown, capillary, decompoundly branched from the very base; branches many-times alternately and dichotomously divided, the smaller naked at the base, alternately compound at the apex; middle articulations 4–5 times as long as broad; upper 2–3 times; uppermost 1$\frac{1}{2}$ times; nodes pellucid; tubes 9 or 10.

Banks's Peninsula, *Lyall.*

22. **P. isogona,** *Harv. in Fl. N. Z.* ii. 231. Very similar to *P. comoides*, and perhaps only a variety of it, but the filaments are more slender; articulations much shorter (1$\frac{1}{2}$ times longer than broad).

Cape Kidnapper and Hawke's Bay, *Colenso;* Blind Bay, *Lyall.*

23. **P. australis,** *J. Agardh.—P. cladostephus*, Mont.;—Fl. N. Z. ii. 232. Frond 6–18 in. long, brown-purple, setaceous, articulate throughout, alternately excessively branched from the base; larger branches decompound; smaller erect, elongate, simple, all clothed with whorled, imbricate, dichotomous, rose-red, 1-tubed branchlets, springing from every node; tips swollen; articulations of the branches 2–3 times as long as broad, with 7 tubes; upper shorter. Conceptacles ovate, sessile on branches at the base of the branchlets. Tetraspores on smaller branchlets.—J. Ag. Sp. Alg. 1044. *P. cladostephus*, Mont. in Voy. au Pôle Sud, t. 13. f. 4 a; Fl. Antarct. 184; Harv. Ner. Aust. 45; Phyc. Austr. t. 154. *P. byssoclados*, Harv. in Hook. Lond. Journ. Bot. iii. 436. *Bindera cladostephus*, Decaisne. *Griffithsia australis*, Agardh.

Probably common, parasitical on *Fuci*, though found hitherto only south of Canterbury; Banks's Peninsula, *Lyall, Raoul.* **Lord Auckland's** group, *Hombron*, etc. (Australia, Tasmania.)

24. **P. decipiens,** *Mont.;—Fl. N. Z.* ii. 230. Frond stout, shrubby, twiggy, coarse, bark brown, eventually black, bushy, grooved, pellucidly articulate, branched from the very base; branches erect, decompound; lateral branches 4-farious, erecto-patent, attenuated upwards; ultimate spinescent, scattered; articulations so short that the branches appear transversely striated; tubes 7.—Mont. in Voy. au Pôle Sud, 131; Fl. Antarct. 184; Harv. Ner.

Austr. 51; J. Ag. Sp. Alg. 1046. *R. rytiphlœoides*, Hook. Lond. Journ. Bot. iv. 537.

Banks's Peninsula and Otago, *Raoul, Lyall.* **Lord Auckland's** group, *Hombron, J. D. H.*

25. **P. cancellata,** *Harv. in Fl. N. Z.* ii. 230. Frond 4–5 in. high, robust, shrubby, brownish-black, globose in outline. Stem as thick as pack-thread, terete, furrowed, excessively branched from the base; branches much divided alternately, spreading all round; branchlets tapering, much more slender, pinnate or 2-pinnate; ultimate spinescent; lower articulations 3 times, upper twice as broad as long; tubes 7–8. Conceptacles small, ovate, sessile.—Harv. Ner. Austr. 51. t. 15.

Banks's Peninsula, *Lyall.* (Tasmania, Australia.)

26. **P. aterrima,** *Hook. f. and Harv. Fl. N. Z.* ii. 230. Frond 4–5 in. high, black when dry, rather rigid. Stem setaceous, furrowed, articulate, naked below, decompoundly branched above, scarcely dichotomous, narrowed to the apex; branches attenuate or secund, obovate in outline; branchlets distant, erect, subulate; axils acute; articulations all very short; tubes 9–10, oblong, hexagonal. Conceptacles sessile, scattered, ovoid-globose, obtuse.—Harv. Ner. Austr. 52; J. Ag. Sp. Alg. 1050.

Parasitical on *Fuci*, abundant, *Colenso*, etc.

27. **P. ramulosa,** *Harv. in Fl. N. Z.* ii. 230. Frond 1 in. high, blackish, flabellately branched. Stem as thick as horsehair, divided from above the base into numerous alternate branches, which are erect, close-set; branchlets numerous, spreading, subulate, simple or again branched, $\frac{1}{12}$–$\frac{1}{6}$ in. long; tips all acute; articulations all about half as long as broad; tubes 7.

Parasitic on *Sargassa*; Parimahu, *Colenso.*

28. **P. nigrescens,** *Grev.*;—*Fl. N. Z.* ii. 231. Frond robust, bushy, 6 in. high, dark brown, rigid below. Stem setaceous, rough below with broken off branches, excessively branched above; branches flaccid, somewhat distichous, repeatedly pinnate, obovate in outline; pinnules elongate, distant, alternate, subulate, the upper again pinnate at their tips; lower articulations short; upper longer; tubes 16. Conceptacles ovoid, sessile.—Harv. Ner. Austr. 44; Phyc. Brit. t. 277; J. Ag. Sp. Alg. 1057.

New Zealand, *Raoul.* (Atlantic and Baltic Oceans.)

40. DASYA, Agardh.

Frond red, filiform or compressed, dendroid; axis articulate, of many radiating cells surrounding a central cavity, the whole coated with still smaller cells; branchlets articulate, 1-tubed. Conceptacles ovate or urceolate, with pyriform spores. Stichidia lanceolate, on the branchlets, containing transverse rows of tetraspores.

Very beautiful bright red seaweeds, natives chiefly of the temperate zones.

1. **D. collabens,** *Hook. f. and Harv. Fl. N. Z.* ii. 222. Frond 1–12

in. high, soft and flaccid, terete, inarticulate, glabrous, decompoundly pinnate; pinnæ ovate-lanceolate. 2–3-pinnate; pinnules distichous, subdivided, clothed with articulate 1-tubed dichotomous branchlets, which are spreading, subulate, forked, contracted at the joints; articulations 2–3 times longer than broad.—Harv. Ner. Austr. 61. t. 21, young state only.

From Akaroa, *Raoul*, southwards.

2. **D. squarrosa,** *Harv. in Fl. N. Z.* ii. 232. Fronds crimson, gregarious, 1–2 in. high, distichous, pinnate. Stems articulate, with 7–8 tubes, pellucid, except at the base, rough below with horizontal hair-like branchlets, smooth above, 1–3-pinnate; branches alternate, on second or third node; branchlets 3-tubed below, 1-tubed above, short, divaricating, cylindric, obtuse, simple or forked; articulations of the branches as long as broad, of the ultimate longer.

Stewart's Island, Port William, *Lyall.*

3. **D. tessellata,** *Harv. in Fl. N. Z.* ii. 233. Frond densely tufted, crimson-lake. Stems 1–2 in. high, setaceous, pellucid, of about 12 cells around the central tube, tessellated with large, transverse, square or oblong cells, pinnate or 2-pinnate; branches spreading, penultimate pinnated, with alternate, distichous, very spreading, short, dichotomous or pectinate, secund branchlets, below 3-tubed; ultimate 1-tubed, subulate, acute or obtuse, divaricating, often changed into sessile stichidia.

Cook's Straits, Blind and Massacre Bays, *Lyall.*

4. **D. Berkeleyi,** *J. Agardh, Sp. Alg.* 1179.—*Polysiphonia punicea,* Mont. Fl. Antarct. i. 182. Frond red-purple, 4–8 in. long, setaceous, very irregularly branched, dichotomous or with a percurrent stem and pinnated branches, which are usually flexuose and zigzag, all furnished with short alternate or spirally disposed branchlets $\frac{1}{12}$–$\frac{1}{6}$ in. long; articulations of the branches 3 times longer than broad, of the branchlets as long as broad; tubes 8. Conceptacles ovoid, sessile; stichidia lanceolate.—*Polysiphonia punicea* and *Heterosiphonia Berkeleyi,* Mont. in Voy. au Pôle Sud, 137 and 128 t. 5. f. 1. and 3. *P. Berkeleyi,* Harv. Ner. Austr. 46.

Lord Auckland's group, *Hombron, J. D. H.* (Fuegia, Falkland Islands, Kerguelen's Land.)

41. POLYZONIA, Suhr.

Frond parasitic, minute, rose-red, filiform, articulate, much distichously branched; branchlets leaf-like, alternate, flattened, angular, toothed or pectinate and secund. Stem of many parallel tubular cells. Conceptacles ovoid, enclosing pyriform spores. Stichidia lanceolate, pedicelled, supra-axillary, often crested, containing one row of large tetraspores.

Most lovely little *Algæ*, always parasitic on the stems of larger ones, growing in deep water, found only in the S. temperate and Antarctic zones.

1 **P. cuneifolia,** *Montagne ;—Fl. N. Z.* ii. 226. Fronds creeping by a filiform rhizome, articulate. Stems erect, 4–5 in. long, undivided, furnished

with long, simple (rarely branched), alternate branches, everywhere covered with distichous, alternate leaflets (branchlets); leaflets vertical, trapezoid, truncate, irregularly cut and toothed, sometimes 2-fid; lower margin entire. Conceptacle sessile on the basal lobe of a deeply cut leaflet. Stichidia supra-axillary, crested with processes, at first simple, then pinnately branched.—Harv. Ner. Austr. 70.

Roots and fronds of large *Algæ*, Otago, Stewart's Island, and Port Preservation, *Lyall.* **Lord Auckland's** group and **Campbell's** Island: abundant, *J. D. H.*

2. **P. adiantiformis,** *Decaisne;—Fl. N. Z.* ii. 226. Minute, vaguely branched; branches spreading; leaflets as in *P. cuneifolia*, but smaller and with much smaller cells.

Parasitic on *Marginaria*. Described by Dr. Harvey from very immature scraps.

3. **P. ovalifolia,** *Hook. f. and Harv. Fl. N. Z.* 226. *t.* 112 B. Habit of *P. cuneifolia*; leaflets horizontal, sessile, obliquely oblong, very obtuse, quite entire or toothed.

Parasitic on *Amphiroa corymbosa, Colenso.* Specimens immature.

4. **P. bipartita,** *Hook. f. and Harv. Fl. N. Z.* ii. 227. *t.* 112 A. Minute; leaflets distichous, alternate, 2-partite; segments flattened, linear, mucronate or obtuse; upper erect, of 3 rows of rectangular cells; lower shorter, horizontal. Stichidia large, axillary, linear-oblong, apiculate, with one row of few large tetraspores.

On *Carpophyllum maschalocarpus, Colenso.*—Very similar in general appearance to *Polysiphonia ceratoclada.*

5. **P. Harveyana,** *Decaisne;—Fl. N. Z.* ii. 227. Stems ½–2 in. high, slender or robust, compressed, closely set with alternate distichous, pectinate, horizontal and recurved leaflets. Stichidia supra-axillary, simple or 2–3-lobed; lobes crested.—*P. Colensoi* and *P. Harveyi*, Harv. Ner. Austr. 71 and 72.

Common, parasitical on various *Algæ*, *J. D. H.*, etc.

42. CHAMPIA, Desvaux.

Frond dull reddish, terete or compressed, branched, tubular, constricted at intervals, internally septate at the constrictions, the septa connected by longitudinal jointed filaments; walls of one or many rows of minute coloured polygonal cells. Conceptacles ovate-ovoid, with a terminal pore, containing a tuft of branched filaments; the branches terminated by ovoid or obconic spores. Tetraspores 3-partite, scattered in the walls of the stem and branches.

1. **C. novæ-Zelandiæ,** *Harv. Fl. N. Z.* ii. 235. Frond 3 in. long, stipitate. Stem $\frac{1}{12}$ in. broad, compressed, subattenuate at the apex; branches opposite or whorled, simple or divided; branchlets often opposite; articulations twice as broad as long, obtuse. Conceptacles conical, ovoid, sessile. Tetraspores scattered through the branches.—*Chylocladia*, Harv. Ner. Aust. 80. *Lomentaria*, J. Ag. Sp. Alg. ii. 739.

Common along the whole coast, *Lyall*, etc.

2. **C. affinis,** *Harv. in Fl. N. Z.* ii. 237. Frond 12 in. and upwards. Stem undivided, obscurely constricted at intervals, clothed throughout with opposite alternate and whorled spreading branches; branches elongate, shorter upwards, tapering at both ends, constricted at regular intervals; branchlets disposed like the branches, sometimes again branching; articulations of the branchlets moniliform, shorter than broad. Conceptacles large, conical.—*Chylocladia,* Harv. Ner. Austr. 79. t. 29. *C. Kaliformis,* Harv. in Hook. Lond. Journ. Bot. iii. 444, excl. synonyms. *Lomentaria,* J. Ag. Sp. Alg. ii. 730.

Stewart's Island, *Lyall.* (Australia, Tasmania.)

3. **C. parvula,** *Harv. in Fl. N. Z.* ii. 236. Fronds densely tufted, 2–4 in. high, bushy, irregularly branched, stemless, flexuous, irregularly branched; branches and branchlets scattered, spreading; articulations of the branches as long or twice as long as broad, sometimes moniliform. Conceptacles large, ovoid.—*Chylocladia,* Grev.; Harv. Ner. Austr. 80; Phyc. Brit. t. 210. *Lomentaria,* Gaill.; J. Ag. Sp. Alg. 729.

Akaroa, *D'Urville, Raoul.* (Atlantic and Mediterranean seas.)

C. obsoleta is mentioned in Harvey's 'Phycologia Australis,' synops. p. 5, as a native of New Zealand, but I do not know on what authority.

43. LAURENCIA, Lamouroux.

Frond red or purple, often fading to green, cartilaginous, cylindric or compressed, linear, pinnate; tips obtuse; structure cellular, solid; cells in 2 series; inner oblong, angular; outer smaller. Conceptacles enclosing tufted spores. Tetraspores 3-parted, imbedded in the branchlets.

A widely diffused genus, especially in the temperate zones.

* *Frond compressed.*

1. **L. elata,** *Harv.;—Fl. N. Z.* ii. 233. Frond 12–18 in. high, blood-red, compressed or flat, repeatedly pinnate; branches $\frac{1}{12}$ in. diameter, compressed, erecto-patent; lower pinnæ longest, all alternate; ultimate linear-elongate. Conceptacles terminal.—Harv. Ner. Austr. 81. t. 33; J. Ag. Sp. Alg. ii. 766.

East coast, *Colenso.* (Australia, Tasmania.)—Very nearly allied to the common *L. pinnatifida.*

2. **L. pinnatifida,** *Lamour. Fl. Antarct.* 184.—Very similar to *L. elata,* but frond broader, more purple, less branched, and the conceptacles are lateral. —Harv. Phyc. Brit. t. 55; Fl. Antarct.

Lord Auckland's group, *J. D. H.* (South Africa, Fuegia, abundant in the north temperate ocean.)

3. **L. botrychioides,** *Harv. in Fl. N. Z.* ii. 234. Frond tufted, small, purplish-brown, erect, 1–3 in. high, rising from creeping matted surculi; rhachis or stem closely pinnatifid, dilated upwards, compressed; branches $\frac{1}{16}$–$\frac{1}{12}$ in. broad, 3–4-pinnate; pinnæ distichous, usually opposite; lower

larger; pinnules very short, cuneate or clavate, crenate, or multifid with the lobes turbinate. Tetraspores in the club-shaped tips of the pinnules.—*L. botryioides*, Harv. Ner. Austr. 82, in part.

Bay of Islands, *J. D. H.*; Parimahu, *Colenso.*

4. **L. distichophylla ?,** *J. Ag. Sp. Alg.* ii. 762 ;—*Fl. N. Z.* ii. 234. Frond 2–3 in. high, crimson rose-red or purple, densely tufted, compressed, distichously decompound, rarely more simple, ovate in outline; rachis straight, dilated upwards; pinnæ usually opposite, spreading, flat; pinnules cylindric, obtuse, simple or again divided. Conceptacles sessile on the branches of the pinnules, ovoid-urceolate, acuminate. Tetraspores fasciate below the tips of ultimate pinnules.

Bay of Islands, *J. D. H.*; East coast, *Colenso*; Watemata, *Lyall.*—Apparently the same as J. Agardh's plant, of which the locality is unknown.

** *Frond terete; branchlets cylindric, narrowed at the base.*

5. **L. cæspitosa,** *Lamour.*;—*Fl. Antarct.* i. 184. Very similar to *L. pinnatifida*, and usually regarded as a variety of that plant, but darker-coloured and more terete.—Harv. Ner. Austr. 82.

Lord Auckland's group, *J. D. H.* (Chili, Falkland, Atlantic.)

6. **L. gracilis,** *Hook. f. and Harv. Fl. N. Z.* ii. 234. Frond tall, very slender, filiform, setaceous, flexuous, excessively and pinnately branched; branches elongate, alternate, simple or again branched; branchlets alternate, horizontal and secund, short or long, cylindric, capitate, $\frac{1}{12}$–$\frac{3}{4}$ in. long.—Harv. Ner. Austr. 84; J. Ag. Sp. Alg. ii. 746.

Hawke's Bay, *Colenso.*

7. **L. papillosa,** *Grev.*;—*Fl. N. Z.* ii. 234. Frond dark purple, fading to green, densely tufted, 4–6 in. high, 2-pinnate, terete; pinnæ close-set, elongate, simple, opposite and alternate, studded with spreading, simple or cleft, obtuse, warted pinnules, which are $\frac{1}{12}$–$\frac{1}{4}$ in. long and depressed or perforated at the apex.—Harv. Ner. Austr. 84; J. Ag. Sp. Alg. ii. 756. *Fucus thyrsoides*, Turn. Hist. Fuc. t. 19.

New Zealand, *Banks and Solander.* (Atlantic, Pacific, and Indian Oceans.)

8. **L. Forsteri,** *Grev.*;—*Fl. N. Z.* ii. 234. Frond pink, fading to orange, terete, filiform, nearly equal throughout, much dichotomously and pinnately branched; branches somewhat fastigiate, the smaller spreading and subdichotomous; branchlets spreading, cylindric, dichotomous, alternate or secund.—Harv. Ner. Austr. 85; J. Ag. Sp. Alg. ii. 744. *Fucus*, Turn. Hist, Fuc. t. 77.

New Zealand, *Forster.* (Australia and Tasmania.)

9. **L. virgata,** *J. Ag. Sp. Alg.* ii. 752;—*Fl. N. Z.* ii. 234. Frond 6–12 in. high, red-purple, about as stout as a pigeon's quill, pyramidal in outline, terete, pinnately branched in all directions; lower branches 4–5 in. long, elongate, racemose, and branchlets more or less regularly opposite and whorled. Conceptacles crowded, spherico-ovoid.

East coast, *Colenso;* Houraki Gulf, Banks's Peninsula, and Otago, *Lyall.* (S. Africa, California.)

44. CHONDRIA, Agardh.

Frond dull red or purplish, filiform, cartilaginous, dichotomously or verticillately branched, opaque; axis articulated, of many parallel tubes, coated with small polygonal irregularly placed cells; branchlets clavate, constricted at the insertions. Conceptacles ovoid, containing pyriform stalked spores. Tetraspores 3-parted, scattered on the ramuli.

A genus of several species, scattered through the temperate and tropical oceans.

1. **C. macrocarpa,** *Harv. in Fl. N. Z.* ii. 223. Frond 4–6 in. long, as thick as a sparrow's quill, cylindric, alternately or pinnately decompound; branches distichous, close or distant, alternate or secund, closely set with alternate pinnæ or curved and distantly compound; branchlets elongate, ½–1 in. long, very obtuse, slightly constricted at the base. Conceptacles large, ovoid at the ends or sessile on the sides of the branchlets.

Foveaux Straits, *Lyall;* Otago, *Lindsay.*

2. **C. flagellaris,** *Harv. in Fl. N. Z.* ii. 224. Fronds dull purple, cartilaginous, setaceous, densely tufted, branched vaguely from the base or alternately decompound; branches elongate, simple, with few lateral, rather spreading branches; branchlets scattered, unequal, filiform, acute, not constricted at the base. Conceptacles large, ovoid, sessile, with a prominent orifice.

Port Nicholson and Paterson's Harbour, *Lyall.*

45. CLADHYMENIA, Harvey.

Frond rose-pink, membranous, flat, thin, with or without a midrib, distichously pinnatifid; structure of 2 strata, internal cells large, polygonal, full of granules, outer minute, coloured. Conceptacles ovoid, enclosing a tuft of pyriform spores. Tetraspores 3-partite, in definite marginal sori.

Genus confined to New Zealand and Australia.

1. **C. Lyallii,** *Hook. f. and Harv. Fl. N. Z.* ii. 235. Frond rose-red, 4–5 in. high, broadly deltoid in outline, filiform at the base, soon dilating to $\frac{1}{10}$–$\frac{1}{6}$ in., then flattened, traversed by an obscure nerve, attenuated to the apex, between gelatinous and membranous, 2–3-pinnatifid; branches spreading, linear-lanceolate; lowest 2-pinnatifid; middle pinnatifid; upper entire or toothed; branchlets linear, filiform, obtuse. Conceptacles stalked, elliptic-oblong.—Harv. Ner. Austr. 87. t. 33.

Bay of Islands, *Lyall.*

2. **C. oblongifolia,** *Harv. in Fl. N. Z.* ii. 235. *t.* 113. Frond red, brown when dry, 1 foot long and more, nearly 1 in. broad, flat, gelatinously membranous, pinnate or 2-pinnate; pinnæ 6–8 in. long, tapering downwards, obtuse, simple or bearing similar pinnules; margins more or less fringed with filiform or compressed, simple or branched processes. Conceptacles sessile

on the terete processes, ovoid. Tetraspores sunk in flattened processes.—Harv. Ner. Austr. 87; J. Ag. Sp. Alg. ii. 771.

East coast, *Colenso;* Paroah Bay and Port Cooper, *Lyall.*—Very similar to, and easily mistaken for *Callophyllis Hombroniana.*

46. ASPARAGOPSIS, Montagne.

Frond filiform, red, inarticulate, panicled; branches thyrsoid, penicillate, pinnately decompound; branchlets setaceous, cellular. Conceptacles ovoid, enclosing a dense tuft of pyriform spores covered with a membrane. Tetraspores unknown.

Natives of the Mediterranean and warmer seas.

1. **A. armata,** *Harv.—A. Delilei,* Fl. N. Z. ii. 233. Rhizome branched, setaceous. Fronds pale or bright rose-red, tufted, 6–12 in. long, much branched; branches linear-lanceolate in outline, virgate, clothed to the base with short setaceous branchlets, and having also at the base a few long, naked, retrorsely spinous, stiff, spreading branches. Conceptacles globose, on cylindric stalks.—Harv. Phyc. Austr. iv. t. 122. *A. Delilei,* Harv. Ner. Austr. t. 35, not of Montagne.

In deep water, Cook's Straits, *Lyall.* (Australia and Tasmania.)

47. DELISEA, Lamouroux.

Frond bright red, cartilaginous, narrow, compressed, 2-edged, much branched alternately; branches distichous, pectinate or coarsely serrate; midrib evident, immersed; structure an axis of roundish-angular, colourless, close-packed cells, surrounded by several series of coloured, minute, cortical cells. Conceptacles ovoid, sessile on the midrib towards the tips of the branchlets, and enclosing a tuft of pyriform spores. Tetraspores immersed in wart-like swellings, spread over the tips of the branches.

A southern genus, found hitherto only in Australia, New Zealand, South Africa, and Kerguelen's Land.

1. **D. elegans,** *Lamour.;—Fl. N. Z.* ii. 233. Frond bright red, 6–12 in. long, excessively and finely branched, compressed, membranous, narrow, pinnately decompound, all the branches and their divisions bordered with distichous, subulate, acute, alternate branchlets. Conceptacles solitary or 2 together, sessile in the axils of the subulate branchlets, opening by a pore.—Harv. Ner. Austr. 89. t. 34; J. Ag. Sp. Alg. ii. 781.

Preservation Harbour and Akaroa, *Lyall.* (Australia and Tasmania.)

48. PTILONIA, J. Agardh.

Frond bright rose-purple, linear, membranous above, flattened and 2-edged, decompound-pinnate, toothed, with a sunk midrib of 3 series of cells; axis of branched, jointed, longitudinal threads, running laterally into an intermediate tissue of rounded-angular cells; cortical cells rounded, minute. Conceptacles subterminal on the pinnules.

A genus of a single most beautiful species.

1. **P. magellanica,** *J. Ag. Sp. Alg.* ii. 774;—*Fl. N. Z.* ii. 235. Root scutate. Frond 2–10 in. long, $\frac{1}{6}$–$\frac{1}{4}$ in. broad, membranous above, cartilaginous below; pinnæ alternate or subopposite in pairs, terminal flattened, toothed; teeth growing out into linear pinnæ; axils rounded.—*Plocamium*, Hook. f. and Harv. Fl. Antarct.; Harv. Ner. Austr. 124. *Thamnophora*, Mont. in Voy. au Pôle Sud, t. 8. f. 2.

East coast, *Lyall*. (Abundant in Fuegia and Kerguelen's Land.)

49. AMPHIROA, Lamouroux.

Frond pale-red, flat, compressed or terete, articulate, dichotomously branched or pinnate or whorled; articulations cartilaginous, coated with calcareous matter. Conceptacles conical, wart-like, opening by a pore, enclosing a tuft of erect pyriform, at length 4-parted spore-threads.

Handsome Corallines, usually found in tide-pools in both temperate, but chiefly tropical seas.

1. **A. corymbosa,** *Decaisne*;—*Fl. N. Z.* ii. 237. Frond 4–5 in. high, densely tufted, 2-pinnate at the top; pinnæ and pinnules close-set; lower articulations cylindric, very short; intermediate and upper deltoid or obcordate, compressed; lateral angles acute or produced, of the branches subsagittate; terminal lanceolate.—Harv. Ner. Austr. 99. t. 38; J. Ag. Sp. Alg. ii. 550.

East coast, *Colenso*; Bay of Islands, *Lyall*, etc. (S. Africa.)

2. **A. elegans,** *Hook. f. and Harv. Fl. N. Z.* ii. 237. Frond 2–3 in. long, slender, irregularly dichotomous; forks distant; lower articulations terete, three times as long as broad; middle and upper sagittate, with acute, subulate, erecto-patent, obtuse lobes. Conceptacles sunk in the lobes of the articulations.—Harv. Ner. Austr. 101. t. 38; J. Ag. Sp. Alg. ii. 546.

Cape Kidnapper, *Colenso*. (Tasmania.)

50. CORALLINA, Linnæus.

Frond pale pink, filiform, jointed, usually pinnately branched; articulations coated with a calcareous deposit. Conceptacles ovoid-turbinate or obovate, opening by a pore, containing a tuft of erect pyriform or clavate transversely-parted tetraspores.

Small tufted perennial calcareous *Algæ*, known as Corallines, and growing in all seas.

1. **C. armata,** *Hook. f. and Harv. Fl. N. Z.* ii. 237. Frond dull purple, short, stout, 1–2 in. high, simple below, flabellately branched at the summit; branches 2-pinnate, fastigiate; articulations of the stem short, crowned with spines, middle broadly obcuneate, compressed, short, with 2–6-

whorled species; upper cuneate, uppermost terete, slender, obtuse, often capitate.—Harv. Ner. Austr. 100. t. 40; J. Ag. Sp. Alg. ii. 566.

East coast, *Colenso.*

2. **C. officinalis,** *Linn.;—Fl. N. Z.* ii. 237. Frond pink, pinnate or 2-pinnate; lower articulations cylindric, twice as long as broad, upper obconic, compressed.—Harv. Ner. Austr. 104; Phyc. Brit. t. 222; J. Ag. Sp. Alg. ii. 562.

Auckland, *Lyall;* East coast, *Colenso.* (Common in all seas.)

51. JANIA, Lamouroux.

Frond pale pink, subterete, filiform, articulated, dichotomously and flabellately branched; articulations coated with calcareous matter, cylindric or compressed. Conceptacles urn-shaped, in the forks of the upper branches, 2–4-horned, with a terminal pore, containing a tuft of at length transversely 4-parted spore-threads.

A common genus in both the temperate and tropical zones.

§ 1. *Frond pinnately compound.*

1. **J. Cuvieri,** *Decaisne;—Fl. N. Z.* ii. 237. Frond 2–4 in. long, plumose, pinnately compound; pinnæ and pinnules slender, close-set pinnules simple or forked; articulations of the branches 1½–2 times as long as broad, broadly obcuneate with prominent angles of the branchlets linear, 2–3 times as long as broad. Conceptacles terminal, urceolate, with slender processes. —Harv. Ner. Austr. 105. *Corallina,* J. Ag. Sp. Alg. ii. 572.

Common on the coasts, *Colenso,* etc. (Australia, Tasmania, S. Africa.)

2. **J. Hombronii,** *Mont.;—Fl. Antarct.* 184. Frond excessively 3-pinnately branched; pinnæ crowded, flabellate; lower articulations terete, upper obcuneate, truncate, as broad as long, compressed, of the branches terete, filiform, 2–3 times as long as broad, or ovate-globose, and bearing conceptacles.—Mont. in Voy. au Pôle Sud, 146; Harv. Ner. Austr. 105. *Corallina,* J. Ag. Sp. Alg. ii. 574.

Lord Auckland's group, *Hombron, J. D. H.*

3. **J. pistillaris,** *Mont.;—Fl. N. Z.* ii. 237. Frond tufted, laxly 2-pinnate; pinnæ and pinnules remote, erecto-patent; lower articulations compressed, oblong-quadrate, dilated above, twice as long as broad, of the pinnules terete, discoid, or thickened at the tip and bearing tetraspores.—Mont. in Voy. au Pôle Sud, 147; Harv. Ner. Austr. 105. *Corallina,* J. Ag. Sp. Alg. ii. 574.

Bay of Islands, in holes of rocks, *Hombron.*—Description from Montagne.

§ 2. *Frond dichotomous, often fastigiate.*

4. **J. micrarthrodia,** *Lamour.;—Fl. N. Z.* ii. 237. Frond 1½–2 in. high, tufted, slender, laxly dichotomously branched; axils patent; branches subarcuate; articulations all about the same proportionate length to breadth, as long or twice as long as broad, those of the stems cylindric, of the branches

subcylindric.—Harv. Ner. Austr. 107; J. Ag. Sp. Alg. ii. 555. *J. tenuissima*, Sond.; Harv. l. c. 106. t. 40.

Probably common, East coast, *Colenso*; Port Cooper, *Lyall*. (Australia.)

5. **J. novæ-Zelandiæ,** *Harv. in Fl. N. Z.* ii. 237. Frond 1–2 in. long, setaceous, dichotomously branched; axils acute; articulations cylindric, 6–12 times longer than broad. Conceptacles urceolate, axillary, crowned with geminate 2–3-articulate filaments,

East coast, *Colenso*; Banks's Peninsula, *Lyall*.

J. gracilis, Lamour., a little-known plant, referred by J. Agardh to *J. Cuvieri*, is stated by Montagne to have been found by Hombron at Akaroa.

52. MELOBESIA, Lamour.

Frond crustaceous, red pink or whitish, free or attached by its surface, never jointed, coated with a calcareous deposit, flattened, orbicular and lobed, or cylindric and branched. Conceptacles scattered, conical, sessile, containing a tuft of transversely-parted oblong tetraspores.

Perennial coral-like *Algæ*, found in all seas, sometimes forming calcareous incrustations on other *Algæ*, shells, and pebbles, etc., at others erect or foliated.

1. **M. calcarea,** *Harv. Fl. N. Z.* ii. 238. Frond shrubby, 2–3 in. high, white, subdichotomous; branches as broad as long, $\frac{1}{12}$–$\frac{1}{6}$ in. broad, distant, lower coalescing, upper free and narrow at the bases.—Harv. Phyc. Brit. t. 291; Ner. Austr. 110. *Lithobia*, J. Ag. Sp. Alg. ii. 523.

Bay of Islands, *J. D. H.* (N. Atlantic, Mediterranean Sea, Galapagos Islands.)

2. **M. Patena,** *Hook. f. and Harv. Fl. N. Z.* ii. 238. Frond horizontal, fixed to the stems of *Algæ*, and by a groove at the base, white red or purplish, $\frac{1}{2}$ in. diameter, thick, flat, concave or undulate, obovate or suborbicular, quite entire, imbricate, concentrically striate, shining. Conceptacles numerous, scattered, depressed.—Harv. Ner. Austr. 111. t. 40.

Parasitic on *Ballia*, *Colenso*. (Tasmania, S. Africa.)

3. **M. antarctica,** *Hook. f. and Harv. in Harv. Ner. Austr.* 111. Frond flat, suborbicular, attached by the lower surface; margins free, quite entire, forming a thin expanded irregularly-lobed crust, pale, surface concentrically undulate. Conceptacles minute, depressed, hemispherical.—J. Ag. Sp. Alg. ii. 514. *M. verrucata*, var. *antarctica*, Fl. Antarct. 482.

Lord Auckland's group, *J. D. H.* (Tasmania, Fuegia, Kerguelen's Land.)

53. DELESSERIA, Agardh.

Frond bright rose-red, leaf-like, simple, dichotomously pinnately or proliferously branched, very membranous, flaccid; midrib evident, substance of one layer of angular cells. Conceptacles hemispherical, on the ribs or nerves, enclosing a tuft of branched filaments bearing spores, which are seated on an elevated tubercle on the floor of the cavity. Tetraspores clustered in sori in the substance of the frond.

A most beautiful genus, common in both temperate zones.

§ 1. *Frond dichotomously branched.*

1. **D. Hookeri,** *Lyall in Fl. N. Z.* ii. 238. *t.* 114, 115. Frond 2 ft. and more long, stipitate, 4–5 in. broad, linear or ovate-lanceolate, or oblong-obovate, lacerate, acute or subacute, crimson-purple, sometimes palmatifid; substance thick, surface-cells small; costa stout, pinnated at distances of $\frac{1}{2}$ in.; nerves thick, erecto-patent, dichotomous; margin erose-toothed. Conceptacles on obovate processes, $\frac{1}{12}$–$\frac{1}{6}$ in. long on the veins; walls thick; orifice prominent. Tetraspores in oblong minute sori near the base of the lamina.

Lyall's Bay or Cook's Straits, Foveaux Straits and Otago, *Lyall.*—The finest species of the genus.

2. **D. dichotoma,** *Hook. f. and Harv. Fl. N. Z.* ii. 239. Frond membranous, rose-purple or blood-red, 4–8 in. high, stipitate; stipes 2–3 in. long, oblong-cuneate or elliptic-lanceolate, obtuse or emarginate, at length 2-fid or laciniate; costa stout, dichotomous, vanishing below the apex. Sori rounded. —Fl. Antarct. 185. t. 71. f. 2; Harv. Ner. Austr. 115; J. Ag. Sp. Alg. ii. 682.

Ruapuke and Chalky Bay, *Lyall.* **Lord Auckland's** group and **Campbell's** Island, *J. D. H.*

3. **D. pleurospora,** *Harv. in Fl. N. Z.* ii. 239. Frond 6–8 in. high, stipitate, flabellate, laciniate; segments dichotomous, cuneate, alternate, acute or subobtuse, spreading; margin quite entire; costa dichotomous, vanishing below the tips of the segments. Sori linear, arranged in confluent lines on each side of the costa.

Preservation Harbour, *Lyall.*—Possibly a form of *D. dichotoma,* but the sori appear very different.

4. **D. Leprieurii,** *Mont.;—Fl. N. Z.* ii. 240. Fronds rising from a creeping rhizome, dichotomous or subfastigiate, dull violet-purple, 1–3 in. long, costate, constricted at the axils as if articulate; segments linear-lanceolate, acute at both ends. Tetraspores in lines radiating from the costa to the margin, on the terminal segments and on axillary leaflets.—Montagne in Ann. Sc. Nat. ser. 3. xiii. 196. t. 5. f. 1; J. Ag. Sp. Alg. ii. 682.

Bay of Islands, *J. D. H.* (Guiana, U. States.)—A very curious plant, doubtfully referred to *Delesseria* by Harvey, more confidently by J. Agardh.

§ 2. *Frond pinnately decompound; segments of the primary first pinnatifid and then pinnate.*

5. **D. quercifolia,** *Bory;—Fl. N. Z.* ii. 239. Frond pale red-purple, 3–6 in. long, broadly ovate, pinnatifid and 2-pinnatifid, irregularly lobed, sinuate and toothed; costa stout, giving off opposite nerves to the tips of the lobes. Sori minute, punctiform, very numerous between the nerves on the lobes.—Bory in Duperr. Voy. Bot. 186. t. 18. f. 1; Harv. Ner. Austr. 114. t. 46; J. Alg. Sp. Alg. ii. 692.

East coast, lat. 42°, *Lyall.* (Fuegia, Falkland Islands.)

6. **D. Davisii,** *Hook. f. and Harv. Fl. N. Z.* ii. 239. Frond with a winged cartilaginous stipes, 5–7 in. long, rose-red, membranous, pinnate or pinnatifid; segments obliquely lanceolate, costate, feather-nerved, nerves

alternate, at length lacerate between the nerves, the ultimate segments erecto-patent, costate.—Fl. Antarct. 470. t. 175; Harv. Ner. Austr. 115; J. Ag. Sp. Alg. ii. 689.

Ruapuke, Preservation Harbour, and Chalky Bay, *Lyall.* (Fuegia.)

§ 3. *Frond proliferous, the branches rising from the midrib.*

7. **D. nereifolia,** *Harv. in Fl. N. Z.* ii. 238. Frond red-purple, stiff and elastic when fresh, when old with a stout stipes, 6–8 in. long, 1 in. broad, linear-oblong, obtuse, costate, pinnately veined; branches or leaflets springing from the often denuded stout costa, oblong, quite entire, flat, obtuse, with broad prominent midribs, and feathered with opposite close-set nerves, between which the membrane is reticulated with articulate lines of cells. Fruit unknown.

Preservation Harbour and Stewart's Island, *Lyall.*

8. **D. oppositifolia,** *Harv. in Fl. N. Z.* ii. 239. Frond lake-red, 1–2 in. high, linear-lanceolate, costate, at length 2–3-pinnately compound, with opposite leaflets springing from the costa; larger leaflets with a cylindric costa coated with small cells; smaller leaflets with an articulate costa.

Stewart's Island, rare, *Lyall.*

9. **D. crassinervia,** *Mont.;—Fl. N. Z.* ii. 239. Frond bright red, linear, 1–2 ft. long, more or less closely pinnated throughout with branches 6–8 in. long; branches emitting leaflets from the midrib, which again are proliferously divided; costa very thick; leaflets ovate-lanceolate, acuminate at both ends. Sori elongate, solitary on each side of the midrib.—Mont. in Voy. au Pôle Sud, 164. t. 8. f. 1 (small state); Fl. Antarct. 184; Harv. Ner. Austr. 115; J. Ag. Sp. Alg. ii. 694.

Ruapuke Harbour and Stewart's Island, *Lyall.* **Lord Auckland's** group and **Campbell's** Island, *J. D. H.* (Fuegia, Kerguelen's Land.)—Perhaps only a large southern state of the European *D. Hypoglossum.*

10. **D. ruscifolia,** *Lamour.;—Fl. N. Z.* ii. 239. Frond crimson, 1–3 in. high, with a branched winged stem, linear-oblong, branched throughout proliferously from the costa; leaflets stipitate, young obovate, old linear, obtuse; costa pinnately veined; nerves pellucid when dry. Sori linear, 2 pairs on each side of the costa.—Harv. Ner. Austr. 115; Phyc. Brit. t. 26; J. Ag. Sp. Alg. ii. 695.

Cook's Straits, *Lyall.* (Australia, Fuegia, S. Africa, N. Atlantic.)

54. HEMINEURA, Harv.

Frond rose-red, leaf-like, delicately membranous, flaccid, 1–3-pinnatifid, entire or serrulate; midrib very slender. Conceptacles horned, pierced by a pore on the costa. Tetraspores intramarginal.

Genus of two species, one Tasmanian and the following.

1. **H. cruenta,** *Harv. in Fl. N. Z.* ii. 240. Fronds 4–5 in. long, $\frac{1}{2}$–1 in. broad; segments narrow, ultimate $\frac{1}{12}$ in. broad, acute, quite entire, but

crisped and waved; costa running through the principal part of the frond, not branching to the lobes, which have isolated costæ of their own; principal costa emitting numerous minute lanceolate leaflets.

Massacre Bay, *Lyall.*

55. NITOPHALLUM, Greville.

Characters of *Delesseria*, but no midrib, or an obscure one confined to the base of the frond.

A genus of most beautiful *Algæ*, which are abundant in the colder zones, and arrive at their greatest stature and beauty in the deep southern bays.

§ 1. *Sori subsolitary below the apices of the segments.*

1. **N. minus,** *Sond.;—Fl. N. Z.* ii. 119. Frond small, tufted, sessile, excessively delicate, 1–2 in. long, bright rose-red, linear, dichotomously pinnate from the base, quite entire, traversed by a few very slender longitudinal veins; branches divaricate, short, obtuse, upper dichotomous, multifid, ultimate obtuse. Sori towards the tips of the segments.—Harv. Ner. Austr. 119. *Cryptopleura*, J. Ag. Sp. Alg. ii. 655.

Parasitic, East coast, *Colenso*; Tauranga, *Davies.* (Australia.)

2. **N. uncinatum,** *J. Ag.?;—Fl. N. Z.* ii. 241. Fronds small, tufted, on a creeping rhizome, delicately membranous, sessile, 1–2 in. long, subpinnately dichotomous from the base, wholly veinless; segments linear, $\frac{1}{10}$–$\frac{1}{8}$ in. broad, upper acuminate, often secund; fruiting ones shorter, often 3-foliolate. Sori solitary towards the tips of the segments.—J. Ag. Sp. Alg. ii. 654.

Blind Bay, Cook's Straits, *Lyall.* (Europe.)—Not found in fruit, and hence doubtful.

§ 2. *Sori more or less intramarginal.*

3. **N. palmatum,** *Harv. in Fl. N. Z.* ii. 240. Frond red-brown, thickish, 6–8 in. long and more, broadly cuneate at the base, and then narrowed into a thickened but not costate stipes 1–2 in. long, above palmately divided into many cuneate erect segments with narrow axils; costa 0, but whole frond traversed with microscopic veinlets; larger segments cuneate, dichotomously palmate, smaller oblong, obtuse, quite entire. Sori oblong, often elongate, usually in interrupted lines towards the apex, but continued downwards within the margin, often irregularly scattered.

Eastern coasts, *Colenso* and *Lyall*, apparently common.—The ordinary form strongly resembles *Rhodymenia palmata*, but Harvey enumerates many varieties, some crisped, some rose-red, some with pinnatifid lobes, and some with proliferous margins.

§ 3. *Sori scattered over the whole surface of the frond or its segments.*

† *Frond nerveless.*

4. **N. punctatum ?,** *Grev.;—Fl. Antarct.* 185. Frond subsessile, delicately membranous, rose-red, nerveless, very variable in size and shape, dichotomously fastigiate or dilated and pinnate or palmate; axils rounded.

Sori large, oblong, scattered over the whole frond.—Harv. Phyc. Brit. t. 202, 203; J. Ag. Sp. Alg. ii. 659.

Campbell's Island, *J. D. H.* (Tasmania, N. Atlantic, and Pacific Oceans.)—Specimens imperfect.

5. **N. denticulatum,** *Harv. in Fl. N. Z.* ii. 241. Frond subsessile or narrowed below into a cuneate stipes, above flabelliform, dichotomously branched, or undivided or deeply divided; margin erose and crisped; tips obtuse; stipes sometimes produced upwards and branched. Sori ovate or oblong, excessively numerous, scattered over the whole frond.

Maketu, *Chapman;* Blind Bay, *Davies;* Cook's Straits and East coast, *Lyall.*

6. **N. variolosum,** *Harv. in Fl. N. Z.* ii. 241. Frond rose-red, 2–3 in. long, shortly stipitate, below narrowed into a thickened cuneate stipes, above dichotomously palmate-partite; segments decompoundly multifid, obtuse, upper narrower, covered with scattered cilia or processes; axils patent. Sori scattered, thickened into convex tubercles, containing few tetraspores.

East coast, *Colenso;* Banks's Peninsula, *Lyall.*

7. **N.? suborbiculare,** *Harv. in Fl. N. Z.* ii. 242. Frond rose-red, delicately membranous, subsessile or with a setaceous stipes, suddenly expanding into an entire suborbicular lamina 1 in. broad, entire crenate or lobed; substance of one layer of angular cells. Sori and conceptacles scattered over the whole frond.

Parasitic, Hawke's Bay, *Colenso;* Cook's Straits, *Lyall.*—A curious plant, of doubtful affinity, perhaps to be referred to *Rhodymenia;* if a new genus, Harvey proposes that it should be called *Abroteia.*

†† *Frond with evident nerves.*

8. **N. multinerve,** *Hook. f. and Harv.;—Fl. N. Z.* ii. 241. Frond rose-red, shortly stipitate, 2–4 in. long, rounded or cuneate at the base, simple or cut into numerous ribbon-like segments, which are traversed by numerous parallel distinct nerves.—Fl. Antarct. 473; Harv. Ner. Austr. 119; J. Ag. Sp. Alg. ii. 666.

Massacre and Chalky bays and Stewart's Island, *Lyall.* (Tasmania, Fuegia.)

9. **N. crispatum,** *Hook. f. and Harv. in Fl. Antarct.* 185. *t.* 71. *f.* 1. Frond rose-red, 3–4 in. long, stipitate, oblong or broadly expanded from a cuneate decompound dichotomous and subpinnately laciniate base, with branched basal nerves; segments cuneate, tips and axils obtuse; margins crenulate and crisped. Sori minute, dot-like, scattered over the whole frond. —Harv. Ner. Austr. 120.; J. Ag. Sp. Alg. ii. 664.

Campbell's Island, *J. D. H.*

10. **N. D'Urvillæi,** *J. Agardh;—Fl. N. Z.* ii. 240. Frond 6–10 in. long, narrowed into a stout cylindric dichotomous stipes, 1–3 in. long, palmately dichotomous, the stipes branching into the segments, which are linear subcuneate, the lower crisped, the upper slightly undulate; terminal lobes linear-oblong, very obtuse; divisions of the costa produced nearly to the tops of the laciniæ. Sori dot-like, thickly scattered over the upper lobes.—

J. Ag. Sp. Alg. ii. 667. *Dawsonia*, Bory, Voy. Coquil. t. 19. f. i. *Aglaophyllum*, Mont. in Voy. Bonite, 111.

Ruapuke Island in Foveaux Straits, *Lyall.* (Chili.)

56. PHACELOCARPUS, Endlicher.

Frond red, slender; midrib strong, compressed, distichously pectinate or subterete, 4-fariously aculeate; structure of three strata, surrounding an articulate tube; medullary of longitudinal densely interlaced filaments; intermediate of larger roundish cells; cortical of minute coloured cells in vertical lines. Conceptacles external, pedicelled or sessile, imperfectly known. Receptacles ovoid or globose, containing numerous cavities in which are zoned tetraspores mixed with unicellular peranemata.

A genus of one species of uncertain affinity.

1. **P. Labillardieri,** *J. Agardh, Sp. Alg.* ii. 648;—*Fl. N. Z.* ii. 242. Root a broad disk. Frond dark-red, 1–3 ft. high, $\frac{1}{20}$–$\frac{1}{12}$ in. diameter, excessively branched; branches irregular, the smaller closely pinnate or 2-pinnate, all of them pectinate with subulate branchlets. Receptacles ovoid, axillary, pedicelled, ovoid.—Harv. Phyc. Austr. iii. t. 163. *Euctenodus* and *Ctenodus*, Kuetzing. *Sphærococcus*, Agardh.

Laminarian zone and deeper water, abundant along all the shores, *Sinclair, Colenso,* etc. (Australia, Tasmania.)

57. CALLIBLEPHARIS, Kuetzing.

Frond dull-red, flat, between cartilaginous and membranous, dichotomously pinnate and fimbriate; structure of two strata; medullary of several series of large roundish angular cells; cortical of minute coloured cellules. Conceptacles sessile, thick-walled, enclosing a tuft of moniliform spore-threads. Tetraspores zoned, scattered amongst the cortical cells.

A fine genus, native of both temperate zones.

1. **C.? tenuifolia,** *Harv. in Fl. N. Z.* ii. 243. Frond 6–8 in. long, 1 in. broad, bright rose-red, delicately membranous, dichotomously pinnate; margins proliferous; segments much attenuated at the base, often subciliate and toothed. Tetraspores scattered, transversely divided.

Chalky Bay, *Lyall.*—Dr. Harvey remarks that in the absence of conceptacles the genus of this plant is doubtful; it may be a *Rhodophyllis.*

58. SARCODIA, J. Agardh.

Frond rose-red, fleshy, cartilaginous when dry, flat, repeatedly dichotomously-branched and bearing marginal segments; structure of 3 strata; medullary of loosely interlaced filiform cells; intermediate of rounded angular cells, gradually passing into the other stratum of minute vertical rounded cells. Conceptacles hemispheric, thick-walled, on the disk and margins of the frond, opening by a pore, enclosing obovate spores arranged on dichotomous filaments which radiate from a central receptacle. Tetraspores scattered, transversely divided.

A genus of a single species.

1. **S. Montagneana,** *J. Ag. Sp. Alg.* ii. 623;—*Fl. N. Z.* ii. 242. Frond 4–8 in. long, cuneate at the base, then expanding into dichotomous lobes ½–1½ in. broad, tips often truncate and proliferous; axils obtuse.—*Rhodymenia Montagneana*, Hook. f. and Harv.; Harv. Ner. Austr. t. 48. *Rhodophyllis Montagneana*, Kuetzing.

Bay of Islands, *J. D. H.*, *Lyall.*

59. GRACILARIA, Greville.

Frond dull red, cartilaginous, cylindric, compressed or rarely flat, irregularly branched; structure of two strata; medullary of large angular colourless cells, surrounded by smaller coloured ones; cortical of minute coloured cells in vertical lines. Conceptacles conical or hemispherical, sessile, thick-walled, at length opening by a pore, enclosing obovate spores arranged on a tuft of filaments. Tetraspores cruciate or 3-parted, scattered beneath the cortical cells.

A large genus, found in both temperate zones.

1. **G. confervoides,** *Grev.;—Fl. N. Z.* ii. 243. Fronds tufted, purple, subcartilaginous, 2 in. to 2 ft. long, as thick as a sparrow- or crow-quill, very variable in branching, etc.; branches long, undivided; branchlets subsecund, filiform close-set, filiform attenuate at both ends. Conceptacles hemispherical, extremely numerous. Tetraspores also abundant, on thickened filiform branchlets.—Harv. Phyc. Brit. t. 65; J. Ag. Sp. Alg. ii. 587.

Probably common, from Hawke's Bay, *Colenso*, to Ruapuke, *Lyall.* (Native of both temperate oceans.)

2. **G. multipartita,** *Harv.*, var. *polycarpa*, J. Ag. Sp. Alg. ii. 601. Frond between cartilaginous and membranous, 4–12 in. high, dull purple, flattened, cleft to the base palmately or dichotomously; segments linear-cuneate, acute, entire, jagged or multifid. Conceptacles scattered over the whole surface of the frond.—Harv. Phyc. Brit. t. 15; J. Ag. Sp. Alg. ii. 601.

Blind Bay, Cook's Straits, *Lyall.* (Native of both temperate and tropical oceans.)

3. **G. coriacea,** *Harv.;—Fl. N. Z.* ii. 243. Closely resembling *G. multipartita*, but much thicker, more coriaceous, minutely wrinkled when dry, and the tips of the segments obtuse. Conceptacles deeply sunk in the frond, depressed at the apex with a large pore.—*Rhodymenia ? coriacea*, Hook. f. and Harv. in Lond. Journ. Bot. iv. 545.

Bay of Islands, Cook's Straits, *Lyall;* Otago, *Lindsay.*

60. MELANTHALIA, J. Agardh.

Frond deep purple, linear, flat, dichotomously flabellately branched, coriaceous, solid, densely cellular; inner substance of colourless elongate angular cells, outer of many series of most minute coloured cells. Conceptacles marginal, thick-walled, hemispherical; spores on moniliform filaments, covering a prominence on the floor of the conceptacle. Tetraspores unknown.

An Australian and New Zealand genus.

1. **M. abscissa,** *Hook. f. and Harv. Fl. N. Z.* ii. 242. Fronds tufted, blackish when dry, 2–3 in. long, fastigiately dichotomously decompound; segments 2-edged; margins obtuse, unequally thickened here and there, lower $\frac{1}{20}$ in. long, upper narrower, subcuneate below the forkings.—J. Ag. Sp. Alg. ii. 613. *Fucus,* Turn. Hist. t. 137. *Chondrococcus,* Kuetzing.

New Zealand, *Banks.*

2. **M. Jaubertiana,** *Mont.;—Fl. N. Z.* ii. 242. Frond similar to *M. abscissa,* but larger, 4–6 in. long; segments quite equal throughout.—J. Ag. Sp. Alg. ii. 613.

Common along all the coasts. (Tasmania.)

61. GELIDIUM, Lamouroux.

Frond dull red or purplish, linear, compressed, pinnate, horny, solid; axis of densely interlaced tenacious fibres; walls of small polygonal cells, the outermost minute in horizontal series. Conceptacles 2-celled, formed of two opposite confluent conceptacles, immersed in swollen branchlets, containing oblong spores which are attached to the walls of the septum. Tetraspores forming sori on dilated branchlets.

Native of both temperate zones.

1. **G. corneum,** *Lamour.;—Fl. N. Z.* ii. 243. Frond 2–6 in. high, between cartilaginous and horny, nearly flat, distichously branched, setaceous or more or less stout; branches 1–2-pinnate, linear, attenuated at each end, pinnules mostly opposite, spreading, obtuse, bearing elliptic conceptacles within their apices.—Harv Phyc. Brit. t. 53; J. Ag. Sp. Alg. ii. 469.

Probably abundant, Hawke's Bay, *Colenso;* Banks's Peninsula, *Lyall.* (All seas.)

2. **G. asperum,** *Grev.;—Fl. N. Z.* ii. 243. Frond very variable in size, two-edged, decompoundly pinnate; pinnæ linear, narrowed at both ends, rather regularly again pinnate, serrulate.—J. Ag. Sp. Alg. ii. 475.

New Zealand, *Baume* (J. Agardh, l. c.) (Australia.)

62. CAULACANTHUS, Kuetzing.

Frond purplish-brown, black when dry, terete, excessively branched; branchlets aculeate, cartilaginous; axis of young parts tubular, of a single flexuous articulate tube, giving off alternate branches which divide excessively, their divisions becoming packed and forming the surface; adult part altogether cellular. Conceptacles unknown. Tetraspores in slightly swollen branchlets, sunk in the cortical threads, transversely divided.

A curious genus, of a N. Atlantic species and the following.

1. **C. spinellus,** *Kuetz.;—Fl. N. Z.* ii. 244. Fronds 1½–1 in. long,

setaceous, forming broad tufts, rigid, irregularly excessively branched; branches very spreading, simple or forked, set with horizontal, often secund spinous branchlets.—J. Ag. Sp. Alg. ii. 434. *Rhodomela ? spinella*, Hook. f. and Harv. in Lond. Journ. Bot. iv. 534.

Common on all the shores, on Corallines, shells, etc., *Colenso.*

63. HYPNEA, Lamouroux.

Frond pink or dull purplish, filiform, much branched; branchlets subulate, cartilaginous; axis fibrous or cellular, dense, surrounded by several series of colourless polygonal cells, which decrease externally, outer series minute, coloured. Conceptacles sessile or immersed, thick-walled, containing stalked spores attached to filaments. Tetraspores zoned, sunk in the surface of swollen branchlets.

Natives of both temperate zones.

1. **H. musciformis,** *Lamour.;—Fl. N. Z.* ii. 244. Frond tufted, dark green and purple, excessively branched, 4–6 in. high; branches filiform, incurved, thickened at the tips; branchlets spreading, sporiferous, pod-like, lanceolate, beaked; those bearing the conceptacles with divaricating spinescent divisions.—J. Ag. Sp. Alg. ii. 442.

New Zealand, *Banks and Solander.* (N. S. and tropical Atlantic, Indian Ocean, Pacific, Australia.)—I have seen no N. Z. specimens.

2. **H.? multicornis,** *Mont. Voy. au Pôle Sud,* 152. *t.* 9. *f.* 1;—*Fl. Antarct.* 187. Frond 2–3 in. long, compressed, excessively corymbosely branched, ultimate branchlets subdistichous, alternate or subsecund, forked; branchlets spreading, recurved, hooked. Fruits in the thickened tips of the hooked branchlets.—*Rhodomela,* Mont.

Lord Auckland's group, *D'Urville.*

64. PTEROCLADIA, J. Agardh.

Frond red, linear, flat, 2-edged; midrib strong or indistinct, pinnate; structure of 3 strata; medullary of densely packed longitudinal fibres; intermediate of larger roundish cells; cortical of moniliform rows of coloured minute cells vertically placed. Conceptacles thick-walled, hemispherical, 1-locular, on one surface of the frond, containing obovate stalked spores; spores inserted on a parietal receptacle which is united to the pericarp by delicate filaments. Tetraspores cruciate, forming sori on dilated branchlets.

Differs from *Gelidium* in the 1-celled conceptacle.

1. **P. lucida,** *J. Agardh;—Fl. N. Z.* ii. 244. Root branching. Frond extremely variable, 6–18 in. long, red-purple, $\frac{1}{12}$–$\frac{1}{4}$ in. broad, pinnately decompound, variable in breadth and branching; often proliferous; pinnules thin, flat, long or short and crisped; midrib strong or faint.—J. Ag. Sp. Alg. ii. 483. *Gelidium,* Sond.; *Fucus,* Br. in Turn. Hist. t. 238.

Common throughout the islands. (W. and S.W. Australia.)

65. APOPHLŒA, Harvey.

Frond red-purple, black when dry, spongy, cylindric, densely dichotomously fastigiate, of 2 strata of cells; axis of dense, slender, anastomosing threads; cortical of vertical, elongate, dichotomous, articulate threads, immersed in a soft gelatine. Conceptacles scattered, immersed in the cortical threads, enclosing parietal fascicled threads. Tetraspores transversely divided.

A curious genus, confined to New Zealand.

1. **A. Sinclairii,** *Harv. in Fl. N. Z.* ii. 244. *t.* 116 B. Root a broad leathery disk. Fronds 1 in. high, very stout, densely fastigiate; branches short, thick, obtuse.—J. Ag. Sp. Alg. ii. 458.

Common on rocks, *Colenso*, etc.

2. **A. Lyallii,** *Hook. f. and Harv. Fl. N. Z.* ii. 244. *t.* 166 A. Frond 5–6 in. high, with a slender stipes thickening upwards, $\frac{1}{4}$–$\frac{1}{3}$ in. diameter, then forking repeatedly; branches flexuous, patent, stout; axils rounded; tips obtuse.

Preservation Harbour and Otago, *Lyall.*

66. PEYSSONNELLIA, Decaisne.

Frond dull red or brownish, flat, broad, fan-shaped, rooting by fibrils from the lower surface, zoned; structure of 2 strata, that of the lower surface of horizontal cylindric cells, cohering in longitudinal filaments, that of the upper surface of similar cells in lines at right angles to the others. Conceptacles superficial, containing either globular spores in moniliform filaments or cruciate tetraspores.

A scarce genus in the northern seas, where one species exists on the Mediterranean and Atlantic coasts; more common in the south temperate zone.

1. **P. rugosa,** *Harv. in Fl. N. Z.* ii. 245. Frond 1–2 in. broad, dark, red-brown, attached by the whole lower surface, between membranous and coriaceous, orbicular, wrinkled over the whole surface. Fructification unknown.

Cape Kidnapper, on the surface of rocks covered with sand, *Colenso.*

67. NEMALION, Duby.

Frond red or purple, cylindric, gelatinous, elastic, dichotomous; structure of 3 strata; medullary of long, subsimple, interlaced filaments, which radiate outwards, forming the second stratum; this consists of a layer of anastomosing, horizontal, fastigiate, dichotomous, moniliform threads, invested in gelatine, the tips of which constitute the outermost stratum or periphery. Spore-mass spherical, of many clavate radiating spore-threads, immersed in the peripheric threads. Tetraspores triangularly divided, in the terminal cells of the peripheric filaments.

Natives of northern and southern temperate zones.

1. **N. ramulosum,** *Harv. in Fl. N. Z.* ii. 245. Frond 6 in. long, $\frac{1}{6}$–$\frac{1}{4}$ in. broad, compressed, once or twice forked, irregularly densely set with spreading branches and branchlets; branches unequal, 1–2 in. long, horizontal, obtuse, simple, or again set with lateral often forked branchlets. Fructification unknown.

Otea, *Lyall.*

68. SCINAIA, Bivona.

Frond red, terete or compressed, dichotomous, between gelatinous and membranous, filled with fluid gelatine, traversed by a fibrous axis, from which slender branched filaments radiate to the periphery, which is formed of a thin stratum of small angular cells. Spore-masses globular, of jointed spore-threads, enclosed in a membrane, suspended on the inner wall of the frond, bearing pyriform spores.

A genus of only one known species.

1. **S. furcellata,** *Bivona;—Fl. N. Z.* ii. 245. Frond tender, rose or lake-red, 2–4 in. long, cylindric, diameter from a sparrow's to a swan's quill, branched, with a semicircular outline; tips and axils obtuse.—J. Ag. Sp. Alg. ii. 422. *Ginnania*, Mont.; Harv. Phyc. Brit. t. 69. *Myclomium*, Kuetzing.

East coast, *Cunningham ?* (Europe, N. America, Chili, Pacific, Tasmania, S. Africa.)

69. WRANGELIA, Agardh.

Frond rose-red or purplish, filiform, articulate, branched, of a single tube, sometimes coated with minute cells; branchlets minute, opposite or whorled round the nodes. Spore-masses usually involucrate, containing several clusters of spores in the uppermost whorls of the branchlets. Tetraspores 3-parted, scattered on the branchlets.

A beautiful small genus, native of both temperate zones.

1. **W. Lyallii,** *Harv. in Fl. N. Z.* ii. 236. Frond flaccid, carmine-red, setaceous, 4–5 in. long, pinnate or 2-pinnate, pellucid above, veined near the base; branchlets elongate, articulate, simple, with whorled or opposite branchlets at the nodes; branchlets pinnate; articulations not coated with cells; walls thick, of the branches many times longer than broad. Tetraspores sessile on the inner faces of the branchlets, 1–2 at each node. Spore-masses minute, terminal, hardly involucrate.

Ruapuke and Preservation Harbours, *Lyall.*

2. **W. (?) squarrulosa,** *Harv. in Fl. N. Z.* ii. 236. Frond rigid, setaceous, articulate throughout, distichously branched; branches opposite and alternate, very compound, loosely divided or closely pinnate or 2-pinnate; pinnæ usually opposite; branchlets minute, squarrose, whorled, clothing the nodes, dichotomously multifid, spinescent; articulations 2–3 times longer than broad, not coated with cells. Tetraspores sessile on lateral branchlets.

Preservation Harbour, *Lyall.*—Dr. Harvey describes two varieties, both gathered on the same spot, one pinnate with the pinnæ opposite and very close; the other vaguely branched, scarcely pinnate. The genus is doubtful, the spore-masses being unknown.

70. RHODYMENIA, Greville.

Frond flat, dull or rose-red or purplish, ribless, membranous, dichotomous, palmate or flabellate; structure of 2 strata of cells; medullary large, oblong, polygonal; cortical minute, in vertical lines, coloured. Conceptacles scattered over the frond or in sori near the tips of the segments, sessile, hemispherical, thick walled, opening by a pore, enclosing a dense tuft of sporiferous filaments in gelatine. Tetraspores cruciate or 3-partite, scattered or in sori.

A large genus in both temperate zones.

§ 1. *Tetraspores scattered over the surface of the frond.*

1. **R. sanguinea,** *Harv. in Fl. N. Z.* ii. 248. Frond 12–14 in. long, purplish blood-red, firmly membranous, with a cylindric stipes and cuneate base, deeply laciniate; segments broad or narrow, cuneate, attenuated; margins simple or foliiferous; leaflets cuneate; axils rounded. Conceptacles very numerous, scattered densely over the whole frond. Tetraspores abundant on the smaller segments.

Foveaux Straits, *Lyall.*

2. **R. lanceolata,** *Harv. in Fl. N. Z.* ii. 248. Segments of the frond 6–8 in. long, quite simple, narrow, $\frac{1}{4}$ in. broad, tapering to the base and acute apex, purple red; fruit as in *R. sanguinea.*

Port Cooper, *Lyall;* Otago, *Lindsay.*—Perhaps only a variety of *R. sanguinea,* with simple longer segments, but the substance is softer, and the cortical layer more developed.

3. **R. prolifera,** *Harv. in Fl. N. Z.* ii. 249. Frond dull red-brown, 4–8 in. long. cuneate at the base, with a short stipes; axils rounded; segments spreading, $\frac{1}{4}$–1 in. wide, linear or subcuneate, sparingly divided, proliferous at the apex; leaflets often stipitate, linear, acute or attenuate at the base, simple or forked. Sori indefinite, remote from the attenuated tips of the segments.

Hawke's Bay, *Colenso.*—Similar to *R. linearis,* but larger, thicker, generally proliferous, and with differently placed tetraspores.

4. **R. ornata,** *Mont.;—Fl. Antarct.* 186. Frond bright purple, subcarnose, oblong lanceolate, very broad, proliferous on the margins; segments or leaflets obovate, substipitate, very large, palmately lobed. Conceptacles scattered over the whole frond.—Mont. Voy. au Pôle Sud, 149. t. 11.

Lord Auckland's group, *D'Urville.*—Unknown to us? Consistence of *Iridæa edulis* (Mont.)

§ 2. *Tetraspores collected in sori towards the tips of the segments.*

5. **R. linearis,** *J. Ag.;—Fl. N. Z.* ii. 248. Frond elongate, 4 in. long, decompound; segments narrow, flabellate, altogether linear, $\frac{1}{12}$–$\frac{1}{6}$ in. broad, terminal obtuse or emarginate; axils subacute. Conceptacles apiculate, never marginal, scattered over the disk of the frond. Tetraspores below the tips of the segments.—J. Ag. Sp. Alg. ii. 379. *R. Palmetta,* Hook. f. and Harv. in Lond. Journ. Bot. vii. 444, not of Greville.

East coast, *Colenso;* Otago, *Lyall.* **Lord Auckland's** group and **Campbell's** Island, *Turnbull, J. D. H.*

6. **R. corallina,** *Grev.;—Fl. N. Z.* ii. 248. Frond bright red or purplish, a foot long, broader than long, flabellately decompound, with a terete stipes, cuneate at the base; segments linear, $\frac{1}{4}$–$\frac{1}{8}$ in. broad, linear or subcuneate; axils rounded. Conceptacles crowded on the disk of the penultimate segments. Tetraspores in sori towards the tips of the fronds.—J. Ag. Sp. Alg. ii. 379. *Sphærococcus*, Bory, Voy. Coquille, t. 16.

East coast and Cook's Straits, *Lyall.* **Lord Auckland's** group, *Hombron.* (Tasmania, Kerguelen's Land, Chili.)

7. **R. dichotoma,** *Hook. f. and Harv. Fl. Antarct.* i. 186. *t.* 72. *f.* 1; *Fl. N. Z.* ii. 248. Frond rose-red, membranous, very broad from a cuneate base; axils rounded; segments spreading, linear or cuneate, obtuse, at length emarginate or 2-fid.—*Callophyllis*, Kuetzing.

Queen Charlotte's Sound, *Lyall;* **Campbell's** Island, *J. D. H.*

Section doubtful.

8. **R. epimenioides,** *Harv. in Fl. N. Z.* ii. 248. Frond rose-red, delicately membranous, many times dichotomous, flabelliform, narrowed at the base into a short cartilaginous stipes, which is produced into the frond as a short costa; segments linear or cuneate, very obtuse, $\frac{1}{2}$ in. wide; axils rounded or subacute. Fructification unknown.

Otago Harbour, *Lyall, Lindsay.*—Very similar to *R. dichotoma*, but the cells of the medullary stratum are larger, thin-walled, and rapidly expand when moistened; it also resembles *Epymenia obtusa*, but has a totally different structure.

71. RHODOPHYLLIS, Kuetzing.

Frond rose-red, flat, membranous, dichotomously or pinnately decompound, usually margined by slender leafy processes; structure of 2 strata; medullary of large, colourless, roundish angular cells; cortical of one or few series of coloured minute cells. Conceptacles marginal, external, containing a compound nucleus of radiating bundles of spore-threads, enclosed in a membrane. Tetraspores zoned, immersed in the cortical cells of the frond on its marginal processes.

A small genus, native of both temperate zones.

1. **R. bifida,** *Kuetzing.* Frond bright red or purplish, delicately membranous, decompound, subflabellate, 1–2 in. long; segments fastigiate, linear or cuneate, $\frac{1}{6}$–$\frac{1}{4}$ in. broad, terminal, obtuse; margins entire, often set with crowded confluent leafy cilia. Conceptacles marginal or scattered over the terminal lobes. Tetraspores in clouded sori.—J. Ag. Sp. Alg. ii. 388.

Lord Auckland's group, *D'Urville.* (Atlantic, Mediterranean and Pacific Oceans.) I have seen no southern specimens.

2. **R. Gunnii,** *Harv. in Fl. N. Z.* ii. 247. Frond decompoundly pinnatifid, rose-red, delicately membranous, 8–12 in. long; segments alternate or subopposite, sub-2-pinnatifid, close set; ultimate toothed; tips obtuse; axils

rounded. Conceptacles marginal. Tetraspores scattered towards the tips of the segments.—*Cladhymenia*, Harv. in Lond. Journ. Bot. iv. 540; Ner. Austr. t. 32 (tetrasp. incorrect); J. Ag. Syst. Alg. ii. 386. *Callophyllis*, Kuetzing.

Chalky Bay and Preservation Harbour, *Lyall*. (Tasmania.)

3. **R. membranacea,** *Hook. f. and Harv. in Fl. N. Z.* ii. 247. *t.* 117. Frond similar to that of *R. Gunnii*, dotted over with red specks, often proliferously fimbriate; segments alternate; primary broad; secondary narrow; ultimate linear-oblong, subacute, narrowed at the bases; axils rounded. Conceptacles marginal. Tetraspores confined to the ultimate segments.—*Rhodymenia* and *Halymenia*, Harv. in Lond. Journ. Bot. vi. 405 and 448. *Stictophyllum*, Kuetzing. *Euthoria*, J. Ag. Sp. Alg. ii. 385.

East coast, *Colenso*; Cook's Straits, *Lyall*. (Tasmania.)

4. **R. (?) angustifrons,** *Harv. in Fl. N. Z.* ii. 247. Fronds densely tufted, 2–3 in. high, membranous, deep red, excessively dichotomously or irregularly branched from a narrow base; segments divaricate, dichotomous, 3-multifid, linear, $\frac{1}{20}$ in. broad; terminal often secund, obtuse or acute; axils rounded. Fructification unknown.

Port Nicholson and Bluff Harbour, *Lyall*.—Genus doubtful.

5. **R. (?) lacerata,** *Harv. in Fl. N. Z.* ii. 247. Frond rose-red, delicately membranous, subdichotomous, narrowed into a cartilaginous filiform stipes; primary segments broad; secondary narrower, dichotomously multifid; ultimate elongate, attenuate, almost subulate, here and there sparingly toothed. Fructification unknown.

Port William, on rocks, *Lyall*.—Genus doubtful.

72. DASYPHLŒA, Montagne.

Frond rose-red or purple, cylindric, dendroid, between gelatinous or cartilaginous and membranous, coated with microscopic hyaline hairs; structure complex; central axis an articulate filament; intermediate of branching radiating moniliform filaments; cortical membranous, of several series of minute coloured cells. Conceptacles immersed in the branchlets, containing tufts of moniliform spore-threads, radiating from a central axis. Tetraspores zoned, in wart-like thick-walled conceptacles.

A genus of N. Australian and New Zealand species.

1. **D. insignis,** *Mont. in Voy. au Pôle Sud*, 102. *t.* 8. *f.* 3;—*Fl. N. Z.* ii. 254. Frond rose-purple, 6–8 in. long, very delicate, subtubular, as slender as a crowquill, broadly ovate in outline, vaguely decompoundly branched; branches spreading, attenuate at both ends.—J. Ag. Sp. Alg. ii. 215.

Akaroa, *D'Urville*.

73. PLOCAMIUM, Lyngbye.

Frond bright pink, between membranous and cartilaginous, flat or compressed, linear, pinnately decompound; pinnules short, acute, alternately secund in pairs or threes or fours, of 2 strata of cells; inner cells oblong, longitudinal; outer small, polygonal coloured. Conceptacles sessile or stalked, hemispherical, opening by a pore, containing a mass of angular spores, on filaments which radiate from the base. Stichidia containing oblong transversely-parted tetraspores.

A genus of most beautiful *Algæ*, found in both temperate zones.

§ 1. THAMNOPHORA.—*Frond usually membranous, flat, ecostate or with a slender costa; branchlets alternately in pairs; conceptacles and stichidia axillary.*

1. **P. procerum,** *J. Agardh;—Fl. N. Z.* ii. 246. Root branching. Frond 1–2 ft. long, linear, almost ribless, pectinate-pinnate, $\frac{1}{12}$–$\frac{1}{4}$ in. diameter, bright pink; pinnæ alternately geminate, lower one and divisions of the upper from a broad base, acuminate, subulate, entire or externally serrulate. Conceptacles axillary, stalked, 1–4 together. Stichidia axillary, tufted, simple, falcate, acute.—Harv. Phyc. Austr. iv. t. 223; Ner. Austr. 122.

New Zealand, *Lyall.* (Australia, Tasmania.)

2. **P. Corallorhiza,** *Harv. Fl. N. Z.* ii. 245. Root branching. Frond 12–18 in. long, clear red, stipitate, membranous, broadly linear, $\frac{1}{2}$ in. broad, subcostate at the base, pinnate or 2-pinnate; segments alternately geminate, lower shortly subulate, upper incised or pinnatifid; outer margin of each segment serrate. Conceptacles pedicelled, and stichidia axillary, crowded.—Harv. Ner. Austr. 121; J. Ag. Sp. Alg. ii. 402.

Dusky Bay, *Forster.*—Perhaps the habitat is erroneous. (South Africa.)

3. **P. costatum,** *Hook. f. and Harv. Fl. N. Z.* ii. 246. Frond narrow, membranous, flat, a foot long, $\frac{1}{12}$ in. wide, decompoundly-pinnate, with a stout percurrent costa; pinnæ lanceolate, alternately geminate, lower subulate, upper elongate, laciniate, falcate or straight; outer margin of the segments serrate. Stichidia axillary, crowded, many times forked.—Harv. Ner. Austr. 122; J. Ag. Sp. Alg. ii. 403. *Thamnophora Cunninghamii*, Grev.

Common in the Bay of Islands, etc., *Cunningham.* (Australia, Tasmania.)

4. **P. angustum,** *Hook. f. and Harv. Fl. N. Z.* ii. 246. Frond rather cartilaginous, 6 in. long, $\frac{1}{20}$ in. broad, with a stout decurrent costa, decompoundly pectinately pinnate; pinnæ alternately geminate, lower subulate, upper elongate, laciniate; outer margin of the segments quite entire. Conceptacles solitary. Stichidia crowded, axillary.—Harv. Ner. Austr. 122; J. Ag. Sp. Alg. ii. 402.

Common from the Bay of Islands to Otago. (Australia, Tasmania.)

5. **P. cruciferum,** *Harv. in Fl. N. Z.* ii. 246. Scarcely to be distinguished by habit or other characters from *P. angustum,* except in the stichidia, which are cruciform or palmately lobed, or even multi-radiate, with lobes short, the whole crowded with tetraspores.

East coast, *Colenso.*

6. **P. abnorme,** *Hook. f. and Harv. Fl. N. Z.* ii. 246. Frond 1–2 in. high, narrow, slender, with an obsolete costa, pinnately decompound; pinnæ

and pinnules alternately geminate, ultimate very narrow; branches subulate, quite entire, acute. Stichidia axillary, subsolitary, lanceolate, simple or forked, sometimes formed of the ultimate pinnules.—Harv. Ner. Austr. 123. t. 42; J. Ag. Sp. Alg. ii. 401.

Bay of Islands, *Cunningham*, etc.; Maketa, *Chapman*.

7. **P. dispermum,** *Harv. in Fl. N. Z.* ii. 246. Frond 6–12 in. high, $\frac{1}{20}$ in. broad, stipitate, costate, decompoundly pectinate-pinnate, subflabellate; pinnæ alternately geminate, lower and laciniæ of the upper narrow, subulate, acute, quite entire. Conceptacles supra-axillary, sessile, scattered. Stichidia extremely minute, axillary, decompoundly branched or lobed; lobes ovoid or oblong, stalked, each with about 2 large tetraspores filling the lobe.

East coast, *Colenso*; Foveaux and Cook's Straits, *Lyall.*

§ 2. Euplocamium.—*Frond cartilaginous, plano-compressed; branches alternately in threes or fours; conceptacles and stichidia scattered, lateral on the branches.*

8. **P. coccineum,** *Lyngbye;—Fl. N. Z.* ii. 246. Fronds tufted, 2–12 in. long, compressed, 2-edged, $\frac{1}{20}$ in. diameter, excessively divided, divisions set throughout with patent alternate branches, and these again with distichous subulate branchlets, pectinate on the inner margin. Conceptacles solitary, sessile on the margins of the upper branches. Stichidia solitary, sessile, branched.—Harv. Ner. Austr. 123; Phyc. Brit. t. 44; J. Ag. Sp. Alg. ii. 395.

Abundant and very variable from the Bay of Islands to **Lord Auckland's** group and **Campbell's** Island. (Atlantic, Pacific, and Antarctic Oceans.)

74. STENOGRAMME, Harvey.

Frond rose-red, membranous, flat, nerveless, dichotomously laciniate; structure of two strata; medullary of several rows of roundish angular cells; cortical of minute coloured cells. Conceptacles confluent into linear rib-like masses, containing minute spores within a thick integument. Tetraspores in superficial convex masses, formed in strings, cuneate.

A beautiful *Alga*, most abundant in New Zealand.

1. **S. interrupta,** *Montagne;—Fl. N. Z.* ii. 249. Root a small disk. Frond 6–12 in. long and broad, flabellate; stem short, at once dilating into the frond; laciniæ $\frac{1}{4}$–$\frac{1}{2}$ in. broad, with obtuse tips and axils.—Harv. Ner. Bor. Am. t. 19 C; Phyc. Brit. t. 157; Phyc. Austr. iv. t. 220; J. Ag. Sp. Alg. ii. 391.

Common along all the coasts, *Colenso*, etc. (Atlantic and Pacific Oceans.)

75. GYMNOGONGRUS, Martius.

Frond dull red-brown or purplish, horny, coriaceous or rather fleshy, flat or filiform, dichotomously branched; structure of two strata; medullary of large colourless angular cells; cortical of moniliform minute coloured cells vertically disposed and set in gelatine. Conceptacles more or less prominent, containing aggregate clusters of spores. Tetraspores developed from cells contained in thick-walled warts.

Natives of both temperate zones.

1. **G. furcellatus,** *J. Ag. Sp. Alg.* ii. 318;—*Fl. N. Z.* ii. 250. Frond tufted, 2–6 in. high, very variable, plano-compressed, dichotomously fastigiate, often secundly proliferous; segments narrow-linear, very variable. Conceptacles often in opposite pairs.—*Sphærococcus*, Agardh. *Gracilaria furcata* and *torulosa*, Hook. f. and Harv. in Lond. Journ. Bot. iv. 545, and vii. 444.

Common on the coasts, as far south as Otago. (Chili, Peru.)

2. **G. vermicularis,** *J. Ag. Sp. Alg.* ii. 323;—*Fl. N. Z.* ii. 250. Frond 2–4 in. long, tufted, purplish, subcompressed, almost terete, $\frac{1}{20}$–$\frac{1}{10}$ in. broad, dichotomously fastigiate, often secundly or fasciculately proliferous; terminal segments obtuse. Conceptacles crowded, rarely solitary, giving the frond a nodose appearance.—*Chondrus*, Grev.; *Fucus*, Turn. Hist. t. 221.

Milford Haven, *Lyall.* (S. Africa, Chili.)

76. CALLOPHYLLIS, Kuetzing.

Frond red, flat, rather broad, between membranous and fleshy, dichotomously and flabellately branched; structure of 2 strata of cells; medullary of large rounded cavities separated by a network of minute cells; cortical of vertical series of minute coloured cells. Conceptacles superficial or immersed, often marginal, thick walled, containing masses of spores. Tetraspores cruciate, dispersed through the cortical layer.

A handsome southern genus of numerous species.

1. **C. variegata,** *Kuetzing*;—*Fl. N. Z.* ii. 250. Frond purple-red, membranous, 6 in. long, decompoundly pinnate; pinnæ close-set, lower simple, linear, upper decompound, dilated, terminal obtuse, doubly-crenate. Conceptacles marginal, immersed.—J. Ag. Sp. Alg. ii. 302. *Rhodymenia*, Mont.; Fl. Antarct. 475. *Sphærococcus*, Bory, Voy. Coquil. t. 14.

Tauranga, *Davies.* **Lord Auckland's** group, *D'Urville.* (Chili, Fuegia.)

2. **C. Hombroniana,** *Kuetzing*;—*Fl. N. Z.* ii. 251. Frond 1–2 ft. long, bright red-purple, decompoundly pinnate, and subflabellately dichotomous from a flattened stipes; pinnæ erecto-patent, lower simple, linear, upper decompoundly dilated with the margins fimbriate, terminal obtuse, toothed. Conceptacles immersed in the fimbriæ.—J. Agardh, Sp. Alg. ii. 303. *Rhodymenia*, Mont. in Voy. au Pôle Sud, 156. t. 1. f. 2; Fl. Antarct. i. 186. t. 72. f. 2.

East coast, *Colenso*; Akaroa, *Raoul*; Foveaux Straits, *Lyall*; Otago, *Lindsay.* **Lord Auckland's** group, *Hombron.*

3. **C. acanthocarpa,** *Harv. in Fl. N. Z.* ii. 251. Very similar to *C. Hombroniana*, but the conceptacles are aculeate.

East coast and Port Cooper, *Colenso*, *Lyall.*

4. **C. coccinea,** *Harv.*;—*Fl. N. Z* ii. 250. Frond membranous, deli-

cate, 4–6 in. long, bright red, 2–3-chotomous or multifid; segments narrow, linear, $\frac{1}{6}$–$\frac{1}{4}$ in. broad, pinnate on the margins; pinnæ like the segments, outer entire on the inner side above the sinus. Conceptacles immersed in the disk of the frond.—J. Ag. Sp. Alg. ii. 301. *Sphærococcus*, Harv. *Chondrococcus*, Kuetzing.

Tauranga, *Davies*. (Tasmania.)

5. **C. erosa,** *Hook. f. and Harv. Fl. N. Z.* ii. 250. *t.* 118. Frond crimson-purple, 6–12 in. long, sublinear, pinnately dichotomously decompound, densely or laxly branched; principal laciniæ $\frac{1}{4}$–$\frac{1}{2}$ in. wide, erect or erecto-patent, elongate, more or less compound; margins with toothed or laciniate lobules or proliferous; axils all rounded. Conceptacles sessile, marginal, spherical.

Port Cooper and Foveaux Straits, *Lyall*; Otago, *Lindsay*.

6. **C. asperata,** *Harv. in Fl. N. Z.* ii. 250. Frond rosy-crimson, 4–5 in. long, delicately membranous, sessile, flabelliform, dichotomously palmate or decompound; segments $\frac{1}{2}$–1 in. broad, broader upwards, crenate and multifid at the fastigiate tip; margin flat or curved, fimbriate or often bearing cilia; surface generally rough, with minute subulate processes. Conceptacles immersed, scattered over the surface, convex.

Port Nicholson, *Lyall*.

77. KALLYMENIA, J. Agardh.

Frond bright rose-red, broad, flat, ribless, between membranous and fleshy; structure of 3 strata of cells, innermost of matted anastomosing filaments; intermediate of large globular colourless cells; cortical of minute vertically-placed coloured cells. Conceptacles immersed often prominent, enclosing compound masses of spores. Tetraspores cruciately parted, scattered amongst the cortical cells.

Beautiful *Algæ*, natives of both cold zones.

1. **K. Harveyana,** *J. Ag. Sp. Alg.* ii. 288;—*Fl. N. Z.* ii. 251. Frond membranous, 6 in. long, bright red, with a very short stipes, broadly obovate or orbicular-cordate; margin entire, undulate. Conceptacles scattered over the whole frond, immersed, not prominent.—*Euhymenia*, Kuetzing.

Blind Bay, Cook's Straits, *Lyall*. (Cape of Good Hope.)

78. GIGARTINA, Lamouroux.

Frond dull purple or red, between cartilaginous and fleshy, flat or cylindric, simple or branched; structure of 2 series of cells; central, a lax network of cylindrical articulate filaments; cortical of moniliform vertical dichotomous filaments imbedded in firm gelatine. Conceptacles globose, opening by a pore, enclosing several dense clusters of spores held together by a network of filaments. Tetraspores cruciately-partite, collected in sori amongst the filaments of the surface.

A large genus, found abundantly in both cold, temperate, and tropical zones.

§ 1. *Frond terete or compressed; conceptacles submarginal on the more or less acuminate pinnules.*

1. **G. pistillata,** *Gmelin;—Fl. N. Z.* ii. 252. Fronds several together, 3–6 in. long, dull purple, horny when dry, compressed, stipitate, flabellately branched, $\frac{1}{20}$–$\frac{1}{2}$ in. diameter; branches forked, naked or pinnate with short horizontal subulate ramuli, which bear the conceptacles at or near their tips. —Harv. Phyc. Brit. t. 232; J. Ag. Sp. Alg. ii. 264. *G. divaricata*, Fl. Antarct. 187; J. Ag. l. c. 280.

Apparently common on the coast, *Colenso*, etc. **Campbell's** Island, *J. D. H.* (N. Atlantic Ocean.)—A very variable plant in New Zealand.

2. **G. Chapmani,** *Hook. f. and Harv. Fl. N. Z.* ii. 251. *t.* 119 B. Frond 2 in. high, dull red, filiform, alternately decompoundly branched, compressed and channelled when dry; branches flexuous, variously divided; branchlets scattered, subulate, divaricate. Fructification unknown.

Maketa, *Chapman.*—Apparently allied to the British *G. acicularis*, but much more slender.

3. **G. Chauvinii,** *J. Ag. Sp. Alg.* ii. 268;—*Fl. N. Z.* ii. 252. Frond 1 ft. and more long, flat, linear, regularly pinnate, between membranous and cartilaginous, dark red or violet, decompoundly pinnate and proliferous from the disk; pinnæ distichous, patent, tips elongate, 3–4 in. long, mixed with shorter, sterile, ovate-lanceolate ones, fertile with numerous marginal conceptacles. Tetraspores forming a submarginal line.—*Sphærococcus*, Bory in Voy. Coquille, t. 20. *Chondroclonium*, Kuetzing.

New Zealand, *D'Urville.* (Peru, Chili, Fuegia.)

§ 2. *Frond thick, flat, subchannelled, pinnately decompound; conceptacles subsolitary on the disks of the pinnæ.*

4. **G. livida,** *J. Ag. Sp. Alg.* ii. 270;—*Fl. N. Z.* ii. 252. Frond purple or livid, a span high, $\frac{1}{4}$–$\frac{1}{3}$ in. broad, narrowed to a terete stipes below, 3-pinnate or more; pinnæ and pinnules distichous, linear or oblong-lanceolate, obtuse, fertile ones more obovate.—*Fucus*, Turn. Hist. t. 254.

New Zealand, *Banks and Solander;* Paroa, Otago, and Jackson's bays, *Lyall, Lindsay.* (Tasmania.)

§ 3. *Frond channelled, dichotomous or subpinnate; conceptacles umbilicate, numerous, submarginal on the convex surface of the frond.*

5. **G. alveata,** *J. Ag. Sp. Alg.* ii. 271;—*Fl. N. Z.* ii. 252. Frond 2–4 in. long, gregarious, linear, above fastigiately dichotomous, $\frac{1}{12}$ in. broad; segments close-set, uppermost revolute; axils spreading, subacute.—*Chondrus*, Grev. *Fucus*, Turn. Hist. t. 239. *Mastocarpus*, Kuetzing.

Probably common, *Banks and Solander*, etc.

6. **G. ancistroclada,** *Montagne in Voy. au Pôle Sud*, 121. *t.* 7. *f.* 4. Root discoid. Frond 2–3 in. high, linear, $\frac{1}{12}$–$\frac{1}{8}$ in. broad, shortly stalked, dark red-purple, rigid, channelled on one side, convex on the other, irregularly 2–3-pinnate; pinnæ alternate or subopposite, often secund; tips strongly incurved or hooked.—Harv. Phyc. Austr. iv. 197; J. Ag. Sp. Alg. ii. 272.

Akaroa, *D'Urville;* Otago, *Lyall.* (Tasmania.)

§ 4. *Frond thick, channelled, dichotomously or pinnately divided into cuneate or oblong segments; conceptacles on tubercles on both surfaces of the fronds.*

7. **G. decipiens,** *Hook. f. and Harv. in Fl. N. Z.* ii. 252. Frond small, cartilaginous, stipitate, flabelliform, flat, dichotomous; segments cuneate, repeatedly forked, ultimate narrow-linear, acute; axils rounded; margins simple or emitting short, stout, simple or pinnate linear branchlets. Fructification sessile, forming scattered oblong spots over the whole frond, or sessile warts immersed in the tips of the branchlets.—*Iridæa*, Lond. Journ. Bot. 1845, 547; J. Ag. Sp. Alg. ii. 257.

New Zealand, *Raoul.*

8. **G. stiriata,** *J. Ag. Sp. Alg.* ii. 277;—*Fl. N. Z.* ii. 252. Frond 3–6 in. long, violet-purple, thick, fleshy, below cylindric, stout, above compressed, slightly channelled, branched below; branches dilating into dichotomous or subpalmate leaflets, everywhere covered with simple or branched papillæ, which are obovate and often foliaceous. Conceptacles wart-like, solitary, or crowded on the papillæ.—*Mastocarpus*, Kuetzing. *Iridæa*, Bory. *I. volans*, Mont.? *Fucus*, Turn. Hist. t. 16.

Paroa Bay, *Lyall.* (Cape of Good Hope.)

9. **G. Radula,** *J. Ag. Sp. Alg.* ii. 278;—*Fl. N. Z.* ii. 252. Frond fleshy, flat, somewhat channelled below, 6 in. to 2 ft. long, simple or divided from the base into many large stipitate, obovate-oblong fronds or leaflets, of a fine red or amethyst-purple colour; margins and surfaces naked or papillose, or covered with small thick leaflets. Conceptacles subsolitary in the papillæ.—*Iridæa,* Bory. *Mastocarpus*, Kuetzing. *Fucus bracteatus*, Turn. Hist. Fuc. t. 25.

Abundant throughout the islands, and in **Lord Auckland's** group and **Campbell's** Island. (Throughout the Antarctic and N. and S. Pacific Oceans.)—A most abundant and variable plant, assuming various shapes as it is split by the waves, etc., often perforated.

79. IRIDÆA, Bory.

Frond dull red, broad, flat, nerveless, between cartilaginous and fleshy; central substance of densely interlaced longitudinal filaments; cortical of closely packed, vertical, coloured, moniliform filaments. Spore-masses immersed in the frond. Tetraspores forming a layer at the base of the filaments of the periphery.

Perennial *Algæ*, often very large, with rough surfaces; when cast up on the beach looking like great red-purple pieces of cloth; found in both temperate regions. The species are excessively variable and difficult to discriminate.

1. **I. micans,** *Bory;—Fl. N. Z.* ii. 252. Frond 1–2 ft. long, ovate-cordate or orbicular-ovate, obtuse, shortly stipitate, dark red-purple, iridescent; surface smooth; margins and fractures ciliated, with short, simple or divided spines.—Bory in Voy. Coquille, 110. t. 13 and 13 *bis;* J. Ag. Sp. Alg. ii. 254.

Akaroa and **Lord Auckland's** Island, *D'Urville.* (Australia and all Antarctic shores.)

2. **I. lanceolata,** *Harv. in Fl. N. Z.* ii. 252. Fronds numerous from the same base; stipes 1–2 in. long, cuneate at the base, then linear, gradually dilated upwards, pinnated below with spreading linear or lanceolate leaflets; above expanding into the lamina, which is 1–2 ft. long, red-purple, crisped and waved, 1–3 in. broad, lanceolate, tapering at both ends, between horny and membranous, rigid when dry.

Otago, *Lyall.*

3. **I. lusoria,** *Harv. in Fl. N. Z.* ii. 252. Frond stipitate, cartilaginous, simple, most frequently split; segments dilated or contracted, very proliferous, vaguely and inordinately produced.—*Rhodymenia,* Grev. in Hook. Comp. Bot. Mag. ii. 329.

East coast, *Cunningham;* Otago, *Lindsay.* (Specimens very imperfect.)—A very doubtful plant.

4. **I. laminarioides,** *Bory in Voy. Coquille,* 105. *t.* 11. *f.* 1. Fronds gregarious, linear-obovate, violet-purple or livid-purple, obovate, narrowed to a channelled stipes 1–2 in. long, usually split into erect or spreading linear-lanceolate segments, which are produced to a point.—Mont. in Voy. au Pôle Sud, 105; J. Ag. Sp. Alg. ii. 253.

Lord Auckland's group, *D'Urville.* (N. and S. Pacific Oceans.)

5. **I. volans,** *Grev.?* "Frond simple, flat, obovate-lanceolate, dirty violet; disk and margins covered with short, flattish, linear branchlets, those of the margin longer."—Mont. in Voy. au Pôle Sud, 104.

Lord Auckland's group, *D'Urville.*—A solitary imperfect specimen is all that Montagne describes from; it is perhaps *Gigartina stiriata.*

80. CHONDRUS, Linnæus.

Frond dull red, cartilaginous, flat or compressed, ribless, flabellately dichotomously cleft; structure of three strata, innermost of dense, interlaced, longitudinal filaments, next of small roundish cells; cortical of vertical, coloured, moniliform filaments. Conceptacles containing radiating filaments, whose lower joints develope into spores, at length opening by a pore. Tetraspores in sori, immersed in the frond; there are also cavities in the substance of the frond containing minute spores.

Usually perennial *Algæ,* common in the north temperate zone.

1. **C. tuberculosus,** *Hook. f. and Harv.;—Fl. Antarct.* i. 188. Frond 2 in. long, cartilaginous, thick, livid, simple and broadly cuneate below, then linear and forked; margins simple or proliferous; segments spreading or divaricate, flat or channelled, broader upwards, obtuse; axils rounded. Conceptacles very numerous on the concave surface of the frond, globose, depressed at the tip, at length opening by a pore; spores minute.—J. Ag. Sp. Alg. ii. 248. *Nothogenia,* Kuetzing.

Lord Auckland's group, *J. D. H.*

2. **C. variolosus,** *Montagne.—Nothogenia,* Fl. Antarct. i. 188. Frond cartilaginous, compressed, 6–8 in. long, linear, red-purple, at length brown

repeatedly dichotomous, subchannelled; segments ½ in. wide, ascending, flabellately fastigiate, obtuse; axils rounded. Conceptacles crowded, hemispheric on both surfaces of the frond.—*Nothogenia*, Mont. in Voy. au Pôle Sud, 109. t. 10. f. 3.

Lord Auckland's group, *D'Urville*, *Hombron*, etc.

81. EPYMENIA, Kuetzing

Frond dull red-purple, caulescent below, expanding into flat forked lamina, which is ribbed at the base. Structure of 2 strata; medullary of oblong colourless cells; cortical of vertical, minute, coloured cells. Conceptacles on proper leaflets, seated on the lamina, of thick-walled hemispherical tubercles, containing a mass of spores. Tetraspores cruciate, dispersed amongst the cortical cells.

A southern genus, found in S. Africa and Australia.

1. **E. obtusa,** *Kuetzing;—Fl. N. Z.* ii. 249. Frond 1 foot long, stipitate, flabellately expanded, repeatedly dichotomous; principal segments cuneate below, costate halfway up; upper linear-cuneate; uppermost subpalmate, very obtuse; disk or costa proliferous, with small leaflets $\frac{1}{6}$–½ in. long. Conceptacles numerous.—J. Ag. Sp. Alg. ii. 220. *Phyllophora,* Grev.;—Fl. Antarct. 187 and 486.

East coast, *Colenso*, *Lyall*. **Lord Auckland's** group, *J. D. H.* (S. Africa, Fuegia.)

2. **E. acuta,** *Harv. in Fl. N. Z.* ii. 249. Frond 5–6 in. high, flabelliform, cuneate at the base, many-times dichotomous, with a stout costa running up every segment to the middle; segments spreading, broadly linear, subacute.

Akaroa, *Lyall.*

82. CHYLOCLADIA, Greville.

Frond rose-red, terete or subcompressed, tubular, decompound, nodose-articulate or constricted at regular intervals, thus divided into sacs, which are full of water, with a few filaments; walls of angular cells. Conceptacles conical, opening by a pore, enclosing cuneate spores, surrounded by a web of filaments. Tetraspores 3-partite, immersed in the walls of the frond.

A northern and southern temperate genus.

1. **C. umbellata,** *Hook. f. and Harv. Fl. N. Z.* ii. 253. *t.* 119 C. Fronds dull purple, 1–2 in. long, terete, $\frac{1}{16}$ in. diameter, trichotomous; primary branches curved and hooked at the summit, emitting on the convex side below the apex, 3 or 4 erect, linear-clavate, slightly curved, lesser branches, which again throw out still smaller from their tips or are crowned with 3–4 clavate or fusiform bodies.

Port Underwood, Cook's Straits, *Lyall.*

2. **C. secunda,** *Hook. f. and Harv. Fl. N. Z.* ii. 253. Frond rose-red, small, 2 in. high, densely tufted, membranous, flaccid; branches setaceous, intricate, generally secund and arched, linear, obtuse, equal; branchlets few,

distant.—*Chrysimenia,* Harv. Ner. Austr. 77. *Dumontia pusilla,* Mont. in Voy. au Pôle Sud, 105. t. 13. f. 3.

Akaroa, *Raoul;* Port Cooper, *Lyall.* (Australia.)

3. **C.? cæspitosa,** *Harv. in Fl. N. Z.* ii. 253. Fronds small, tufted; stipes arcuate, rooting, giving off erect, oppositely pinnate or whorled, divided branches; branchlets spindle-shaped, opposite or 4-nate, a little constricted at the base, attenuated at the apex, acute, the larger curved.

Port Nicholson, *Lyall.*—Perhaps an immature state of the Australian *C. clavellosa.*

83. CHRYSYMENIA, J. Agardh.

Frond more or less compressed, tubular, not articulate, full of fluid and traversed by a few filaments; walls of several series of cells; innermost long and distended; outer smaller; outermost minute. Conceptacles enclosing a dense tuft of spores. Tetraspores 3-parted, immersed in the branchlets.

A small genus, common to both temperate zones.

1. **C.? polydactyla,** *Hook. f. and Harv. Fl. N. Z.* ii. 253. *t.* 119 A. Frond filiform, solid, firmly membranous, 3–4 in. long, twice as thick as a hog's bristle, excessively subdichotomously branched, dark red-purple; larger branches flexuous; smaller straight, twiggy, alternate or secund, all densely clothed with alternate, secund or fascicled, fusiform, hollow branchlets, which are attenuate at both ends. Fructification unknown.

Stewart's Islands, *Lyall.*—A curious and pretty little plant of doubtful affinity.

84. HALYMENIA, J. Agardh.

Frond bright or dark red, terete, compressed or flat, between membranous and gelatinous, dichotomous or pinnatifid; structure of 2 strata; medullary of few, lax, branched filaments in gelatine; cortical membranous, of minute coloured cellules. Spore-masses immersed in the medullary stratum. Tetraspores cruciate, scattered amongst the cortical cells.

Beautiful seeweeds, natives of both temperate zones.

1. **H. latissima,** *Hook. f. and Harv. Fl. Antarct. t.* 73. *f.* 1. Frond bright red, tender, between gelatinous and membranous, 2–14 in. long, 4–6 broad, cuneate at the base, broadly oblong- or ovate-lanceolate, simple or 2-fid, or margin laciniate-pinnatifid; segments ovate-lanceolate. Spore-masses extremely numerous, scattered over the whole frond, very slightly prominent. —J. Ag. Sp. Alg. ii. 204.

Lord Auckland's group and **Campbell's** Island: on rocks, *J. D. H.*

2. **H. novæ-Zelandiæ,** *Mont. in Voy. au Pôle Sud,* 107. *t.* 12. *f.* 2; —*Fl. N. Z.* ii. 253. Frond rose-purple, flat, between fleshy and membranous, rigid when dry, spirally convolute, dichotomously pinnate; pinnæ narrowed at the base, spreading, subterete, lanceolate, acute, quite entire, again pinnulate.—J. Ag. Sp. Alg. ii. 207. *H. Urvilleana,* Mont. l. c. t. 12, et Phyc. Antarct. 8.

Akaroa, *D'Urville.*—Harvey thinks that this, which is unknown to us, may possibly be his *Nemastoma Daviesii.*

85. PRIONITIS, J. Agardh.

Frond dark red, coarse, rigid, dense, compressed or flat, linear, without a rib, dichotomous or pinnate, proliferous on the disk or margin; structure of 3 strata; medullary broad, of densely interlaced filaments; intermediate of roundish cells; cortical of minute, coloured, vertically disposed cells. Spore-masses simple, imbedded in the substance or marginal processes of the frond, opening by a pore, enclosing roundish spores in a gelatinous pellicle. Tetraspores oblong, cruciate, scattered in the cortical layer.

A small genus, natives of the North and South Pacific.

1. **P. Colensoi,** *Hook. f. and Harv. Fl. N. Z.* ii. 254. *t.* 120 A. Frond rigid, dark brown, 6–8 in. long when dry, ½ in. broad, irregularly pinnate from the base; branches 3–4 in. long, subopposite, spreading, narrow below, dilated beyond the middle, then linear-oblong, obtuse or attenuate; margins below simple, above fimbriate or lacerate, with minute lobes, which are toothed at their dilated tips.

Cape Turnagain, on rocks, *Colenso.*

86. GRATELOUPIA, J. Agardh.

Frond dull purple or greenish, membranous, flexible, solid, pinnate; structure of densely woven, anastomosing, branching filaments; those of the periphery moniliform, short, densely compacted. Spore-masses globular, immersed in the periphery, opening by a pore. Tetraspores cruciate, in subdefined sori, vertically placed amongst the peripheric filaments.

A small northern and southern temperate genus.

1. **G.? Aucklandica,** *Mont. in Voy. au Pôle Sud,* 115. *t.* 10. *f.* 1; *Fl. Antarct.* 187. Frond 4–5 in. long, cartilaginous, filiform, terete, as thick as a sparrow-quill, vaguely excessively branched; branches and branchlets crowded and fascicled, subcompressed, narrowed at the base, obtuse, membranous, sometimes tubular, solitary or several together, springing from a curious tubercle on the frond.—J. Ag. Sp. Alg. ii. 185.

Lord Auckland's group, *D'Urville.*

87. NEMASTOMA, J. Agardh.

Frond dull red-purple, compressed or flattened, between fleshy and gelatinous, dichotomous or subpinnate; structure of two strata; medullary of longitudinal interwoven filaments; cortical of radiating, fastigiate, dichotomous, moniliform filaments, imbedded in gelatine. Spore-masses immersed in the cortical filaments, enclosing roundish spores, imbedded in gelatine. Tetraspores cruciate, scattered amongst the cortical filaments.

A northern and southern temperate genus.

1. **N. intestinalis,** *Harv. in Fl. N. Z.* ii. 255. Frond, delicate, red-purple, 6–8 in. long, $\frac{1}{4}$-$\frac{1}{8}$ broad, cylindric or compressed, tubular, linear, vaguely branched or irregularly divided into simple or forked branches, which are constricted at the base and attenuated, or 2-horned at the tip; medullary stratum very fluid. Spore-masses scattered over the whole frond, suspended below the periphery, enclosed in a dense plexus of filaments.

Preservation Harbour, on rocks, *Lyall.*

2. **N. ? attenuata,** *Harv. in Fl. N. Z.* ii. 255. Nearly related to *N. intestinalis*, but the only fragment known is larger, 15 in. long, $\frac{1}{3}$ in. wide below, tapering upwards, compressed; medullary stratum lax.

Jackson's Bay, *Lyall.*

3. **N. pinnata,** *Hook. f. and Harv. Fl. N. Z.* ii. 255. *t.* 120 B. Frond soft, pale red, 4–5 in. long, and as broad in outline, undivided or forked, $\frac{1}{4}$–$\frac{1}{2}$ in. broad, closely regularly set, with long, distichous, horizontal, opposite or alternate pinnæ, which are $\frac{1}{6}$–$\frac{1}{4}$ in. broad, attenuated to both ends; pinnæ again pinnate, with similar but much smaller ciliiform pinnules; medullary stratum compact.

Akaroa, *Lyall.* (A single specimen.)

4. **N. Daviesii,** *Harv. in Fl. N. Z.* ii. 255. Frond purple-red, between gelatinous and membranous, 6–8 in. long, stipitate, foliaceous, ovate-lanceolate, $\frac{1}{2}$–$\frac{1}{12}$ in. broad, pinnate, with ovate-lanceolate lobes, 2–4 in. long and $\frac{1}{2}$–1 in. wide; pinnæ tapering to both ends, crisped and curled, spreading, toothed and ciliate. Spore-masses abundant, scattered over the whole frond.

Tawanga, *Davies*; Otago, *Lindsay.*

5. **N. endiviæfolia,** *Harv. in Fl. N. Z.* ii. 255. Frond cuneate at the base, excessively variable, vaguely dichotomously palmate or laciniate, narrow-linear and multifid, with acute forked tips, or broad and subpinnately branched, with dichotomous, acute or truncate, entire or toothed segments; axils rounded.

Port Nicholson and Blind Bay, *Lyall.*—Harvey thinks this so variable that it may be a form of *Daviesii.*

6. **N. prolifera,** *Harv. in Fl. N. Z.* ii. 255. Upper half of a frond, 8 in. long, cuneate, 1 in. broad, widening upwards to 2 in. then forking, gelatinous, everywhere beset on the surface and margins with filiform or cuneate-multifid leaflets, less than 1 in. long.

Akaroa, on stones, *Lyall.*

88. CATENELLA, Greville.

Frond dull purple, cylindric, constricted at intervals, subtubular, branched; structure of 2 strata, medullary of interlacing anastomosing filaments, passing into a cortical of vertically-placed moniliform minute filaments cohering at the periphery. Conceptacles formed in minute ovoid or subglobose branchlets. Tetraspores transversely divided, scattered through the periphery.

A genus of a single species.

C. Opuntia, *Grev.;—Fl. N. Z.* ii. 254. Roots tufted, matted, creeping. Frond ½–1 in. high, sparingly branched, forming strings of oblong internodes, varying in length and breadth, terete or compressed a little. —Harv. Phyc. Brit. t. 88; J. Ag. Sp. Alg. ii. 352.

Bay of Islands, *J. D. H.* (Europe, Australia.)

89. DUMONTIA, Lamouroux.

Frond dull red-brown, cylindric, tubular, full of watery fluid, with a central stem and alternate simple or forked branches; walls membranous; inner surface of longitudinal filaments formed of elongated cells, outer of minute closely-packed roundish cells. Spores obovate, in minute clusters on the inner wall of the frond or in conceptacles. Tetraspores roundish, amongst the cortical cells.

A small genus of annual, often tall *Algæ*, found in both temperate zones.

1. **D. filiformis,** *Grev.;—Fl. N. Z.* ii. 254. Frond tender, membranous, 2–18 in. long, usually undivided, but sometimes divided from the base, narrowed at both ends, pinnated with numerous alternate long, simple, or forked branches, each like the main stem, yellow-greenish or purplish, 1 line to ½ in. diameter. Conceptacles scattered over the whole frond.—Fl. Antarct. i. 189; Harv. Phyc. Brit. t. 95, et Suppl. t. 357; J. Ag. Sp. Alg. ii. 349.

Tidal rocks, Cape Turnagain, *Colenso*. **Campbell's** Island, *J. D. H.* (Europe, Australia.)—A very variable plant; in Europe the fronds are sometimes compressed, curved, and twisted.

2. **D. cornuta,** *Hook. f. and Harv. Fl. Antarct.* i. 189. Frond tufted, 2–3 in. long, irregularly constricted here and there, ¼ in. broad, brownish-red, naked below, above with crowded fascicled or 4-farious branches; branches spreading, forked, flexuous; branchlets recurved, ultimate attenuated at both ends with acute apices.—J. Ag. Sp. Alg. ii. 350.

Campbell's Island, *Lyall.*

3. **D. pusilla,** *Montagne in Voy. au Pôle Sud*, 105. *t.* 13. *f.* 2. Frond 1–2 in. long, cylindric, loosely branched; branches forked at the apex, clothed with short 2-nate branchlets. Tetraspores globose, scattered or arranged in lines.

On *Gigartina Ancistroclada*, Akaroa, *D'Urville.*

90. SPYRIDIA, Harvey.

Frond rose-red, filiform, pinnately decompound, articulated, formed of a wide tube, coated below and at the nodes, or all over, with a layer of minute coloured cells; branches and branchlets covered with minute articulated setæ. Conceptacles naked or involucrate, terminating short branches, containing masses of spores, each mass enclosed in a distinct membrane. Tetraspores sessile, naked on the articulated setæ.

A small genus, found in both temperate, but chiefly in tropical seas.

1. **S. opposita,** *Harv. in Fl. N. Z.* ii. 256. Root a small disk. Frond dark red, rather rigid; stem 4–6 in. long, not evidently articulate, forked at the base, compressed, 4-angled; angles prominent; branches spreading, distichously pinnate; pinnæ unequal, transparently articulate; joints short, as long as broad; pinnules few, short, pectinate, with capillary distichous or decussating ultimate branchlets. Conceptacles imperfectly 3-lobed, not involucrate.—Harv. Phyc. Austr. iii. t. 158.

Chalky Bay, *Lyall.* (Australia, Tasmania.)

91. CERAMIUM, Roth.

Frond red, filiform, dichotomously branched, jointed; joints opaque, of a single cell, the nodes coated with a stratum of minute coloured cells, which sometimes extend over the whole internode. Spore-masses globular, sessile, with a pellucid limb containing angular spores, subtended by shortened (involucral) branchlets. Tetraspores external or immersed.

Annual small *Algæ*, beautiful when laid out on paper, abundant in all oceans.

§ 1. *Frond unarmed, coated at the articulations only with a zone of minute cells.*

1. **C. diaphanum,** *Roth;—Fl. N. Z.* ii. 256. Frond 2–4 in. long, subsetaceous, dichotomously and sublaterally branched; branches erecto-patent, terminal erect with converging tips; lower articulations 3–4 times longer than broad, interstices pellucid. Spore-masses lateral on the branches, with few short involucral branchlets. Tetraspores whorled round the nodes in a single series.—Harv. Phyc. Brit. t. 193; J. Ag. Sp. Alg. ii. 125.

Probably abundant, Port Cooper, Akaroa and Otago, *Lyall.* **Lord Auckland's** group, *J. D. H.* (Atlantic and Southern Oceans.)

2. **C. virgatum,** *Hook. f. and Harv. Fl. N. Z.* ii. 256. Frond erect, straight, sparingly branched; branches strict, virgate, simple or again branched; branches dichotomously multifid, appressed; tips incurved; axils very narrow; articulations of the branches as long as broad, of the branchlets very short. Spore-masses subterminal; involucre of many branchlets.—J. Ag. Sp. Alg. ii. 137.

East coast, parasitical on *Carpophyllum, Colenso.*

§ 2. *Frond unarmed; internodes more or less completely coated with small cells.*

3. **C. rubrum,** *Agardh;—Fl. N. Z.* ii. 256. Frond 2–12 in. long, robust, very variable in form and ramification, irregularly dichotomous; branchlets hooked inwards; lower articulations twice as long as broad, upper shorter than broad. Spore-masses globose, with 3–4 involucral branchlets or lateral branchlets. Tetraspores whorled in the articulations.—J. Ag. Sp. Alg. ii. 127; Harv. Phyc. Brit. i. 181.

Abundant throughout the islands, and in **Lord Auckland's** group and **Campbell's** Island. (Common in all temperate seas.)

4. **C. vestitum,** *Harv. in Fl. N. Z.* ii. 256. Very similar to *C. rubrum*, but articulations so coated with cells that they are scarcely visible, and the tips of the branchlets are straighter.

Stewart's Island, *Lyall*.—Probably a state of *C. rubrum*.

5. **C. cancellatum,** *Agardh ;—Fl. N. Z.* ii. 226. Frond 3–6 in. high, compressed, distichously pinnate; pinnæ subcorymbose, dichotomously pinnulate; terminal segments obtuse, spreading; articulations completely coated with cells, lower as long as broad. Spore-masses subterminal; involucral branchlets elongate, divided. Tetraspores in irregular lines along the margins of the segments.—J. Ag. Sp. Alg. ii. 136. *Pteroceras*, Kuetzing.

Abundant on the coasts, *Colenso*, etc. **Lord Auckland's** group, *J. D. H.*

§ 3. *Frond with spinules, especially on the outer sides of the terminal branchlets.*

6. **C. uncinatum,** *Harv. in Fl. N. Z.* ii. 257. Frond capillary, regularly dichotomously fastigiate; axils patent; tips incurved; articulations not coated with cells, lower 3–4 times longer than broad; nodes swollen, armed with a long and a short uncinate spinule.

Cape Turnagain, *Colenso;* Cook's Straits, *Lyall*.

92. CENTROCERAS, Kuetzing.

Frond red, filiform, branched, jointed; joints of a single cell, coated with oblong longitudinal cellules, placed in longitudinal series. Fructification of *Ceramium*.

A genus of one variable species, abounding in the Atlantic and Pacific Oceans, differing from *Ceramium* in the longitudinal cortical cells.

1. **C. clavulatum,** *Mont. ;—Fl. N. Z.* ii. 257. Frond densely tufted, dark red, very variable, 2–4 in. long, rather rigid, capillary, dichotomously branched; nodes swollen, their bases sunk in the top of that below them, when young crowned by minute spinous processes; young plants covered with deciduous hairs.—J. Ag. Sp. Alg. ii. 148; Harv. Ner. Bor. Am. 211. t. 33 C.

Common on all the coasts, *Colenso*, etc. (Tasmania, and common in tropical and subtropical seas.)

93. PTILOTA, Agardh.

Frond bright or dull red, linear, slender, compressed, 2-edged, distichously branched, pectinate-pinnate; pinnules sometimes articulate; axis of one long articulate tube, more or less coated with many series of small cells. Spore-masses involucrate, full of angular spores, the involucre of many incurved branchlets. Tetraspores sessile or stalked, 3-partite, solitary or crowded on the pinnules.

A beautiful genus, common in both temperate zones, but more abundant and larger in the southern.

1. **P. formosissima,** *Mont. in Voy. au Pôle Sud*, 97. *t.* 9. *f.* 3 ;—*Fl. N. Z.* ii. 257. Frond bright red, cartilaginous, 6–16 in. long, and as broad

in outline, flabellate, 2-edged, $\frac{1}{12}$–$\frac{1}{6}$ in. broad, decompoundly pinnate; pinnæ alternate; pinnules pinnatifid, ultimate serrate or pectinate. Spore-masses sessile, with 4 pectinate involucral branchlets. Tetraspores marginal, pedicelled, globose.—Fl. Antarct .190. t. 77; J. Ag. Sp. Alg. ii. 102. *Euptilota*, Kuetzing.

Throughout the islands, and in **Lord Auckland's** group, abundant.

2. **P. pellucida,** *Harv. in Fl. N. Z.* ii. 257. Frond bright rosy-red, 2–3 in. long, ovate in outline, filiform, distichously decompound, primary branches 2–3-pinnate, not coated with cells above; pinnæ articulate, opposite, unequal, one undivided, the other pinnate or pectinate on one side; ultimate branchlets filiform, constricted at the base, subacute; articulations 1$\frac{1}{2}$–2 times as long as broad.

Otago and Stewart's Island, *Lyall.*—Harvey suggests the possibility of this being the same as the *Callothamnion Ptilota,* Fl. Antarct. t. 189, found off the Crozet Islands.

94. GRIFFITHSIA, Agardh.

Frond rose-red, filiform, dichotomously branched, articulate, of one tube; branchlets often whorled; nodes transparent. Spore-masses involucrate, enclosing numerous angular spores. Tetraspores globose, 3-parted, involucrate.

A considerable genus, found in both temperate zones.

1. **G. setacea,** *Agardh;—Fl. N. Z.* ii. 258. Frond 3–6 in. long, setaceous, rather rigid, straight, irregularly dichotomous, crimson; axils acute; branchlets sometimes opposite; articulations cylindric, 5–6 times longer than broad; involucres lateral, pedicelled.—Harv. Phyc. Brit. t. 184; J. Ag. Sp. Alg. ii. 84.

East coast, *Colenso;* Foveaux Straits, *Lyall.*—A state with more slender branches and longer articulations has been found at Bluff Harbour and Port Cooper, *Lyall.* (N. and S. temperate oceans.)—A beautiful plant, stains paper crimson; when put in fresh water the articulations burst with a crackling noise, and discharge the red colour.

2. **G. antarctica,** *Hook. f. and Harv. Fl. Antarct.* ii. 488;—*Fl. N. Z.* ii. 258. Frond tufted, flaccid, stout, dichotomous; branches elongate, erect, and branchlets often secund, naked, constricted at the nodes; articulations of the branches 4–6 times, of the branchlets about 3 times longer than broad; fruiting branchlets lateral, short, crowned with an umbellate involucre.—J. Ag. Sp. Alg. ii. 87.

East coast, *Colenso;* Foveaux Straits and Stewart's Island, *Lyall.* (Tasmania.)

95. BALLIA, Harvey.

Frond bright or rose-red, fading to green or dirty brown, filiform, rigid, dendroid; stem and branches covered with short capillary fibres; branchlets pellucid, pinnately decompound, or pectinately pinnate. Spore-masses enclosed in an involucre of many incurved filiform branchlets, containing numerous angular spores. Tetraspores on the short capillary fibres of the branches.

A small genus of beautiful plants.

1. **B. callitricha,** *Mont.;—Fl. N. Z.* ii. 257. Root a spongy disk. Fronds densely tufted, 2–10 in. long, with a stout rigid woolly stipes, setaceous above, rose-purple, 3–4-pinnate; pinnæ and pinnules capillary; branches opposite, very close-set, plumosely decompound, lanceolate in outline, ultimate undivided, subulate, very acute; articulations ovoid or cylindric, about twice as long as broad.—J. Ag.; Fl. Sp. Alg. ii. 75. *B. Brunonis*, Harv. Fl. Antarct. 182. *B. Hombroniana*, Mont. in Voy. au Pôle Sud, t. 12. f. 1. *Sphacelaria callitricha*, Agardh.

Abundant throughout the islands, and in **Lord Auckland's** group and **Campbell's** Island. (All the southern temperate oceans.)

2. **B. scoparia,** *Harv. Phyc. Austr.* iii. 168.—*Callithamnion*, Fl. N. Z. ii. 259. Habit and woolly root and stipes of *B. callitricha*, but dull in colour, most densely fastigiately branched, the branches springing from all sides; branches pinnately branched.—*Callithamnion*, Fl. Antarct. t. 189. f. 3; J. Ag. Sp. Alg. ii. 35. *Phlebothamnion*, Kuetzing.

East coast, *Colenso;* Preservation Harbour and Foveaux Straits, *Lyall.* (Fuegia, Falkland Islands, Australia.)

96. CALLITHAMNION, Lyngbye.

Frond rose-red or purplish, filiform or capillary, branched, articulate, of one series of long tubular cells, those of the stem sometimes coated by decurrent cells, the branches always of single articulate tubes, pellucid at the nodes. Spore-masses solitary or geminate, axillary or lateral, containing numerous angular spores. Tetraspores sessile or stalked on the branchlets.

A large genus, found in all seas.

§ 1. *Frond short, densely tufted, excessively branched from the base, with no distinct stem; articulations without cortical cells.*

1. **C. Rothii,** *Lyngb.;—Fl. N. Z.* ii. 260. Frond deep red or purple, $\frac{1}{4}$–1 in. high, densely tufted; filaments excessively slender, short, erect, dichotomous; branches few, long, straight, equal, alternate or dichotomous; articulations twice as long as broad. Tetraspores clustered on short terminal subcorymbose branchlets.—Harv. Phyc. Brit. t. 120 B; J. Ag. Sp. Alg. ii. 17.

Tidal rocks, caverns, etc., probably common, Hawke's Bay and Cape Kidnapper, *Colenso.* (N. Atlantic Ocean.)

§ 2. *Frond pinnate; pinnules (upper or all) opposite.*

2. **C. flaccidum,** *Hook. f. and Harv. Fl. N. Z.* ii. 258. Fronds flaccid, membranous, rosy, 2–4 in. long, loosely branched, excessively slender; branches densely crowded, opposite and alternate, pinnate; pinnules all opposite or alternate, short, simple, sometimes secund, incurved; primary articulations many times, secondary 6–10 times longer than broad. Tetraspores numerous, secund on the pinnules.—Fl. Antarct. 490. t. 188. f. 1; J. Ag. Sp. Alg. ii. 31.

Otago Harbour, *Lyall.* (Fuegia, Tasmania.)

3. **C. Plumula,** *Lyngb.;—Fl. N. Z.* ii. 258. Frond flaccid, rose-red, 2–5 in. long, distichously branched; branches alternate or irregular, slender, each node with a pair of opposite horizontal or recurved pectinate branchlets, the segments of which in luxuriant specimens are sometimes again pectinate; articulations 3–4 times as long as broad. Tetraspores on the pectinate branchlets.

D'Urville's Island, *Lyall.* (Atlantic Ocean, Tasmania, Fuegia.)

4. **C. applicitum,** *Harv. in Fl. N. Z.* ii. 258. Frond minute, $\frac{1}{6}$–$\frac{1}{4}$ in. long, rose-red, capillary, prostrate, attached by its whole length, distichously pinnately branched; branches pinnate or 2-pinnate; pinnæ all opposite, close-set; articulations of branches cylindric, 3–4 times as long as broad, of the pinnæ hexagonal, of the pinnules square.

Creeping on *Amphiroa, Colenso.*

5. **C. pectinatum,** *Mont. in Voy. au Pôle Sud,* 90;—*Fl. Antarct.* 191. Habit and size of *C. applicitum,* or even smaller; pinnæ horizontal; pinnules few, 5–6, very short, simple or nearly so; articulations terete, primary twice as long as broad. Tetraspores axillary or at the base of the pinnules.

Lord Auckland's group, on *Algæ, D'Urville.*

§ 3. *Frond pinnate throughout or above; pinnules alternate.*

* *Internodes without a coating of small cells (or lowest coated at the base only in* C. puniceum, brachygonum, *and* byssoideum).

6. **C. hirtum,** *Hook. f. and Harv. Fl. Antarct.* 192. *t.* 78. *f.* 2;—*Fl. N. Z.* ii. 258. Fronds tufted, blackish or dark brown, stout, intricate, 2–3 in. long, 4-fariously branched, densely clothed throughout with 4-farious branchlets; primary branches robust, veined below, jointed above; secondary elongate; pinnules simple, incurved, obtuse; articulations 1$\frac{1}{2}$ as long as broad.—J. Ag. Sp. Alg. ii. 58.

Various places, in the **Middle** Island, probably common, *Lyall.* **Lord Auckland's** group, *J. D. H.*

7. **C. gracile,** *Hook. f. and Harv. Fl. Antarct.* 191. *t.* 78. *f.* 1. Frond excessively slender, rose-red, erect from a branched creeping rhizome, laxly 2–3-pinnate; pinnæ opposite and alternate, elongate, branched; pinnules short, erecto-patent, alternate, simple or branched at the tip; primary articulations pellucid, 4–5 times as long as broad. Tetraspores minute, spherical.—J. Ag. Sp. Alg. ii. 48.

Campbell's Island, *J. D. H.*

8. **C. cryptopterum,** *Kuetzing, Sp. Alg.* 646.—*C. micropterum,* Hook. f. and Harv. Fl. Antarct. 192 (not of Montagne). Frond small, rose-red, from a scutate root, erect, $\frac{1}{6}$–$\frac{1}{4}$ in. long, sparingly distichously and alternately divided, obovate in outline, 2–3-pinnate; pinnæ flexuous, alternately multifid; pinnules erecto-patent, obtuse; internodes pellucid, 2–3 times longer than broad. Tetraspores minute, scattered, elliptic, appressed.—J. Ag. Sp. Alg. ii. 55.

Lord Auckland's group, *J. D. H.*

9. **C. puniceum,** *Harv. in Fl. N. Z.* ii. 259. Frond fine lake-purple,

2–3 in. long, tender, excessively branched, inextricable; stem and branches decompoundly pinnate; upper divisions pinnate; pinnules subulate, acute; primary articulations pellucid, veined, 8–10 times, of the branchlets 3–4 times longer than broad; lower only coated with small cells. Tetraspores solitary, lateral.

Tauranga, *Davies*.

10. **C. brachygonum,** *Harv. in Fl. N. Z.* ii. 259. Frond flaccid, capillary, carmine-red, densely tufted, 1–2 in high. Stem subsimple, set on all sides with virgate branches, which again branch and bear distichous, dense, plumose pinnæ; plumules narrow, erecto-patent; lower pinnate; upper longer and more compound; middle one long and 2-pinnate; articulations short, thick-walled; lowest sparingly coated with cells. Tetraspores globose, in series on the inner faces of the branchlets.

Blind Bay, *Lyall*; Tauranga, *Davies*.

11. **C. byssoideum,** *Arnott;—Fl. N. Z.* ii. 260. Frond flaccid, gelatinous, capillary, much divided into inextricable branched filaments, or with a few stems thicker than the rest; branches linear-lanceolate, set with long, slender, flexuous, pinnate plumules; articulations pellucid, of the stouter stems veined, 8 times as long as broad, of the branchlets 4 times. Tetraspores elliptic, subsolitary near the bases of the branchlets.—Harv. Phyc. Brit. t. 262; J. Ag. Sp. Alg. ii. 40. *Phlebothammia*, Kuetzing.

Foveaux Straits and Otago Harbour, *Lyall*; Maketu, *Chapman*. (N. Australia.)

** *Internodes coated with small cells.*

12. **C. Colensoi,** *Harv. in Fl. N. Z.* ii. 259. Frond dark purple, robust, dendroid, 2–3 in. high; main stem as thick as a sparrow's quill below, laterally branched; branches densely clothed all round with minute, compound, squarrose, imbricate branchlets; ultimate divaricate, subulate, acute, alternate; articulations coated with small cells, 1½ times longer than broad; ultimate pellucid.

East coast and Hawke's Bay, *Colenso*.

13. **C. consanguineum,** *Harv. in Fl. N. Z.* ii. 259. Frond densely tufted, flaccid, rose-purple, 2–3 in. high, capillary, excessively branched; branches spreading, the smaller distichous upwards, alternately decompoundly pinnate and alternately plumulate; plumules short, flabellate, naked below, pinnate above, terminal; pinnæ close-set; articulations opaque, with cortical cells, 2–3 times as long as broad; upper pellucid. Tetraspores lateral, subsolitary, triangularly divided.

Port Nicholson, *Lyall*.

SUBORDER III. CHLOROSPERMEÆ.

97. CAULERPA, Lamouroux.

Rhizome prostrate, creeping, throwing up erect green branches (fronds) of horny-membranous texture. Structure continuously tubular, filled with a spongy network of anastomosing fibres and grumous fluid. Fructification unknown.

A fine genus, most abundant in tropical seas; the species are often much branched, like Pine-branches, or pinnate, like Mosses.

1. **C. sedoides,** *Agardh;—Fl. N. Z.* ii. 261. Rhizome slender, glabrous, matted, glossy, several in. long, $\frac{1}{16}$–$\frac{1}{10}$ in. diameter. Fronds crowded, 2–6 in. high, linear, clothed throughout their length with short, pyriform, loosely imbricate, more or less distichous, opposite sacs, $\frac{1}{12}$–$\frac{1}{6}$ in. long.—Harv. Ner. Austr. ii. t. 72. *Chauvinia*, Kuetzing. *Ahnfeldtia*, Trev.

Rocks near low-water mark, *Colenso;* Cook's Straits, *Lyall.* (Australia, Mauritius, Indian Ocean.)

2. **C. hypnoides,** *Agardh;—Fl. N. Z.* ii. 260. Rhizome robust, $\frac{1}{6}$–$\frac{1}{4}$ in. diameter, covered with cylindric, dichotomous scales. Frond bright green, erect, stalked, lanceolate, 8–12 in. high, pinnate; stipes, rachis, and pinnæ densely clothed with very short, forked, subulate branchlets; pinnæ distichous, simple or forked, 1–2 in. long, close-set, patent.—Harv. Ner. Austr. ii. t. 84. *Chauvinia*, Kuetzing.

Deep tide-pools, East coast, *Colenso.* (Australia, Tasmania.)

3. **C. Brownii,** *Endlicher;—Fl. N. Z.* ii. 260. *t.* 121 A. Rhizome stout, clothed with cylindric scales. Frond green, erect, 10–12 in. high, slender, branched, clothed throughout with erecto-patent, 4-fariously, loosely imbricate, curved, cylindrical, slender, simple, acute, mucronulate or obtuse processes $\frac{1}{8}$ in. long.

East coast, *Colenso;* Chalky and Lyall's Bay, *Lyall.* (Tasmania and Australia.)

4. **C. furcifolia,** *Hook. f. and Harv. Fl. N. Z.* ii. 260. *t.* 121 B. Rhizome as in *C. Brownii.* Frond slender, 4–6 in. high, clothed throughout with suberect, 4-fariously imbricate, curved, forked (rarely simple), cylindrical, slender, acute processes.—*C. Selago*, Hook. f. and Harv. in Lond. Journ. Bot. iv. 550, not of Agardh.

East coast, *Colenso.* (Tasmania.)

5. **C. articulata,** *Harv. in Fl. N. Z.* ii. 261. Rhizome unknown. Branches slender, filiform, 4–5 in. long, bright green, simple or forked, constricted at intervals of $\frac{1}{12}$–$\frac{1}{6}$ in., clothed throughout with distichous branchlets, which are opposite, attached to the middle of each internode, $\frac{1}{2}$ in. long, cylindric, obtuse, constricted at the base.

East coast, rare, *Colenso.*

98. **CODIUM,** Stackhouse.

Frond dark green, sponge-like, consisting of a matted mass of 1-celled branching tubuli, filled with semifluid endochrome. Fructification of lateral vesicles on the branchlets at the circumference, containing myriads of antherozoids.

Natives of both temperate and tropical zones.

1. **C. tomentosum,** *Agardh;—Fl. N. Z.* ii. 261. Frond from a spongy base, cylindric, 6–12 in. long, ¼–½ in. diameter, cylindric, dichotomous or palmatipartite.—Harv. Phyc. Brit. t. 93.

Probably abundant, East coast, *Colenso;* Tauranga, *Davies;* Port Nicholson, *Lyall*. **Lord Auckland's** group, *J. D. H.* (Abundant in all seas.)

2. **C. adhærens,** *Agardh;—Fl. N. Z.* ii. 261. Frond forming green velvety patches, 1 in. to 1 ft. broad, on rocks; substance gelatinous, dense. —Harv. Phyc. Brit. t. 35 A.

Banks' Peninsula, *Lyall;* Cape Kidnapper, *Colenso.* (Generally diffused in warmer seas, but not hitherto found in Australia.)

99. VAUCHERIA, De Candolle.

Frond bright green, tufted, capillary, cylindric, tubular, full of a green granular mass. Fructification globose or oblong stalked bodies (spores), full of green matter, attached to the surface of the frond.

Usually annual freshwater *Algæ*, always small.

1. **V. Dilwynii,** *Agardh;—Fl. N. Z.* ii. 195. Frond decumbent, irregularly branched, farinose when highly magnified, minutely tubercled. Spores globose, sessile, solitary.—Grev. Alg. Brit. t. 19.

Damp ground, *Colenso.* (Europe, Kerguelen's Land, etc.)

Colenso and Lyall send fragments of other species of this genus, but these cannot be determined except in a fresh state.

100. BRYOPSIS, Lamouroux.

Frond bright or yellow-green, filiform, tubular, membranous, cylindric, glistening, branched; branches imbricate or distichously pinnate, full of green granuliferous fluid.

Small annual *Algæ*, found in all seas.

1. **B. plumosa,** *Agardh;—Fl. N. Z.* ii. 195. Frond glossy green, feathery, lubricous, 1–4 in. high, more or less branched or simple; branches naked below, above pinnate, with subopposite, slender, distichous branchlets. —Harv. Phyc. Brit. t. 111.

Tidal pools, probably common, Cook's Straits, Akaroa, and Otago Harbour, *Lyall.* (N. Atlantic, Australia.)

101. PORPHYRA, J. Agardh.

Frond pink or green-purple, flat, membranous, leaf-like, without midrib. Fructification red or purple. Spores arranged in fours, dispersed over the whole frond.

A considerable genus, found in all seas.

1. **P. laciniata,** *Agardh ;—Fl. N. Z.* ii. 264. Fronds clustered, 4–18 in. long, fixed by the base or centre, glossy, fine purple, deeply and irregularly cleft ; margin cut and lobed.—Harv. Phyc. Brit. t. 92.

Common on rocks and stones in the sea (all latitudes).—This is much used as "Laver," in England, in scrophulous cases.

2. **P. vulgaris,** *Agardh ;—Fl. N. Z.* ii. 264. Frond 1–2 ft. long, simple, lanceolate, often much waved, undivided.—Harv. Phyc. Brit. t. 211.

With the last and probably not different from it.

3. **P. capensis,** *Kuetzing ;—Fl. Antarct.* 193. Frond tufted, small, violet-purple, orbicular, crisped and undulate, 1–2 in. broad.—*P. Columbina*, Mont. in Voy. au Pôle Sud, 33.

Lord Auckland's group and **Campbell's** Island, *D'Urville, J. D. H.* (S. Africa, etc.)

102. ULVA, Linn.

Frond bright green, flat, membranous, leaf-like, without midrib, sometimes inflated or saccate when young. Fructification of green spores, arranged in fours, dispersed over the whole frond.

Perhaps the commonest marine plants in the world, annual ; the species are probably all forms of one.

1. **U. latissima,** *Linn. ;—Fl. N. Z.* ii. 265. Frond tufted, 6–18 in. long, delicately membranous, pale green, broadly obovate or oblong, flat, or variously lobed or waved.—Harv. Phyc. Brit. 171.

Abundant on the shore everywhere, as far south as **Campbell's** Island. (All seas.)

2. **U. rigida,** *Agardh ;—Fl. N. Z.* ii. 265. Frond subcartilaginous, contracted to a flat thickish stipes, deeply lobed ; lobes curved, undulate ; cells subquadrate, orbicular.—*Phycoseris rigida* and *lobata*, Kuetzing.

East coast, *Colenso.* (Atlantic, Chili, etc.)

3. **U. crispa,** *Lightf. ;—Fl. N. Z.* ii. 265. Fronds forming an irregular stratum, short, crisped and plaited.—Fl. Antarct. 498. *Prasiola crispa* and *antarctica*, Kuetzing.

Probably common on damp ground near the sea. (Europe, etc.)

4. **U. bullosa ?,** *Roth ;—Fl. N. Z.* ii. 265. Fronds forming green strata, small, obovate and saccate when young, at length breaking up into ragged bullate laminæ.—Kuetzing, Tab. Phyc. t. 28. f. 1.

On stones, under water, *Colenso.*—Specimens young, and perhaps not referable to this species. (Europe, etc.)

U. reticulata, Forst., a tropical species, found in the Red Sea, Sumatra, etc., is mentioned by Montagne (Voy. au Pôle Sud ; Bot. Crypt. 33) as a native of **Lord Auckland's** group, but I suspect some mistake ; it may be recognized by its flat perforated frond divided into linear-reticulated lobes.

103. ENTEROMORPHA, Link.

Frond green, tubular, hollow, cylindric or compressed, usually branched; structure reticulated. Fructification of roundish spores in threes or fours, clustered in the reticulations.

An excessively common genus of annual *Algæ*, found in all seas, and in brackish water also. The two first species appear to pass into one another, and often to be difficult to distinguish from *Ulva*.

1. **E. compressa,** *Grev.;—Fl. N. Z.* ii. 264. Frond very variable, 6–12 in. long, elongate; branches cylindrical or compressed, firm, capillary 2–$\frac{1}{4}$ in. diameter; branches simple or nearly so, attenuated at the base, obtuse.—Harv. Phyc. Brit. t. 335.

Abundant on the seashores. (Common in all seas.)

2. **E. intestinalis,** *Grev.;— Fl. N. Z.* ii. 264. Frond 2 feet and more long, $\frac{1}{12}$–3 in. diameter, simple, tapering downwards, inflated, often floating, often waved and curled, full of green-coloured fluid, fading to yellow and white.—Harv. Phyc. Brit. t. 154.

Abundant on seashores, with *E. compressa.* (Common to all seas.)

3. **E. clathrata,** *Grev.*—Var. *ramulosa*, Fl. N. Z. ii. 265. Frond 5 in. to 2 feet long, very slender, much branched, highly reticulated, curled and twisted, forming a dense interwoven mat, beset on all sides with short spine-like branchlets, hence harsh to the touch.—*E. ramulosa,* Smith; Harv. Phyc. Brit. t. 245.

Var. *erecta.* Branches all attenuated to a fine point.—*E. erecta,* Harv. Phyc. Brit. t. 43. *Zignoa,* Trevis; Mont. in Voy. au Pôle Sud, t. 30.

Var. *ramulosa,* Otago Harbour, *Lyall.* Var. *erecta,* **Lord Auckland's** group, *D'Urville.* (Europe, etc.)

104. BANGIA, Lyngbye.

Frond generally pink or purple, capillary, tubular, simple or branched, of numerous radiating cells in transverse rows, enclosed in a hyaline sheath. Fructification of purple or green spores, one in each cell of the frond.

A considerable genus of very small fresh- and salt-water *Algæ*, often forming a scum or stratum on rocks, walls, etc.

1. **B. ciliaris,** *Carmichael;—Fl. N. Z.* ii. 264. Filaments minute, simple, straight, $\frac{1}{2}$ in. long, variable in breadth, compressed, purple, sometimes expanding and leafy in the middle; cells in 2 rows except where expanded, globular.—Harv. Phyc. Brit. t. 322.

Parasitic on leaves of *Zostera,* Cook's Straits, *Lyall.* (Europe, also on *Zostera.*)

2. **B. lanuginosa,** *Harv. in Fl. N. Z.* ii. 264. Filaments curved, cylindric, bright purple, soft, gelatinous, of one series of lenticular cells, which are shorter than broad.

Parasitic on *Chordaria, Colenso.*

105. BATRACHOSPERMUM, Roth.

Frond immersed in gelatine, filiform, branched, usually moniliform and nodose, longitudinally striate, composed of colourless jointed filaments agglutinated together, beset with distant whorls of moniliform branchlets. Spores in globular masses seated on the whorls.

Delicate filamentous freshwater *Algæ*, growing in subalpine lakes and running streams.

1. **B. moniliforme,** *Roth;—Fl. N. Z.* ii. 261. Vaguely branched, variously coloured (green purple violet brown or blackish); nodes moniliform, distinct, globose, those of the branches confluent.—Kuetz. Sp. Alg. 535.—*Conferva gelatinosa*, Linn.; Dillwyn, Conferv. t. 32.

Freshwater streams, probably common, *Colenso*; Canterbury, *Lyall*. (Europe, Africa, America.)

106. CLADOPHORA, Kuetzing.

Frond green, filiform, dichotomously branched, articulate, uniform throughout; articulations of single long or short tubes, filled with granular endochrome, that at maturity developes antherozoids.

A large and widely-diffused genus.

§ 1. *Filaments forming dense intricately-matted masses.*

1. **C. herpestica,** *Kuetzing;—Fl. N. Z.* ii. 262. Filaments tufted, creeping, ultra-setaceous, bladder-like, rigid, irregularly branched; uppermost branches fascicled; articulations about 15 times longer than broad.—*Conferva*, Mont. in Voy. au Pôle Sud, 6.

Sandy soil, Bay of Islands, *Hombron, J. D. H.*; Cape Kidnapper, *Colenso*.

2. **C. Lyallii,** *Harv. in Fl. N. Z.* ii. 262. *t.* 121 C. Filaments pale yellow-green, forming dense wide-spreading matted tufts, stout, rigid, decumbent below, then erect, $\frac{1}{4}$–$\frac{2}{3}$ in. long, very irregularly branched; branches spreading, secund or opposite or alternate; branchlets few, 3–4 articulations, which are constricted at the nodes, and $1\frac{1}{2}$ times as long as broad.

Stewart's Island, *Lyall*.

3. **C. pacifica,** *Kuetzing.—Conferva*, Mont. in Voy. au Pôle Sud, 7. t. 14. f. 2; Fl. Antarct. 192. Filaments 1–2 in. high, dull green, olive-green below, forming an intricate spongy compact mass, capillary, rigid, excessively branched; branches erect, lower simple, upper dichotomous; branchlets subsecund, acute, spiniform, straight or hooked; articulations as long or twice as long as broad.

Lord Auckland's group, *D'Urville, Hombron*. Unknown to us.

§ 2. *Filaments tufted, but not forming spreading matted masses, more or less regularly branched.*

4. **C. pellucida,** *Kuetzing;—Fl. N. Z.* ii. 262. Filaments bright green, a span long and under, firm, subcartilaginous, long, simple and naked below, above branched; branches opposite alternate or whorled, uppermost

fascicled; primary articulations 8–16 times, of the branches 8–10 times, of the branchlets 4–8 times longer than broad.—Harv. Phyc. Brit. t. 174.—*C. catenifera*, Kuetz. l. c.

Waitemata Harbour, *Lyall.* (Europe, Australia, S. Africa.)

5. **C. verticillata,** *Hook. f. and Harv. Fl. Antarct.* i. 193. Filaments 3–8 in. long, tufted, rigid, setaceous, very slender, straight, sparingly divided; lateral branches elongate, simple, naked, erect, fastigiate, opposite or 3–4-nate, slightly dilated at the tips; primary articulations 6–8 times longer than broad, upper as long or twice as long as broad.—Kuetz. Sp. Alg. 388.

Port William, in a cave, *Lyall.* **Lord Auckland's** group, *J. D. H.*

6. **C. Colensoi,** *Harv. in Fl. N. Z.* ii. 262. Filaments not densely tufted, dark green, slender, flexuous, sparingly decompound from the base; branches slender, curved, flexuous, naked below, above with short secund very erect appressed branchlets; articulations 3–4 times longer than broad, full of green matter, cylindric, not contracted at the nodes, uniform throughout.

Hawke's Bay, *Colenso*, rare.

7. **C. Daviesii,** *Harv. in Fl. N. Z.* ii. 263. Filaments 4–5 in. long, bright green, densely tufted, rigid, stout, much branched; branches alternate, usually naked below; axils very broad and rounded, above with short crowded subcorymbose generally secund branchlets of 2 or 3 articulations, the latter seldom twice as long as broad, uniform throughout the frond.

Tauranga, *Colenso.*

8. **C. gracilis,** *Griffiths;—Fl. N. Z.* ii. 263. Filaments 10–15 in. long, pale green, very slender, much branched; branches erecto-patent, stout, flexuous; branchlets secund, pectinate, elongate, slender, of many articulations, subtomentose; articulations 2–4 times as long as broad, contracted at the nodes.—Kuetzing, Sp. Alg. 403; Harv. Phyc. Brit. t. 18.

Port William, *Lyall.* (Europe, Tasmania.)

9. **C. crinalis,** *Harv. in Fl. N. Z.* ii. 263. Filaments densely tufted below, deep green, 1–2 in. high, slender, rigid, shining when dry, decompound, but not much branched; branches and branchlets few, scattered, short or long, very straight and erect, obtuse; articulations 4–8 times longer than broad, upper shorter, constricted at the nodes.

New Zealand, *Colenso.*

10. **C. virgata,** *Kuetzing.—Conferva*, Agardh;—Fl. Antarct. 192. Filaments excessively branched below; branches elongate, fastigiate, ultra-setaceous, distinctly dilated at the tips; lower articulations 2–4 times longer than broad, upper about equal, collapsed when dry.

Lord Auckland's group, *D'Urville.* (S. Africa.)

107. CONFERVA, Linn.

Frond of green, slender, jointed filaments; joints of a single cell, floating or attached. Fructification of clusters of antherozoids contained in the joints.

A large salt and freshwater genus, abundant in all climates.

1. **C. Darwinii,** *Kuetz. Sp. Alg.* 380.—*Chætomorpha*, Fl. N. Z. ii. 263. Frond 2–8 in. long, very stout, gradually dilated upwards; articulations often geminate; lower rather longer than broad; upper shorter, $\frac{1}{8}$ in. broad, constricted at the nodes.—*C. clavata*, var. *Darwinii*, Fl. Antarct. 493. t. 192. f. 1.

East coast, *Lyall, Colenso;* Otago, *Lindsay.* (Tasmania, Fuegia?)

2. **C. ærea,** *Dillwyn, Conf. t.* 80;—*Fl. N. Z.* ii. 263. Filaments numerous, erect, pale green below, paler above, gradually dilated upwards; lowest articulation longest, 5 times as long as broad, the middle ones half as long; uppermost as long as broad.—Harv. Phyc. Brit. t. 99 B. *Chætomorpha*, Kuetz. Sp. Alg. 379.

Houraki Gulf, *Lyall.* (Australia, Europe.)

3. **C. valida,** *Hook. f. and Harv. Fl. N. Z.* ii. 263. Filaments pale green, loosely matted, flaccid; articulations $1\frac{1}{2}$–3 times as long as broad, contracted at the nodes. *Chætomorpha*, Kuetz. Sp. Alg. 379.

Port William, *Lyall.* (Tasmania.)

108. TYNDARIDEA, Bory.

Filaments simple, green-yellow or brown, finally inosculating by transverse tubes. Endochrome of subglobular masses, which, after conjugation, unite and form a roundish globule lodged in one of the articulations, or in the connecting tube.

In streams, pools, ditches, and wet places, often forming spreading masses of coloured scum.

1. **T. anomala,** *Ralfs;—Fl. N. Z.* ii. 263. Joints equal or shorter than long, constricted here and there. Cell-wall often very thick.—E. Bot. Suppl. t. 2899; Hassall, Brit. Freshwater Alg. t. 38. f. 2, 3. *Zygogonium*, Kuetz. Sp. Alg. 447.

Freshwater, *Colenso.* (Europe.)—Apparently the same as the British plant, but impossible to determine from comparison of dried specimens.

2. **T.? byssoidea,** *Harv. in Fl. N. Z.* ii. 264. Filaments blackish, gelatinous, arachnoid, greenish under the microscope; articulations about 3 times as long as broad.

Freshwater lake, Kapiti, Cook's Straits, *Lyall.*

109. CALOTHRIX, Agardh.

Frond of capillary, green internally, short, tufted or fascicled, rather rigid filaments, fixed, without a mucous coat, and without a waving motion in water; tube continuous, transversely closely striated, at length dissolving into lenticular sporules.

Fresh- and salt-water small *Algæ.*

1. **C. scopulorum,** *Weber and Mohr;—Fl. N. Z.* ii. 265. Filaments $\frac{1}{10}$ in. long, minute, erect, curved, flexuous, simple, forming a continuous, dirty-green, velvety stratum, attenuated to the point, slimy at the base, indistinctly striate.—Harv. Phyc. Brit. t. 58 B.

Mud and rocks near high-water mark, *Colenso*. (Widely diffused.)

110. OSCILLATORIA, Vaucher.

Frond of capillary interlacing filaments floating in mucus; filaments simple, continuous, presenting a curious oscillating motion under the microscope; tube transversely closely striated.

A large genus, abounding chiefly in fresh water, forming green or coloured slimy masses on moist rocks etc., and a scum on ponds, etc.

Species of this genus are no doubt as common in New Zealand as in Europe and everywhere, but they cannot be determined except in a fresh state. Fragments of several have been collected by Colenso, Lyall, and others.

111. TOLYPOTHRIX, Kuetzing.

Frond of free capillary filaments, articulate at the base; branches fascicled or radiating, not opposite. Sheaths hyaline, open at the apex.

A considerable European freshwater genus, probably found all over the globe.

1. **T. irregularis,** *Berk. in Fl. N. Z.* ii. 265. Filaments verdigris-green, irregular, $\frac{1}{2000}$–$\frac{1}{2250}$ in. diameter, compressed, sometimes constricted or torulose, often attenuated above and below, furnished at the base with a minute, hyaline, elliptic connecting joint; rings very narrow.

Tidal mud, amongst patches of *Vaucheria*, *Colenso*.

112. NOSTOC, Vaucher.

Frond gelatinous or coriaceous, dull green or brownish, amorphous, traversed by moniliform filaments, which finally break up as spores.

Gregarious *Algæ*, forming jelly-like masses on the damp ground, moss, rocks, and sometimes in water, often resembling *Collemas*.

1. **N. verrucosum,** *Vaucher;—Fl. N. Z.* ii. 266. Frond $\frac{1}{2}$–1 in. across, gregarious, confluent, subglobose, plaited, blackish green, at length hollow filaments short, curled, fragile.—Agardh, Syst. Alg. 21.

Canterbury plains, in freshwater streams, *Lyall*. (Europe, etc.)

113. ACHNANTHES, Bory.

Frond like a little flag, rectangular, attached to a short or long stipes by one corner, composed of a few frustules, which separate without cohering at the angles.

Very minute *Algæ*, parasitic, like minute flags on the stems of other *Algæ*.

1. **A. brevipes,** *Agardh.* Frond with a very short stipes and 2–5 frustules, which have 2 coloured spots.—Mont. in Voy. au Pôle Sud, 1.

Lord Auckland's Island, on *Polysiphonia glomerulata, D'Urville.* (Common in many seas.)

114. SCHIZONEMA, Agardh.

Frond of microscopic, brownish or greenish, flat filaments, which are excessively fragile, breaking up into rectilinear frustules; frustules in longitudinal series, enclosed in the simple or branched mucous or membranous frond.

Very minute *Algæ,* parasitic on other *Algæ,* etc., or occurring in floating mass.

1. **S. crispum,** *Montagne;—Fl. N. Z.* ii. 266. Threads tufted, green, very crisp, $\frac{1}{3}$–$\frac{1}{4}$ in. long, the tips penicellate; branchlets obtuse; frustules nearly parallelograms.

Lord Auckland's group, on *Polysiphonia glomerulata, D'Urville.*

ADDITIONS, CORRECTIONS, &c.

Two valuable contributions to New Zealand botany have been published since the first part of this Handbook appeared, viz. Mr. Colenso's 'Essay on the Botany of the North Island,' which was printed for the Commissioners of the New Zealand Exhibition of 1865, and is full of interesting matter, which the author alone was able to supply, especially regarding the altitudinal and latitudinal ranges of the species; the other is Dr. Mueller's work 'On the Vegetation of Chatham Islands,' founded chiefly on Mr. W. Travers's collections.

For additional matter regarding the flora of New Zealand, I am especially indebted to Dr. Haast, Dr. Hector and Mr. Buchanan, Mr. Travers, Logan, Mair, Colenso, T. Kirk, Captain Rough, Drs. Knight and Lindsay.

Page

lv. In the Key to the Artificial Classification, under Classes XXI.–XXIII., the following—*Myriophyllum*, *Loranthaceæ*, *Araliaceæ*, *Umbelliferæ*, *Corneæ*, *Sicyos*—should form a section, with "Perianth double, superior. Corolla polypetalous."

lxii. **Corneæ.**—The petals are *usually* valvate, not always (see *Griselinia*).

lxvii. The Orders 5, *Jungermannieæ*, and 6, *Marchantieæ*, are treated as divisions of one Order, **Hepaticæ,** at p. 497.

Class I. DICOTYLEDONS.

Order I. RANUNCULACEÆ.

1. In the generic character of *Clematis*, *read* Sepals 3–8, and Stamens 5–20.

2. **Clematis hexasepala.**—Southern Alps, *Haast*.

3. The Key to the species of *Ranunculus* might be improved by bringing together the species with silky carpels and white flowers—*R. Lyallii*, *Traversii*, *Buchanani*, and perhaps *insignis*.

4. **R. Lyallii,** var. *araneosa*.—Whole plant more or less covered with scattered flaccid hairs.

Page

4. **R. Traversii.**—F. Mueller considers this identical with *R. Lyallii*, but though I have received numerous specimens of the latter I find none of them with the characters of this.

5. After *R. pinguis* add—

No. 4 *bis*. **R. Godleyanus,** *Hook. f., n. sp.* Erect, very stout, a foot high, everywhere quite glabrous except the receptacle of the fruit. Stem 1 in. diameter at the base. Leaves all radical, with broad thick petioles 2–5 in. long and 1 in. broad; blade 6–7 in. long, broadly oblong, rounded at the apex, narrowed into the petiole, coarsely crenate or doubly crenate, thickly fleshy or coriaceous, with radiating reticulated nerves. Scapes stout, ascending, as long or longer than the leaves, bearing at the middle 2 or more sessile oblong bracts, and several branching or simple flowering peduncles that are 2-bracteolate at the axils. Flowers numerous, corymbose, 1–1½ in. across. Sepals broadly obovate-oblong, spreading. Petals 5, broadly obcuneate-obovate, 2-lobed at the apex, with 3 naked glands at the base, deep yellow. Achenes very numerous, forming a broad-oblong head, spreading, the lower recurved, subulate, narrowed into the slender curved styles, nearly glabrous; receptacle pilose.

Middle Island: Whitcombe's Pass, on the edge of a lagoon, alt. 4200 ft., *Haast*, March, 1866.—A noble plant, and worthy of the distinguished colonist whose name it bears. It approaches *R. pinguis*, and may prove a form of that very variable plant, but so far as Haast's excellent specimens go, it is abundantly distinct in its glabrousness, broad oblong leaves, cuneate at the base, not at all lobed, and smaller more slender achenes.

5. The following species was omitted in the Handbook:—

No. 4 *ter*. **R. Aucklandicus,** *A. Gray, Bot. U. S. Expl. Exped.* i. 8. Silky-strigose; rhizome short, thick; stem solitary, erect, 6–10 in. high, 2-leaved, 1-flowered. Leaves long-petioled, petiole strigose-hirsute, orbicular or reniform, 3-fid to or beyond the middle with closed sinus, both surfaces equally silky-strigose, 1–1½ in. diameter, thick, but not fleshy; lobes cuneiform, 2–3-lobed, again toothed or incised. Flower not seen. Receptacle of achenes cylindric, ¼ in. long, papillose, minutely hairy; achenes ovate compressed, not margined, smooth and glabrous with a short straight subulate style.

Lord Auckland's group, *Wilkes's Expedition.*—Near *R. pinguis* and possibly a form of it, if I am correct in considering *R. Munroi* a variety of that species.

5. **R. geraniifolius.**—Southern Alps, alt. 2500 ft., *Haast.*

5. **R. Buchanani.**—Erase the Macaulay river and Waimakeriri habitats, which belong to the following species:—

5. No. 7 *bis*. **R. chordorhizos,** *Hook. f., n. sp.* Everywhere glabrous, short, stout. Rhizome as thick as the finger, 2–4 in. long, almost woody, with numerous perpendicular root-fibres 6–10 in. long and as thick as whipcord. Leaves with stout petioles, 1–1½ in. long; blade coriaceous and fleshy, suborbicular, 3-partite to the base; segments broadly obovate-spathulate, almost petiolulate, 3-lobed; lobes crenate; upper surface pitted when dry. Peduncles few, thick, 1-flowered, from a scape that is much shorter than the petioles. Flowers not seen. Head of achenes globose. Receptacle

Page

small, globose. Achenes glabrous, turgid, not margined nor angled, with a slightly curved subulate style as long as the achene.

Middle Island: Alps of Canterbury, Macaulay river, alt. 5000 ft.; Waimakeriri district, on limestone-gravel and Mount Somers range, alt. 2–4000 ft., *Haast.*—A very remarkable species.

6. **R. sericophyllus.**—Browning's Pass, 5–7000 ft., *Haast.*

6. **R. Haastii.**—Dr. Haast assures me that this is not at all a variable plant. On the other hand, Mr. Travers sends me a specimen from shingle beds, Saddle Hill, Wairau, alt. 5000 ft., with a 1-flowered scape; it is in fruit only, and shows the remarkable flat-beaked achenes well.

7. **R. macropus.**—Found at Tapaitoitoi, 300 miles N. of Poverty Bay, *Kirk.*

8. **R. gracilipes.**—Dr. Hector informs me that this species has creeping stems.

ORDER II. MAGNOLIACEÆ.

10. **Drimys axillaris.**—Both Colenso and Buchanan agree with Raoul and others in considering there are two species of *Drimys* in New Zealand; of these, the southern one, also found in mountain districts of the Northern Island, is *D. colorata*, Raoul, differing according to Buchanan, especially in the leaves being hardly coriaceous, green on both surfaces, and not glaucous nor spotted, in the midrib not hairy, in the flowers more abundant, in the long peduncles, and in the bark being pale, not black-brown.

ORDER III. CRUCIFERÆ.

10. In the ordinal character, *for* Stamens . . . 2 longer than the others, *read* shorter.

13. **Braya novæ-Zelandiæ.**—I have received better specimens of this from Dr. Hector; the pod is laterally much compressed, and the plant evidently closely allied to *Notothlaspi*, but differs in the valves not being winged, and in the short funicles. Considering its very curious habit, I think it should be regarded as a new genus and called *Pachycladon* in reference to the remarkably stout branches of the rhizome or stock.

5. PACHYCLADON, Hook. f.

Pod laterally compressed, elliptic or linear-oblong; valves convex, subcymbiform, keeled, not winged; nerves very obscure; septum incomplete; style very short; stigma capitate, 2-lobed. Seeds 3–5 in each cell, obovoid; funicles short; cotyledones incumbent.

1. **P. novæ-Zelandiæ.**—*Braya novæ-Zelandiæ*, p. 13.

15. **Notothlaspi rosulatum.** — Smells deliciously of orange-blossoms, *Haast.*

ORDER IV. VIOLARIEÆ.

18. **Hymenanthera crassifolia.**—Banks's Peninsula, *Haast.* Chatham

Page

Island, *W. Travers* (a very coriaceous-leaved variety, distinguished by F. Mueller as var. *Chathamica*). Dr. Mueller is disposed to unite the New Zealand, Norfolk Island and Australian species, which are all very variable plants.

ORDER V. PITTOSPOREÆ.

19. **Pittosporum Colensoi.**—*For* "leaves smaller" *read* "leaves larger."

19. 2 *bis.* **Pittosporum Buchanani,** *Hook. f., n. sp.* A rambling-branched shrub, 10–12 ft. high; young shoots and leaves silky-pubescent. Leaves scattered, 2–5 in. long, of the male plant elliptic-oblong, acuminate, flat, quite entire, glabrous, of the female more oblong, shining green above, paler below. Peduncles axillary, solitary, slender, curved, silky-pubescent, $\frac{1}{3}$–$\frac{2}{3}$ in. long, 1- rarely 2-flowered, the male usually longest, sometimes the peduncle is short and gives off 1 or 2 slender pedicels; bracts at the base of the peduncle small, acute. Sepals oblong, obtuse. Petals linear, dark purple; of the female smaller. Capsule as in *P. tenuifolium.*

Northern Island: from near Tongariro (cultivated at Wellington), *Buchanan.*

It is with great reluctance that I add another *Pittosporum* to the series, already very difficult of discrimination, that consists of *tenuifolium, Colensoi,* and *fasciculatum.* The present seems to be nearer to the last named species, but differs in the almost solitary flowers, slender silky peduncles, and obtuse oblong sepals.

19. **P. patulum.**—Mr. Travers sends a fruiting specimen from the Upper Wairau, together with young foliage, which is linear, very narrow, and pinnatifid.

20. **P. rigidum.**—Acheron Island, Dusky Bay, *Hector and Buchanan.*

21. **P. umbellatum.**—Peduncles sometimes only $\frac{1}{2}$ in. long.

ORDER VI. CARYOPHYLLEÆ.

22. **Gypsophila tubulosa,** appears to be rapidly spreading over New Zealand, and is, doubtless, an imported plant.

Stellaria gracilenta.—"Leaf-margin not revolute" (*Knight*), but I find it always much thickened or revolute to the midrib in the common form of the plant.

25. **Colobanthus Billardieri.**—Chatham Island, *W. Travers.* Distinguished as a variety by F. Mueller, on account of the short pedicels and rigid channelled leaves.

25. **C. acicularis.**—Hurumui mountains, *Travers.*

25. **Spergularia rubra,** var. *marina.*—This is a rather large flowered state of the plant; common in S. America, S. Africa, and Australia.

ORDER IX. HYPERICINEÆ.

29. **Hypericum gramineum.**—Colenso considers this to be a rare and local plant, but I have specimens from very distant localities (Auckland, Waipura river, and Akaroa).

ORDER X. MALVACEÆ.

Page

30. **Plagianthus betulinus.**—Chatham Island, *W. Travers.*

31. **Hoheria populnea.**—Colenso informs me that the var. *a.* is not found south of Bream Bay.

31. **HIBISCUS.**—In the last line of generic character, *for* cells 3- or many-ovuled, *read* cells 2-ovuled.

ORDER XI. TILIACEÆ.

33. **Aristotelia fruticosa.**—Hector and Buchanan appear to regard this as a form of *A. racemosa.* It is an extraordinarily variable plant.

ORDER XII. LINEÆ.

35. **Linum monogynum.**—Var. *a.* Chatham Island, *W. Travers.*

ORDER XIII. GERANIACEÆ.

35. In the Ordinal character, line 1, *read* Leaves opposite or alternate.

At the end of *Geraniaceæ* insert—

37. 5. **Geranium Traversii,** *Hook. f., n. sp.* All over hoary and almost silvery with fine white pubescence. Stems 1–2? feet long, rather stout. Leaves long-petioled, orbicular, 1½ in. diameter, 7-lobed to the middle; lobes cuneate, 3-fid, both surfaces equally hoary; stipules broad, almost orbicular, cuspidate. Peduncle 4 in. long, 1-flowered, slender, with 2 ovate-lanceolate acuminate bracts in the middle. Flower large, 1 in. diameter. Sepals broadly ovate, cuspidate, silky. Petals orbicular-ovate, quite entire, much larger than the sepals, white.

Chatham Island: waste ground, *W. Travers.*—F. Mueller refers this to *G. dissectum,* from all states of which it appears to me to differ widely in the much larger size, almost coriaceous leaves, hoary pubescence, long 1-flowered peduncle, large white flowers, silky sepals, and entire petals. It is a beautiful plant. I have but one small specimen.

37. **Pelargonium australe.**—Common on the Alps of the Lake district, Otago, *Hector and Buchanan.*

ORDER XIV. RUTACEÆ.

40. **Melicope simplex.**—Leaves 3-foliolate, with winged petioles in young state, *Kirk.*

ORDER XV. MELIACEÆ.

41. **Dysoxylum spectabile.**—The Middle Island habitat is probably erroneous, *Colenso.*

ORDER XVI. OLACINEÆ.

41. **PENNANTIA.**—There are now three species of this genus known, the present an Australian and Norfolk Island one.

ORDER XVIII. RHAMNEÆ.

Page

44. **Discaria Toumatou.**—This becomes 15 feet high in subalpine localities, *Haast.*

ORDER XXI. CORIARIEÆ.

46. **Coriaria ruscifolia.**—Chatham Island, *W. Travers.* For an elaborate report on the poisonous properties of this plant, see Dr. L. Lindsay in Brit. and Foreign Medic. Chirurg. Review, July, 1865.

47. **C. thymifolia.**—Hector says that this is the poisonous *Tutu* and annual herbaceous *Tutu* of Otago, that it grows 4–5 ft. high, and is difficult to distinguish by leaf from *C. ruscifolia.*

47. **C. angustissima.**—Hector informs me that this is the "ground *Tutu,*" of Otago, and is subalpine and rare.

ORDER XXII. LEGUMINOSÆ.

48. Three lines from top, for *Edwardsia* read *Sophora.*

49. **Carmichælia grandiflora.**—Flowers very odoriferous, often small, but not so small as *C. odorata.* Beak of pod sometimes $\frac{1}{6}$–$\frac{1}{4}$ in. long.

50. **C. odorata.**—It is difficult to distinguish this from *C. grandiflora.*

51. **Notospartium Carmichæliæ.**—Ascends to 1500 ft., *Haast.*

51. **Swainsonia novæ-Zelandiæ.**—Hurumui valley, *Travers.*

53. **Sophora tetraptera,** var. *α.* Chatham Island, *W. Travers.* I have seen no specimen.

ORDER XXIII. ROSACEÆ.

54. **Potentilla anserina,** var. *β.* Chatham Island, *W. Travers.*

55. **Geum parviflorum.**—An extremely variable plant. Haast sends Mount Torlesse specimens with very slender flowering branches, and flowers not $\frac{1}{4}$ in. diameter.

57. **Acæna inermis.**—Hurumui Mountains, *Travers.*

ORDER XXIV. SAXIFRAGEÆ.

58. **Donatia novæ-Zelandiæ.**—Dusky Bay, on the hills, *Hector and Buchanan;* Ramsay glacier, 4200 ft., *Haast.*

61. **Weinmannia racemosa.**—Becomes a large tree, with close-grained hard brittle wood, *Hector and Buchanan.*

ORDER XXVI. DROSERACEÆ.

63. **Drosera stenopetala.**—Milford Sound, *Hector.*

Page

ORDER XXVII. HALORAGEÆ.

64. In the Key to the genera of *Haloragea*, in the second line, *for* stigmas *read* style.

68. **Callitriche verna.**—This is the form or species called *C. stagnalis*, Scopoli (*C. platycarpa*, Kuetzing), and is *C. verna* β, of Linnæus, distinguished by the broad thick wings of the fruit, and broad short leaves.

ORDER XXVIII. MYRTACEÆ.

70. **Leptospermum ericoides.**—Mr. Kirk sends a remarkable variety? of this species, with silky branchlets, and differing greatly in size and habit from *L. ericoides*, which grows near it. The species of this genus are, however, so variable that I do not venture to make a new one of this. He thus describes it:—

"Young leaves and branches more or less pubescent. Leaves sessile, scattered and fascicled, linear-lanceolate, margins recurved, pungent, much dotted. Flowers minute, sessile or shortly peduncled. Petals much crumpled. Capsules very small, with persistent calyx-lobes."

Northern Island: Apatawapa and Coxe's Creek, *Kirk*.

71. **Metrosideros lucida.**—Often attains the size of a large tree, *Hector*.

74. **Myrtus pedunculata** and **M. obcordata.**—Both abundant in Otago, *Hector and Buchanan*, etc.

ORDER XXIX. ONAGRARIEÆ.

75. **FUCHSIA.**

There are three tolerably distinct forms or species of the genus in New Zealand, which may be thus distinguished:—

Petioles shorter than the leaves. Stamens as long or longer than the calyx-lobes . . . . 1. *F. excorticata.*
Petioles as long as or longer than the leaves.
Stems procumbent, filiform. Stamens shorter than the calyx-lobes 2. *F. procumbens.*
Stems branching low, woody. Stamens as long as or longer than the calyx-lobes . . . . 3. *F. Colensoi.*

Of these, **F. excorticata** forms a tree with ovate-lanceolate leaves, silvery below. Flowers $\frac{3}{4}$–$1\frac{1}{2}$ in. long. It is common throughout the Island.

F. procumbens is very slender, creeping or prostrate. Leaves rounded-ovate or -cordate. Flowers not $\frac{1}{3}$ in. long. I have it from the Northern Island, *A. Cunningham* and *Colenso*.

F. Colensoi is intermediate in size. Leaves very variable, ovate orbicular or cordate. Flowers as large as in *F. excorticata*.—Northern Island, *Colenso*. Middle Island, common, Canterbury plains, *Travers;* Otago, *Lindsay*, *Hector*.

81. **Epilobium pallidiflorum.**—*E. tetragonum*, F. Mueller, Veg. Chatham Island. Chatham Island, *W. Travers*.

ORDER XXX. PASSIFLOREÆ.

Page

81. The seeds are inadvertently described as *exalbuminous* instead of *albuminous*.

81. Line 4 from bottom, *for* tetramerous *read* pentamerous.

ORDER XXXII. FICOIDEÆ.

84. **Mesembryanthemum australe.**—Chatham Islands, *W. Travers.*

84. **Tetragonia trigyna.**—This Dr. Mueller assures me is the true *T. implexicoma*, Miquel, of Australia. Littleton Harbour, *Travers.* Styles 2–4. Stamens 12–16.—Var. *Chathamica*, F. Muell. Veg. Chatham Island. Leaves broader. Calyx-lobes 4–5, broader, more unequal. Styles 3–4, shorter. —The flowers are usually unisexual in this genus.

ORDER XXXIII. UMBELLIFERÆ.

86. **Hydrocotyle heteromera.**—Colenso sends a small state with leaves only $\frac{1}{8}$ in. broad.

88. **Pozoa trifoliolata.**—This is a Middle Island plant, the Totara-nui of Captain Cook being in that Island.

90. **Eryngium vesiculosum.**—Colenso assures me that this is a rare and local plant, and has not been found at Auckland, where I supposed Dr. Sinclair had gathered it.

91. **Oreomyrrhis Colensoi.**—Waipura, on limestone rocks, *Haast.*

91. **O. ramosa.**—Glacier gully, Wairau Mountains, *Travers.*

91. **O. andicola,** *Endl.* A native of Australia and Chili; is erroneously stated in Benth. Fl. Austral. iii. 377, to be a native of N. Zealand.

92. **Aciphylla Lyallii.**—Waiau range, *Travers.* Buchanan regards this as a var. of *A. Colensoi*, and *A. Munroi* as a var. of *A. squarrosa.*

93. **A.? Traversii,** *F. Mueller, Veg. Chatham Islands,* 18 (*Gingidium*). Dr. Mueller enumerates the above as a probably new Chatham Island species, allied to *A. Munroi;* his description is incomplete; the leaves have a denticulate sheath and 2 or 3 rigid longitudinally-striated segments; bracts linear, undivided, passing into the sheath without denticulations.

93. **Ligusticum Dieffenbachii.**—(*Gingidium*, F. Muell. Veg. Chatham Island, 17.) Tall, erect, sheaths of petioles produced into 2 obtuse teeth. Leaves 3-pinnatisect; segments long-linear, flat, flaccid, 7-nerved, mucronate; rachis imperfectly jointed. Bracts 5-partite. Umbels compound, numerous, panicled; involucral leaves few, narrow-lanceolate. Calyx-teeth of male-florets short.—*Mueller, l. c.*

Chatham Island: under maritime cliffs, *W. Travers.* (Specimen imperfect.)

95. **L. Lyallii.**—Common on the shores of the West coast of Otago, *Hector and Buchanan.*

Page

95. **L. Haastii.**—Waiau range, alt. 4000 ft., and Hurumui Mountains, *Travers.*

95. **L. filifolium.**—Fruiting specimens have been sent by Haast, and by Travers from the Waiau and Hurumui Mountains.—Fruit $\frac{1}{3}$ in. long, narrow-oblong, much compressed, pale, styles very short; mericarps flattened, with 5 nearly equal prominent acute ridges; commissural suture flat.—This plant hence exhibits the broad flat lateral wings of *Angelica*, and short styles of *L. brevistyle.* It is intermediate between *Angelica* and *Ligusticum.*

96. **L. carnosulum.**—Fruiting specimens have been received from Haast, also gathered on Mount Torlesse. They are very stout, almost woody, rigid. Fruit sunk in the axils of the involucral leaves, almost sessile, $\frac{1}{3}$ in. long; styles rigid, subulate; mericarps incurved, ridges acute, not prominent; commissural suture rounded.

96. **L. piliferum.**—Travers sends specimens from the Waiau Mountains, alt. 4000 ft., with shorter hairs terminating the segments of the leaflets.

96. **L. aromaticum.**—*For* 4–6500 ft. *read* 2–5000 ft.

97. **L. imbricatum.**—Waiau range, Saddle hill, on shingle, alt. 5500 ft., *Travers.*

98. **Angelica decipiens.**—Waiau range, *Travers.* I have an imperfect specimen apparently of this plant from Colenso.

98. **A. rosæfolia.**—Canterbury, *Travers.*

ORDER XXXIV. ARALIACEÆ.

101. **Panax anomalum** is not found at the Bay of Islands, according to Colenso.

101. **P. Edgerleyi.**—Leaflets in young plant deeply sinuate-pinnatifid. (See below, note on *P. Colensoi.*) Used by the natives to scent themselves (*Hector*).

101. **P. lineare.**—Dusky Bay, *Hector and Buchanan.* A shrub 6–10 ft. high.

101. **P. crassifolium.**—Chatham Island, *W. Travers.* Referred to *Hedera* by A. Gray (Bot. U. S. Expl. Exped. 719), and to *Pseudopanax* by Seemann (Journ. Bot. 1865, p. 178).

102. **P. longissimum.**—Referred to *Pseudopanax crassifolium* by Seemann (Journ. Bot. 1864). It is *P. coriaceum*, Regel, Gartenflora, 1859, 45. Mr. Logan has sent me specimens clearly showing that it is only the young state of *P. crassifolium.*

102. **P. Lessonii.**—Otago, West coast, *Hector.* Is referred to *Hedera* by A. Gray. Bream Bay, where it was gathered by D'Urville, is in the Northern Island. Also referred to *Pseudopanax* by C. Koch and Seemann.

102. **P. Colensoi.**—Mr. Logan informs me that the plant with pinnatifid leaflets, figured in Fl. N. Z. t. 21, is the young state of *P. Edgerleyi*, and not of this species.

Page

102. **P. arboreum.**—Colenso informs me that this is found chiefly on sea-coasts and banks of rivers.

104. **Meryta Sinclairii.**—Dr. Seemann having referred this plant to its proper genus in the 'Bonplandia,' his name should be attached to it.

104. **Polyscias pinnata,** *Forst.*, a Pacific Island Araliaceous plant, was erroneously described as a native of New Zealand.

ORDER XXXVI. LORANTHACEÆ.

107. **Loranthus tetrapetalus.**—Otago, *Hector and Buchanan.*

108. **Viscum salicornioides.**—Colenso observes that this is a very rare and local plant in the Northern Island, being confined to two spots, near the Bay of Islands, and at Wellington, and found only on *Dracophyllum*, *Gaultheria*, and *Leptospermum.*

ORDER XXXVII. CAPRIFOLIACEÆ.

109. **Alseuosmia macrophylla.**—"Flowers full bright crimson, berries crimson," *Kirk, mss.*

109. **A. quercifolia.**—Mr. Colenso affirms that this is found nowhere between the Bay of Islands and Te-Hawera. "Tube of flower crimson, lobes tipped with green, berry crimson," *Kirk, mss.*

109. **A. linariifolia.**—"Flowers crimson, lobes paler. Berry crimson," *Kirk, mss.*

ORDER XXXVIII. RUBIACEÆ.

113. **Coprosma petiolata.**—Chatham Island, *W. Travers.*

ORDER XXXIX. COMPOSITÆ.

122. In Key of *Compositæ*, 12. *Ozothamnus*, omit the word *not* from the diagnosis.

124. **Olearia operina.**—Edges of sounds on the West coast of Otago, from Milford Sound southwards, growing sometimes 10–12 ft. high, with trunk 6–8 in. diameter. Found only at the level of the sea. *Hector and Buchanan.*

Chatham Island, *W. Travers.* A form with lax bracts on the scapes, thus connecting it with *O. angustifolia.*

124. **O. semidentata.**—Chatham Island, *W. Travers.*

125. **O. Colensoi,** var. *β.*—Otago, West coast, found at the level of the sea. 10–12 ft. high, on the mountains, alt. 3–4000 ft., where it forms impenetrable thickets, and has simply serrated leaves. *Hector and Buchanan.*

126. **O. insignis**—Wairau district, in rocky places, *Rough.*

126. After 9. *Olearia dentata*—

9*a.* **O. Traversii,** *F. Muell. Veget. of Chatham Island*, 19 (*Eurybia*). Arboreous, 30–35 ft. high. Branches and branchlets opposite, covered with silky down, as are the panicles, involucres, and leaves beneath. Leaves plane,

Page

opposite, oblong- or ovate-lanceolate, acuminate, quite entire, 1½–2½ in. long, glabrous and shining above; midrib obscure beneath, as are the veins, which form an acute angle with the midrib. Panicles numerous, cymose, axillary and terminal, much branched. Flower-heads very numerous, on slender pedicels, small, ¼ in. long. Involucral scales few, hardly imbricate, linear-oblong, obtuse, very silky. Florets of the ray with a short ray. Achene silky. Pappus rufous when dry.

Chatham Island, *W. Travers*, where it forms the principal wood, and is called "Ake Ake," and "Bastard Sandalwood-tree."

126. No. 11 *bis*. **O. lacunosa,** *Hook. f.*, *n. sp.* Branchlets, petioles, leaves below, and panicle covered with pale rusty buff tomentum. Leaves shortly petioled, 4–6 in. long, narrow-linear or linear-oblong, ⅓ to ⅔ in. broad, acute or acuminate, quite entire, glabrous and reticulated above, coriaceous; midrib stout; lateral nerves at right angles to the midrib, close set, very prominent, dividing the under surface of the leaf into numerous rather deep pits. Panicle on a short slender peduncle, corymbose; branches elongate and branchlets slender. Heads small, ⅛ in. diameter; involucre turbinate, scales few, pubescent, linear-oblong; florets very small. Achene compressed, pubescent.

Middle Island: Rotoroa Lake, alt. 5–7000 ft., *Travers*; Harper's Pass, alt. 3000 ft., *Haast*. A very singular and most distinct species, quite unlike any other. The very numerous short stout parallel nerves at right angles to the midrib afford a distinctive character.

126. **O. Haastii.**—Otago Alps, *Hector and Buchanan*. Leaves longer and narrower.

127. **O. nummularifolia,** var. *cymbifolia*.—A short stout erect woody shrub, 1 ft. high and upwards. Leaves close-set, spreading or deflexed, ¼ in. long, shortly petioled, oblong, obtuse, quite entire, glabrous and very convex above with strongly revolute margins all round, hence boat-shaped with the concavity downwards, viscid and pitted above, beneath and petiole covered with buff or white tomentum; nerves invisible. Heads solitary or few, on slender stiff erect powdery pedicels, as long or longer than the leaves, ½ in. long; involucre cylindric, scales oblong, glutinous, obtuse, appressed, imbricate. Achene narrow, elongate, angled, pubescent near the top.

Alps of Canterbury, Mount Fenzl and Mount Potts, alt. 3–4000 ft., *Haast*; Hürumui Mountains, *Travers*. A very distinct-looking variety, but hardly a different species.

127. **O. Forsteri.**—Mount Somers range. Leaves broader, subcordate, *Haast*.

127. **O.** (*Eurybia*) **cydoniæfolia,** *DC. Prod.* v. 267, is erroneously stated to be a native of N. Zealand.

130. **Celmisia holosericea.**—West coast of Otago, from the sea-level to 3–4000 ft., *Hector and Buchanan*.

133. **C. Lyallii.**—The "blunt-leaved Spaniard" of the Canterbury Alps.

133. **C. viscosa.**—For *Sinclair and Haast* read *Hector and Buchanan*.

134. **C. petiolata,** page 135, **C. Hectori,** and **C. sessiliflora.** After Mount Brewster read *Haast* (not *Hector and Buchanan*), who found the same species on Mount Alta.

134. **C. Traversii.**—Mr. Travers has sent me another specimen, with the leaves (and sheath) 12 in. long and the blade 2½ across.

Page

135. 21 *bis*. **C. ramulosa,** *Hook. f., n. sp.* Stem woody, procumbent, branched, slender; branches ascending, 2–4 in. long, densely clothed with closely imbricating short leaves. Leaves very numerous, $\frac{1}{4}$–$\frac{1}{3}$ in. long, linear-oblong, obtuse, subcoriaceous, with broad membranous sheathing bases, glabrous above, clothed beneath with snow-white cobwebby wool; margins strongly revolute. Scape $\frac{1}{3}$ in. long, slender, terminal, tomentose, with one or two bracts. Head campanulate, nearly 1 in. diameter; involucral scales linear-oblong, erect, pubescent and glandular. Rays $\frac{1}{2}$ in. long.

Middle Island: Dusky Bay mountains, alt. 3500 ft., *Hector and Buchanan*. A remarkable species, with a woody prostrate stem as thick as a crowquill and subfrutescent habit, approaching *Olearia*.

136. **Vittadinia australis,** *A. Rich.*—Mueller and Bentham consider this to be identical with the Australian species, and the former observes that it is also found in New Caledonia.

141. **C. lanata.**—The *Leptinella potentillina*, F. Mueller (Veg. Chatham Islands, 28), only differs from this in the eglandular florets, and from *L. dioica* in the more deeply-cut leaves and unisexual heads.

142. **C. atrata.**—This should be inserted before 3. *C. plumosa*, as in Key.

142. **C. minor.**—Otago, *Hector and Buchanan*.

142. *After* 8. *C. pectinata*, insert, **Cotula** (*Myriogyne*) **Featherstonii,** F. Muell. Veg. Chatham Islands, 27 (*Leptinella*). Branched, prostrate, downy. Leaves obovate or oblong cuneate, 3-toothed at the apex, narrowed to the sessile base. Head axillary and terminal, peduncled, 2-sexual. Involucral scales 10–12, sub-2-seriate, unequal, lanceolate. Receptacle depressed, hemispherical; female florets numerous, stipitate; corolla minute, conic, pale, glandless; male florets 4-toothed, glandless. Achene obovate-cylindric, striate, slightly glandular. *F. Muell.*

Chatham Island: damp maritime rocks, *W. Travers*.

143. **C. dioica.**—Erase "but eglandular" in the last line of the description, the florets being eglandular both in this and *C. perpusilla*.

144. See previous page.

146. **Ozothamnus glomeratus.**—Colenso informs me that this is not found north of the East Cape.

146. **O. depressus.**—Canterbury, descending nearly to the sea, *Haast*.

148. **Raoulia tenuicaulis.**—Otago mountains, *Hector and Buchanan*. Leaves sometimes obtuse.

150. **R. grandiflora.**—The Mount Brewster specimens were collected by *Haast* (not *Hector and Buchanan*, who, however, gathered this species in Mount Alta).

155. **HAASTIA.**—I accidentally omitted to mention under this genus that it is named in honour of my friend Dr. Julius Haast, F.L.S. and F.G.S., the distinguished explorer and geological surveyor of Canterbury, whose botanical discoveries have so greatly enriched this volume.

Page

156. **H. Sinclairii.**—Mount Alta, and in the lake district Otago, on dry débris, *Hector and Buchanan.*

157. **Erechtites scaberula.**—After "in various places," *read* "not so common," etc.

159. **Senecio bellidioides,** var. γ; this is *S. Traversii,* F. Muell. in Trans. Bot. Soc. Edinb.

159. **S. latifolius.**—Harper's range, Hurumui Mountains, alt. 2600 ft., *Haast.* Lower leaves petiolate. Achenes pilose.

160. **S. lautus.**—Syn. *S. radiolatus,* F. Muell. Veg. Chatham Island.—Chatham Island, *W. Travers,* a broad membranous-leaved form, also found in New Zealand. *S. angustifolius,* Forst. Prod. p. 91, seems to be a variety of this.

161. **S. odoratus.**—Banks's Peninsula, *Haast.* This is distinct from the Australian *S. odoratus* in the ligulate ray-flowers, and must here be kept as *S. Banksii,* according to Bentham (Fl. Austral. iii. 671).

161. **S. bifistulosus.**—A branching shrub 1–2 ft. high. Dusky Bay, alt. 2500 ft., *Hector and Buchanan.*

161. 13 *bis.* **S. Huntii,** *F. Muell. Veget. of Chatham Island,* 23. *t.* iii. A shrub 20 ft.; branchlets, leaves above more or less, and inflorescence glandular-pubescent. Leaves 2 in. long, 2⅓ in. broad, elliptic-lanceolate, sessile, acute, beneath covered with thin appressed fulvous tomentum; margins downy, subrevolute, quite entire; midrib rather thick below; nerves obscure. Panicle terminal, very large, broadly and shortly conical, much branched, 3–5 in. broad. Heads crowded, on slender almost hirsute glandular pedicels, 2⅓ in. diameter. Pedicels with a few linear bracts. Involucre broadly campanulate; scales linear-oblong, obtuse, membranous, glandular and villous at the tips. Rays few, ¼ in. long, revolute. Anthers without tails; pappus-hairs slender, scabrid, white. Achene (young) glabrous, grooved.

Chatham Island: waste places, *W. Travers.* A very distinct species, but described from single specimens.

162. **S. elæagnifolius.**—Probably only a form of *S. rotundifolius.* The achenes are sometimes glabrous, the petioles slender, and tails of anthers very short indeed.

162. **S. rotundifolius.**—Syn. *S. Reinoldi,* Endl.—Abundant on the west coast of Otago, on the east found at 2000–3000 ft.; also in Stewart's Island. It varies extremely in the foliage from orbicular to elliptical.

163. **S. cassinioides.**—Cameron river, alt. 3000–4000 ft, *Haast.* Lake district, Otago, alt. 1000–2000 ft., *Hector and Buchanan.*

163. **Brachyglottis repanda.**—Syn. *Senecio Georgii,* Endl.

163. **TRAVERSIA.**—I accidentally omitted to state that this remarkable genus is named in honour of my correspondent W. T. Locke Travers, Esq., F.L.S., of Canterbury, who has contributed so largely to our knowledge of the botany of the Middle Island of New Zealand.

Page
165. **Taraxacum Dens-leonis.**—Chatham Island, *W. Travers.* The indigenous form.

166. **Sonchus oleraceus.**—Chatham Island, *W. Travers* (fruits unknown).

ORDER XLI. **CAMPANULACEÆ.**

169. **Wahlenbergia gracilis,** *A. Rich.*—Mueller points out that A. De Candolle's specific name of *gracilis* has two years' priority over Richard's, and therefore *A. DC.* should be substituted for *A. Rich.*

171. **Lobelia rugulosa,** *Graham in Edinb. New Phil. Journ.* 1829; *DC. Prod.* vii. 366, is, I suspect a synonym of *Pratia angulata.*

171. **L. anceps.**—Chatham Island, *W. Travers.*

173. **Selliera radicans.**—Limestone Cliffs, Weka Pass, *Haast.*

ORDER XLII. **ERICEÆ.**

175. Under observations on *Gaultheria rupestris, for* "ground-parrots" *reaa* "tree-parrots."

177. **Cyathodes? pumila,** *Hook. f., n. sp.* Small, densely tufted; branches 1–2 in. long, slender, suberect, fascicled. Leaves densely imbricate, $\frac{1}{8}$–$\frac{1}{5}$ in. long, incurved, shortly petiolate, linear-oblong, apiculate; margins broadly scarious towards the apex, flat above, glaucous below; midrib stout, margin thickened. Flowers solitary, axillary, sessile, longer than the leaves; pedicels covered with 8 or more broad concave ciliated bracts. Sepals oblong, obtuse, ciliated. Corolla tubular, twice as long the sepals; lobes densely bearded within. Style short, 5-toothed at the apex. Ovary 5-celled, cells 1-seeded.

Middle Island: Hurumui Mountains, *Travers.* This may be a *Pentachondra;* it is impossible to say without the fruit.

178. After *L. fasciculatus,* add—

Leucopogon Richei, *Brown.* A slender erect, much-branched shrub, 4 ft. high and upwards; branches somewhat fascicled. Leaves $\frac{1}{2}$ in. long, linear-lanceolate or rather dilated towards the tips or above the middle, caute or acuminate; margins recurved, pale or glaucous below, with 3–5 faint nerves. Spikes subterminal, shorter than the leaves, many-flowered, glabrous or puberulous. Flowers small, $\frac{1}{8}$ in. diameter. Calyx-lobes oblong, obtuse. Corolla-lobes linear. Bracts persistent, small. Drupe fleshy, 3–5-celled (not seen in New Zealand specimens).

Chatham Island, *W. Travers.* A native of extratropical Australia, but not hitherto found in New Zealand proper.

179. **Epacris pauciflora.**—Rare and local, according to *Colenso.* I have seen it from the Bay of Islands, Nelson, and some intermediate localities.

180. **Archeria Traversii,** var. *australis.* Leaves $\frac{1}{2}$–$\frac{3}{4}$ in. long, elliptic-lanceolate or oblong, obtuse, very coriaceous, quite entire, flat or concave above, ribbed with few parallel nerves below when dry. Racemes terminal, $\frac{2}{3}$ in. long, stout, erect, pubescent, 8–12 ft. Flowers $\frac{1}{5}$ in. long, secund; pedicels short, pubescent; bracts linear, as long as the pedicels, caducous.

Page

Calyx-lobes broadly oblong, very obtuse, striate on the back, not half as long as the campanulate red corolla. Ovary glabrous; style very short. Capsule depressed, $\frac{1}{10}$ in. diameter.

Middle Island: Otago, common on the west coast, *Hector and Buchanan.* Much stouter than *A. Traversii.* Leaves broader and flowers larger. I have specimens of *A. Traversii* from Browning's Pass, alt. 3000–3500 ft.

181. **D. latifolium,** var. *ciliolatum.*—Edges of the leaves ciliolate.—Chatham Island, *W. Travers.*

181. **D. strictum.**—All sounds on the west coast, *Hector and Buchanan.* A small tree in Canterbury, *Haast.*

181. After *Dracophyllum strictum*, add—

1. **D. Traversii,** *Hook. f.*—Much the largest species of the genus. Leaves 1 ft. long and upwards, $1\frac{1}{2}$–2 in. broad at the base, very coriaceous, ensiform, gradually tapering from the base to the apex; margin quite entire and smooth to the touch, slightly concave, minutely striated. Panicle terminal, strict, sessile, linear-oblong, very dense; branches short, suberect, crowded, stiff, puberulous. Flowers crowded, sessile, small, $\frac{1}{6}$ in. diameter; bracts and sepals nearly orbicular, obtuse, glabrous. Corolla very broadly campanulate; tube no longer than the sepals; lobes reflexed. Ovary glabrous; scales oblong, retuse; style stout and exserted.

Middle Island: Nelson Province, *Travers.*

182. **D. Urvilleanum,** var. *scoparium.*—Chatham Island, *W. Travers;* Otago, *Hector and Buchanan.*

Order XLIII. MYRSINEÆ.

184. After *Myrsine Urvillei*—

M. Chathamica, *F. Muell. Veg. Chatham Island*, 38. Shrubby, erect, much-branched. Leaves $1\frac{1}{2}$–$2\frac{1}{2}$ in. long, obovate-oblong, obtuse or emarginate, pale and reticulated on both surfaces. Flowers unknown.

Chatham Island, *W. Travers.* There are no flowers in my specimen. The fruit is described by F. Mueller as purplish, spherical, 1-seeded, size of a large pea. Calyx-lobes ciliolate; pedicels very short. Perhaps only a large state of *M. Urvillei.*

184. **M. divaricata.**—Common on both the east and west coasts of Otago, *Hector and Buchanan.*

Order XLIV. PRIMULACEÆ.

185. **Samolus littoralis.**—Chatham Island, *W. Travers.* Mueller correctly observes that the specific name of *S. repens*, Persoon, has the priority over *S. littoralis*, Br.

Order XLVI. JASMINEÆ.

187. **Olea montana.**—Mr. Colenso assures me that the Bay of Island habitat is erroneous.

ORDER XLVIII. LOGANIACEÆ.

Page

188. Before *Logania*, insert—

1. **MITRASACME.**—Small annual, rarely perennial, herbaceous plants. Leaves opposite, often connate, quite entire, exstipulate. Flowers minute, axillary and umbelled, or solitary and terminal. Calyx 2–4-lobed or partite. Corolla 4-lobed or partite; lobes valvate. Stamens 4, filaments short or slender. Ovary 2-celled; styles 2, distinct, or connate by the capitate stigmas; ovules several or numerous, attached to axile placentæ. Capsule turgid, 2-celled, dehiscing loculicidally between the styles at the summit. Seeds numerous, roughish; albumen fleshy.

A considerable genus of small plants, natives chiefly of Australia. A few are Indian.

1. **M. novæ-Zelandiæ,** *Hook. f., n. sp.* A small moss-like glabrous densely-tufted perennial. Stems and branches very short, filiform, interwoven. Leaves $\frac{1}{12}$–$\frac{1}{10}$ in. long, densely imbricate, linear-oblong, acuminate, with a bristle at the apex, connate by their broad bases, quite glabrous, coriaceous, rather concave, nerveless. Flowers terminal, sessile, hidden by the uppermost leaves. Sepals like the leaves. Corolla shortly and broadly tubular; lobes 4, short, obtuse. Stamens with short filaments and broad didymous anthers. Stigmas free.

Middle Island: Dusky Bay, *Hector and Buchanan*, on the hills, alt. 3500 ft. Allied to the Tasmanian *M. montana.*

189. **Logania tetragona.**—Sounds of the west coast of Otago, *Hector and Buchanan.*—Stipules 0. Flowers solitary, terminal. Calyx-tube and lobes below pubescent, glandular, and almost hispid. Corolla funnel-shaped, $\frac{1}{3}$ in. diameter; tube short; lobes 5, very large and spreading. Throat quite glabrous. Stamens 5; filaments short, inserted within the tube; anthers short, 2-cleft halfway up. Capsule enclosed in, and half as long as the calyx-lobes; valves obovate, retuse.

189. 3. **L. ciliolata,** *Hook. f. n. sp.* Stems prostrate, much branched; branches ascending, 1–3 in. high, glabrous, densely covered with closely-imbricate leaves, hence tetragonous and $\frac{1}{3}$ in. diameter. Leaves densely packed, spreading, linear or linear-oblong, obtuse, very coriaceous, connate in pairs, ciliate along all the margins, nerveless, convex at the back. Flowers solitary in the axils of uppermost leaves, $\frac{1}{6}$–$\frac{1}{4}$ in. diameter. Calyx deeeply 4-cleft; lobes glabrous, ciliated on the margins. Corolla with a short funnel-shaped tube; lobes 4, orbicular, spreading; mouth glabrous. Stamens 4; filaments short, inserted within the mouth of the corolla; anthers oblong-sagittate. Ovary narrow, seated in a cup-shaped disk; style slender, stigma minute. Capsule not seen.

Middle Island: slopes above Browning's Pass, alt. 4–6000 ft., *Haast.*—So similar in most characters to *L. tetragona* that it is difficult to suppose that intermediates may not occur, but the leaves are longer, ciliate to the apex; the branchlets and calyx glabrous; the flowers very much smaller and 4-merous, and the anthers shorter.

ORDER XLIX. GENTIANEÆ.

190. **Gentiana pleurogynoides.**—Chatham Island, *W. Travers.*

Page

192. **Myosotis uniflora.**—Shingle beds, Lake district, Otago, *Hector and Buchanan*. A monstrous state, with proliferous tips to the branches.

194. After **Myosotis Traversii,** insert—

10. **M. albo-sericea,** *Hook. f. n. sp.* Root perennial. Whole plant covered with white appressed silky hairs. Stem tufted, short, woody at the base. Leaves densely tufted, about 1 in. long, very narrow, linear-spathulate, acute, $\frac{1}{12}$ in. broad where most dilated, gradually dilated from the petiole, uniformly silky on both surfaces. Flowering stems 3–4 in. high, slender, ascending, with scattered leafy bracts to above the middle, upper part naked; lobes linear, hairs straight, appressed. Flowers apparently yellow, on short peduncles. Raceme short, naked. Calyx $\frac{1}{8}$ in. long. Corolla tube dilated, twice as long as the calyx; lobes $\frac{1}{3}$ in. across. Anthers included, narrow linear-oblong.

Middle Island: Dunstan gorge on the Clutha river, Otago, *Hector and Buchanan*. A most distinct plant, of which I have but one specimen.

125. **Exarrhena macrantha.**—The flowers are dark orange and yellow, very sweet-scented.

197. **Myosotidium nobile.**—Chatham Island, *Travers*.

ORDER LI. CONVOLVULACEÆ.

198. **Convolvulus Tuguriorum.**—Chatham Island, *W. Travers*.

198. **Convolvulus erubescens.**—Ascends to 5500 ft. on the Barefells pass, *Travers*.

ORDER LII. SOLANEÆ.

200. **Solanum aviculare** and **S. nigrum.**—Chatham Island, *W. Travers*.

ORDER LIII. SCROPHULARINEÆ.

201. **Calceolaria Sinclairii.**—Corolla white, spotted with purple, *Colenso*.

203. **Mazus Pumilio.**—The position of the bract is very variable.

237. **Veronica salicifolia.**—Chatham Island, *Travers*.

208. **V. Traversii.**—Chatham Island, *Travers*.

209. **V. elliptica.**—Chatham Island, *W. Travers*.

209. **V. lævis.**—Hurumui Mountains, 3000–4000 ft., *Travers*.

211. **V. pimeleoides,** var. *minor*.—Smaller, leaves lanceolate, acute, $\frac{1}{6}$ in. long.

Middle Island: Shingle beds near Lake Heron, *Haast*.

211. **V tetragona.**—Otago, Greenstone Valley, *Hector and Buchanan*.

218. **Ourisia macrophylla.**—Otago, west coast, *Hector and Buchanan*.

218. **O. sessilifolia.**—Dusky Bay, alt. 3500 ft., *Hector and Buchanan*.

219. **O. cæspitosa,** var. *gracilis*.—Stems more slender. Leaves much smaller, $\frac{1}{10}$–$\frac{1}{6}$ in. long. Scapes slender, 1-flowered; bracts opposite, small; pedicels very slender. Sepals linear-oblong, obtuse, not dilated at the apex. Corolla smaller.

Page

Middle Island: Kowai river and Mount Torlesse, alt. 4000 ft., *Haast;* Otago, Lake district, alpine and Maranoa river, *Hector and Buchanan.*—Very different-looking from *C. cæspitosa,* but there are quite intermediate forms.

219. **O. glandulosa.**—Lake Guyon range, Waiau, 3500 ft. on rocky ground, *Travers.*

ORDER LIV. GESNERIACEÆ.

221. **Rhabdothamnus Solandri.**—Colenso observes that this is an excessively local plant, and not found by him south of Bream Bay.

ORDER LVI. VERBENACEÆ.

225. The **Teucridium sphærocarpum,** Mueller, is now referred by that author to *Spartothamnus* and to the Natural Order *Myoporineæ.*

225. **Myoporum lætum.**—Chatham Island, *W. Travers.*

ORDER LVIII. PLANTAGINEÆ.

228. **Plantago Raoulii.**—Colenso says that this is by no means common in the Northern Island.

P. major.—Mr. Colenso is disposed to regard this as indigenous, it having a native name. I cannot doubt its having been introduced.

ORDER LIX. NYCTAGINEÆ.

229. **Pisonia Brunoniana,** Endl.—Seemann, Journ. Bot. i. 244, observes that the New Zealand plant is identical with *Ceodes umbellifera,* Forst. (*Pisonia umbellifera,* Seemann); whereas the *P. Brunoniana,* Endl., is different, and the same with *P. inermis,* Forst., a native of the Malayan and Pacific islands and of Australia.

ORDER LX. CHENOPODIACEÆ.

231. **Chenopodium triandrum.**—Hector observes that the utricle is fleshy.

232. **Atriplex Billardieri.**—Outer East coast, *Colenso.*

ORDER LXIII. POLYGONEÆ.

235. **Polygonum minus,** var. *decipiens.*—Chatham Island, *W. Travers.*

236. **Muhlenbeckia appressa.**—Add syn. *Polygonum Forsterii,* Endl.

237. **M. ephedroides.**—Canterbury, Waipura, *Haast.*

ORDER LXIV. LAURINEÆ.

238. **Tetranthera calicaris.**—Add syn. *T. Tangao,* R. Cunn. mss.; Meissn. in DC. Prod. 35, pt. i. 191.

ORDER LXV. MONIMIACEÆ.

240. **Hedycarya dentata.**—Common in Otago, *Hector and Buchanan.*

Page

ORDER LXVII. **THYMELEÆ.**

243. **Pimelea buxifolia.**—Kaweka Mountain, Hawke's Bay, *Colenso*

244. **P. arenaria.**—Chatham Island, *W. Travers.*

244. **P. Urvilleana.**—Otago, *Lindsay.*

ORDER LXIX. **EUPHORBIACEÆ.**

248. **Euphorbia glauca.**—Chatham Island, *W. Travers.*

248. **Carumbium polyandrum.**—This species had, unknown to me, been already described by Joh. Mueller in the 'Ratisbon Flora' (1864, p. 434), and under the same name of *C. polyandrum.*

ORDER LXX. **CUPULIFERÆ.**

250. **Fagus cliffortioides.**—*For* 5–7000 ft. *read* 2–4800 ft.—This is the *Cliffortioides cordata* of Forster's Prodromus.

ORDER LXXI. **URTICEÆ.**

251. **Epicarpurus microphyllus.**—Otago, *Lindsay.*

251. **Urtica incisa.**—Chatham Island, *W. Travers* (according to F. Mueller) I have seen no specimens.

251. **Urtica australis.**—Chatham Island, *W. Travers.*

ORDER LXXII. **CHLORANTHACEÆ.**

253. **Ascarina lucida.**—Middle Island, Totara-nui, *Banks and Solander.* Common in all the sounds on the west coast of Otago, with stems 6–12 in. diameter, *Hector.*

ORDER LXXIII. **PIPERACEÆ.**

254. **Piper excelsum.**—Add syn. *Macropiper excelsum,* Miquel. Chatham Island, *W. Travers.* Mr. Colenso observes that the leaves are not eaten, but the pulp of the ripe fruit was.

256. 1. **DAMMARA.**—The inflorescence is monœcious in various species, perhaps in all.

ORDER LXXIV. **CONIFERÆ.**

257. **Podocarpus ferruginea.**—Common in the Canterbury province, skirting the plains, *Haast.* Hector and Buchanan state that this is common on the west coast of Otago, on river flats; it is one of the "Black Pines" of settlers. "Miro" of Middle Island. According to these gentlemen's notes there are three forms of this tree in Otago, viz. —

1. A large tree, common near Dunedin, with very small leaves.
2. A large-leaved light-coloured tree ("common Black Pine").
3. A tree on the west coast, with very large dark-coloured leaves

Page

258. **P. Totara.**—Trunk sometimes 10 ft. diameter. Hector and Buchanan consider the species of the west coast of Otago different from the eastern; having dense foliage and more obtuse leaves, its trunk does not attain more than 2 ft. diameter.

258. **P. spicata.**—Add syn. *P. Matai*, Lambert. Like *P. ferruginea*, this is also called "Black Pine" in Otago, is the "*Mai*" or "*Matai*" of the Southern Island. Wood good, white, tough.

258. **P. dacrydioides.**—Hector and Buchanan mention as "*Dacrydium?*" the "White Pine" of Mataura, which differs from the "Kahi-Katea" of the 'New Zealand Flora,' in being a taller stronger tree, with upright foliage and bright-red berries. Elsewhere, the same observers remark that *P. dacrydioides* has two distinct varieties:—

1. A tall tapering tree, common in the Northern Island.
2. A dense round-headed tree, common near Dunedin, with the other.

The remark in the 'Handbook' that Otago specimens have hard close wood arose from an error of Mr. Buchanan; it should apply to *D. Colensoi.* Colenso says the leaves of *P. dacrydioides* are certainly not used for eel baskets, but those of *P. spicata* may be.

258. **Dacrydium cupressinum.**—Hector and Buchanan speak confidently of the west coast Otago *D. cupressinum* being different from that of the east coast, having bright-green plumose more upright branches drooping at the points. Wood close and heavy, like that of *D. Colensoi.*

259. **D. Colensoi.**—Sounds and exposed islets of the west coast of Otago, *Hector and Buchanan.* Of this also there are two forms, a more slender western and stiffer-branched eastern.

259. **Phyllocladus trichomanoides.**—Bark used to dye yellow in Otago.

260. **P. alpinus.**—Regarded by Hector and Buchanan as a form of *P. trichomanoides.*

Class II. MONOCOTYLEDONS.

Order I. ORCHIDEÆ.

262. **Earina mucronata.**—Chatham Island, *W. Travers.*

267. **Caladenia? bifolia.**—Referred to *Chiloglottis* (as *C. Traversii*) by Mueller in his 'Chatham Island Florula,' and with much reason; it differs slightly from that genus in the slender habit, less arched upper sepal, almost sessile lip, and small glands. It is certainly much nearer *Chiloglottis* than *Caladenia*, and on a revision of the former genus its character should probably be modified so as to admit it.

Chatham Island, *W. Travers.*

268. **Pterostylis Banksii.**—Chatham Island, α and β, *W. Travers.*

Order II. IRIDEÆ.

269. **Libertia ixioides.**—Chatham Island, *W. Travers.*

Order IV. PANDANEÆ.

Page
275. **Freycinetia Banksii.**—Northern parts of the Middle Island.

Order V. TYPHACEÆ.

276. **Typha angustifolia.**—Middle Island.—Mr. Kirk doubts the "Raupo" of the natives being identical with this. It may be the *T. latifolia*, which he is the first to discover in the island.

After *T. angustifolia*, add—

276. 2. **T. latifolia,** *Linn.* A larger coarser plant than *T. angustifolia.* Male and female catkins contiguous.

Northern Island: swamps between Cape Horn and Titirangi Manukeau Harbour, *Kirk.*

277. **Sparganium simplex.**—Scape sometimes with 1 or 2 branches.

Order VI. NAIDEÆ.

278. **Lemna minor.**—Northern Island, *Kirk.*

279. **Potamogeton heterophyllus.**—The true plant has been found by Mr. Kirk in the Auckland district, and in the Waikato by Mr. Travers. Mr. Kirk informs me that he has also found *P. oblongifolius*, but neither *gramineus* nor *compressus.*

280. After **Zannichellia**, insert—

6. ZOSTERA, L.

Slender, creeping, grassy-leaved marine plants. Spathes adnate to the inner face of the leaves at their bases. Stamens and ovaries inserted in 2 rows on the face of the flat thin spadix. Flowers monœcious. Perianth 0. Male flower a sessile 1-celled anther with tubular pollen. Female flower an ovoid ovary, with 1 filiform style and 2 stigmata, 1 cell and 1 pendulous ovule. Fruit a utricle. Seed pendulous, albumen 0; radicle large, with an inflexed cotyledon lodged in a slit in its side.

1. **Z. marina,** *Linn.* Leaves 6–14 in. long, variable in length and breadth, $\frac{1}{12}$–$\frac{1}{3}$ in. with numerous faint nerves. Flowers not seen.

Northern Island: Waitemata, *Kirk.* A narrow-leaved form, perhaps *Z. angustifolia*, Roth. (Temperate seas, northern and southern)

Order VII. LILIACEÆ.

281. **Rhipogonum scandens.**—Chatham Island, *W. Travers.*

281. **Cordyline.**—The New Zealand species have been referred to a new genus *Dracænopsis*, by Planchon, but I think on insufficient grounds.

281. **Cordyline australis.**—I am indebted to Mr. Buchanan for excellent sketches of this plant, at all stages of growth, and full descriptions, made in

Page

various parts of the Otago province, where it is common. Its average height is 20–30 ft. and the trunk 10–18 in. diameter, though sometimes reaching 3 ft. diameter; it branches repeatedly and soon decays. The leaves form large tufts at the ends of the branches, amongst which the dense terminal panicles form conspicuous white drooping masses as large as the human head. The plants vary much in amount of branching, appear to grow rapidly, often root from the side of the trunk when this is inclined or prostrate, and are with difficulty extirpated.—Whether this is the same as the Bay of Islands plant, which, I think, has an erect loose panicle, is doubtful.

As this sheet was going to press, I received the following very full and valuable communication regarding the *Cordylines* from Dr. Hector, F.R.S.

"There are, I believe, 7 distinct *Cordylines* in the island.

"1. C. AUSTRALIS. Ti-rahau.

"2. C. BANKSII. Ti Ngahere.

"3. C. PUMILIO. Ti-rauriki. Sessile; blue flowers.

"4. C. STRICTA? Ti-parae. Leaves rigid, with fine serratures on the margin. Stem about 1 in. diameter and 7–9 long, droops, and throws off heads irregularly. Flowers and berries white, but in a sparse scape like *C. Pumilio*." (This seems to agree with *C. Banksii.—J. H.*)

"5. C. INDIVISA. This is the broad-leaved deep-green *Ti*, with red veins, a single head, and long elegant flowers, that Forster found in Dusky Bay. The leaf has a slight resemblance to the true *Toii* of Colenso, which has led to the confusion no doubt.

"6. C. sp.? Toii. A large tree with many heads and huge broad massive leaves, yellowish with yellow and red veins and ponderous inflorescence with long bracts and black shiny seeds. This is the *Ti* that the natives use for mats, etc. The portion of the description of the 'Handbook' which refers to *C. indivisa*, and which you got from Colenso, applies to this plant." (I have no Dusky Bay specimens of Forster's plant, but Colenso's agrees well with Forster's figure in the British Museum.—*J.H.*)

"7. C. sp.? Ti-tawhiti. The *Ti* which is cultivated by the natives in the Upper Whanganui district; it has a long dark-green flexible leaf and thick flexible pulpy stem, which they propagate by layers. It grows rapidly. I have seen young plants only of this, and do not feel so certain of it as of the others."

281. **C. indivisa.**—Common in Bligh's Sound, *Hector and Buchanan.*

282. **C. Pumilio.**—In last line of description, *for* bracts *read* pedicel.

283. **Dianella intermedia.**—I have seen no specimens from the south of Nelson Province.

283. **ASTELIA.**—Mr. Kirk, of Auckland, sends me the following valuable notes on this genus, but, not being in all cases accompanied with specimens, I am not positive of the identifications.

"1. A. CUNNINGHAMII. Berries fine purple, black, handsome. Flowers March, April (before *A. Banksii*); fruit January–March.

"2. A. sp. Leaves 1–9 ft. long, margins flat, 2–4½ in. broad, erect. *Female* scape 6–20 in. high, stout, 1½ in. diameter. Flower dark-green.

Page

Perianth campanulate. Berry deep orange, 3-celled, with 3-lobed stigma, clasped by the perianth, which also becomes orange-yellow. Seeds large, angled.—A striking plant, not uncommon in swampy gullies in Auckland district. Flowers October, November; fruit July."—(This must be either *A. nervosa* or a near ally of that plant. Mr. Colenso has sent imperfect specimens? from Bream Bay.)

"3. A. SOLANDRI. Flowers January and February; fruit May, bright-crimson.

"4. A. BANKSII. Flowers April, May; fruit March, rich purple, black when fully ripe, yellowish when immature only.

"5. A. sp. Leaves 2–6 ft. long, pale, glabrous or slightly silky, margin not recurved. *Male:* Panicle silky, bracts long, slender. Perianth silky; segments narrow, acuminate, recurved. Filaments subulate; anthers oblong. *Female:* Panicle more slender, very silky; branches ascending, crowded. Perianth campanulate, small; segments short, lanceolate, acuminate, externally clasping the fruit. Berry large, globular, 3-celled, deep crimson. Seeds angled.—Hilly forests, rarely epiphytal. Flowers March, April; fruit February. Easily distinguished from *A. Cunninghamii* and *A. Banksii* by its green leaves, slender female panicle, *large* crimson flower, and terrestrial habit."—(I do not see how this is to be distinguished by the above description from *A. Banksii*, for the terrestrial habit is not constant; the flowers are described as small (not large), and the slenderness of the female panicle and colour of berry are not very good characters.

284. An **Astelia,** referred to *A. Cunninghamii* by Mueller (Chatham Island, *W. Travers*), seems identical with the Oahu *A. Menziesii* in the form and structure of the bract, fruiting panicle, conic-ovoid fruit, and in the terete hardly angled seeds, with a brittle testa. It is certainly not *A. Cunninghamii*, which has sharply-angled seeds, thick hard testa, and oblong-ovoid berry.

284. **A. nervosa.**—Dusky Bay, *Hector*. Probably two species are included under this, one with the berry sunk in the calyx tube, the other with a much larger, partially exserted berry.

286. **Anthericum Hookeri.**—Descends to the sea level on the east coast of Otago, *Hector*.

286. **Phormium tenax.**—Chatham Island, *W. Travers.*—Hector and Buchanan distinguish two Otago plants:—

1. Very robust. Flowers dark red. Capsules erect, 3-gonous.

2. Slender. Leaves drooping, greenish, narrow. Capsules much twisted, terete, 4 in. long. $\frac{3}{4}$ in. diameter This is more common on the west coast. In another communication, Dr. Hector alludes to a S.W. coast species with globular capsules.

ORDER VIII. PALMEÆ.

288. **Areca sapida.**—Middle Island, abundant in Banks's Peninsula and on the west coast of the Nelson and Canterbury provinces. Chatham Island, *W Travers.* The flowers are dingy purplish in a specimen now flowering in the Royal Gardens, Kew. (October, 1866.)

Page

ORDER IX. JUNCEÆ.

289. **Juncus vaginatus.**—Middle Island; Otago, *Lindsay.*

290. **J. planifolius.**—Chatham Island, *W. Travers.*

291. **J. novæ-Zelandiæ.**—Canterbury, Mount Harper, *Haast.*

291. **Rostkovia gracilis.**—Canterbury Alps, slopes above Browning's Pass, alt. 5000 ft., *Haast.*

292. **Luzula campestris.**—Chatham Island, *W. Travers.*

ORDER XI. CYPERACEÆ.

297. After **Cyperus ustulatus,** add—

2. **Cyperus tenellus,** Linn. fil. A small, very slender, tufted annual. Culms numerous, capillary, 2 in. high. Leaves small, setaceous. Spikelets 1–3, near the apex of the culm, large for the size of the plant, $\frac{1}{6}$–$\frac{1}{4}$ in. long, oblong, obtuse, very much flattened. Scales 12–20, oblong-lanceolate, obtuse or apiculate, green with red-brown at the base, and hyaline margins, deeply grooved. Nut minute, 3-gonous, almost perfectly smooth. Stigmas 3.

Northern Island: abundant in the Newton and Dedwood districts, Auckland, *Mr. Kirk.* (West Australia, South Africa.)—A beautiful little plant, resembling an *Alepyrum,* probably introduced both into Swan River and New Zealand.

301. **Eleocharis gracilis.**—Mr. Carruthers, of the British Museum, has done me the service of comparing a series of New Zealand specimens with Brown's *E. gracilis,* and informs me that none agree, but that most of them may be referred to obtuse-scaled forms of Brown's *E. acuta,* from which however they further differ in the scales not being keeled. Its best character is the truncate mouth of the sheath with a foliaceous mucro. My var *gracillima,* again, is a very different plant, with an oblique mouth to the sheath and broad pale small spikelets. The two species may be diagnosed as follows:—

2. **E. acuta,** *Br.,* var. *platylepis.* Culms tufted, stout or slender, 4–18 in. high; mouth of sheath truncate with a foliaceous mucro. Spikelet very variable in length, ovoid oblong or cylindric, acute or obtuse; scales numerous, broad, rounded at the apex, convex, dark brown with broad scarious margins. Nut obtusely 3-gonous or compressed; bristles 4–8, longer than the nut.—*E. gracilis,* Fl. N. Z. and Tasman., not of Brown; *E. subsphacelata,* Steudel.

Northern, Middle, and **Chatham** Islands. (Australia, Tasmania.)

2. **E. gracillima,** *Hook. f.* Culm very slender, 4–10 in. high, from a stout or slender creeping scaly rhizome; mouth of sheath oblique, scarious. Spikelet short, ovate; scales few, oblong, obtuse or subacute, pale and scarious. Nut obtusely 3-gonous or compressed; bristles 8, longer than the nut. —*E. gracilis,* var. *gracillima,* and var. *radicans,* Fl. N. Z.

Northern, Middle, and **Chatham** Islands: not so common as *E. acuta.*

303. **Desmoschœnus spiralis.**—Chatham Island, *W. Travers.*

Page

307. **Lepidosperma tetragona.**—Otago, *Lindsay.*

313. **Carex stellulata** is a native of the Australian Alps.

313. **C. appressa.**—Chatham Island, *W. Travers.*

315. **C. Forsteri.**—Chatham Island, *W. Travers,* according to Dr. Mueller. I have seen no specimens.

316. **C. trifida.**—Chatham Island, *W. Travers.*—It is not found in the Northern Island, Totara-nui being in the Middle Island.

316. **C. Neesiana.**—Add Middle Island, in which is Totara-nui of Banks.

317. **C. Lambertiana?**—Chatham Island, *W. Travers,* a form with several female spikes; specimens indifferent. Add Middle Island, in which is Banks's Totara-nui.

Order XII. GRAMINEÆ.

321. **Hierochloe redolens.**—Add syn. *H. Banksiana,* Endl.; and *H. racemosa,* Trin.

323. **Paspalum tenax,** *Trin. Diss.* ii. 122, described as a New Zealand plant, is quite unknown to me.

322. **Agrostis pilosa.**—Chatham Island, *W. Travers.*

331. **Arundo conspicua.**—Add syn. *A. Richardii,* Endl.; and to habitats, Chatham Island, *W. Travers.* To this also should probably be referred *Kampmannia Zelandiæ* and *Dichelachne procera,* Steudel.

331. **Phragmites communis,** *Fries,* a common European reed, also found in Australia and Tasmania, has been, as Dr. Mueller informs me, found at Canterbury by Dr. Haast. It is a large coarse grass, 4–8 ft. high, with a long dense purple panicle, as in *Arundo conspicua,* but the lower flower of each spikelet is male; the glumes are silky and awns short.

332. **Danthonia Cunninghamii.**—*For* uplands near Otago, *read* Chain hills, Otago.

334. **Deschampsia cæspitosa.**—Chatham Island, *W. Travers* (fid. *F. Mueller*).

337. In Key to Poa, under 9. *P. Lindsayi, for* 2–4 in., read 3–6 in., and *for* Leaves flat, *read* Leaves flat or involute.

337. **Poa imbecilla.**—Add syn. ? *P. hypopsila,* Steud.

338. **P. foliosa** is often difficult to distinguish from *Festuca littoralis.*

341. **Festuca littoralis.**—Chatham Island, *W. Travers.* A form with pubescent empty glumes.

Class III. CRYPTOGAMIA.

Order I. FILICES.

347. After Tribe VII. enter **Marattieæ.**

Page

348. **Gleichenia dicarpa.**—Chatham Island, *W. Travers* (*F. Mueller*).

348. After 4. *G. flabellata*, add—

5. **G. dichotoma,** *Willd.* Stipes zigzag, repeatedly 2–3-chotomous, ultimate branches bearing a pair of forked pinnæ; a pair of pinnæ also arises from the base of the forked branches; pinnæ lanceolate, acuminate, pinnatifid; segments linear, obtuse or emarginate, glaucous below. Capsules 10–12, exposed.—Hook. Sp. Fil. i. 12. *G. Hermanni*, Br.

Northern Island: Hot springs, Karapiti, *Hochstetter;* and at Rotamahaua, *Mair.* A most abundant tropical fern all over the globe. Curiously enough, Forster (as stated in the Handbook) gives New Zealand as a habitat of this plant, where, however, he never saw it in all probability, and erroneously describes the roots as eaten.

349. **Cyathea dealbata.**—Chatham Island, *W. Travers.*

349. **C. medullaris.**—Chatham Island, *W. Travers* (*C. Cunninghamii*, F. Muell. Veg. Chatham Island, 657).

350. **C. Cunninghamii.**—Colenso assures me that this was not found at the Bay of Islands, but given by himself to A. Cunningham.

351. **Dicksonia squarrosa.**—Chatham Island, *W. Travers.*

354. **Hymenophyllum dilatatum.**—Frond sometimes 24 in. long, *Logan.*

354. **H. crispatum.**—This name should give place to **H. Javanicum,** *Sprengel*, which is of earlier date.

354. **H. polyanthos.**—Frond sometimes 10 in. long, *Logan.*

355. **H. flabellatum.**—Chatham Island, *W. Travers.* Mr. Logan sends specimens 9 in. long.

355. **H. æruginosum.**—This is considered distinct from the Tristan d'Acunha plant, in Hook. Synops. Fil., and Mr. Colenso's name of *H. Franklinianum* confirmed to it. I must confess my inability to distinguish it. Mr. Kirk informs me that it is found at Hunua, 25 miles from Auckland.

355. **H. Lyallii.**—This is referred to *Trichomanes*, in Hook. Synops. Fil. It is found at Hunua-Huia, Auckland.—*Kirk's MSS.*

355. After *H. Lyallii*, add—

16. **H. ciliatum,** *Swartz.* Rhizome 1–2 in. long. Frond oblong, acuminate, 3-pinnatifid, 2–6 in. long, 1–2 in. broad, main rachis broadly winged throughout, ciliated; lower pinnæ oblong or rhomboid, central part undivided. Segments numerous, spreading, linear, simple or forked. Involucres 2–12 on a pinnule, at the apices of the segments, immersed, suborbicular; valves divided halfway down, ciliated.—Hook. Synops. Fil. 63.

Middle Island: Nelson, *Travers.* Closely allied to *H. æruginosum.* (America, Africa.)

356. **Trichomanes reniforme.**—This has been discovered in Australia.

356. **T. strictum.**—Mr. Baker regards this as the true common tropical *T. rigidum*, Swartz, and *T. elongatum* as a variety of it.

357. **T. venosum.**—Chatham Island, *W. Travers.*

Page

357. **T. Malingii.**—Francis Joseph Glacier, alt. 7000 ft., *Haast.*

359. **Lindsæa linearis.**—Colenso informs me that this is very local in the Northern Island.

360. **Adiantum Cunninghamii.**—Chatham Island, *W. Travers.*

362. **Hypolepis distans.**—Mr. Colenso informs me that Mr. Edgerly was never at Cape Maria Van Diemen, but gathered this plant at Hokianga.

362. **Cheilanthes tenuifolia,** var. *Sieberi.*—This is usually kept as a distinct species.—*C. Sieberi,* Kunze.

363. **Pteris aquilina,** var. *esculenta.*—Chatham Island, *Captain Anderson.*

363. **P. scaberula.**—Chatham Island, *W. Travers.*

363. **P. incisa.**—Chatham Island, *Captain Anderson* (*F. Muell.*).

366. **Lomaria procera.**—Chatham Island, *W. Travers* (*L. Capensis,* F. Muell. Veg. Chatham Island, 74).

367. **L. pumila.**—Mr. Logan thinks this is connected by intermediate forms with *L. lanceolata.*

367. After 6. *L. vulcanica,* add—

6*a*. **L. dura,** *Moore in Gard. Chron.* 1866, 290. Rhizome erect. Frond tufted. Barren fronds lanceolate, 8–10 in. high, 1½–2 broad, erect, rigid, pinnatifid to the rachis; sinus acute; segments alternate, contiguous, lanceolate, falcate, finely serrulate, lower decreasing in size, obtuse, veins evident. Fertile fronds as rigid and broad as the sterile; segments becoming involute, densely covered with fruit.—*L. rigida,* J. Sm. Ferns, Brit. and Foreign, 290.

Chatham Island (fide *Smith* and *Moore*).—I have seen no specimen.

368. **L. discolor.**—Chatham Island, *W. Travers.*

370. **Doodia caudata.**—Colenso thinks there must be a mistake in describing this as fragrant, and that *Polypodium pustulatum* is referred to.

371. **Asplenium obtusatum.**—Chatham Island, *W. Travers* (*A. marinum,* F. Muell. in Veg. Chatham Island, 66). A var. with pinnatifid lobes.

372. **A. falcatum.**—Chatham Island, *W. Travers.* This is the *A. Forsterianum,* Colenso in Tasman. Phil. Journ. ii. 171, according to Moore, but Mr. Colenso refers his plant to *A. caudatum.* The rhizome is tufted, not creeping as described by error.

372. **A. Hookerianum.**—I have another variety of this plant from Mount Mania, Whangarei, sent by Mr. Mair; it has a small slender flaccid oblong frond, 1–2 in. long, with narrow linear-obovate substipitate entire ultimate pinnules ⅛ in. long. I have seen but one frond.

373. **A. bulbiferum.**—Chatham Island, *W. Travers.*

374. **A. flaccidum.**—Chatham Island, *W. Travers.*

375. **A. aculeatum,** var. *vestitum.*—Chatham Island, *W. Travers.*

376. **A. oculatum.**—Chatham Island, *W. Travers.* A var. with brownish scales.

Page

376. **A. coriaceum.**—Chatham Island, *W. Travers.*

377. **Nephrodium molle.**—I am indebted to Mr. Mair for a specimen of this.

378. **N. decompositum.**—Chatham Island, *W. Travers.*

378. Add—

6. **N. unitum,** *Brown.* Glabrous. Rhizome creeping. Stipes glabrous or slightly chaffy at the base. Frond ovate-oblong, pinnate, 1–3 ft. high, rather rigid; pinnules rather distant, spreading or ascending, narrow ensiform, very shortly stipitate, pinnatifid halfway down or lower; segments broadly ovate, acute, quite entire; veins free. Sori numerous, halfway between the margin and costa, or nearer the former; involucre glabrous.

Northern Island, *Sinclair;* Rotamahaua, in a hot water swamp, *H. Mair.* (Tropical Australia, Asia, Africa, and America.) This is no doubt the plant alluded to by Mrs. Jones, and mentioned in the note at bottom of page 378 of this Handbook.

380. **Polypodium Grammitidis.**—Chatham Island, *W. Travers.*

380. **P. tenellum.**—Middle Island, Cape Farewell, *Travers.*

381. **P. rugulosum.**—Chatham Island, *W. Travers.*

382. **P. pennigerum.**—Chatham Island, *W. Travers.*

382. **P. pustulatum.**—Chatham Island, *W. Travers.*

382. **P. Billardieri** (*P. scandens,* Forst., var. *Billardieri,* Mueller, Veg. Chatham Island, 69).—Chatham Island, *W. Travers.*

383. **Gymnogramme leptophylla.**—Middle Island, *Lyttleton,* in crevices of rocks, *Travers.*

383. **Nothochlæna distans.**—Middle Island, Maiku valley, *Travers.*

385. **Lygodium articulatum.**—This is rare in and confined to the Northern part of the Middle Island.

385. **Schizæa dichotoma.**—Colenso informs me that this never grows in marshy places, but only at the base of Kaudi-trees.

385. **S. bifida.**—I have confounded two plants under this name, both of which occur intermixed in Cunningham's herbarium, as *S. propinqua;* they are:—

2. **S. bifida,** *Swartz.* Stipes often 2-fid, more or less flattened, concave on one surface with or without a prominent rib, $\frac{1}{20}$–$\frac{1}{10}$ in. broad, slightly rough. Fruiting limb $\frac{1}{2}$–$\frac{3}{4}$ in. long, broad, of 8–16 pairs of ascending pinnules, $\frac{1}{3}$–$\frac{1}{2}$ in. long, whose edges are fringed with very slender flexuous hairs.

Northern Island, Bay of Islands, *A. Cunningham,* etc., and perhaps elsewhere. (Australia, Pacific Islands, India, S. America.)

3. **S. fistulosa,** *Labillardière.* Stipes simple, terete, deeply grooved, $\frac{1}{40}$–$\frac{1}{20}$ in. diam., smooth. Fruiting limb $\frac{1}{2}$–$\frac{3}{4}$ in. long, narrow, of 8–12 pairs of spreading pinn $\frac{1}{8}$–$\frac{1}{4}$ in. long, whose edges are torn and toothed.

Var. *β. australis.* Much smaller; frond 1–3 in. high.—*S. australis,* Gaudichaud; *S. pectinata,* Homb. and Jacq.

Northern Island, Bay of Islands, *A. Cunningham,* etc. **Middle** Island, Canterbury, *Haast;* Otago, *Hector and Buchanan.* Var. *β.* **Lord Auckland's** group, *D'Urville's Exped.,* etc. (Australia, Chili, Falkland Islands.)

Page

386. **Ophioglossum vulgatum.**—Sometimes 3 and even 4 scapes arise from the root.

387. **Botrychium cicutarium.**—Chatham Island, *W. Travers.*

389. **Lycopodium varium.**—Chatham Island, *W. Travers.*

389. **L. Billardieri.**—Chatham Island, *W. Travers.*

389. **L. densum.**—Chatham Island, *W. Travers.*

391. **L. volubile.**—Chatham Island, *W. Travers.*

391. **Tmesipteris Forsteri.**—Colenso distinguishes between *T. Forsteri*, with acute or acuminate segments, and *T. truncatum*, Br., with truncate emarginate segments, both of which grow in the islands, and which he affirms are distinct in habit and habitat. In the herbarium I am quite unable to distinguish the numerous intermediate forms.

391. **Psilotum triquetrum.**—Rangitoto Island, near Auckland, on *Metrosideros tomentosa, Colenso.*

Order IV. MUSCI.

437. **Mielichoferia longiseta.**—Mitten informs me that the New Zealand differs from the American plant, and is this—

1. **M. tenuiseta,** *Mitten.* Stems ½–1 in. long. Leaves glossy, on the sterile branches oblong-lanceolate acute; nerve percurrent; apex serrate; cells narrow elongate; perichætial ovate, acute, serrulate, ½-nerved. Fruit-stalk elongate, red. Capsule pyriform; operculum large, mamillate; teeth (inner) narrow, smooth, united into a short membrane. Inflorescence monœcious, *Mitten.*

Middle Island: Alps of Canterbury, *Sinclair and Haast.*—Differs from the Andean plant in the narrower leaves and smooth teeth.

Order V. HEPATICÆ.

The following is a new arrangement of the New Zealand genera of *Hepaticæ*, by Mitten, which will be of great use in aiding the student in determining the genera of this most difficult Order :—

A. Foliosæ.—Leaves distinct.

* *Leaves succubous; base with the lowest angle on the upper side of the stem.*

a. *Perianth leafy.*

1. *Fruit terminal.*

3. Plagiochila. Perianth compressed; stems erect or ascending; stipules 0.
4. Leioscyphus. Perianth compressed; stems procumbent, stipulate.

2/1. Temnoma. Perianth above 3-gonous, truncate.
2/2. Chandonanthus. Perianth tubular, many-plicate; mouth open; stems erect or ascending.
5. Lophocolea. Perianth 3-quetrous; angles often alate; mouth 3-lipped, closed; stems procumbent, stipulate.
5/1. Trigonanthus. Perianth 3-gonous; mouth contracted; stems procumbent, stipulate near the fruit.

2. Jungermannia. Perianth tubular; mouth contracted, dentate; stipules 0 or present on the stem.
2/3. Solenostoma. Perianth obovate, 5-plicate above, with a tubular beak.

2. *Fruit lateral.*

10. Adelanthus. Stems erect, nodding; stipules 0.
8. Chiloscyphus. Stems procumbent, stipulate.
9. Psiloclada. Stems procumbent; leaves and stipules deeply cleft.

b. *Perianth a descending fleshy bag.*

1. *Fruit terminal.*

11/1. Tylimanthus. Stems erect or ascending; leaves nearly entire; stipules 0.
12/2. Acrobolbus. Stems procumbent; leaves 2-fid; stipules small or 0.
11/3. Lethocolea. Stems procumbent; leaves entire; stipules 0.
11/4. Balantiopsis. Stems procumbent, stipulate.

2. *Fruit lateral.*

11/5. Marsupidium. Stems erect or ascending; stipules 0.
12. Saccogyna. Stems procumbent, stipulate.

** *Leaves vertical; base crossing the stem transversely.*

a. *Perianth none.*

1. *Fruit terminal.*

1. Gymnomitrium.

b. *Perianth leafy.*

1. *Fruit terminal.*

16. Isotachis. Leaves and stipules nearly equal; perianth tubular; mouth connivent.
6. Scapania. Leaves complicate; stipules 0; perianth compressed in plane with leaves; mouth truncate.
7. Gottschea. Leaves with adherent lobe; perianth overlaid by involucral leaves.

2. *Fruit lateral.*

15. Lembidium. Stipulate. Perianth from the lower part of stem, 3-gonous.

*** *Leaves incubous; base with the lowest angle on the under side of the stem.*

† *Without an inferior lesser lobe.*

a. *Perianth leafy.*

1. *Fruit terminal.*

18. Sendtnera.

2. *Fruit lateral.*

18/1. Leperoma. Leaves and stipules deeply cleft; calyptra adnate with the involucral leaves; fruit near the top of the stem.
13. Lepidozia. Leaves and stipules usually deeply cleft; perianth near the base of the stem, 3-gonous.
14. Mastigobryum. Leaves and stipules entire or with their apices truncate, dentate; perianth in the lower part of the stem, 3-gonous.

†† *With an inferior lesser lobe (lobule).*

a. *Lobule plane.*

1. *Fruit terminal.*

20. Radula. Perianth compressed in plane with leaves; mouth truncate; stipules 0.

Page 22. LEJEUNIA. Perianth obovate, 3–6-plicate; mouth a tubular beak.
17. TRICHOCOLEA. Calyptra and involucral leaves combined; leaves capillary, multifid.

2. *Fruit lateral.*

18/2. MASTIGOPHORA. Perianth ventricose, subcampanulate.
21. MADOTHECA. Perianth contracted at the mouth, compressed, plicate.

b. *Lobule inflated-galeate.*

23. FRULLANIA. Perianth 3–6-plicate or terete, with a tubular beak.
19. POLYOTUS. Involucral leaves overlying each other, adnate below.

B. **Frondosæ.**—Without distinct leaves.

a. *Perianth complete.*

24. FOSSOMBRONIA. Perianth on upper side of frond; leaves angular.
25. NOTEROCLADA. Perianth near the apex; leaves rounded.
26. PETALOPHYLLUM. Perianth on upper side of frond; frond continuous.
27. ZOOPSIS. Perianth lateral; frond continuous, with alternate lateral projections, tipped with cilia.
28. PODOMITRIUM. Perianth from the under side of a continuous frond.
29. STEETZIA. Perianth on upper side of a continuous frond.

b. *Perianth none.*

30. SYMPHYOGYNA. Calyptra on upper side of often stipitate frond; nerve narrow.
31. METZGERIA. Calyptra on under side of continuous frond; nerve narrow.
32. SARCOMITRIUM (ANEURA). Calyptra lateral; frond composed almost entirely of thickened nerve.

C. **Carnosæ.**—Fronds fleshy, with oblique scales on under side.

a. *Fruits imbedded in substance of the frond.*

40. RICCIA.

b. *Fruits terminal on the under side of frond.*

38. TARGIONIA. Involucre 2-valved.

c. *Fruits many on the under side of a stalked peltate receptacle.*

33. PLAGIOCHASMA. Perianths opening laterally.
34. MARCHANTIA. Perianths opening downwards.
35. DUMORTIERA. Perianth 0; involucres opening by a slit.
36. REBOULIA. Perianth 0; involucres opening by 2 valves.
37. FIMBRIARIA. Perianth split into bands cohering at their apices.

D. **Anthocerotæ.**—Frond fleshy, without scales beneath.

39. ANTHOCEROS.

Mr. Mitten has further favoured me with the following valuable remarks on the genera:—

3. **PLAGIOCHILA.** Add after 16. *P. deltoidea*:—

507. **P. læta,** *Mitten.* Stem nearly simple, elongate, ascending, curved, with innovations below the perianth. Leaves patent, nearly orbicular, convex above; dorsal margin recurved towards the apex; ventral ciliate; cells small, orbicular with narrow interstices. Perianth oblong-obovate; mouth truncate, sparingly toothed.

Page

507. **Northern** Island, *Colenso*. Differs from *P. deltoidea* in the more oblong perianth and rounder softer leaves.

2/1. **TEMNOMA,** *Mitten* (from the truncate mouth of the perianth).—To this are referred *Jungermannia pulchella* and *quadrifida* (p. 504), together with some other not New Zealand plants.

2/2. **CHANDONANTHUS,** *Mitten*.—From the open plaited mouth of the perianth, includes *Jungermannia squarrosa* (p. 503) of New Zealand, and others.

5. **LOPHOCOLEA.**—Characterized by its 3-gonous closed perianth. *L. Colensoi* (p. 509) has been found in fruit, and proves to be a *Chiloscyphus*.

5/1. **TRIGONANTHUS,** *Spruce*, approaches *Lophocolea*, but the perianth is different and the stipules usually wanting on the barren stems. The New Zealand species is *J. dentata* (p. 503).

2/3. **SOLENOSTOMA,** *Mitten*, includes *Jungermannia inundata* (p. 502), and *rotata* (p. 503). The perianth in this genus adheres to the base of the uppermost leaves.

8. **CHILOSCYPHUS.**—**Lophocolea Colensoi,** p. 509, should be transferred here; it resembles *C. coalitus*, but is larger, more succulent, the dorsal angles of the leaves are more widely separate; perianth short, campanulate; lips toothed, not exserted beyond the leaves.

11/1. **TYLIMANTHUS,** *Mitten*, here proposed for *Gymnanthe saccata* (excl. syn. *Urvilleana*) (p. 520) and another,—the name *Cymnanthe* being too near *Gymnanthes* and *Gymnanthus*.

11/2. **ACROBOLBUS,** *Lehm. and Lindb.*, includes *Gymnanthe unguiculata* and *G. lophocoleoides* (p. 519). (In description of *unguiculata*, *for* leaves obscurely 12-lobed *read* 2-lobed.)

11/3. **LETHOCOLEA,** *Mitten*, consists of *Gymnanthe Drummondii*, p. 519.

11/4. **BALANTIOPSIS,** *Mitten*, includes *Gymnanthe*, *G. diplophylla* (p. 519), and the following.

3. **B. erinacea,** *Mitten*. Lobes of the leaf free at the base, inferior broadly ovate, notched at the apex with 2 teeth, both margins ciliated; superior lobe nearly orbicular, notched at the apex and ciliate. Stipule unequally 6-partite; laciniæ toothed and ciliate.—*Mitten*.

New Zealand, *Lyall*.

11/5. **MARSUPIDIUM,** *Mitten*.—To this genus belong *Gymnanthe setulosa* (p. 519), and the true *G. Urvilleana*, which was confounded with *G. saccata* at p. 520.

2. **M. Knightii,** *Mitten*. Stem short, simple, curved. Leaves largest towards the middle of the stem, patent or appressed, orbicular or subquadrate; anterior margin entire, ending in a spine; apex subtruncate; posterior margin sparingly toothed. Sac oblong on a very short branch, to which it is

attached at about the middle of its lèngth. Fruitstalk basal. Capsule elongate.

New Zealand, *Knight.*

3. **M. Urvilleana.**—*Gymnanthe*, Tayl. in Fl. Antarct. 153. Rhizome creeping. Stems tufted, ½–2 in. high, erect, tips nodding. Leaves spreading or appressed and imbricate, nearly round, with an indistinct notch or with one or two short teeth at their tops. Involucre attached to the lower parts of the stem by its side, rooting. Capsule on a long stalk, ovoid.—*Plagiochila Urvilleana*, Mont. in Voy. au Pôle Sud, t. 16. *Jung. abbreviata*, Hook. f. and Tayl. in Lond. Journ. Bot. 1844, 374; Syn. Hepat. 647 (*Plagiochila*).

Northern Island, *Colenso.* **Lord Auckland's** group, *D'Urville's Expedition, J. D. H., Col. Bolton.*—Less than *G. saccata*, and in its mode of growth more like some *Plagiochilæ*, e. g. *P. ansata* and *P. circinalis.*

15. **LEMBIDIUM,** *Mitten.*—This genus is established for the *Micropterygium nutans* (p. 526), a very peculiar plant, destitute of flagella; the stem is clothed with a vesicular coat.

18/1. **LEPEROMA,** *Mitten,* is proposed for *Sendtnera ochroleuca*, *S. attenuata*, which may be a variety of the same, and *S. Scolopendra* (p. 528).

18/2. **MASTIGOPHORA.**—Quite different from *Sendtnera* in its lateral fruit, perianth not overlaid, and free calyptra; the habit is more that
508. of *Madotheca.*

In character of *Lophocolea*, after, Fruit terminal, add, rarely lateral. Lips
519. of perianth 3, closed.

Gymnanthe lophocoleoides.—Mr. Mitten has furnished me with a description of the involucre, which is about as long as the uppermost leaves, with scattered pale rootlets.

5 **G. setulosa.**—Add, Antheridia covered by minute scale-like leaves, on slender branches from the lower part of the stem.

520. **G. saccata.**—*G. Urvilleana* (*Plagiochila*, Mont.) proves to be a different species. See above, under *Marsupidium.*

521. **Lepidozia capilligera.**—Mr. Mitten informs me that the perianth is elongated, its mouth toothed.

523. **L. albula.**—Mr. Mitten suspects that this is nothing but *L. ulothrix.*

524. **Mastigobryum monilinerve.**—Add—

Middle Island: Dusky bay, *Menzies;* Otago, *Hector.*

525. **M. decrescens.**—Mr. Mitten has little doubt but that this is *M. novæ-Hollandiæ.*

526. **ISOTACHIS.**—Place "almost trifarious" after "Leaves."

529. Add—

5. **P. allophylla,** and 6. **P. reticulata,** from p. 538, where they are described as *Frullaniæ.*

Page

531. Erase the description of **Radula dentata,** and substitute:—Stems short, ½ in. long, pinnate. Leaves roundly ovate, with a short point; lobule nearly square, appressed; leaves of the branches with 3 or 4 strong teeth on the dorsal margin; lobule saccate. Perianth narrow, very long, compressed above. Capsule cylindric.—*Lejeunia dentata,* Mitten in Fl. N. Z. ii. 159.

Northern Island, *Stephenson in Hb. Mitten;* Auckland, *Sinclair.*—The figure in Fl. N. Z. was taken from fragments of the branches.

531. **LEJEUNIA.**—The stipules are sometimes absent.

537. After **Frullania pentapleura,** add—

F. rostellata, *Mitten.* Small, creeping, appressed, irregularly pinnate. Leaves orbicular-ovate; lobule compressed, decurved, galeate, half the length of the leaf; stipules broadly obovate, with a small notch. Involucral leaves spreading, obtuse, united below with the stipule, which is deeply cleft with lanceolate segments. Perianth broadly obovate, with a long tubular beak, covered with short laciniæ, flattened on the upper side, obtusely keeled on the under.—*Mitten.*

Northern Island, on a *Sticta, Knight.*

538. Remove *F. allophylla* and *F. reticulata* to *Polyotus,* p. 531.

539. **FOSSOMBRONIA.**—For Antheridia on the "under" *read* "upper" surface.

540. Last line, for *Lyallii* read *Lyellii.*

545. **Marchantia tabularis.**—In fourth line of observations, etc., before "midrib" insert "dark."

ORDER VII. LICHENES.

Since the printing of the Lichens for this work, several papers on this Order have appeared, to which I can here only refer. They are:—Dr. Nylander's 'Lichenes Novæ-Zelandiæ' (Journ. Linn. Soc. Lond. Bot. ix. 244), and Dr. L. Lindsay's 'Observations on New Lichens and Fungi, collected in Otago, New Zealand' (Trans. Royal Soc. Edinb. xxiv. 407. t. 29, 30), and his paper 'On a New Species of *Melanospora* from Otago' (Trans. Bot. Soc. Edinb. viii. 426).

I am indebted to Dr. Lauder Lindsay for the following list of Lichens found in Chatham Island by W. Travers, Esq.:—

Sticta orygmæa, Ait.
Sticta Urvillei, Ait., var. *flavicans.*
Ramalina scopulorum, Ait.
Cladonia aggregata, Eschw.
Stereocaulon mixtum, Nylander.

ORDER VIII. FUNGI.

602. **Agaricus (Pleurotus) euphyllus,** *Berk.* Pileus 3 in. across, finely striate and coarsely wrinkled longitudinally when dry, reniform, glabrous, pale chestnut; stipes 0 or obsolete; gills broad; interstices smooth.

New Zealand, *Sinclair.*

Page

611. **Hydnum Sinclairii,** *Berk.* Black. Pileus thin, coriaceous, somewhat zoned, radiately striate and rugose; margin irregularly lobed; stipes thin, subconfluent; processes slender, acute.

New Zealand, *Sinclair.*—Allied to *H. zonatum*, but differing in the black spines and pileus.

625. **Ustilago urceolorum,** *Tulasne in Ann. Sc. Nat. Ser.* 3. vii. 36. *t.* 4. *f.* 7–10. Spores black-brown, thick, oblong, angular, at length agglutinated and forming a black powder; integument thick smooth or minutely papillose.

Northern Island: on *Carex ternaria, Colenso.* (Europe, etc.)

625. **Æcidium disseminatum,** *Berk.* Spots 0 or effuse. Peridia scattered, short; margin lobed. Spores white?

Middle Island: on leaves of *Hypericum japonicum.*—Apparently quite distinct from the N. American *Cæoma Hypericastrum,* Link.

625. **Peziza (Lachnea) scutellata,** *Linn.* Flat, vermilion-red, pale externally, hispid towards the margin with black straight hairs.

Northern Island: on dead wood, *Colenso.* (Europe, etc.)

632. **Hypoxylon Colensoi,** *Berk.* Stroma of dense club-shaped masses, pitchy-black externally, somewhat stratified internally, rugose-punctate. Perithecia oblong.

Northern Island, *Colenso.*—Specimen old and imperfect, but evidently allied to *H cœnopus,* and hence representing a peculiar type. In an early state it resembles states ot *Antennaria scoriadea.*

Order IX. ALGÆ.

720. After **TOLYPOTHRIX,** insert—

CHROOLEPUS, Agardh.

Frond of minute, erect, rigid, subsolid filaments, opaque, falling to powder; joints often contracted.

An obscure genus, belonging to the tribe *Byssoideæ,* which should follow *Oscillatorieæ* the Key, p. 646.

1. **C. aureus,** *Harvey.* Filaments forming soft cushion-like tufts, flexuose, irregularly branched, yellow-green brick-red or orange; articulations twice as long as broad.

Northern Island: on rocks and trees, *Colenso.* (Europe, etc.)—I am indebted to Mr. Berkeley for this identification; he observes that the specimens are full of dumb-bell crystals of oxalate of lime.

LIST OF THE PRINCIPAL NATURALIZED, OR APPARENTLY NATURALIZED, PLANTS OF NEW ZEALAND.

THE rapidity with which European weeds, and especially the annuals of cultivated grounds, are being introduced into and disseminated throughout New Zealand, is a matter of much surprise to all observers, and not only to professed naturalists. It is a point of very great significance in reference to all inquiries relating to their superior powers of propagation and establishing themselves, which the plants as well as animals of some countries display, as contrasted with those of others; and when, as in the case of New Zealand, the result is the actual displacement and possible extinction of a portion of the native flora by the introduced, the facts may well arouse the interest of the most listless colonist. It is impossible for me here to enter into this subject, which, novel as it is, yet suggests a thousand curious reflections. I have touched lightly on it in an article "On the Replacement of Species in the Colonies and elsewhere" in the 'Natural History Review' of January, 1864; and I can here do no more than again call attention to the fact, that now is the time for certifying the dates of the introduction of many plants which, though unknown in the islands a quarter of a century ago, are already actually driving the native plants out of the country, and will, before long, take their places, and be regarded as the commonest native weeds of New Zealand.

The following list is compiled from many sources, and has for the most part appeared in the 'Flora of New Zealand.' Important additions have been made by various collectors, but especially by Mr. Kirk, of Auckland, who has favoured me with a list of upwards of eighty species (marked A in this catalogue) from within sixty miles of that city. The majority of these plants are British, and will be found described in any British Flora.

I am informed that the late Mr. Bidwill habitually scattered Australian seeds during his extensive travels in New Zealand; if this be true, it is remarkable how few Australian plants have naturalized themselves in the islands, considering both this circumstance and the extensive commerce between these countries.

I have added the duration to the species, to show the great contrast in this respect between the indigenous and introduced plants. Of the indigenous plants described in this work, nearly all (as in other oceanic islands) are perennial (♃); of the introduced plants, now to be ennumerated, fully one-half are annual (☉), and thirteen are biennial (♂).

CLASS I. DICOTYLEDONES.

RANUNCULACEÆ.

Ranunculus parviflorus, *L.* (see p. 8). Europe, fields, etc. ☉

CRUCIFERÆ.

Nasturtium officinale, *L.* (p. 15). Europe, aquatic. (Watercress.) ♃
Erysimum officinale, *L.* Europe, waste places. ⊙
Senebiera Coronopus, *Poir.* (p. 15). Ditto, ditto. ⊙
Senebiera pinnatifida, *DC.* (p. 15). Ditto, ditto. ⊙
Capsella Bursa-pastoris, *L.* (p. 15). Ditto. (Shepherd's Purse.) ⊙
A. Lepidium ruderale, *L.* Ditto, ditto. ⊙
Lepidium sativum, *L.* Ditto. (Garden Cress.) ⊙
Alyssum maritimum, *Willd.* Ditto. (Sweet Alyssum.) ♃
Cochlearia Armoracia, *L.* Ditto. (Horseradish.) ♃
A. Sinapis arvensis, *L.* Ditto, fields. (Charlock.) ⊙
Brassica Rapa, *L.* Ditto, ditto. (Rape-seed.) ♂
Brassica Napus, *L.* Ditto, ditto. (Turnip.) ♂
Brassica oleracea, *L.* Ditto, ditto. ♂
Brassica campestris, *L.* Ditto, ditto. (Swedish Turnip.) ⊙ or ♂
Raphanus sativus, *L.* Ditto, ditto. (Radish.) ⊙

CARYOPHYLLEÆ.

Gypsophila tubulosa, *Boiss.* (p. 22). Levant. ⊙
Silene quinquevulnera, *L.* Ditto, waste places. ⊙
Stellaria media, *With.* Ditto, ditto. ⊙
A. Arenaria serpyllifolia, *L.* ⊙
A. Sagina apetala, *L.* Europe, waste places. ⊙
Cerastium vulgatum, *L.* Ditto. ⊙
Cerastium viscosum, *L.* Ditto. ⊙
Polycarpon tetraphyllum, *L.* Ditto. ⊙
Spergula arvensis, *L.* Ditto. ⊙

HYPERICINEÆ.

A. Hypericum humifusum,* *L.* Europe, heathy places. ♃

MALVACEÆ.

A. Malva rotundifolia, *L.* Europe, waste places. ♃

GERANIACEÆ.

A. Geranium molle, *L.* Europe, pastures. ⊙
Erodium cicutarium, *L.* (p. 38). Ditto, waste places. ⊙

LEGUMINOSÆ.

A. Ulex europæus, *L.* Europe. (Furze, Gorse, or Whin.) ♃
A. Trifolium repens, *L.* Ditto, meadows and waste places. (Dutch Clover.) ♃
A. Trifolium pratense, *L.* Ditto, meadows. (Purple Clover.) ♃
A. Trifolium procumbens, *L.* Ditto, waste places. ⊙

* Inserted on Mr. Kirk's authority. It much resembles *H. Japonicum* (p. 29).

A. Melilotus arvensis, *L.* Europe, waste places. ♂
A. Medicago lupulina, *L.* Ditto, ditto. ⊙
A. Medicago maculata, *L.* Ditto, pastures. ⊙
A. Medicago denticulata, *Willd.* Ditto, ditto. ⊙
A. Lotus corniculatus, *L.* Ditto, ditto. (Bird's-foot Trefoil.) ♃
A. Lotus major, *Scop.* Ditto, ditches, etc. ♃
A. Vicia hirsuta, *L.* Ditto, fields, etc. ⊙
A. Vicia sativa, *L.* Ditto, ditto. ⊙ or ♂
A. Vicia tetrasperma, *L.* Ditto, ditto. ⊙

ROSACEÆ.

Alchemilla arvensis, *Sm.* Europe, waste places. ⊙
A. Rubus discolor, *W. and N.* Ditto, hedges, etc. (Bramble.) ♃
A. Rubus rudis, *Weihe.* Ditto, ditto. ♃
A. Rosa micrantha, *Smith.* Ditto, ditto. ♃
A. Rosa rubiginosa, *L.* Ditto, ditto. (Sweetbriar Rose.) ♃
A. Rosa canina, *L.* Ditto. (Dog Rose.) ♃

LYTHRARIEÆ.

Lythrum hyssopifolium, *L.* Ditto, waste places. ⊙

ONAGRARIEÆ.

Œnothera stricta, *L.* S. America. (Evening Primrose.) ⊙

CACTEÆ.

Opuntia vulgaris, *Mill.* S. America. (Common Cactus or Prickly Pear.) ♃

CURCUBITACEÆ.

Curcubita* sp. Cultivated by the natives (" Hue " of natives), *Colenso.* ⊙

UMBELLIFERÆ.

A. Petroselinum sativum, *L.* Europe, cultivated places. (Parsley.) ♂
A. Fœniculum vulgare, *L.* Ditto, ditto. (Fennel.) ♃
A. Daucus Carota, *L.* Ditto, ditto. (Carrot.) ♂
A. Pastinaca sativa, *L.* Ditto, waste places. (Parsnip.) ♂
A. Scandix Pecten, *L.* Europe, fields and waste places. ⊙
Chærophyllum cerefolium, *Cranz.* ⊙

CAPRIFOLIACEÆ.

A. Sambucus nigra, *L.* Europe, coppices, etc. (Elder.) ♃

RUBIACEÆ.

Sherardia arvensis, *L.* Europe, fields, etc. ⊙

VALERIANEÆ.

Fedia olitoria, *L.* Europe, cornfields. (Corn Salad.) ☉

COMPOSITÆ.

Conyza ambigua, *DC.* Europe, waste places. ☉
Erigeron canadensis, *L.* N. America and Europe, ditto. ☉
A. Bellis perennis, *L.* Europe. (Daisy.) ♃
Bidens pilosa, *L.* (p. 138). Warm Asia, etc. ☉
A. Anthemis nobilis, *L.* Europe, waste places. (Chamomile.) ♃
A. Achillea millefolia, *L.* Ditto, pastures. (Yarrow.) ♃
Wollastonia biflora, *DC.?* India. ☉
A. Chrysanthemum leucanthemum, *L.* Europe, fields, etc. (Ox-eye Daisy.) ♃
Siegesbeckia orientalis, *L.* India, tropical weed. ☉
A. Senecio vulgaris, *L.* Europe, weed. (Groundsel.) ☉
Eclipta erecta, *L.* India, tropical weed. ☉
B. Carduus lanceolatus, *L.* Europe, roadsides. (Common Thistle.)
Centaurea calcitrapa, *L.* Ditto, fields, etc. ☉
A. Centaurea solstitialis, *L.* Ditto, ditto. ☉
Lapsana pusilla, *L.* Ditto, ditto. ☉
A. Tragopogon minus, *L.* Ditto, meadows, etc. ♂
A. Thrincia hirta, *Roth.* Ditto, dry pastures, etc. ♃
A. Apargia autumnalis, *Willd.* Ditto, ditto. ♃
A. Hypochæris radicata, *L.* Ditto, meadows, etc. ♃
A. Crepis virens, *L.* Ditto, pastures, etc. ☉
Taraxacum Dens-leonis, *Desf.* (p. 165). Ditto, waste places. ♃
Sonchus arvensis, *L.* (p. 166). Ditto, cornfields, etc. (Sow-thistle.) ♃
A. Cichorium Intybus, *L.* Ditto, ditto. (Wild Chicory.) ♃
Xanthium spinosum, *L.* Ditto, ditto. (Burweed.) ☉

PRIMULACEÆ.

Anagallis arvensis, *L.* Ditto, ditto. (Pimpernel, or Poor Man's Weather-glass.) ☉

GENTIANEÆ.

A. Erythræa Centaurium, *Pers.* Europe, pastures. (Centaury.) ☉

BORAGINEÆ.

A. Lithospermum arvense, *L.* Europe, waste places. ☉
Cynoglossum micranthum? (p. 197). India. ♂

CONVOLVULACEÆ.

Ipomœa chrysorrhiza (Convolvulus, *Forst.*). The originally cultivated Kumarah of the islanders, probably a variety of the following. ♃

Ipomœa Batatas, *Lam.* The common Sweet Potato of the tropics, now much cultivated in New Zealand. ♃

SOLANEÆ.

Solanum nigrum, *L.* (p. 200). Europe, waste places. ⊙
Solanum tuberosum, *L.* S. America. (Common Potato.) ♃
Physalis peruviana, *L.* Ditto. (Cape Gooseberry, Tipareh.) ♃
Capsicum annuum, *L.* America, cult. everywhere. (Capsicum.) ⊙
Lycopersicum esculentum, *Mill.* Ditto, ditto. (Tomato.) ⊙
A. Datura Stramonium, *L.* Europe, waste places. ⊙
A. Lycium Barbarum, *L.* Ditto, ditto. ♃

SCROPHULARINEÆ.

A. Verbascum Thapsus, *L.* Europe, waste places. (Mullein.) ♂
A. Verbascum phœniceum, *L.* Ditto, ditto. ♂
A. Veronica arvensis, *L.* Ditto, fields, etc. ⊙
A. Veronica serpyllifolia, *L.* Ditto, roadsides, etc. ♃
A. Veronica agrestis, *L.* Ditto, fields. ⊙
A. Veronica Buxbaumii, *Ten.* Ditto, ditto. ⊙
Veronica officinalis, *L.* Ditto, woods and dry places. ♃
Veronica Anagallis, *L.* Ditto, ditches, etc. ♃
A. Digitalis purpurea, *L.* Ditto, banks, etc. (Foxglove.) ♃
A. Linaria Elatine, *Mill.* Ditto, cornfields, etc. ⊙

VERBENACEÆ.

A. Verbena officinalis, *L.* Europe, waste places. (Vervain.) ♃

LABIATÆ.

Mentha aquatica, *L.* Europe, watery places. ♃
A. Mentha viridis, *L.* Ditto, ditto. (Spearmint.) ♃
A. Stachys arvensis, *L.* Ditto, waste places. ⊙
A. Marrubium vulgare, *L.* Ditto, ditto. (Horehound.) ♃
A. Prunella vulgaris, *L.* Ditto, ditto. ♃

PHYTOLACCEÆ.

A. Phytolacca decandra, *L.* N. America. (Virginian Poke-weed.) ♃

PLANTAGINEÆ.

Plantago major, *L.* Europe, fields, etc. (Plaintain.) ♃
Plantago lanceolata, *L.* Ditto, ditto. ♃

POLYGONEÆ.

Polygonum aviculare, *L.* (p. 235). Europe, waste places. (Cow Grass of Colonists.) ⊙
Polygonum minus, *Huds.* Ditto, ditto. ⊙
Rumex obtusifolius. *L.* Ditto, ditto. (Dock.) ♃

A. Rumex conglomeratus, *Murr.* Ditto, watery places. ♃
Rumex crispus, *L.* Ditto, roadsides, etc. ♃
Rumex Acetosa, *L.* Ditto, meadows. (Sorrel.) ♃
Rumex Acetosella, *L.* Gravelly places. ♃

CHENOPODIACEÆ.

A. Chenopodium album, *L.* Europe, waste places. (Goosefoot.) ⊙
A. Chenopodium viride, *L.* ⊙
Chenopodium urbicum, *L.* (p. 231). Ditto, ditto. ⊙
Chenopodium ambrosioides, *L.* (p. 231). Ditto, ditto. ⊙

AMARANTHACEÆ.

Euxolus viridis, *Moq.* Tropics, weed of cultivation. ⊙

EUPHORBIACEÆ.

A. Euphorbia Peplus, *L.* Europe, waste places. (Spurge.) ⊙
Euphorbia Helioscopia, *L.* Ditto, ditto. ⊙
Jatropha Curcas, *L.* America, waste places. (Physic-nut.) ♃
Ricinus Palma-Christi, *L.* Ditto, ditto. (Castor Oil.) ♃
Poranthera ericifolia, *Rudge.* Australia. ♃

URTICEÆ.

Urtica urens, *L.* Europe, waste places. (Common Nettle.) ♃

Class II. **MONOCOTYLEDONES.**

DIOSCOREÆ.

Dioscorea alata, *L.* India. Occasionally cultivated during late years by the natives. (Yam.) ♃

AROIDEÆ.

Colocasia antiquorum, *Schott,* (esculenta, *Schott*). Asia. Two principal varieties are cultivated by the aborigines, a small (Taro), and a large, introduced by Europeans (Taro-hoia). ♃
Alocasia Indica, *Schott.* India, recently introduced and cultivated by the natives. (Edoes.) ♃

CYPERACEÆ.

Cyperus tenellus, *L.* (p. 745). South Africa. ⊙

GRAMINEÆ.

A. Phleum pratense, *L.* (p. 321). Europe, meadows. (Timothy Grass.) ♃
Alopecurus agrestis, *L.* (p. 321). Ditto, fields. (Foxtail Grass.) ⊙
Phalaris canariensis (p. 321). Ditto, waste places. (Canary Grass.) ⊙

Holcus mollis, *L.* (p. 333). Ditto, pastures, etc. ♃
Holcus lanatus, *L.* Ditto, ditto. ♃
Anthoxanthum odoratum, *L.* (p. 321). Ditto, meadows. (Sweet Vernal Grass.) ♃
Panicum colonum, *L.* (p. 324). S. Europe, waste places. ⊙
Panicum glaucum, *L.* (p. 324). Europe, waste places. ⊙
Panicum sanguinale, *L.* (p. 324). Ditto, ditto. ⊙
Cynodon Dactylon, *L.* (p. 331). Ditto, ditto. (Doab Grass.) ♃
Agrostis vulgaris, *With.* Ditto, meadows. (Bent Grass.) ♃
Avena sativa, *L.* (p. 336). Ditto, cultivated places. (Oats.) ⊙
Eleusine indica, *Gærtn.* (p. 331). India, cultivated places. ⊙
A. Poa trivialis, *L.* Europe, meadows, etc. ♃
Poa annua, *L.* (p. 340). Ditto, waysides, etc. ⊙
Briza minor, *L.* Ditto, meadows. ⊙
Briza maxima, *L.* Ditto, ditto. (Quaking Grass.) ⊙
Cynosurus cristatus, *L.* Ditto, dry pastures. ♃
A. Dactylis glomerata, *L.* Ditto, waysides, etc. (Cock's-foot Grass.) ♃
Festuca bromoides, *L.* (p. 341). Ditto, ditto. ⊙
A. Bromus sterilis, *L.* Ditto, ditto. ⊙
A. Bromus erectus, *Huds.* Ditto, fields, etc. ♃
A. Bromus commutatus, *Schrad.* Ditto, ditto. ⊙ or ♂
Bromus mollis, *L.* (p. 341). Ditto, meadows. ⊙ or ♂
Bromus racemosus, *L.* (341). Ditto, ditto. ⊙ or ♂
Lolium temulentum, *L.* Ditto, cornfields, etc. (Rye Grass.) ⊙
Lolium perenne, *L.* Ditto, ditto. ♃ or ♂
Triticum sativum, *L.* ⊙
Hordeum sativum, *L.* (p. 343). Ditto. (Barley.) ⊙
A. Lepturus incurvatus, *Trin.* Seashores. (Sands, Waitemata, *Kirk.*) ⊙
Anthistiria australis, *Br.* (p. 325.) ♃

It is possible that some of the above, which are annual ⊙ in Europe, may become perennial ♃ in New Zealand, owing to the mildness of the winters.

ALPHABETICAL LIST OF NATIVE AND VERNACULAR NAMES.

The native names here enumerated are, with few exceptions, supplied by my friend Mr. Colenso, whose intimate knowledge of the botany and language of New Zealand guarantees their accuracy. For a few chiefly in use by natives of the Otago district I am indebted to Dr. Hector, and for still others to a MS. obligingly lent me by Dr. Lauder Lindsay. And, lastly, whilst this sheet was in the printer's hands, I have received the valuable Maori-Latin Index to the Handbook of this Flora, edited by Dr. Hector, and printed under the authority of the New Zealand Government at Wellington (1866)

Aka, *Col.* *Metrosideros scandens.*
Akakura. *Metrosideros scandens.*
Ake, *Col.* *Dodonæa viscosa.*
Ake-ake, *Hector.* *Olearia avicenniæfolia.*
Ake-piro, *Col.* *Olearia furfuracea.*
Akerautangi, *Mantell.* *Dodonea viscosa.*
Akewharangi, *Geolog. Surv.* *Olearia Cunninghamii.*
Arbor vitæ. *Libocedrus Doniana.*
Aruhe, *Col.* *Pteris aquilina.*
Aster, native. *Celmisia*, various sp.
Aute, *Col.* *Broussonetia papyrifera.*
Auta-taranga, *Col.* *Pimelea arenaria.*
Avens. *Geum urbanum*, var. *strictum.*
Beech, native. *Fagus.*
Birch of New Zealand. *Fagus.*
Black pine of Otago, *Hector.* *Podocarpus spicata* and *ferruginea.*
Bluebell, native. *Wahlenbergia*, several sp.
Bracken. *Pteris aquilina.*
Broad leaf, *Hector.* *Griselinia lucida.*
Bocarro, or Bokako, *Hector.* *Elæocarpus Hookerianus.*
Bramble. *Rubus australis.*
Broom, native. *Carmichælia*, various sp.
Broom, pink. *Notospartium Carmichæliæ.*
Bulrush. *Typha angustifolia.*
Burr. *Acæna.*
Bur-reed. *Sparganium simplex.*
Buttercup. *Ranunculus*, various sp.
Celery, native. *Apium australe.*
Cohou-cohou, *Raoul.* *Pittosporum obcordatum.*
Cowage. *Bidens pilosa.*
Cotton plant of Otago. *Astelia*, various sp.
Cranesbill. *Geranium.*
Cutting grass. *Gahnia*, various sp.
Cypress. *Libocedrus Doniana.*
Daisy, native. *Lagenophora*, various sp.
Daisy-trees. *Olearia*, various sp.
Dandelion. *Taraxacum Dens-leonis.*
Dock. *Rumex flexuosus.*
Dodder. *Cuscuta densiflora.*
Duckweed. *Lemna.*
Eketera, *D'Urville.* *Lepidium oleraceum.*
Everlasting. *Gnaphalium*, various sp.
Flax, native. *Linum monogynum.*
Flax, New Zealand. *Phormium tenax.*
Groundsel. *Senecio*, various sp.
Haekaro. *Pittosporum umbellatum.*
Hakeke, *Col.* *Polyporus* sp.
Hange-hange, *Col.* *Geniostoma ligustrifolium.*
Harakeke, *Col.* *Phormium tenax* and *Colensoi.*
Harori, *Col.* *Agaricus adiposus.*
Hauama, *Col.* *Entelea arborescens.*
Hawhato, *Col.* *Cordiceps Robertsii.*
Hawthorn, native. *Discaria Toumatou.*
Heru-heru, *Col.* *Leptopteris hymenophylloides.*
Heath, native of Otago. *Leucopogon Frazeri.*

Hinahina, *Geolog. Surv.* *Melicytus ramiflorus.*
Hinatoli, *Geolog. Surv.* *Epilobium.*
Hinau, *Cunn.* *Elæocarpus dentatus.*
Hinau, *Raoul.* *Elæocarpus Hookerianus.*
Hiri turiti, *Hector.* Epiphytic Orchids, various.
Hohere, *Cunn.* *Hoheria populnea* and *Plagianthus Lyallii.*
Hohoeka, Middle Island, *Lyall.* *Panax crassifolia.*
Horahora, *Geolog. Surv* *Astelia.*
Horoeka, *Col.* *Panax crassifolia.*
Horoweeka, *Hector* *Panax*, various sp.
Horopito, *Col.* *Drimys axillaris.*
Houka, *Geolog. Surv.* *Cordyline australis.*
Houma, *Hector.* *Sophora tetraptera.*
Horu-horu, *Lindsay.* *Brassica Rapa.*
Houhere, *Col.* *Hoheria populnea.*
Houi, *Col.* *Hoheria populnea.*
Hue, *Col.* *Cucurbita* sp.
Hunangamoho, *Col.* *Apera arundinacea.*
Hune, *Col.* Pappus of seeds of *Typha angustifolia.*
Hutiwai, Middle Island, *Lyall.* *Acæna Sanguisorbæ.*
Hutu, *Geolog. Surv.* *Ascarina lucida.*
Ice-plant, native. *Tetragonia expansa* and *Mesembryanthemum australe.*
Ini-ini, *Hector.* *Melicytus ramiflorus.*
Iron-wood, *Hector.* *Metrosideros lucida.*
Ivy-tree. *Panax Colensoi.*
Kaha-kaha, *Col.* *Astelia Solandri.*
Kahikatea, *Col.* *Podocarpus dacrydioides.*
Kahikatoa, *Col.* *Leptospermum scoparium.*
Kahikomako, *Col.* *Pennantia corymbosa.*
Kaho, *Cunn.* *Linum monogynum.*
Kaikaiatua. *Rhabdothamnus Solandri.*
Kaikatea. *Podocarpus dacrydioides.*
Kaikomako, *Col.* *Pennantia corymbosa.*
Kai-ku, *Col.* *Parsonsia heterophylla.*
Kaiwhiria, *Col.* *Hedycarya dentata.*
Kaiwiria, *Geolog. Surv.* *Panax simplex.*
Kakaha, *Hector.* *Astelia nervosa.*
Kakaho, *Col.* *Arundo conspicua.*
Kakaramu, *Geolog. Surv.* *Coprosma lucida.*
Kakika, *Geolog. Surv* *Senecio glastifolius.*
Kalamou, *Hector.* *Coprosma lucida.*
Kaneree, *Cot.* *Vitex littoralis.*
Kapia, *Geolog. Surv.* Gum dug up of *Dammara australis.*
Kapook, *Hector.* *Griselinia lucida.*
Karaka, *Col.* *Corynocarpus lævigata.*
Karamu, *Col.* *Coprosma fœtidissima.*
Karamu, *Col.* *Coprosma lucida* and *robusta.*
Karangu, *Col.* *Coprosma lucida.*
Kareao, *Col.* *Rhipogonum parviflorum.*
Karengo, *Col.* *Laminaria* sp
Karetu, *Col.* *Hierochloe redolens.*
Karito, *Lindsay.* *Typha angustifolia.*
Karmahi, *Hector.* *Weinmannia silvicola* and *racemosa.*
Karo, *Col.* *Pittosporum cornifolium* and *crassifolium.*
Karo, Middle Island, *Lyall.* *Pittosporum tenuifolium.*
Katute, *Hector.* *Dicksonia antarctica.*
Kaudi. *Dammara australis.*
Kauere. *Vitex littoralis.*
Kauri, *Col.* *Dammara australis.*
Kawaka, *Col.* *Libocedrus Donianus.*
Kawa-kawa, *Col.* *Piper excelsum.*
Kiekie, *Col.* *Freycinetia Banksii.*
Kihii, *Geolog. Surv.* *Pittosporum crassifolium.*
Kohe-kohe, *Col.* *Dysoxylum spectabile.*
Koheriki, *Col.* *Angelica rosæfolia.*
Kohi, *Hector.* *Schefflera digitata.*
Kohia, *Col.* *Passiflora tetrandra.*
Kohoho, *Col.* *Solanum aviculare.*
Kohoukohou, *Geolog. Surv.* *Pittospórum obcordatum.*
Kohuhu, *Col.* *Pittosporum tenuifolium.*
Kohu-kohu, *Geolog. Surv.* *Pittosporum obcordatum.*
Kohu-kohu, *R. Cunn.* *Scleranthus biflorus.*
Kohu-kohu, *Lindsay.* *Stellaria media.*
Kohutuhutu, *Col.* *Fuchsia excorticata.*
Kokaho. *Arundo conspicua.*
Koke. *Passiflora tetrandra.*
Kokihi, *Col.* *Tetragonia trigyna.*
Kokomuka. *Veronica* sp.
Konine. *Fuchsia excorticata.*
Kopa-kopa, *Lindsay.* *Plantago* sp.
Kopata, *Lyall.* *Pelargonium clandestinum.*
Kopata, *Col.* *Geum urbanum* var. *strictum.*
Koporokaiwhiri. *Hedycarya dentata.*
Kopoupou, *D'Urville.* *Scirpus lacustris.*
Kopupungawha, *Lindsay.* *Typha angustifolia.*
Kopura, *Col.* *Hepaticæ*, various.
Korari, *Col.* *Phormium tenax* and *Colensoi.*
Korau, *Lindsay.* *Brassica Rapa.*
Korau, *Col.* *Cyathea medullaris.*
Koreirei, *Col.* Roots of *Typha angustifolia.*
Korikori, *Lindsay.* *Ranunculus*, various sp.
Koroi, *Col.* Fruit of *Podocarpus dacrydioides.*
Korokio, *Lindsay.* *Veronica* sp.
Korokio-taranga, *Col.* *Corokia buddleioides.*
Koromeek, *Hector.* *Panax simplex.*
Koro-miko, *Bidwill.* *Veronica parviflora.*
Koromiko, *Geolog. Surv.* *Veronica elliptica.*
Korumeek, *Hector.* *Veronica salicifolia.*
Koromiko-taranga, *Col.* *Veronica.*

Koromuti, *Hector*. *Panax simplex*.
Koropuku, *Col*. *Gaultheria depressa*.
Kotohituk, *Hector*. *Fuchsia excorticata*.
Kotukutuku, *Col*. *Fuchsia excorticata*.
Kouka, *Lindsay*. *Cordyline australis*.
Koware, *Lindsay*. *Typha angustifolia*.
Kowhai, *Col*. *Sophora tetraptera*, var. *grandiflora*.
Kowhaingutu-kaka, *Cunn*. *Clianthus puniceus*.
Kowhara-whara, *Col*. *Astelia Cunninghamii* and *Banksii*.
Kowhiti-whiti, *Lindsay*. *Nasturtium officinale*.
Kumara, *Col*. *Ipomœa Batatas*.
Kumarahou, *Col*. *Pomaderris elliptica*.
Ku-papa, *R. Cunn*. *Passiflora tetrandra*.
Kuri-kuri, Middle Island, *Lyall*. *Aciphylla squarrosa*.
Kuriwao, *Hector*. *Rhipogonum scandens*.
Lawyer. *Rubus australis*, *Parsonsia*, etc.
Leather plant. *Celmisia*, various sp.
Luma-luma, *Hector*. *Coprosma fœtidissima*.
Maanawa, *Col*. *Avicennia tomentosa*.
Maha-maka, *A. Cunn*. *Ackama rosæfolia*.
Mahimahi, *Col*. *Elæocarpus Hookerianus*.
Mahoe, *Col*. *Melicytus ramiflorus*.
Mahuri, *Raoul*. *Alternanthera sessilis*.
Mai, *Cunn*. *Podocarpus spicata*.
Maikaika, *Col*. *Arthropodium cirrhatum*.
Maikaika, *Lyall*. *Thelymitra pulchella*.
Maikaika, *Col*. *Orthoceras Solandri*.
Maikaika, *Geolog. Surv*. *Metrosideros robusta*.
Maire, *Col*. *Olea* sp. (in the south of N. Isl.).
Maire, *Col*. *Santalum Cunninghamii* (in the north of N. Island).
Mairehau, *Col*. *Phebalium nudum*.
Maire raunui, *Col*. *Olea Cunninghamii*.
Maire-tawhake, *Col*. *Eugenia Maire*.
Makaika, *Col*. *Orthoceras Solandri*.
Makaka, *Geolog. Surv*. *Carmichælia australis*.
Makamaka, *Col*. *Ackama rosæfolia*.
Makomako, *Col*. *Aristotelia racemosa*.
Mamaku, *Col*. *Cyathea medullaris*.
Mamangi, *Geolog. Surv*. *Coprosma spathulata*.
Mamuk, *Hector*. *Cyathea medullaris*.
Manawa, *Cunn*. *Avicennia tomentosa*.
Manawau, *Geolog. Surv*. *Dacrydium Colensoi*.
Mangeao, *Col*. *Tetranthera calicaris*.
Mange-mange, *Col*. *Lygodium articulatum*.
Mangrove. *Avicennia officinalis*.
Manoao, *Col*. *Dacrydium Colensoi*.
Manouea, Middle Island, *D'Urville*. *Leptospermum ericoides*.
Manuka, *Col*. *Leptospermum scoparium*.
Manuka-rau-riki, *Col*. *Leptospermum ericoides*.
Maori Parsnip, *Hector*. *Ligusticum Lyallii*.
Mapara, *Col*. Hardwood of *Dacrydium cupressinum*.
Mapau, *Col*. *Myrsine australis*.
Mapauriki, *Cunn*. *Pittosporum tenuifolium*.
Maru, *Col*. *Sparganium simplex*.
Mataii, *Col*. *Podocarpus spicata*.
Matangoa, *Huegel*. *Cardamine stylosa*.
Matata, *Geolog. Surv*. *Rhabdothamnus Solandri*.
Matipo, Middle Island, *Lyall*. *Myrsine Urvillei*.
Matuakumara, *Col*. *Geranium dissectum*.
Maukoro, *Lindsay*. *Carmichælia australis*.
Mawhai, *Col*. *Sicyos angulatus*.
Meeka-meek, *Hector*. *Hymenophyllum* sp.
Miko. *Areca sapida*.
Mingi, *Lindsay*. *Coprosma myrtillifolia*.
Mingi, *Lindsay*. *Cyathodes acerosa*.
Mingimingi, *Col*. *Leucopogon fasciculatus*.
Mint. *Mentha*.
Miro, *Col*. *Podocarpus ferruginea*.
Mistletoe. *Loranthaceæ*, various.
Moka, *Hector*. *Aristotelia racemosa*.
Moko-piko, *Bidwill*. *Libocedrus Donianus*.
More, *Lindsay*. *Dammara australis*.
Muka, *W. Mantell*. Dressed fibre of *Phormium tenax*.
Muka-muk, *Hector*. *Hymenophyllum* sp.
Naéréoré, Middle Island, *D'Urville*. *Scleranthus biflorus*.
Nahinahi, *Lindsay*. *Convolvulus soldanella*.
Nani, *Lindsay*. *Brassica campestris*.
Nao, *D'Urville*. *Linum monogynum*.
Neinei, *Col*. *Dracophyllum latifolium*.
Neinei, *Lyall*. *Carmichælia australis*.
Nettle. *Urtica*, various sp.
Ngaio, *Col*. *Myoporum lætum*.
Ngawha, *Lindsay*. *Typha angustifolia*.
Nightshade. *Solanum nigrum*.
Nikau, *Col*. *Areca sapida*.
Noté-noho, *D'Urville*. *Arenaria media*.
Oho, *Hector*. *Panax Lessonii*.
Oehiakura, *Lindsay*. *Dicksonia squarrosa*.
Ohoeka, *Lindsay*. *Panax crassifolia*.
Oioi, *Col*. *Leptocarpus simplex*.
Onga-onga, *Col*. *Urtica ferox*, and others.
Orange-leaf of Otago. *Coprosma lucida*, *etc*.
Orewa, *Geolog. Surv*. *Sapota costata*.
Oru, *Cunn*. *Colensoa physaloides*.
Pahautea, *Col*. *Libocedrus Bidwillii*.
Pakue, *Hector*. *Dicksonia squarrosa*.
Panake, *Col*. *Convolvulus sepium*.
Palm, *Nelson*. *Areca sapida*.

Penahi, *Col.* *Convolvulus sepium.*
Panapana, *Col.* *Cardamine hirsuta.*
Papaauma, *Col.* *Coprosma grandifolia.*
Papaii, *Col.* *Aciphylla squarrosa.*
Para, *Col.* *Marattia salicina.*
Para-para, *Col.* *Pisonia Sinclairii.*
Parataniwha, *Col.* *Elatostemma rugosum.*
Parerarera, *Lindsay.* *Plantago* sp
Paretao, *Col.* *Asplenium obtusatum.*
Paruwhatitiri, *Col.* *Ileodictyon cibarium.*
Patate, Middle Island, *Lyall.* *Schefflera digitata.*
Patete, *Geolog. Surv.* *Melicope ternata.*
Pate, *Col.* *Schefflera digitata.*
Patotara, *Col.* *Botrychium Virginicum.*
Patotara, *Col.* *Leucopogon Fraseri.*
Pa-totara, Middle Island, *Lyall.* *Cyathodes Oxycedrus.*
Peka-peki, Middle Island, *Lyall.* *Erechtites quadridentata.*
Pekepekekiore, *Col.* *Hydnum clathroides.*
Pellitory. *Parietaria debilis.*
Pepper-tree. *Drimys axillaris.*
Pepper, native. *Piper excelsum.*
Pére, *Col.* *Alseuosmia Banksii.*
Perei, *Col.* *Gastrodia Cunninghamii.*
Piamanuka, *Col.* Manna exudation of *Leptospermum scoparium.*
Pikiarero, *Col.* *Clematis hexasepala.*
Pinakitere, *Lindsay.* *Geranium dissectum.*
Pine, red. *Dacrydium cupressinum.*
Pine, black. *Podocarpus ferruginea.*
Pingae or Pingao, } *Desmoschœnus spiralis.*
Pirikahu, *Col.* *Acæna Sanguisorbæ.*
Piri-piri, *Col.* *Acæna Sanguisorbæ.*
Piri-piri, *Col.* *Bolbophyllum pygmæum.*
Piri-piri, *Cunn.* *Haloragis tenella.*
Piri-piri,*Lindsay.* *Pittosporum cornifolium.*
Piri-piri Whata, *Cunn.* *Carpodetus serratus.*
Pirita, *Col.* *Rhipogonum parviflorum.*
Piri-ta, *Col.* *Tupeia antarctica*
Pitau, *Lindsay.* *Cyathea medullaris.*
Piu-Piu, *Col.* *Polypodium pennigerum.*
Plantain. *Plantago*, various sp.
Pohue, *Col.* *Convolvulus sepium.*
Pohuehue, *Lindsay.* *Polygonum complexum.*
Pohuihui, *Geolog. Surv.* *Passiflora tetrandra.*
Pohutukawa, *Col.* *Asplenium flaccidum.*
Pohutukawa, *Col.* *Metrosideros tomentosa.*
Pokaka, *Col.* *Elæocarpus dentatus* and *Hookerianus.*
Poko-poko-nui-ha-ura. *Clematis parviflora.*
Ponja, of Chatham Island. *Cyathea Cunninghamii.*
Ponga, *Col.* *Cyathea dealbata.*
Pondweed. *Potamogeton.*
Popero, *Lindsay.* *Solanum aviculare.*
Popoiahakeke, *Col.* *Polyporus* sp.
Popi-hui, *Hector.* *Libertia micrantha.*
Popohui, *Hector.* *Arthropodium cirrhatum.*
Porokaiwhiri, *Col.* *Hedycarya dentata.*
Poroporo, *Col.* *Solanum aviculare* and *nigrum.*
Pororua, *Col.* *Sonchus oleraceus.*
Poukatea, *Raoul.* *Griselinia littoralis.*
Puawhananga, *Lindsay.* *Clematis indivisa.*
Puheritaiko, *Lyall.* *Senecio rotundifolius.*
Puhou, Southern Island, *Lyall.* *Coriaria ruscifolia.*
Puka, *Col.* *Polygonum australe* and *Meryta Sinclairii.*
Pukapuka, *Col.* *Brachyglottis repanda.*
Pukariao, *Geolog. Surv.* *Brachyglottis repanda.*
Pukatea, *Col.* *Atherosperma novæ-Zelandiæ* and *Griselinia lucida.*
Pukurau, *Col.* *Lycoperdon Fontainesii.*
Pungapunga, *Geolog. Surv.* Pollen of *Typha angustifolia.*
Punui, *Col.* *Cyathea Cunninghamii.*
Puri-puri-ki-pili, *Hector.* *Hedycarya dentata.*
Puriri, *Col.* *Vitex littoralis.*
Putawa, *Col.* *Boletus.*
Puwha, *Col.* *Sonchus oleraceus.*
Puwhananga, *Col.* *Clematis indivisa.*
Rahu-rahu, *Lindsay.* *Pteris aquilina.*
Raka-pika, *R. Cunn.* *Metrosideros florida.*
Ramarama, *Col.* *Myrtus bullata.*
Raugiora, *Col.* *Brachyglottis repanda.*
Rata, *Col.* *Metrosideros robusta*
Rata, Middle Island, *Lyall.* *Metrosideros florida.*
Ratapiki, *R. Cunn.* *Metrosideros florida.*
Rauhuia, *Col.* *Linum monogynum.*
Raukawa, *Col.* *Panax Edgerleyi.*
Raupeti, *Col.* *Solanum nigrum.*
Raupo, *Col.* *Typha angustifolia.*
Rau-Raua, *Edgerley.* *Panax Edgerleyi.*
Rautahi, *Col.* *Carex ternaria.*
Rawiri, Northern Island, *Cunn.* *Leptospermum ericoides.*
Red Pine. *Dacrydium cupressinum.*
Reedmace. *Typha angustifolia.*
Rengarenga, *Col.* *Arthropodium cirrhatum.*
Rewarawa, *Geolog. Surv.* *Dysoxylum spectabile.*
Rewarewa, *Col.* *Knightia excelsa.*
Ribbon-wood of Otago. *Hoheria populnea.*
Rimu, *Bidwill.* *Dacrydium laxifolium.*
Rimu, *Col.* *Dacrydium cupressinum.*
Rimurapa, *Col.* *D'Urvillea utilis.*
Ririwaka, *Col.* *Scirpus maritimus.*
Rohutu, *Col.* *Myrtus pedunculata.*

Roi, *Col.* *Pteris aquilina* (root of).
Roniu, *Col.* *Brachycome odorata.*
Roniu, *Col.* *Brachycome radicata.*
Rue, black. *Podocarpus spicata.*
Rush. *Juncus.*
Runa, *Lindsay.* *Rumex* sp.
Sarsaparilla, native. *Rhipogonum scandens.*
Spinach, New Zealand. *Tetragonia expansa.*
Sorrelwood. *Oxalis magellanica.*
Sow-thistle. *Sonchus oleraceus.*
Spear-grass. *Aciphylla squarrosa.*
Sterile wood of Otago. *Coprosma fœtidissima.*
Sundew. *Drosera.*
Supple-jack. *Rubus australis, Parsonsia, Lygodium*, etc.
Tamingi, *Geolog. Surv.* *Epacris pauciflora.*
Tanekaha, *Col.* *Phyllocladus trichomanoides.*
Tangeao, *Col.* *Tetranthera calicaris.*
Taraire, *Col.* *Nesodaphne Taraire.*
Taramea, *Col.* *Aciphylla Colensoi.*
Tarata, *Col.* *Pittosporum eugenioides.*
Tarata, *R. Cunn.* *Pittosporum crassifolium.*
Tarata, *Lindsay.* *Pittosporum tenuifolium.*
Taro, *Col.* *Caladium esculentum.*
Tataka, *Mantell.* *Melicope ternata.*
Tatara-hake, *Col.* *Coprosma acerosa.*
Tataramoa, *Col.* *Rubus australis.*
Tauhinu, *Col.* *Pomaderris ericifolia.*
Tauhinu, *Geolog. Surv.* *Podocarpus nivalis.*
Taupata, *Col.* *Coprosma retusa.*
Tawa, *Col.* *Nesodaphne Tawa.*
Tawaapou, *Col.* *Sapota costata.*
Tawai, *Col.* *Fagus Menziesii.*
Tawai, *Bidwill.* *Fagus fusca.*
Tawai, *Cunn.* *Weinmannia racemosa.*
Tawairauriki, *Geolog. Surv.* *Fagus Solandri.*
Tawaiwai, *Lindsay.* *Phyllocladus trichomanoides.*
Tawari, *Col.* *Ixerba brexioides.*
Taweku, *Lindsay.* *Coriaria ruscifolia.*
Tawhai, *Col.* *Fagus Menziesii* and *Solandri.*
Tawhai-rau-nui, *Col.* *Fagus fusca.*
Tawhara, *Col.* *Freycinetia Banksii.*
Tawhero, *Col.* *Weinmannia sylvicola.*
Tawhero (Southern Island), *Lyall.* *Weinmannia racemosa.*
Tawhiwhi, *Col.* *Pittosporum tenuifolium.*
Tawiri Karo, *Lindsay.* *Pittosporum cornifolium.*
Tea-tree of settlers. *Leptospermum scoparium.*
Tepuaotereinga, *Taylor.* *Dactylanthus Taylorii.*
Thorn, native. *Discaria Toumatou.*
Thyme, wild, of Otago. *Samolus littoralis.*
Ti, *Col.* *Cordyline australis* and *indivisa.*
Tikapu, *Col.* *Cordyline indivisa.*
Tikoraha, *Col.* *Cordyline stricta.*
Tikupenga, *Lindsay.* *Cordyline stricta.*
Tikumu, *Col.* *Celmisia coriacea.*
Tikoraka, *Geolog. Surv.* *Cordyline pumilio.*
Tingahere, *Geolog. Surv.* *Cordyline stricta.*
Tiparae, *Geolog. Surv.* *Cordyline Banksii.*
Tipau, *Col.* *Myrsine Urvillei* and *salicina.*
Tirauriki, *Geolog. Surv.* *Cordyline pumilio.*
Titawhiti, *Geolog. Surv.* *Cordyline* sp.
Titi-rangi (Middle Island), *Lyall.* *Veronica speciosa.*
Titoki, *Col.* *Alectryon excelsum.*
Titongi, *Col.* *Alectryon excelsum.*
Toa-toa, *Col.* *Phyllocladus trichomanoides.*
Toa-toa, *D'Urville.* *Haloragis alata.*
Toe-toe. *Arundo conspicua.*
Toe-toe, *Col.* *Cyperus ustulatus.*
Toe-toe-whatu-manu, *Lyall.* *Cyperus ustulatus.*
Toi, *Col.* *Barbarea vulgaris.*
Topitopi, *Mantell.* *Alectryon excelsum.*
Toro, *Col.* *Persoonia Toro.*
Toromiro. *Podocarpus ferruginea.*
Torotoro, *Lindsay.* *Metrosideros scandens.*
Totara, *Lindsay.* *Cyathodes oxycedrus.*
Totara, *Col.* *Leucopogon Frazeri.*
Totara, *Col.* *Podocarpus totara.*
Totarakirikotukutuku, *Mantell.* *Libocedrus Doniana.*
Totera, *Cunn.* *Fuchsia procumbens.*
Toumatou, *Raoul.* *Discaria australis.*
Towai, *Col.* *Weinmannia racemosa.*
Towai, *Hector.* *Cordyline indivisa.*
Towai, *Raoul.* *Epicarpurus microphyllus.*
Traveller's-joy. *Clematis.*
Tree-fern, grey. *Cyathea medullaris.*
Tree-fern, silver. *Cyathea dealbata.*
Tuakura, *Lindsay.* *Dicksonia squarrosa.*
Tukirunga, *Col.* *Dicksonia antarctica.*
Tukarehu, *Lindsay.* *Plantago* sp.
Tumatakuru, *Col.* *Discaria Toumatou.*
Tumingi, Middle Island, *Lyall.* *Leucopogon fasciculatus.*
Tupak Grass of Otago. *Carex appressa.*
Tupa-kihi, *Col.* *Coriaria ruscifolia.*
Tupari, *Hector.* *Olearia operina* and *O. Lyallii.*
Turawera, *Col.* *Pteris tremula.*
Turutu, *Lindsay.* *Dianella intermedia.*
Turutu, *Lyall.* *Libertia ixioides.*
Tute, a wild, *Hector.* *Olearia operina.*
Tutu, *Col.* *Coriaria ruscifolia.*
Tutuheuheu, *Mantell.* *Coriaria thymifolia.*
Tutu-nawai, *Col.* *Polygonum prostratum.*
Tutupapa, *Col.* *Coriaria thymifolia.*
Upoko-tangata, *Col.* *Cyperus ustulatus.*
Ureure, *Col.* Fruit of *Freycinetia Banksii.*

Vegetable-sheep. *Raoulia eximia* and others.
Waekahu, *Lindsay*. *Lycopodium volubile*.
Wae-wae-kaka, *Col*. *Gleichenia semi-vestita*.
Wae-wae-koukou, *Col*. *Lycopodium volubile*.
Wae-wae-matuku, *Col*. *Gleichenia semivestita*.
Warikauri, *Geolog. Surv*. Kauri gum.
Wawa paka, *Hector*. *Panax arboreum*.
Wawapaku, *Col*. *Panax anomala*.
Wainatua, *Col*. *Euphorbia glauca*.
Wainatua, *Lindsay*. *Rhabdothamnus Solandri*.
Wakaka, *Lyall*. *Carmichælia australis*.
Water-lily, New Zealand. *Ranunculus Lyallii*.
Wawa-paku, *Col*. *Panax anomala*.
Waupaku, *Lindsay*. *Panax arborea*.
Weki, *Col*. *Dicksonia squarrosa*.
Wekiponga, *Col*. *Dicksonia antarctica*.
Whakapiopio, *Lindsay*. *Metrosideros scandens*.
Whakatangitangi, *Lindsay*. *Metrosideros* sp.
Whanako. *Cordyline australis*.
Wharaeki. *Phormium Colensoi*.
Wharangi, *Col*. *Melicope ternata*.
Wharangi-piro, Middle Island, *Lyall*. *Olearia Cunninghamii*.
Wharangi-pirou, *Col*. *Melicope ternata*.
Whawhako, *Geolog. Surv*. *Eugenia Maire*.
Wha-whi, Chatham Island. *Plagianthus betulinus*.
Whau, *Col*. *Entelea arborescens*.
Whau-whau, *R. Cunn*. *Panax Lessoni*.
Whau-whau, *Lindsay*. *Plagianthus Lyallii*.
Whau-whau-paku, *Col*. *Panax arborea*.
Whau-whi, *Col*. *Hoheria populnea*.
Whau-whi, *Lyall*. *Plagianthus Lyallii*.
Whau-whi, *Geolog. Surv*. *Plagianthus betulinus*.
Wheki, *Lindsay*. *Dicksonia squarrosa*.
Whinau, *Lindsay*. *Elæocarpus Hinau*.
Whi, *Col*. *Juncus maritimus*.
Whitau, *Mantell*. Same as *Muka* (dressed flax).
Wild Irishman. *Discaria Toumatou*.
Wild Spaniard. *Aciphylla*.
Wiwi, *Col*. *Juncus maritimus* and *effusus*.
Wiwi, *Lyall*. *Isolepis nodosa*.

INDEX OF GENERA AND SPECIES.

The synonyms are printed in italics.

THE END.

PRINTED BY J. E. TAYLOR AND CO.,
LITTLE QUEEN STREET, LINCOLN'S INN FIELDS.

Printed in the United States
By Bookmasters